Proceedings of the International
Inhibitors, Munich, Nov. 1970

        Important

We wish to point out clearly that
to this volume which had not been

This especially applies to:

<u>H. Tschesche et al.</u>:   On t
            p. 29

<u>T. Ikenaka et al.</u>:   Struc
            Trypsi

<u>D. Čechová et al.</u>:   Amino
            Cow Co

<u>V. Dlouhá-Keil et al.</u>: On the
            Basic Pa
            p. 95 (r

<u>H. Tschesche et al.</u>:   Studies
            Secretor

Proceedings of the International Research Conference
on Proteinase Inhibitors, Munich, Nov. 1970

Proceedings

of the
International Research Conference

on

Proteinase Inhibitors

Munich, November 4—6, 1970

edited by

H. Fritz · H. Tschesche

With 246 figures and 95 tables

Walter de Gruyter · Berlin · New York
1971

The quotation of registered names, trade names, trade marks, etc. in this book does not imply, even in the absence of a specific statement, that such names are exempt from laws and regulations protecting trade marks, etc. and therefore free for general use.

ISBN 3 11 003776 9

Library of Congress Catalog Card Number 70-168659

© Copyright 1971 by Walter de Gruyter & Co., Berlin

All rights reserved, including those of translation into foreign languages. No part of this book may be reproduced in any form — by photoprint, mikrofilm, or any other means — nor transmitted nor translated into a machine language without written permission from the publisher.

Printed in Germany by Walter de Gruyter & Co., Berlin.

Cover designed by U. Hanisch, Berlin.

Preface

It was the aim of the First International Research Conference on Proteinase Inhibitors, held in Munich on November 4—6, 1970, to review the present status of the field, to discuss the recent advances in our knowledge on isolation, structure and mechanism of interactions between proteinases und protein proteinase inhibitors and to promote the research by mutual exchange of experience. We are beginning to understand the physiological significance of this class of proteins, which was substantiated by the primary and tertiary structures elucidated and the theory of protein-protein interaction presented.

The present volume records all of the lectures and those papers and discussion remarks that contributors submitted for publication.

We wish to express our gratitude to all participants for their contributions and for their open and cordial discussions. We are especially grateful to contributors which were unable to attend the Conference. We have to apologize for the limitations to the number of participants for this First Conference. The widespread and general interest and the anticipated future progress will perhaps stimulate the organization of a second conference. We would therefore appreciate any relevant suggestion submitted.

On this occasion we wish to express our thanks to the following organizations we are indebted to for financial support which made possible the arrangement of this Conference: Behringwerke AG, Marburg; C. H. Boehringer & Sohn, Ingelheim; Boehringer Mannheim GmbH, Tutzing; Farbenfabriken Bayer AG, Wuppertal-Elberfeld; Farbenfabriken Hoechst AG, Frankfurt; E. Merck AG, Darmstadt; Novo Industrie, Mainz. The organization of the program was further supported by Bayer-Leverkusen, München; Beckman Instruments, München; Röhm & Haas GmbH, Darmstadt; Schering AG, Berlin.

Munich, May 1971

The Organizing Committee

HANS FRITZ, KARL HOCHSTRASSER, EUGEN WERLE
Universität München

HARALD TSCHESCHE
Technische Universität München

Contents

List of Contributors x
List of Participants xi

Proteinase Inhibitors in Human Medicine

N. Heimburger, H. Haupt, H. G. Schwick: Proteinase Inhibitors of Human Plasma 1
G. F. B. Schumacher: Sex Hormones and Serum Proteins 21
M. H. Coan, J. Travis: Interaction of Human Pancreatic Proteinases With Naturally Occurring Proteinase Inhibitors 294
E. Werle: Proteinase Inhibitors in Clinical Studie 23

Specific Isolation Methods

H. Fritz, B. Brey, M. Müller, M. Gebhardt: Specific Isolation and Modification Methods for Proteinase Inhibitors and Proteinases 28
G. Feinstein: Isolation of Chymotrypsin Inhibitors by Affinity Chromatography through Chymotrypsin-Sepharose 38
B. Kassell, M. B. Marciniszyn: A Simple Method of Purification of the Basic Trypsin Inhibitor of Bovine Organs 43

Inhibition Mechanism: Theories and Methods

D. Shotton: The Molecular Architecture of the Serine Proteinases 47
M. Laskowski, Jr., R. W. Duran, W. R. Finkenstadt, S. Herbert, H. F. Hixson, Jr., D. Kowalski, J. A. Luthy, J. A. Mattis, R. E. McKee, C. W. Niekamp: Kinetics and Thermodynamics of Interaction Between Soybean Trypsin Inhibitor (Kunitz) and Bovine β Trypsin 117
H. Tschesche, R. Obermeier: Mass Spectral Determination of Peptide Bond Hydrolysis Equilibria in Protein Proteinase-Inhibitors 135
M. Laskowski, Jr.: Comments to the Modification Reaction 141
R. E. Feeney: The Non-Bond Splitting Mechanism of Action of Inhibitors of Proteolytic Enzymes — the Conservative Interpretation 162
H. Tschesche, H. Klein, G. Reidel: On the Mechanism of Temporary Inhibition 299

Inhibitors of Plant Tissues

T. Ikenaka, T. Koide, S. Odani: Structure and Chemical Modification of Kunitz Soybean Trypsin Inhibitor . . 108

I. E. Liener: Chemical Studies on the Site of Interaction between Trypsin and Kunitz Soybean Trypsin Inhibitor (STI) 156

Y. Birk, A. Gertler: Chemistry and Biology of Proteinase Inhibitors from Soybeans and Groundnuts 142

F. C. Stevens: Lima Bean Protease Inhibitor: Amino Acid Sequence and Active Sites against Trypsin and Chymotrypsin 149

K. Hochstrasser, E. Werle: Sequential Studies on Proteinase Inhibitors Isolated by Use of Trypsin Resins . . 169

R. E. Feeney: Comparative Biochemistry of Avian Egg White Ovomucoids and Ovoinhibitors 189

C. A. Ryan, L. K. Shumway: Studies on the Structure and Function of Chymotrypsin Inhibitor I in the Solanaceae Family 175

Basic Bovine Inhibitor and Related Inhibitors

R. Huber, D. Kukla, A. Rühlmann, W. Steigemann: The Atomic Structure of the Basic Trypsin Inhibitor of Bovine Organs. (Kallikrein Inactivator) 56

M. Laskowski, Jr.: Comments to the Reactive Site 64

M. Laskowski, Sr., S. L. Schneider, K. A. Wilson, L. F. Kress, J. H. Mozejko, S. R. Martin, U. Kucich, M. Andrews: Naturally Occurring Trypsin Inhibitors: Further Studies on Purification and Temporary Inhibition 66

M. Rigbi: Studies on the Reactive Site towards Chymotrypsin and Trypsin of the Basic Trypsin Inhibitor of Bovine Pancreas 74

M. Laskowski, Jr.: Modification of Soybean Trypsin Inhibitor by α-Chymotrypsin 88

B. Kassell, Tsun-Wen Wang: The Action of Thermolysin on the Basic Trypsin Inhibitor of Bovine Organs . 89

V. Keil-Dlouhá, J.-M. Imhoff, B. Keil: On the Mechanism of the Interaction between Basic Pancreatic Trypsin Inhibitor and Trypsin 95

D. Čechová, V. Jonáková, F. Šorm: Amino Acid Sequence of Trypsin Inhibitor from Cow Colostrum 105

Pancreatic Secretory Inhibitors

L. J. Greene, M. H. Pubols: Isolation of a Secretory Trypsin Inhibitor from Human Pancreas 196

L. J. Greene, O. Guy: Disulfide Bridges of the Bovine Pancreatic Secretory Trypsin Inhibitor — Kazal's Inhibitor 201

H. Tschesche, E. Wachter, S. Kupfer, R. Obermeier, G. Reidel, G. Haenisch, M. Schneider: Studies on Structure and Activity of Pancreatic Secretory Trypsin Inhibitors from Pig, Sheep, Dog and Cat 207

L. J. Greene, D. C. Bartelt: Porcine Pancreatic Secretory Trypsin Inhibitor I, Amino Acid Sequence 223

Seminal Inhibitors, Inhibitors and Fertilization

E. Fink, G. Klein, F. Hammer, G. Müller-Bardorff, H. Fritz: Protein Proteinase Inhibitors in Male Sex Glands 225

L. J. D. Zaneveld, K. L. Polakoski, R. T. Robertson, W. L. Williams: Trypsin Inhibitors and Fertilization 236

G. F. B. Schumacher: a_1-Antitrypsin in Seminal Fluid 245

H. Ingrisch, H. Haendle, E. Werle: Inhibition by the Trypsin-Plasmin-Inhibitor from Sperm Plasma of the Dispersion of the Corona Radiata and Zona Pellucida by a Trypsin-like Enzyme from Spermatozoa 244

G. F. B. Schumacher, J. R. Swartwout, F. P. Zuspan: Fertility Experiments in Mice and Rabbits with the Trypsin-Kallikrein Inhibitor from Bovine Lung 247

G. F. B. Schumacher: Alpha$_1$-Antitrypsin in Uterine Secretions 253

H. Haendle, H. Ingrisch, E. Werle: A Trypsin-Chymotrypsin Inhibitor in Human Female Cervix Secretion. 256

Inhibitors from Dog Submand. Glands, Ascaris Lumbricoides and Leeches

H. Fritz, E. Jaumann, R. Meister, P. Pasquay, K. Hochstrasser, E. Fink: Proteinase Inhibitors from Dog Submandibular Glands — Isolation, Amino Acid Composition, Inhibition Spectrum 257

R. J. Peanasky, G. M. Abu-Erreish: Inhibitors from *Ascaris Lumbricoides*: Interactions With the Host's Digestive Enzymes 281

H. Fritz, M. Gebhardt, R. Meister, E. Fink: Trypsin-Plasmin Inhibitors from Leeches — Isolation, Amino Acid Composition, Inhibitory Characteristics 271

List of Contributors

G. M. Abu-Erreish, D. C. Bartelt, Y. Birk, B. Brey, M. H. Coan, D. Čechová, R. W. Duran, V. Dlouhá-Keil, R. E. Feeney, G. Feinstein, E. Fink, E. R. Finkenstadt, H. Fritz, M. Gebhardt, A. Gertler, L. J. Greene, O. Guy, M. Haendle, G. Haenisch, F. Hammer, H. Haupt, N. Heimburger, S. Herbert, H. F. Hixson, Jr., K. Hochstrasser, R. Huber, T. Ikenaka, J. M. Imhoff, H. Ingrisch, V. Jonáková, E. Jaumann, B. Kassell, R. E. McKee, B. Keil, U. Kucich, G. Klein, H. Klein, T. Koide, D. Kowalski, L. F. Kress, D. Kukla, S. Kupfer, M. Laskowski, Sr., M. Laskowski, Jr., I. E. Liener, J. A. Luthy, M. B. Marciniszyn, S. R. Martin, J. A. Mattis, R. Meister, J. H. Mozejko, M. Müller, G. Müller-Bardorff, C. W. Niekamp, R. Obermeier, S. Odani, P. Pasquay, R. J. Peanasky, M. H. Pubols, G. Reidel, M. Rigbi, A. Rühlmann, C. A. Ryan, M. Schneider, S. L. Schneider, G. F. B. Schumacher, H. G. Schwick, D. Shotton, L. K. Shumway, F. Šorm, W. Steigemann, F. C. Stevens, J. R. Swartwout, J. A. Travis, H. Tschesche, E. Wachter, T.-W. Wang, E. Werle, K. A. Wilson, L. J. D. Zaneveld, F. P. Zuspan

List of Participants

ACHER, R., Laboratoire de Chimie Biologique, 96, Boulevard Raspail, Paris VIe, France

BIRK, Y., The Hebrew University of Jerusalem, Rehovot Campus, P. O. B. 12, Israel

ČECHOVÁ, D., Czechoslovak Academy of Science, Institut of Organic Chemistry and Biochemistry, Flamingovo náměstí 2, Praha 6, CSSR

KEIL-DLOUHÁ, V., Institut de Chimie des Substances Naturelles, 91 Gif-Sur-Yvette, France

FEENEY, R. E., University of California, 209 Roadhouse Hall, Davis, California 95616, U. S. A.

FEINSTEIN, G., Tel-Aviv University, Dept. of Biochemistry, Tel-Aviv, Israel

FINK, E., Institut für Klinische Chemie und Klinische Biochemie, 8 München 15, Nußbaumstr. 20, W.-Germany

FRITZ, H., Institut für Klinische Chemie und Klinische Biochemie, 8 München 15, Nußbaumstr. 20, W.-Germany

GREENE, L. J., Brookhaven National Laboratory, Upton, L. I., N. Y. 11973, U. S. A.

HEIMBURGER, N., Behringwerke AG, 355 Marburg, W.-Germany

HUBER, R., Max Planck-Institut für Eiweiß- und Lederforschung, 8 München 15, Schiller Straße 46, W.-Germany

HOCHSTRASSER, K., Institut für Klinische Chemie und Klinische Biochemie, 8 München 15, Nußbaumstr. 20, W.-Germany

IKENAKA, T., Osaka College of Science, Toyonaka, Japan

KASSELL, B., The Medical College of Wisconsin, Dept. of Biochemistry, 561 North Fifteenth Street, Milwaukee, Wisconsin

LASKOWSKI, M., Jr., Purdue University, Dept. of Chemistry, Lafayette, Indiana 47907, U. S. A.

LASKOWSKI, M., Sr., Roswell Park Memorial Institute, 666 Elm Street, Buffalo, New York 14203, U. S. A.

LIENER, I. E., University of Minnesota, Dept. of Biochemistry, 140 Gortner Laboratory, St. Paul, Minnesota 55101, U. S. A.

PEANASKY, R. J., The University of South Dakota, Dept. of Biochemistry, Vermillion, South Dakota 57069, U. S. A.

RIGBI, M., The Hebrew University of Jerusalem, Dept. of Biological Chemistry, Jerusalem, Israel

RYAN, C. A., Washington State University, Dept. of Agricultural Chemistry, Pullman, Washington 99163, U. S. A.

SCHUMACHER, G. F. B., The University of Chicago, Dept. of Obstetrics and Gynecology, 5841 Maryland Avenue, Chicago, Illinois 60637, U. S. A.

SHOTTON, D., The Molecular Enzymology Laboratory, Department of Biochemistry, University of Bristol, Bristol BS 9 1 AE, England

STEVENS, F. C., The University of Manitoba, Dept. of Biochemistry, Winnipeg 3, Manitoba, Canada

TRAVIS, J., The University of Georgia, Dept. of Biochemistry, Athens, Georgia 30601, U. S. A.

TSCHESCHE, H., Organisch-Chemisches Laboratorium der Technischen Universität München, Lehrstuhl für Organische Chemie und Biochemie, 8 München 2, Arcisstr. 21, W.-Germany

WERLE, E., Institut für Klinische Chemie und Klinische Biochemie, 8 München 15, Nußbaumstr. 20, W.-Germany

ZANEVELD, L. J. D., The University of Georgia, Dept. of Biochemistry, Athens, Georgia 30601, U. S. A.

Proteinase Inhibitors of Human Plasma

N. Heimburger, H. Haupt and H. G. Schwick

Behringwerke AG, Marburg (Lahn)

Human plasma may be assumed to contain far more than a hundred single proteins. It is generally agreed that each protein has a specific function. The most simple function might be the transport of inorganic ions and organic substances. Transport functions involve binding, and plasma contains such proteins which bind generally or specifically, e. g. transferrin acts as carrier of iron only, whereas albumin combines with such different substances as fatty acids, heme, bilirubin, copper and hormones. In special cases, not only a single, but several proteins are engaged in the same function: Thyroxin is bound by three proteins, prealbumin, albumin but only specifically by the so-called Thyroxin binding globulin.

Finally there are proteins in plasma which do not reveal any biological activity as single proteins until after interaction with one or more plasma factors. The interaction may have the character of a cascade reaction, as typified not only in the activation of coagulation and fibrinolysis but also in the kallikrein and complement system. The final products of the activation chains are enzymes. It is not surprising therefore, that human plasma contains proteins which can bind and neutralize these enzymes. These proteins are generally termed proteinase inhibitors (PI). It is thought that their function is to limit the action of the proteolytic enzymes to the biological requirements.

First observations concerning the antitryptic activity of serum were obtained in 1894 by Fermi and Pernossi [1]. In 1897 three papers appeared concerning the proteinase inhibitors of serum; namely from Camus and Gley [2], Pugliese and Coggi [3] and Hahn [4]. In 1900 Landsteiner [5] reported the concentration of the antitryptic activity by salt fractionation procedures in the albumin fraction. In the following years many workers tried to find out the chemical character of the PI's: Delezenne and Pozersky [6], Schwartz [7], Bauer [8] and Jobling [9] assumed the antitryptic factors to be lipids because they were destroyed by treatment with chloroform. Their hypothesis, however, was not generally accepted [10, 11, 12]. At the same time Hedin [13], Wiens [14], Meyer [15] and Eisner [16] occupied themselves with the biological nature of the PI and their mechanism of action. They came to the conclusion that the "antifermente" had the character of antibodies. As a result of the discovery of the antitryptic activity of serum, its concentration and partition was studied during various diseases, e. g. in Pneumonia, an increase in antitryptic activity in the beginning of sickness, an elevated level in the acute phase and a reduction in the third phase was found (Ascoli and Bezzola, 1903 [17]). Deviations of normal values were observed in various infectious diseases [18, 19, 20]. Also in cancer the trypsin inhibitors were found to be increased, but did not prove to be specific [21].

Independently from the antitryptic activity the antithrombin efficiency of human serum was observed in 1905 by Morawitz [22].

Separation of the inhibitors begun by LANDSTEINER [5] in 1900 was continued by SCHMITZ [23] in 1938; he used the methods of NORTHROP and KUNITZ [24] which already had been proved in the isolation of the peptide inhibitor of the pancreas. By this technic a TCA-soluble inhibitor which inactivated trypsin in molar ratios, but not chymotrypsin, was isolated. His results were confirmed by GROB [25] and DUTHIE and LORENZ [26] who used methods according to SCHMITZ [23]. But the procedures used resulted in partially denatured inhibitors which could not be exactly characterized. This is the reason why the PI's isolated by the above mentioned scientists cannot be attributed to one of the now known inhibitors.

In 1939, SMITH and LINDSLEY [27], using the recently introduced electrophoresis, localized the main antitryptic activity of serum in the albumin zone. More than ten years later two inhibitors were demonstrated in serum by SHULMAN [28]. JACOBSSON [29] in 1953 reported their electrophoretic separation. Some time later the same author [30] published that 90% of the antitryptic activity is localized in the α_1-globulin zone and 10% in the α_2-globulin zone and antiplasmin activity in the slower α_2-zone only. A 12-fold concentration of the α_1-trypsin-inhibitor was reported by MOLL and co-workers [31] in 1958 before BUNDY and MEHL succeeded in its isolation in a native state [32] in 1959.

Human plasma is now known to contain at least six PI's. Starch gel-electrophoresis and fibrin-containing agar plates, as used in the sandwich technic [33, 34], have proved to be very helpful in the discovery, identification and isolation of these PI's. This technic is demonstrated in Fig. 1. To prove the presence of PI, plasma is separated electrophoretically in starch gel. The gel is then cut into two discs; the upper one is colored with amido black and the lower one is used to cover agar-plates containing heat-inactivated fibrin. As soon as the proteins have entered the fibrin-agar film, troughs parallel to the migration direction are cut and filled with enzyme solutions. The substrate plates are then incubated at 37° C for about 20 hours. During this time the proteases enter the gel. Diffusion is evidenced by lysis of the fibrin. Only in the electrophoretic positions of the inhibitors fibrin remains free

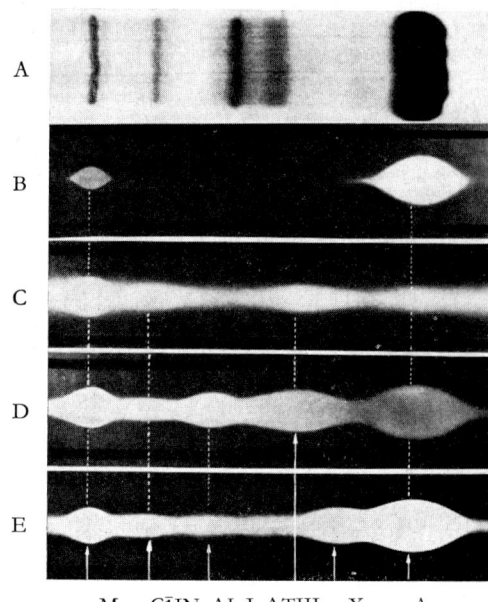

Fig. 1. Identification of six proteinase inhibitors in human plasma.
Starch gel electrophoresis used in combination with fibrin agar plates according to the sandwich technic. A. Electrophoregram of plasma stained with amido-black; fibrin plates after being covered with and lysed by elastase (B), plasmin (C), trypsin (D) and chymotrypsin (E).

from attack. By this technic plasma can be demonstrated to contain inhibitors against elastase, plasmin, trypsin, chymotrypsin, in total six, from which five proved to be polyvalent. The PI's were determined according to their electrophoretic mobility and to their, first recognized or most important, physiological function, as far as known, at the time of isolation.

Many scientists have been involved in the isolation and identification of the PI's, and in most cases, these occurred independent of each other. In some cases proteinase inhibitors were well characterized chemically, but their inhibitor function was at that time still unknown, e. g. α_1-antitrypsin (α_1 A) was isolated in 1955 in our laboratories by SCHULTZE and co-workers [35] as α_1-3.5-glycoprotein. Its identity as the main trypsin inhibitor of plasma was not recognized until 1962 [36] when fibrin-agar electrophoresis [33] was used to control our processing of α_1-3.5-glycoprotein. Then, we became aware too, that the inhibitor is irreversibly denatured when the pH of the solution drops below pH 5. We designated the inhibitor as α_1-antitrypsin [36] and recognized it to be identical with the inhibitor isolated three years previously by BUNDY and MEHL [32].

α_2-Macroglobulin (α_2M) also was not identified as an inhibitor until 9 years after its isolation [37] and chemical characterization in our laboratories [35, 38, 39].

α_1-Antichymotrypsin (α_1X) was also isolated and characterized by our laboratories [40] and identified three years later as the most specific inactivator present in serum [34].

Antithrombin III (AT III) was purified by ABILDGAARD [41] in 1967 and at the same time also in our laboratories [42]. However, before this, a concentrated, but impure preparation of AT III was achieved by LOEB [43], HENSEN and LOELIGER [44] and MARKWARDT and WALSMAN [45].

The inter α-trypsininhibitor (IαI) was first described as π protein by STEINBUCH [46]. Four years later in 1965 a protein with the chemical properties of the π protein was isolated and characterized as an inhibitor [47].

The inactivator efficiency of human serum for C1-esterase was observed by RATNOFF and LEPOW [48] in 1957. Partial purification and some properties were later reported in 1961 by the same scientists [49]. About this time α_2-neuraminoglycoprotein was isolated by SCHULTZE and co-workers [50]. Its function was unknown. However, in 1969 [51] it was shown to be identical with C$\bar{1}$-inactivator.

Isolation

Fractionation charts for all the six PI's from human serum are shown in tables 1—4. Concerning the methods used, it can be stated generally, that crude purification of the PI was obtained by their different solubilities in ammonium sulfate, fractionation with Rivanol, ethanol precipitation and specific adsorption steps.

A high degree of purification was obtained with zone electrophoresis and sieve gel-filtration. α_2-M isolation was started with plasma. To prevent kallikrein contact-activation, plasma should be drawn in plastic tubes and/or with peptide inhibitors. Otherwise α_2-M partially saturated with kallikrein will be obtained [52]. For the isolation of C$\bar{1}$ INA and IαI freshly recalcified serum is used. Processing is started with adsorption on DE-32; during the buffer gradient elution IαI is obtained in a fraction separate from C$\bar{1}$ INA and can then be highly purified by ammonium sulfate precipitation and Sephadex G-150 gel-filtration.

α_1A, α_1X and AT III are enriched in the mother liquid of albumin after crystallization with ammonium sulfate [53]. All three inhibitors are characterized by their solubility in a 50% saturated ammonium sulfate solution. α_1A and α_1X were separated from each other by their different precipitability with Rivanol. Further purification of α_1X was achieved by zone electrophoresis at pH 5.2, and of α_1A by ethanolfractionation and electrophoresis too.

The most important steps in the isolation of AT III are the adsorption on $Ca_3(PO_4)_2$ and the zone electrophoresis. The last fractionation step of all procedures required gel-filtration in order to separate the monomer forms of the inhibitors from the polymers generated during the isolation procedure.

The following abbreviations are used: α_1A, α_1-Antitrypsin; α_2M, α_2-Macroglobulin; α_1X, α_1-Antichymotrypsin; AT III, Antithrombin III; IαI, Inter α-Trypsininhibitor; C$\bar{1}$ INA, C$\bar{1}$-Inactivator.

Table 1. Fractionation Flow chart for α_1-Antitrypsin and α_1-Antichymotrypsin

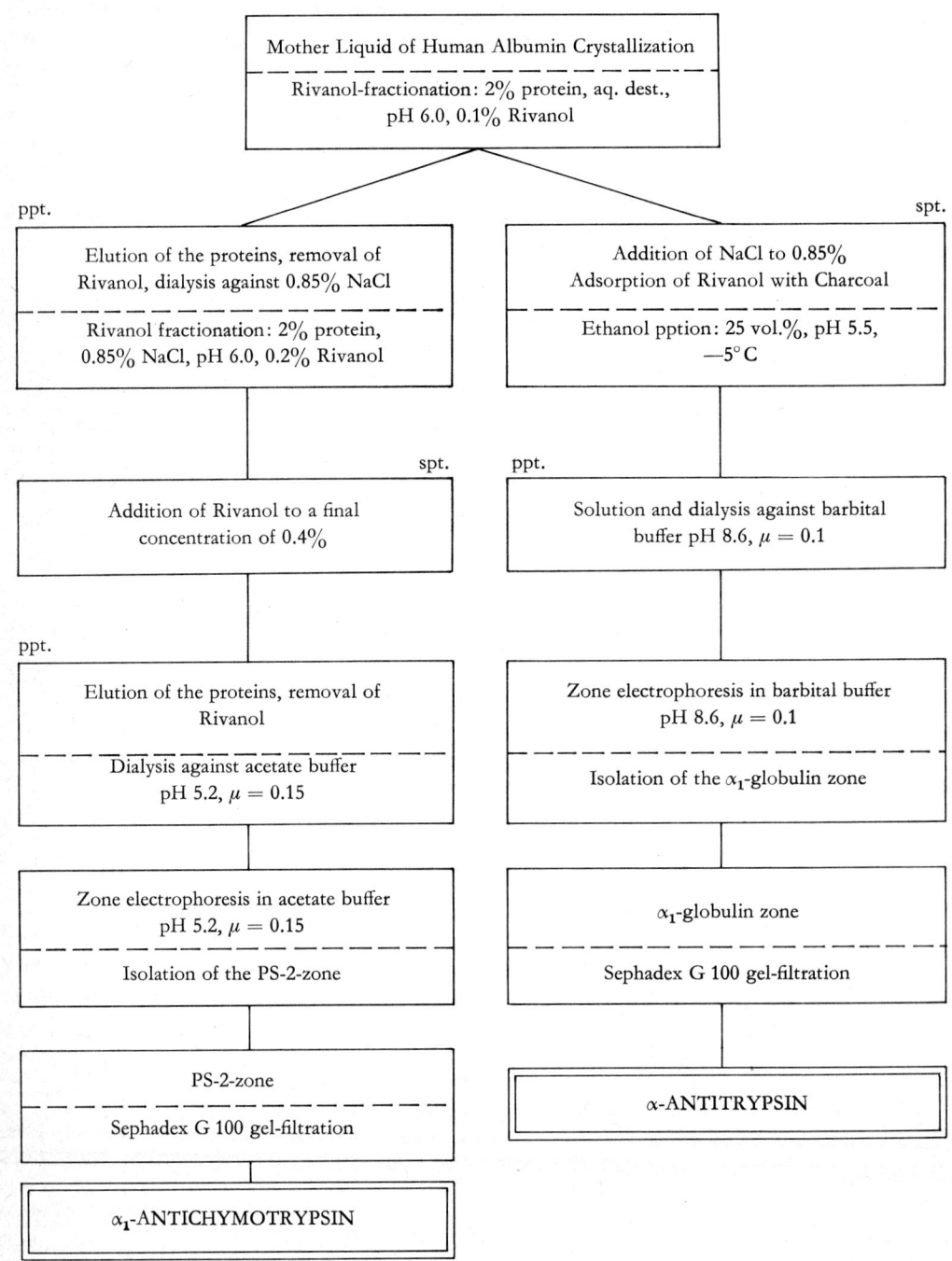

ppt, precipitation/spt, separation

Table 2. Fractionation Flow Chart for C$\bar{1}$-Inactivator and Inter-α-Trypsininhibitor

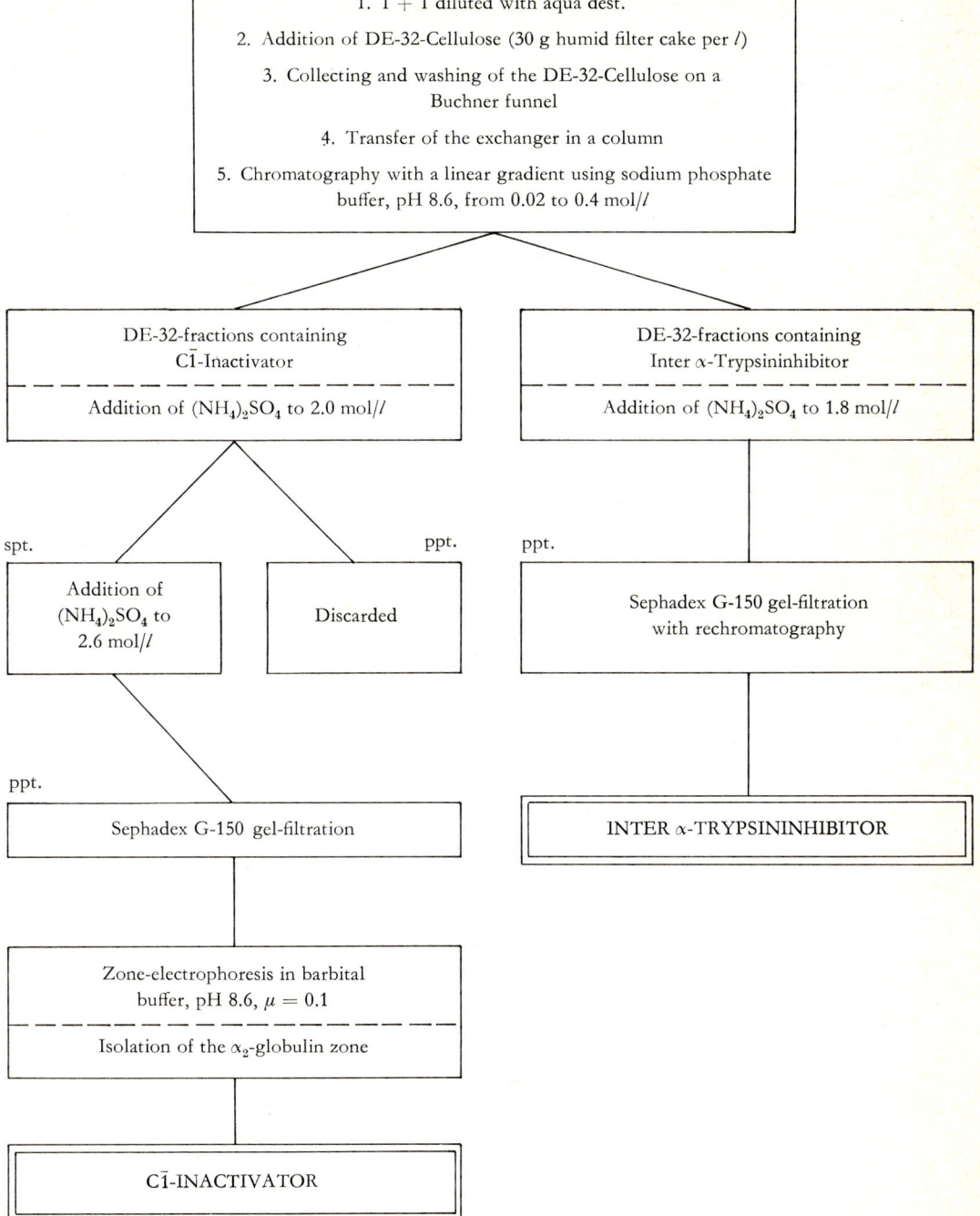

ppt, precipitation/spt, separation

Table 3. Fractionation Flow Chart for Antithrombin III

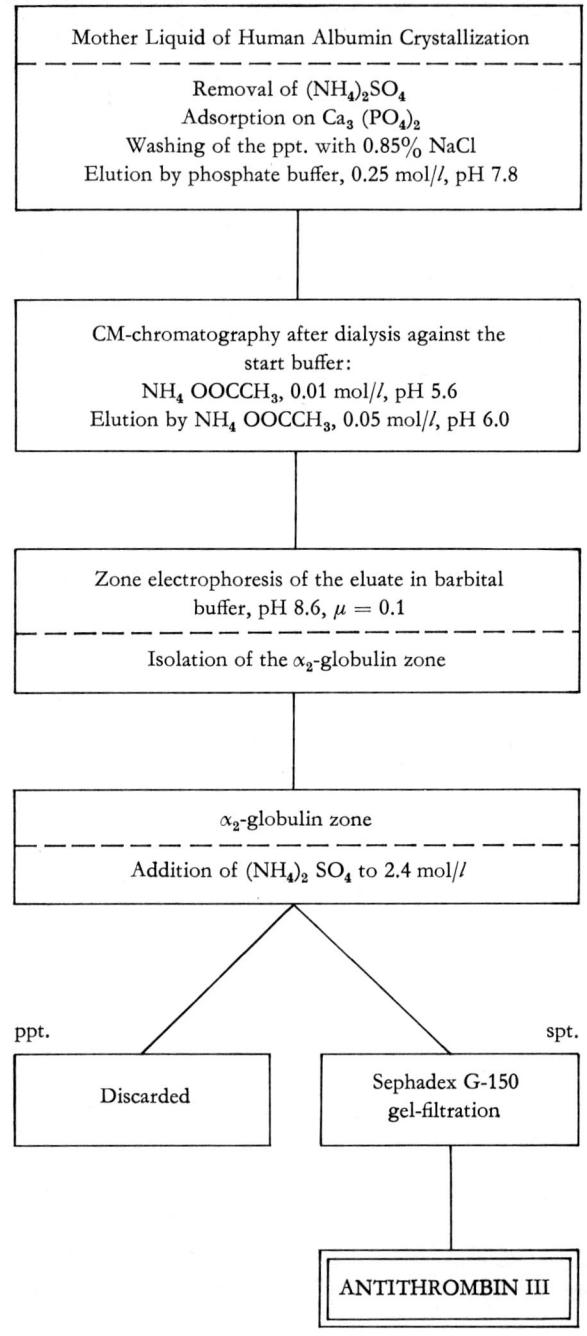

ppt, precipitation/spt, separation

Table 4. Fractionation Flow Chart for α_2-Macroglobulin

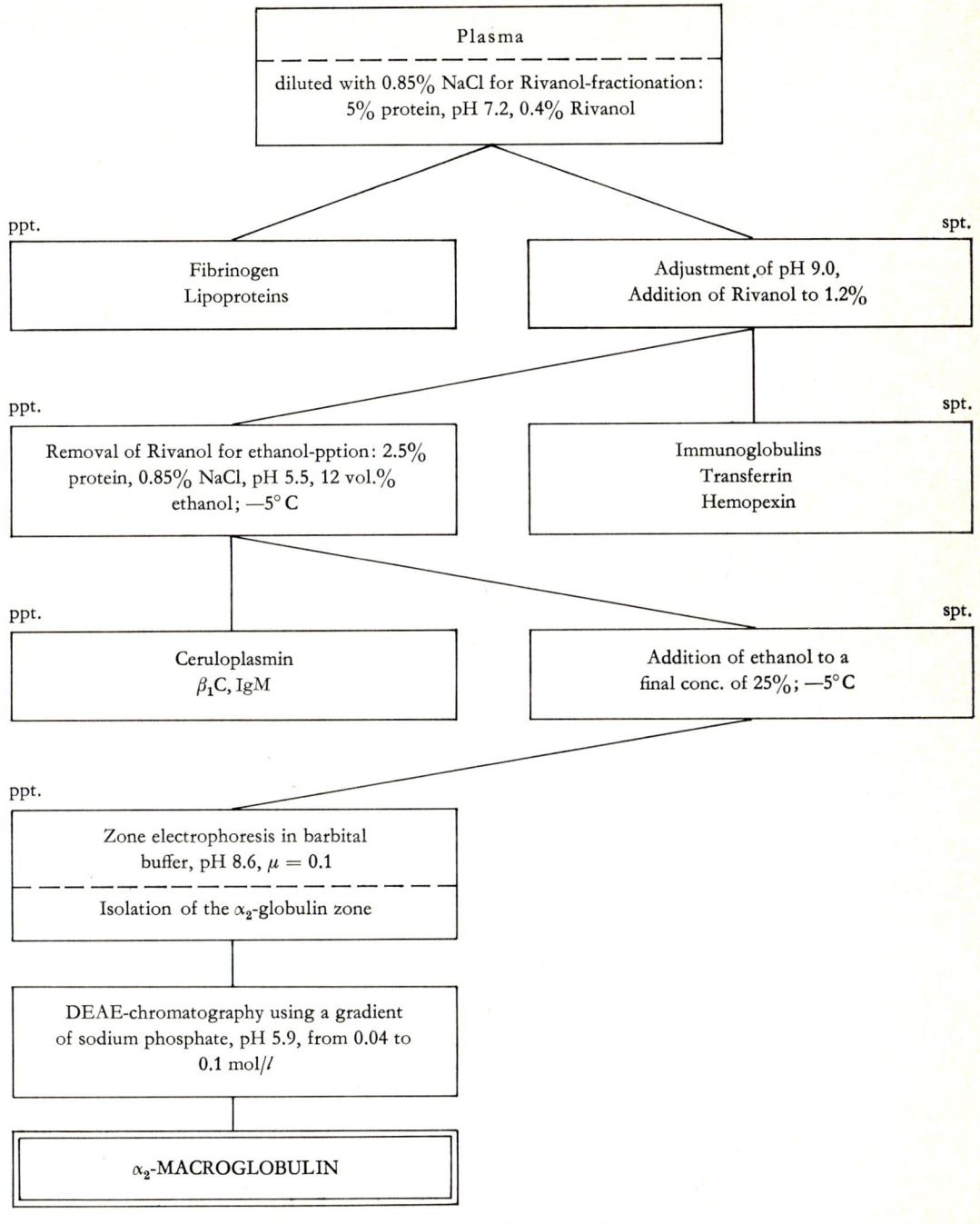

ppt, precipitation/spt. separation

Special Sensitivity

The PI proved to be extremely sensitive, during fractionation, to fractionation methods generally used in protein chemistry; e. g. as illustrated by the following examples: $\alpha_1 A$ tends to form high molecular partially-inactivated aggregates when precipitated by ethanol. Despite this we use this fractionation step because good elimination of all other α-globulins is achieved therewith.

In Fig. 2A separation can be seen of the polymers and dimers from the monomer form on Sephadex G-100 and their electrophoretic characterization. Aggregation of AT III occurs already during dialysis against water and is accompanied by inactivation (Fig. 2B). It becomes more evident also when the ion strength of plasma or serum is reduced. $\alpha_2 M$ is inactivated when it has been precipitated by ammonium sulfate about or above the neutral point as evidenced by its increased mobility in polyacrylamide gels (Fig. 2C). IαI can be aggregated by heating [54] to 56°C for 30 minutes. Although the aggregates formed are of too high a molecular weight for polyacrylamide gel sieving, they remain still active as trypsin inhibitors (Fig. 3). By treatment of plasma with sulfhydryl reagents, $\alpha_2 M$ is preferentially inactivated and probably C$\overline{1}$ INA as well. Addition of ammonia, in a concentration so that a pH of 8.5 is not exceeded, causes a relative specific, and approximately quantitative inactivation of $\alpha_2 M$ in plasma. This inactivation is irreversible in contrast to that produced by aliphatic amines [55, 56].

Finally it should be mentioned that a progressive denaturation can be frequently observed when the pH is below 5. The preceding examples show the difficulties encountered in the isolation of the inhibitors of serum in a native state, particularly because of their high sensitivity. Proved

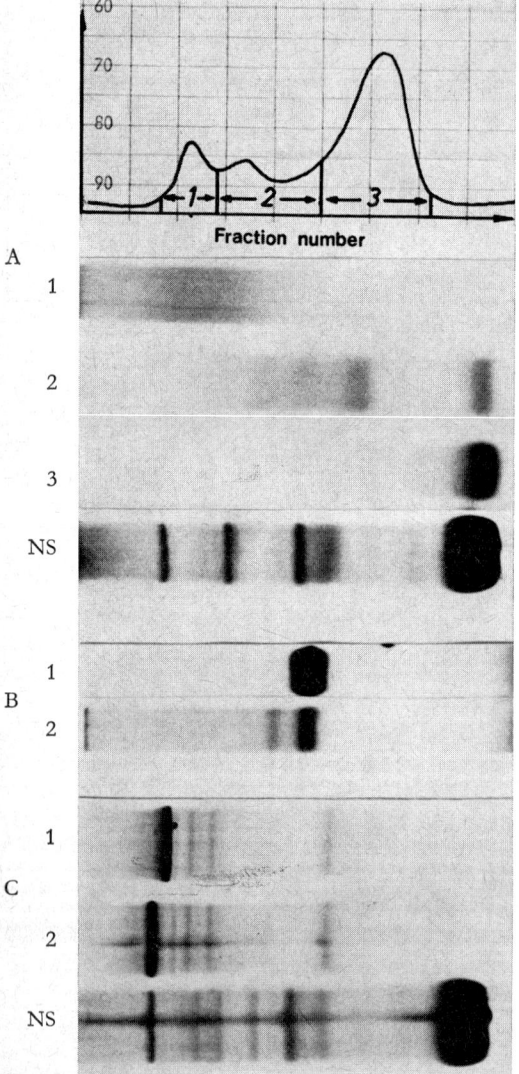

Fig. 2. Denaturation of the proteinase inhibitors by normal fractionation methods as demonstrated by starch gel electrophoresis.

A) Aggregation of α_1-antitrypsin following ethanol separation of the aggregates on Sephadex G-100 and characterization of the fractions (below).
B) Aggregation of antithrombin III in an ion-poor environment:
   1. before, 2. after dialysis against aq. dest.
C) Inactivation of α_2-macroglobulin as evidenced by its increased mobility after $(NH_4)_2 SO_4$ pption. in alkaline medium: 1. 50% ppt at pH 8.0 and 2. at pH 5.0 (NS = normal serum)

fractionation methods cause an alteration in structure, which is of basic importance for the interaction with the specific enzymes.

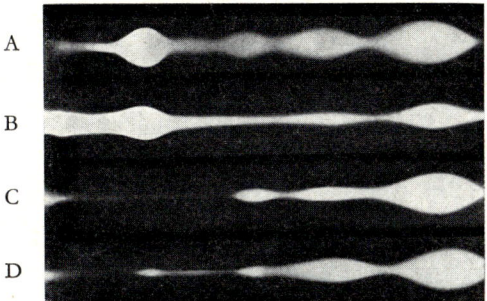

Fig. 3. Stability of the inhibitors under various conditions. Lysis of fibrin agar plates being covered with starch gel electrophoretograms of
A) normal plasma,
B) plasma heated at 56°C for 30 min.,
C) plasma treated with thioglycollate (final conc. 0,1M),
D) plasma adjusted to pH 8.5 using NH_4OH, (troughs contain trypsin).

Chemical Characterization

All six PI's are characterized by their electrophoretic mobility in polyacrylamide gels in which migration is a function of the net charge, the form and the molecular seize (Fig. 4): the low molecular α_1-globulins ($\alpha_1 A$, $\alpha_1 X$) migrate in front of all, the high molecular α_2-globulins ($C\bar{1}$ INA, $\alpha_2 M$) remain near the start point. In between the inter-α-globulins, IαI and AT III are found, the latter nearer to the front due to its lower molecular weight.

Tab. 5 shows the most relevant data of the PI's Plasma is rich in $\alpha_1 A$ and $\alpha_2 M$. Ten weight per cent of the plasma proteins are inhibitors,

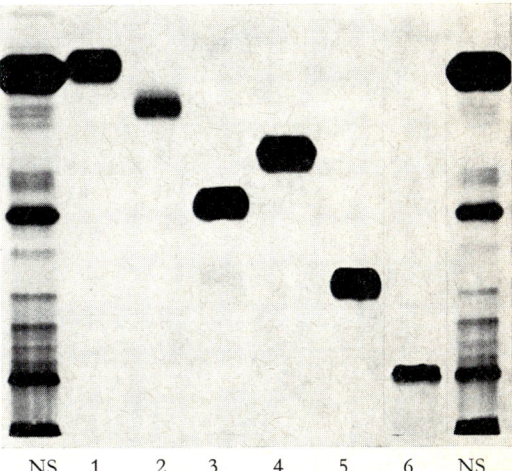

Fig. 4. Polyacrylamide gel-electrophoresis. Characterization of the isolated inhibitors.

NS = normal serum; 1. α_1-antitrypsin; 2. α_1-antichymotrypsin; 3. inter α-trypsininhibitor; 4. antithrombin III; 5. $C\bar{1}$ inactivator; 6. α_2-macroglobulin.

considering the molar concentration, $\alpha_1 A$ is the predominant. The molar concentration of the others gives a total of 40% of $\alpha_1 A$. It should be noted, that PI's are all glycoproteins. Especially

Table 5. Concentration and characterization of the Inhibitors in Human Serum

Inhibitors	mg/100 ml C. Mean ± SD	MW	μMol	Peptide Content (%)	Carbohydrate Content (%)
1. α_1-Antitrypsin	290.0 ± 45.0	54.000	54.0	86	12.2
2. α_1-Antichymotrypsin	48.7 ± 6.5	69.000	7.0	73	24.6
3. Inter α-Trypsininhibitor	50.0	160.000	3.1	90	8.4
4. Antithrombin III	29.0 ± 2.9	65.000	4.5	85	13.4
5. $C\bar{1}$-Inactivator	23.5 ± 3.0	104.000	2.3	65	34.7
6. α_2-Macroglobulin	260.0 ± 70.0	820.000	3.3	92	7.7
	701.2				

Table 6. % of Carbohydrate Residues in Proteinase Inhibitors

	α_1-Antitrypsin	α_1-Antichymotrypsin	Inter α-trypsin inhibitor	Antithrombin III	$\overline{C1}$-Inactivator	α_2-macroglobulin
Hexoses	5.0	9.9	3.1	6.2	10.8	3.2
Fucose	0.2	0.7	0.1		0.4	0.1
Acetylhexosamines	3.6	7.4	3.2	4.1	9.2	2.7
Sialic acid	3.4	6.6	2.0	3.1	14.3	1.7
	12.2	24.6	8.4	13.4	34.7	7.7

Table 7. Amino Acid Composition of the Proteinase Inhibitors (% Amino Acid Residue)

	α_1-Antitrypsin	α_1-Antichymotrypsin	Inter-α-trypsininhibitor	Antithrombin III	$\overline{C1}$-Inactivator	α_2-macroglobulin
Lysine	8.41	5.30	5.63	6.65	4.43	5.75
Histidine	3.33	1.80	2.20	1.12	1.59	2.85
Ammonia	1.45	1.26	1.44	1.21	2.49	1.23
Arginine	2.33	3.49	5.18	4.70	2.72	3.86
Aspartic acid	9.75	7.97	9.50	8.60	5.92	7.08
Threonine	5.66	4.65	4.64	3.78	5.88	5.36
Serine	3.52	3.97	4.87	4.38	4.34	5.51
Glutamic acid	12.92	10.27	11.97	11.70	7.64	12.29
Proline	3.26	2.20	3.69	3.24	3.53	3.96
Glycine	2.38	1.34	2.90	1.75	0.93	2.75
Alanine	3.55	3.28	3.37	3.53	2.49	3.50
Half cystine	0.00	0.00	1.69	0.93	0.55	1.52
Valine	4.74	3.65	5.86	4.32	3.59	6.93
Methionine	2.14	2.28	2.00	2.17	1.79	1.63
Isoleucine	3.89	3.38	3.91	3.57	2.26	3.16
Leucine	9.90	9.05	7.40	7.23	7.09	7.77
Tyrosine	2.14	2.36	4.01	3.13	2.64	4.75
Phenylalanine	7.63	5.58	5.00	5.72	4.14	4.63
Tryptophan	0.55	0.80	1.69	1.99	1.31	1.30
Carbohydrates	12.20	24.60	8.40	13.40	34.70	7.70
Total	99.75	97.23	95.35	93.12	100.03	93.53

rich in carbohydrates are $\overline{C1}$ INA and α_1X with a content of 35, respectively 25%.

Tab. 6 shows the carbohydrate composition of all six inhibitors. The function of the carbohydrates is still unknown. By treatment with neuraminidase only the electrophoretic mobility, but not the biologic activity, has been affected (Fig. 5). It can be seen from the amino acid composition (Tab. 7), that α_1A and α_1X are free from cysteine [57], and that the $\overline{C1}$ INA contains only a few cysteine residues [58]. The absence of structure stabilizing disulfide bridges might explain the rapid inactivation below pH of 5.0. A connection between the amino acid composition and the binding function of the inhibitors for different enzymes cannot be seen from the amino acid composition. But comparing the analysis of the 6 PI's, it is evident that $\overline{C1}$ INA contains the lowest amount of aspartic and glutamic acid; in addition the basic amino acids

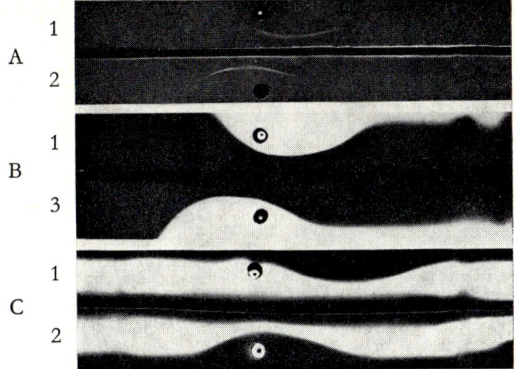

Fig. 5. Characterization of antithrombin III before (1) and after (2) treatment with neuraminidase.
A) Immunoelectrophoresis using anti α_1-antitrypsin.
B) Fibrin agar-electrophoresis with trypsin in the trough.
C) Fibrinogen agarose-electrophoresis with thrombin in the trough.

are found to be also diminished. Accordingly the net charge of C$\bar{1}$ INA is mainly due to the high content of sialic acid (Tab. 6).

We found with C$\bar{1}$ INA that inactivation in acid medium was accompanied by an increase of the sedimentation and diffusion coefficient as well [58]. It is therefore assumed that the inactivation has been caused by an alteration in tertiary structure. These changes are evaluated by titrating the inhibitors with suitable proteolytic enzymes. Immunological methods are applied in our laboratories for the detection of PI's in purification procedures; they have been of special value in the isolation of the inhibitors with α_1-globulin mobility, which are so similar in many properties. Contaminants as low as 5% are still evident with this method and cannot be seen by the use of other methods.

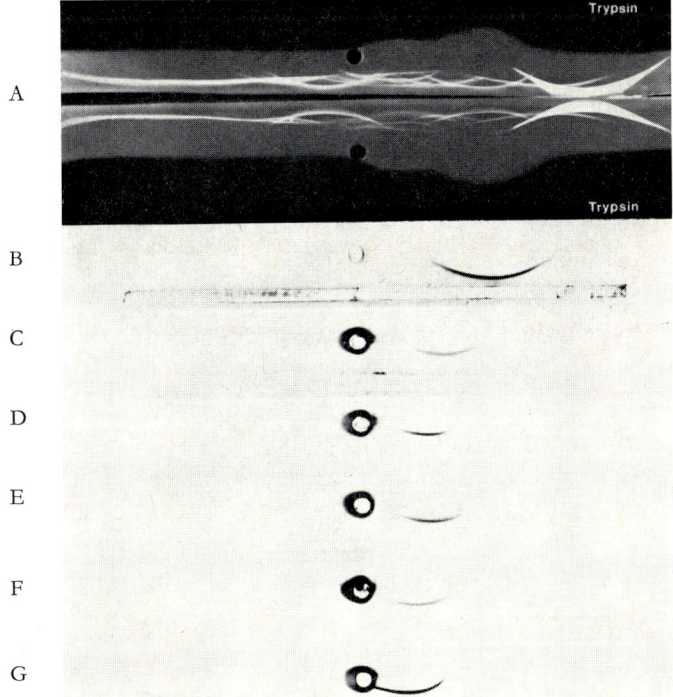

Fig. 6. Identification of six proteinase inhibitors in human serum by immunoelectrophoresis using monospecific antisera.
Electrophoresis using fibrin agar (A) and agar only (B-G).
Troughs containing antisera to A) human serum proteins; B) α_1A; C) α_1X; D) AT III; E) IαI; F) C$\bar{1}$ INA; G) α_2M. The upper plate was lysed by trypsin to indicate the electrophoretic position of the proteinase inhibitors.

By using the purified PI for immunizing rabbits monospecific-antisera were obtained. These antisera are helpful to quantitate and to characterize the inhibitors as for example in immunoelectrophoresis as illustrated in Fig. 6. Finally it should be pointed out that the immunological quantitation of the inhibitors is much easier than the usual enzymatic determinations. We have found good agreement between these methods. This combination might be of special diagnostic value in recognizing molecular diseases.

Biological Characterization

In Tab. 8 inhibitors and enzymes are compiled according to their interaction. The predominant inhibitors $\alpha_1 A$ and $\alpha_2 M$ also appear to be the most polyvalent; only $\alpha_1 X$ proves to be specific against chymotrypsin. Trypsin and chymotrypsin are the enzymes which are inactivated by most of the inhibitors. The inhibition was shown to be stoichiometric and irreversible [57]. We found lower values for AT III [42] which are probably due to a partial inactivation of the preparations tested. The affinity of trypsin to $\alpha_2 M$ is 7 fold stronger than to all other trypsin inhibitors [59] and also in contrast to these 1 mol $\alpha_2 M$ combined with 2 mols trypsin [60]. Lower combining ratios were found with $\alpha_2 M$ isolated from contact-activated plasma [57], which contained already bound kallikrein and plasmin. The binding sites of $\alpha_2 M$ for trypsin are probably non-equivalent [59]. But it might be of special interest that 80% of the tryptic activity was found preserved in the complex with trypsin, as could be proved only when low-molecular substrates were used [61]. Physiologically more important are the inhibitors which combine with the proteases of blood, i. e. plasmin, thrombin and kallikrein. Of special interest is that the same inhibitors which combine with plasmin neutralize plasma kallikrein also. The mechanism of action as well

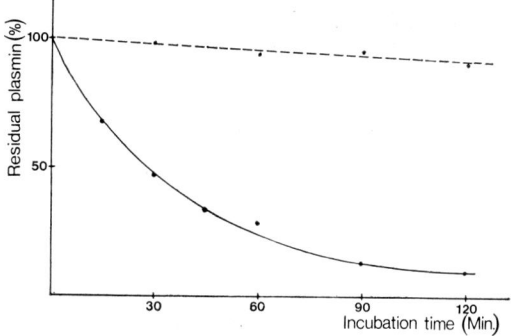

Fig. 7. Progressive inhibition of human plasmin stabilized with methylamine by α_1-antitrypsin.
Residual plasmin (-) was determined from a mixture containing 1 Remmert and Cohen unit plasmin and 0,5 mg α_1-antitrypsin per ml at various incubation times at 37°C. (---) Controls without α_1-antitrypsin.

Table 8. Specificity of human plasma inhibitors

Inhibitors	Enzymes							
	Trypsin	Chymo-trypsin	Plasmin	Plasma-Kallikrein	Pancreas-Kallikrein	Thrombin	C$\overline{1}$-Esterase	Elastase
1. α_1-Antitrypsin	+	+	+	?	+	−	−	+
2. α_1-Antichymotrypsin	−	+	−	?	−	−	−	−
3. Inter α-Trypsininhibitor	+	weak	−	?	−	−	−	−
4. Antithrombin III	+	−	weak	−	−	+	−	−
5. C$\overline{1}$-Inactivator	weak	weak	+	+	−	−	+	−
6. α_2-Macroglobulin	+	+	+	+	−	+	−	+

? not yet determined

Table 9. Ratio of „potential" plasmin to plasmin inhibitors in human serum

Plasmin inhibitors	MW	mg/l	μMol	Partition (%)
1. α_1-Antitrypsin (Progressive Antiplasmin)	54.000	2.900	54	91
2. α_2-Macroglobulin (Immediate Antiplasmin)	820.000	2.650	3.3	5.5
3. C$\bar{1}$-Inactivator	104.000	235	2.3	3.5
Plasmin-Inhibitor Capacity			59.6	100.0
Proactivator Plasminogen	90.000	200	2.2	

The molar ratio of "potential" plasmin to plasmin-inhibitors in human serum is 1:27

as the time-dependence of the inactivation is similar: C$\bar{1}$ INA and α_2M combine immediately with plasmin and α_1A inactives plasmin in a strongly time-dependent reaction [39] (Fig. 7). NORMAN, therefore classified the inhibitors, that he found in the α_1- and α_2-globulin region, as slow and immediate reacting antiplasmins [62]. The antiplasmin capacity of serum, as determined by NORMAN using the caseinolytic technic, is in good agreement with our immunological determined values (Tab. 9), 27 times more plasmin inhibitors were found in human serum than plasminogen (proactivator-plasminogen). For the calculation, molar ratios of interaction could be assumed, because α_2M also combines only with one mol plasmin as well. The combining ratios of the inhibitors for plasma and pancreas kallikrein have still not yet been determined. Until now, two of the six PI's have been identified as inactivators of plasma kallikrein: α_2M [63] and the C$\bar{1}$ INA [64]; both neutralize kallikrein in a time-dependent raction and compete against one another for the enzyme. α_2M-bound kallikrein has been proved also to be active against synthetic esters. Using these low-molecular weight substrates and different kallikrein concentrations, 90% of the applied enzyme activity was inactivated by the C$\bar{1}$ INA and 53% by α_2M; this ratio was not affected by the addition of 4 times more α_2M [63]. The physiological functions induced by plasma kallikrein, namely uterus contraction and vascular permeability, are completely suppressed by both α_2M and C$\bar{1}$ INA. The interactions between plasma kallikrein and α_1A are still not yet known. But α_1A was found by WERLE, FRITZ and co-workers to be identical with the progressive serum inhibitor of kallikrein from pancreas [65]. The inactivation has been characterized as a strongly time-dependent process, comparable to the plasmin inhibition by the same protein: no inhibition occurred at 0°C, but the inactivation ratio per minute increased with increasing temperature. The kallikrein inhibitor-capacity of human blood was estimated to be 150—200 times that of the potential kallikrein [65].

Two thrombin inhibitors can be identified in human plasma: α_2M [66, 67] and AT III [42, 68]. The main inactivator is certainly the AT III, as can be shown when normal plasma and AT III free plasma are compared (Fig. 8); in this graph it can be seen also, that the inactivation of thrombin is a time- and temperature-dependent reaction. The thrombin inactivation ratio per minute is determined mainly by the concentration of AT III. It is of diagnostic value because it is of relevance to various disease states. Of therapeutic importance was the observation, that the inactivation of thrombin can be catalytically accelerated by heparin. AT III is involved in this mechanism also (Fig. 9). Concerning the role of AT III we could prove, for the first time, that AT III is identical with the heparin co-factor [42], as was previously assumed by several workers.

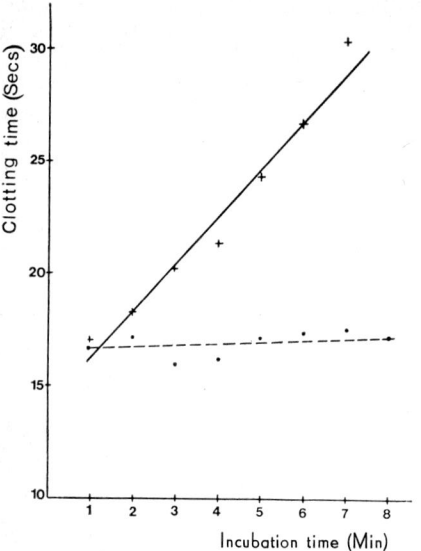

Fig. 8. AT III assay in heat-defibrinated plasma before (—) and after absorption with a specific antibody against AT III (----) (as performed in a coagulation system).

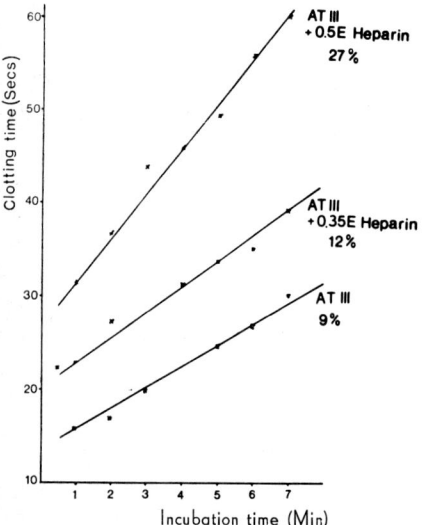

Fig. 9. Catalytic acceleration by heparin of the inactivation of thrombin by antithrombin III.

Fig. 10. shows the interaction of AT III and heparin and dissociation of the complex by the heparin-antidote protamine sulfate.

Our results concerning the mechanism of the thrombin inhibitors, including $\alpha_2 M$, are in accordance with those reported by ABILDGAARD [69, 70]. He used a test system, which closely approaches physiological conditions. ABILDGAARD determined the release of the fibrinopeptides A and B by thrombin, in pre-

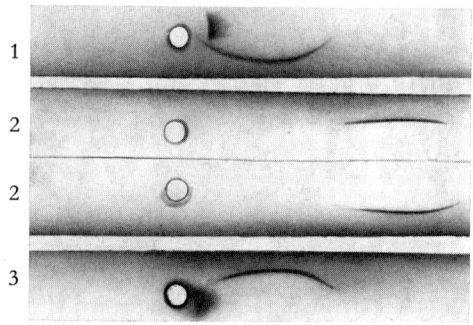

Fig. 10. Immunoelectrophoretic characterization of antithrombin III in human plasma.
1. Normal plasma; 2. plasma + 500/ng heparin/ml;
3. plasma + 500/ng heparin + 500/ng protamine sulfate/ml. Troughs containing anti-antithrombin III-serum.

sence of the inhibitors, by the quantification of the generated N-terminal residues. Using this method the binding of thrombin by $\alpha_2 M$ was characterized to be a strongly time-dependent reaction [70]. Despite this, thrombin was demonstrated to be inactivated by $\alpha_2 M$ even in the presence of fibrinogen [70].

Using the quantitative technic of radial immunodiffusion, 20—30% lower values are found in serum than in plasma of the same subject [71, 72, 73]. This may be due to a partial consumption during blood clotting. Tab. 10 illustrates the partition of the total progressive anti-thrombin capacity in human plasma; furthermore it is evident that plasma contains 4.5 times more inhibitors than the (pro-)thrombin that can be maximally generated. All calculations were performed under the assumption of molar interactions.

Only one PI, endowed with the ability to neutralize $C\bar{1}$ esterase of complement, is found in plasma, the so-called $C\bar{1}$ inactivator [74]. Therefore it is of special physiological importance and

Table 10. Partition of the progressive thrombin-inhibitor capacity
in human plasma
(quantified immunologically)

Antithrombins	MW	mg/l	μMol	Partition (%)
Antithrombin III	65.000	388	6.0	67
α_2-Macroglobulin	820.000	2.400	2.9	33
Antithrombin-Capacity			8.9	100
Thrombin	35.000	71*	2.0	

The molar ratio of "potential" thrombin to antithrombins is 1:4.5.
* calculated: 150 NIH-U thrombin per ml plasma and 2100 NIH-U per mg highly purified thrombin were assumed.

probably also in connection with its ability to inhibit plasma kallikrein and plasmin as well. Finally it should be mentioned, that the predominant and most polyvalent inhibitors of plasma, α_1A and α_2M, can also combine with elastase [75]. Its physiological importance is still not yet clear.

The mechanism of action of the proteinase inhibitors

The interactions between enzymes and inhibitors are characterized by the formation of a complex, in which the active sites of the enzymes are covered or, as in one exception, the active site is penetrated by molecules of a definite size. Enzyme-inhibitor complexes can be characterized by several and different methods. A simple technic, provided that monospecific antisera are available, is immunoelectrophoresis (Fig. 11 and 12). The complexes have a mobility different to the free inhibitors. Complex formation can also be quantitated immunologically [72] (Fig. 13). The AT III titre in serum is 20—30% lower than in plasma [71—73]. This phenomenon may be

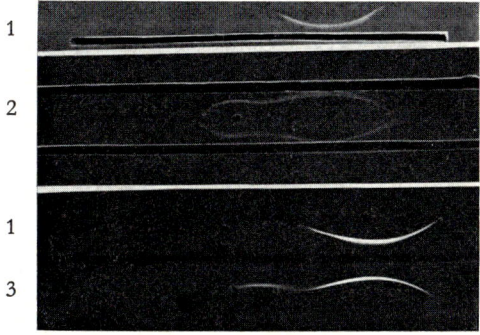

Fig. 11. Characterization of α_1-antitrypsin-enzyme complexes by a monospecific antiserum against α_1-antitrypsin. 1. α_1-antitrypsin; 2. in mixture with plasmin; 3. in mixture with trypsin.

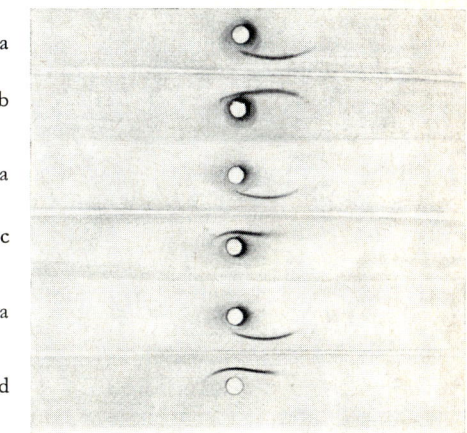

Fig. 12. Immunoelectrophoretic characterization of the thrombin-antithrombin complex.
a) Normal heat-defibrinated plasma; b-d) after addition of thrombin: b. 200 NIH-U/ml; c. 400 NIH-U/ml; d. 800 NIH-U/ml. Troughs contain anti-human antithrombin III serum from rabbits.

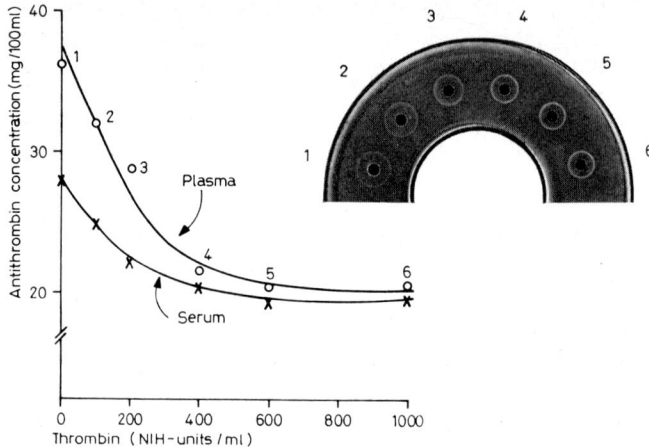

Fig. 13. Antithrombin level in serum and heat-defibrinated plasma after incubation with thrombin.

explained by the fact that AT III has combined, to some extent, with the thrombin generated during clotting. It is obvious that thrombin-saturated antithrombin is no longer fully accessible to its homologous antibody. This might explain that after addition of thrombin in excess, only one half of the original amount of AT III is to be found in plasma and in serum as well.

According to FINKENSTADT and LASKOWSKI [76, 77] the enzyme-inhibitor reaction is comparable to an enzyme-substrate reaction, with respect to the first phase: however, after the cleavage of the peptide chain, the sequence of which is specific for each protease, a co-valent bond is formed between enzyme and inhibitor. According to their specificity lysine- and arginine-residues are involved in the interaction of trypsin and plasmin with their inhibitors. To find out the sensitive peptide bond in the inhibitors which is to be cleaved, i. e. the reactive center, the method recommended by FRITZ and co-workers [78] was used. The lysyl- and arginyl-residues were modified using maleic anhydride and 2.3-butanedione reagent respectively. The treated inhibitors were compared to controls dissolved in the reaction medium only.

The polymaleoyl derivatives of α_1A and AT III are characterized by an increased mobility and

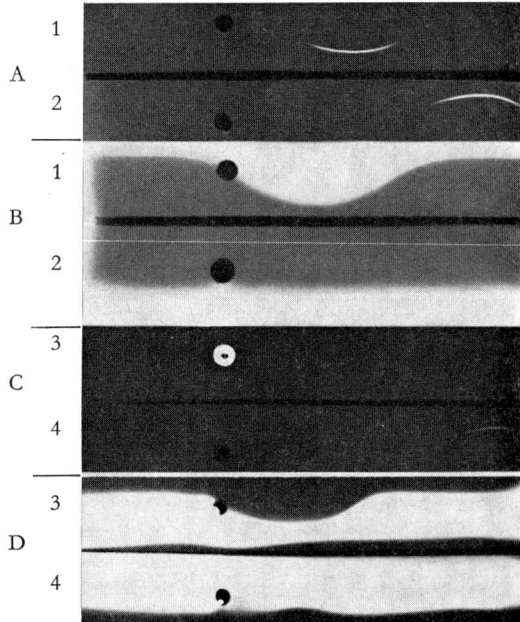

Fig. 14. Identification of lysine as the reactive center in α_1-antitrypsin for trypsin, and in antithrombin III for thrombin.

Immunoelectrophoresis with antisera against α_1-antitrypsin (A) and α_2-macroglobulin (C). B) Fibrin agar-electrophoresis with trypsin and D) fibrinogen agarose-electrophoresis with thrombin in the trough. 1. α_1A; 2. α_1A + maleic anhydride; 3. AT III; 4. AT III + maleic anhydride.

a loss of their activity to inactivate trypsin and thrombin respectively (Fig. 14). The antitryptic activity of α_2M, however, is not sensitive to acylation of the ε-amino groups. After addition of 2.3-butanedione reagent only the activity is markedly reduced, and at the same time and in a comparable degree, its precipitability with the homologous antibody (Fig. 15). Therefore we assumed arginyl-residues to be the reactive centres of α_2M. The same arginine sequences or others may function simultaneously as antigenic sites. The IαI is also characterized by an arginine residue in the reactive center and no longer reacts with its antibody when the guanidino groups have been masked.

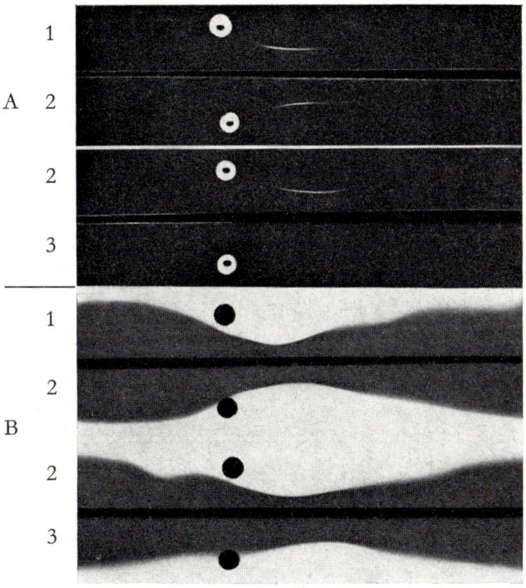

The interaction of α_2M with proteases has, however, a special characteristic. As already mentioned, the activity of trypsin, plasmin and thrombin to catalyze hydrolysis of low-molecular weight substrates remains intact in complex with α_2M [79]. It has been demonstrated that α_2M-bound enzymes are protected also against other inhibitors. Only one exception has been found: the activity of α_2M-bound trypsin could be completely abolished by the polyvalent pancreas trypsin inhibitor [80]. It is generally assumed today that the active site of the enzymes bound to α_2M is still accessible to molecules of a definite size. The physiological role of α_2M-bound enzymes have been the object of many speculations. I would like to point out, that α_2M-bound trypsin can function as an activator of plasminogen (Fig. 16). Using the gel diffusion

Fig. 15. Identification of arginine as the reactive center in α_2-macroglobulin for trypsin.
A) Immunoelectrophoresis with an anti-α_2-macroglobulin antiserum.
B) Fibrin agar-electrophoresis with trypsin.
1. α_2M; 2. α_2M + maleic anhydride; 3. α_2M + maleic anhydride + butanedione-(2.3) reagent.

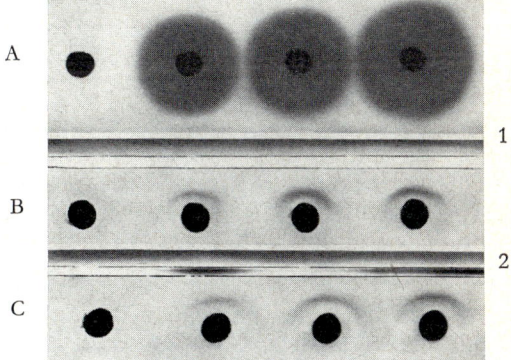

Fig. 16. Plasminogen-activator properties of the trypsin-α_2-macroglobulin complex.
A) Confirmation, using heated fibrin-agar plates, of the fibrinolytic activity of α_2M (500 mg%), and also in trypsin solutions containing 3, 6 and 9 mg% respectively (from the left to the right).
B+C) From the left: α_2M alone and admixtures with equal volumes of trypsin in the concentration as shown in A.
1. Bovine plasma; 2. human plasma.

with heated fibrin-agar plates, it can be demonstrated that a complex, which does not lyse fibrin, causes a liberation of fibrinolytic activity when it comes in contact with plasminogen. The activity is certainly low but perhaps therefore within the physiological range.

The physiological importance of the inhibitors

It is generally agreed that inhibitors control the proteolytic-regulated processes in blood, namely blood clotting, fibrinolysis, kinin liberation and the action of complement. For this reason, variations in the titre of the PI's have proved to be of diagnostic importance.

Antitryptic activity has been found considerably increased during a variety of processes such as metastatic carcinoma, acute diseases and inflammation. Linearly increasing concentrations of $\alpha_1 A$ to double the normal values at parturition, have been observed during pregnancy [81]. During the diagnostic investigations, electrophoretic polymorphism [82] and an $\alpha_1 A$-deficiency state have been found [83]. The various $\alpha_1 A$-patterns observed could be explained by the assumption of 7 co-dominant alleles in man. Of special clinical interest is the finding that hereditary $\alpha_1 A$-deficiency is associated with pulmonary emphysema [83, 84]. A disturbed equilibrium between not yet exactly defined enzymes and the inhibitor is given as a causal explanation [85]. Halflife was found to be 5—6 days in patients homozygous for hereditary deficiency of $\alpha_1 A$ [86]. In this study $\alpha_1 A$ disappeared relatively quickly from the blood into the extracellular fluid compartments. In one patient, extremely poor in $\alpha_1 A$, it was found also in the sputum after infusion of plasma.

Plasma, as already mentioned, contains only one inhibitor which inactivates C1 esterase. Its physiological role became evident, when a causal connection between angioneurotic oedema and $\overline{C1}$ INA was recognized in cases of inherited deficiency of this protein [87, 88].

AT III is partially consumed when coagulation has occurred in vitro, as well as in vivo. Using immunological methods it can be shown to be complexed to thrombin [71, 72]. A reduction of AT III, therefore, signalizes a hypercoagulable, respectively prethrombotic state. In cases of disseminated intravascular coagulation, for example, markedly decreased AT III values have been seen. Therefore, the drop below normal values in women using progestagen-oestrogen pills merits special attention [73]. In a family with an inherited AT III deficiency a high incidence of thrombo-embolic disease was associated with a reduction in AT III to 50% of the normal values [89].

A deficiency in $\alpha_2 M$ has not yet been observed. It was found to be increased in diabetes mellitus and in several conditions associated with growth, namely in pregnancy, in childhood [81, 90], in regeneration- and tumour-growth [71]. In cell cultures it substitutes for calf serum [91]. The reasons for this have not yet been discovered. However, the general transporting function of $\alpha_2 M$ may be responsible. It combines not only with proteolytic enzymes but also with cationic aspartate aminotransferase [92]. $\alpha_2 M$ has a strong affinity to proteohormones, e. g. growth hormone [93], it is precipitated by phytohemagglutinins [94]. An insulin carrier function, originally suggested, has not been confirmed [95]. But $\alpha_2 M$ is identical with the single protein of serum which can bind zinc [96].

The interactions of the inhibitors with the various enzymes in blood are compiled in Fig. 17. With exception of AT III all inhibitors can be seen to bind and to inactivate several proteolytic enzymes. It may be assumed, therefore, that the PI's have a regulatory function on the equilibrium between the different enzyme systems, $\alpha_2 M$, for example, on the equilibrium between coagulation, fibrinolysis and liberation of kinins. This polyvalency of the inhibitors is of biological importance; furthermore it appears necessary to consider the fact, that by only one stimulus, e. g. the contact with foreign surfaces several enzyme systems are activated.

Finally it should be mentioned that the interaction between the enzymes of blood and their inhibitors can be affected by products of the lipidcatabolism. We have recently found that AT III binds fatty acids. Two complexes, probably different in their oleate content, can be seen using polyacrylamide gel electrophoresis (Fig. 18). The binding reaction is associated with a loss of the antithrombin activity; simultaneously the precipitability of AT III with its

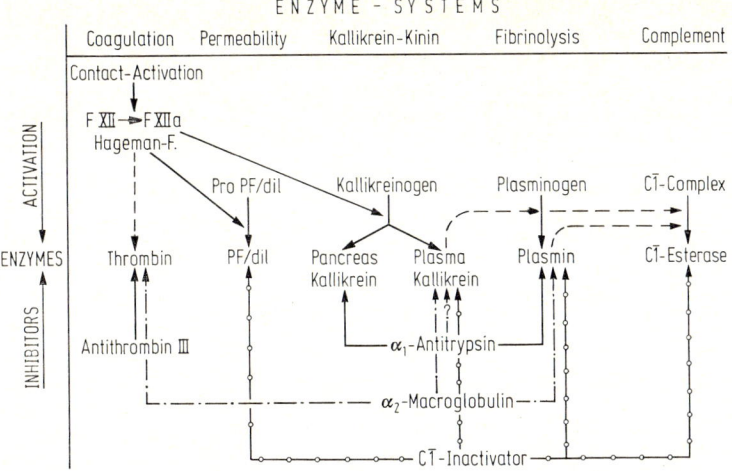

Fig. 17. The interrelationship between the various enzyme systems of blood and the regulatory function of the inhibitors.

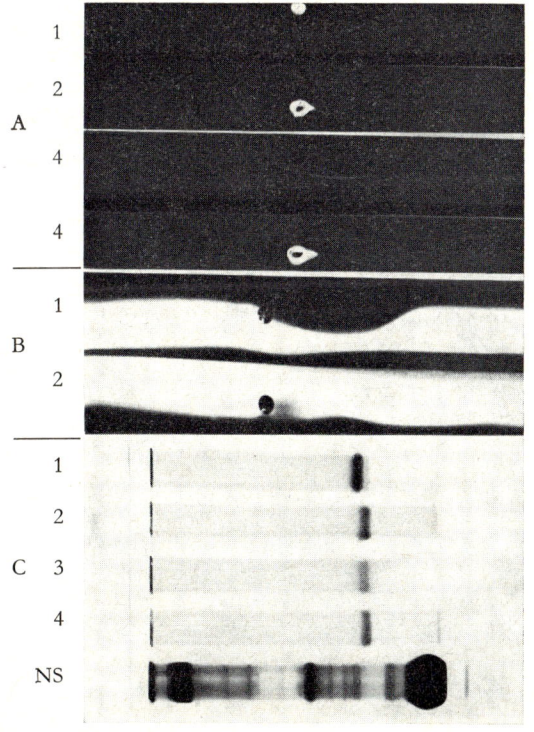

homologous antibody is reduced. Oleate is not displaced from AT III by heparin. This interrelationship between the proteolytic enzyme systems and the lipid-catabolism may prove to be of particular physiological importance.

Fig. 18. Characterization of the antithrombin III (AT III) after treatment with oleate and heparin.

A) Immunoelectrophoresis with anti-antithrombin III serum;
B) fibrinogen agarose-electrophoresis with thrombin in the trough;
C) polyacrylamide gel-electrophoresis.
1. AT III; 2. AT III + sodium oleate; 3. AT III + heparin; 4. AT III + sodium oleate + heparin; NS) normal serum.

References

[1] FERMI, C. and L. PERNOSSI, Zschr. Hyg. **18**, 83 (1894).
[2] CAMUS, L. and E. GLEY, C. R. Soc. Biol. **49**, 8 (1897).
[3] PUGLIESE and COGGI, Bulletino Scienze med. (1897) (zitiert bei 21).
[4] HAHN, M., Berl. klin. Wschr. **34**, 499 (1897).
[5] LANDSTEINER, K., Zbl. Bakt. **27**, 357 (1900).
[6] DELEZENNE, C. and E. POZERSKY, C. R. Soc. Biol. **55**, 690 (1903).
[7] SCHWARTZ, O., Wien. Klin. Wschr. **22**, 1151 (1909).
[8] BAUER, J., Z. Immun. Forsch. Orig. **5**, 186 (1910).
[9] JOBLING, J. W. and W. PETERSEN, J. Exp. Med. **19**, 459 (1914).
[10] MEYER, K., Berl. Klin. Wschr. **46**, 1890 (1909).
[11] COBLINER, S., Biochem. Z. **25**, 494 (1910).
[12] TRALE, F. H. and E. BACH, Proc. R. Soc. Med. **13**, pt. 3, Pathol. 43 (1919/20).
[13] HEDIN, S. G., J. Physiol. **30**, 195 (1903/04).
[14] WIENS, Münch. Med. Wschr. 2637 (1907).
[15] MEYER, K., Folia serolog. VII, 472 (1911).
[16] EISNER, G., Zschr. f. Immunitätsforschung Orig. I, 650 (1908/09).
[17] ASCOLI, M. and C. BEZZOLA, Berl. Klin. Wschr. **40**, 391 (1903).
[18] KOLACZEK and MÜLLER, Deutsch. Med. Wschr. 7 (1907)*.
[19] BITTORF, Deutsch. Arch. f. Klin. Med. 91 IX*.
[20] WIENS, Deutsch. Arch. f. Klin. Med. 91 XX*.
[21] BRIEGER, L. and J. TREBING, Berl. Klin. Wschr. **45**, 1041 (1908).
[22] MORAWITZ, P., Ergebn. Physiol. **4**, 307 (1905).
[23] SCHMITZ, A., Hoppe Seyler's Z. Physiol. Chem. **255**, 234 (1938).
[24] NORTHROP, J. H. and M. KUNITZ, J. Gen. Physiol. **19**, 991 (1936).
[25] GROB, D., J. Gen. Physiol. **26**, 405 (1943).
[26] DUTHIE, E. S. and L. LORENZ, Biochem. Z. **44**, 167 (1949).
[27] DE SMITH, L. S. and C. H. LINDSLEY, J. Bacteriol. **38**, 221 (1939).
[28] SHULMAN, N. R., J. Exp. Med. **95**, 593 (1952).
[29] JACOBSSON, K., Scand. J. Clin. Lab. Invest. **5**, 97 (1953).
[30] JACOBSSON, K., Scand. J. Clin. Lab. Invest. Suppl. **14**, (1955).
[31] MOLL, F. C., S. F. SUNDEN and J. R. BROWN, J. Biol. Chem. **233**, 121 (1958).
[32] BUNDY, H. F. and J. W. MEHL, J. Biol. Chem. **234**, 1124 (1959).
[33] HEIMBURGER, N. and H. G. SCHWICK, Thrombos. Diathes. Haemorrh. **7**, 432 (1962).
[34] HEIMBURGER, N. and H. HAUPT, Clin. Chim. Acta **12**, 116 (1965).
[35] SCHULTZE, H. E., I. GÖLLNER, K. HEIDE, M. SCHÖNENBERGER and G. SCHWICK, Z. Naturforschung **10b**, 463 (1955).
[36] SCHULTZE, H. E., K. HEIDE and H. HAUPT, Klin. Wschr. **40**, 427 (1962).
[37] BROWN, R. K., W. H. BAKER, A. PETERKOFSKY and D. L. KAUFFMAN, J. Am. Chem. Soc. **76**, 4244 (1954).
[38] SCHÖNENBERGER, M., R. SCHMIDTBERGER and H. E. SCHULTZE, Z. Naturforschung **13b**, 761 (1958).
[39] SCHULTZE, H. E., N. HEIMBURGER, K. HEIDE, H. HAUPT, K. STÖRIKO and H. G. SCHWICK, Proc. 9th Congr. Europ Soc. Haemat., Lisbon 1963. S. Karger, Basel/New York, 1315 (1963).
[40] SCHULTZE, H. E., K. HEIDE and H. HAUPT, Naturwiss. **49**, 133 (1962).
[41] ABILDGAARD, U., Scand. J. Clin. Lab. Invest. **19**, 190 (1967).
[42] HEIMBURGER, N., First Intern. Symp. on Tissue Factors in the Hemeostasis of the Coagulation-Fibrinolysis System, Florence, Mai 1967, p. 353.
[43] LOEB, J. Arch. Sci. physiol. **10**, 129 (1956).
[44] HENSEN, A. and E. A. LOELIGER, Thrombos. Diathes. haemorrh. (Stuttg.) IX Supp. 1 (1963).
[45] MARKWARDT, F. and P. WALSMAN, Hoppe-Seyler's Z. physiol. Chem. **317**, 64 (1959).
[46] STEINBUCH, M., Nature (Lond.) **192**, 1196 (1961).
[47] HEIDE, K., N. HEIMBURGER and H. HAUPT, Clin. Chim. Acta **11**, 82 (1965).
[48] RATNOFF, O. D. and I. H. LEPOW, J. Exp. Med. **106**, 327 (1957).
[49] PENSKY, J., L. R. LEVY and I. H. LEPOW, J. biol. Chem. **236**, 1674 (1961).
[50] SCHULTZE, H. E., K. HEIDE and H. HAUPT, Naturwiss. **49**, 133 (1962).
[51] PENSKY, J. and H. G. SCHWICK, Science **163**, 698 (1969).
[52] HARPEL, P. C., J. Exp. Med. **132**, 329 (1970).
[53] HEIDE, K. and H. HAUPT, Behringwerk-Mitteilungen, Heft 43, 161 (1964).
[54] STEINBUCH, M. and R. AUDRAN, C. R. Acad. SC. Paris, **260**, 7058 (1965).
[55] MEHL, T. W., W. O'CONNEL and J. DEGROOT, Science **145**, 821 (1964).

* zitiert bei [21]

[56] STEINBUCH, M., L. PEJAUDIER, M. QUENTIN and V. MARTIN, Biochem. Biophys. Acta **154**, 228 (1968).
[57] SCHWICK, H. G., N. HEIMBURGER and H. HAUPT, Zschr. inn. Med. **21**, 1 (1966).
[58] HAUPT, H., N. HEIMBURGER, T. KRANZ and H. G. SCHWICK, Europ. J. Biochem. **17**, 254 (1970).
[59] GANROT, P. O., Arkiv Kemi **26**, 577 (1967).
[60] MEHL, T. W., W. O'CONNELL and J. DEGROOT, Science **145**, 821 (1964).
[61] GANROT, P. O., Clin. Chim. Acta **13**, 518 (1966).
[62] NORMAN, P. S., J. Exp. Med. **108**, 31 (1958).
[63] HARPEL, P. C., J. Exp. Med. **132**, 329 (1970).
[64] GIGLI, J., J. W. MASON, R. W. COLMAN and K. F. AUSTEN, J. Immunol. **104**, 574 (1970).
[65] FRITZ, H., B. BREY, A. SCHMAL and E. WERLE, Hoppe-Seyler's Z. Physiol. Chem. **350**, 1551 (1969).
[66] STEINBUCH, M., C. BLATRIX, F. L. JOSSO, Proc. XIth Congr. Int. Soc. Haemat. (Sidney) 7, 1966.
[67] LANCHANTIN, G. F., M. L. PLESSET, J. A. FRIEDMANN and D. W. HORT, Proc. Soc. exp. Biol. (NY) **121**, 444 (1966).
[68] ABILDGAARD, U., Scand. J. clin. Lab. Invest. **19**, 190 (1967).
[69] ABILDGAARD, U., Scand. J. clin. Lab. Invest. **20**, 207 (1967).
[70] ABILDGAARD, U., Thrombos. Diathes. haemorrh. **21**, 173 (1969).
[71] SCHWICK, H. G. and N. HEIMBURGER, 12. Tagung der Deutschen Arbeitsgemeinschaft für Blutgerinnungsforschung in Deidesheim, 5. April 1968.
[72] HEIMBURGER, N., XII. Hamburger Symposion über Blutgerinnung, 16. u. 17. Mai 1969.
[73] FAGERHOL, M. K. and U. ABILDGAARD, Lancet **30**, 1175 (1970).
[74] RATNOFF, O. D., J. PENSKY, D. OGSTON and G. B. NAFF, J. exp. Med. **129**, 315 (1969).
[75] HEIMBURGER, N. and H. HAUPT, Klin. Wschr. **44**, 1196 (1966).
[76] FINKENSTADT, W. R. and M. LASKOWSKI, JR., J. Biol. Chem. **240**, PC 962 (1965).
[77] FINKENSTADT, W. R. and M. LASKOWSKI, JR., J. Biol. Chem. **242**, 771 (1967).
[78] FRITZ, H., E. FINK, M. GEBHARDT, K. HOCHSTRASSER and E. WERLE, Hoppe-Seyler's Z. Physiol. Chem. **350**, 933 (1969).
[79] HAVERBACK, B. J., B. DYCE, H. F. BUNDY, S. K. WIRTSCHAFTER and H. A. EDMONDSON, J. clin. Invest. **41**, 972 (1962).
[80] GANROT, P. O., Arkiv Kemi **26**, 583 (1967).
[81] GANROT, P. O. and B. BJERRE, Acta obst. et gynec. scandinav. **46**, 1 (1967).
[82] FAGERHOL, M. K. and C.-B. LAURELL, Clin. Chim. Acta **16**, 199 (1967).
[83] ERIKSSON, S., Acta med. scand. **177**, Suppl. 175 (1965).
[84] LAURELL, C.-B. and S. ERIKSSON, Scand. J. clin. Lab. Invest. **15**, 132 (1963).
[85] KUEPPERS, F. and A. G. BEARN, Proc. Soc. exp. Biol. **121**, 1207 (1966).
[86] MAKINO, S. and C. E. REED, J. Lab. Clin. Med. **75**, 742 (1970).
[87] DONALDSON, V. H. and P. R. EVANS, Amer. J. Med. **35**, 37 (1963).
[88] ROSEN, F. S., P. CHARACKE, J. PENSKY and V. H. DONALDSON, Science **148**, 957 (1965).
[89] EGEBERT, O., Thrombos. Diathes. haemorrh. **13**, 516 (1965).
[90] GANROT, P. O., Clin. Chim. Acta **15**, 113 (1967).
[91] HEIMBURGER, N., unpublished.
[92] BOYDE, T. R. and J. F. PRYME, Clin. Chim. Acta **21**, 9 (1968).
[93] HADDEN, D. R. and T. E. PROUT, Nature **202**, 1342 (1964).
[94] GAHNE, B., Hereditas **51**, 365 (1964).
[95] ADHAM, N. F., P. WILDING, J. MEHL and B. J. HAVERBACK, J. Lab. clin. Med. **71**, 271 (1968).
[96] PARISI, A. F. and B. L. VALLEE, Biochemistry **9**, 2421 (1970).

Discussion Remarks: **Sex Hormones and Serum Proteins**

GEBHARDT F. B. SCHUMACHER, Chicago

Dr. HEIMBURGER has mentioned that proteinase inhibitors such as alpha$_1$-antitrypsin and alpha$_2$-macroglobulin increase under certain conditions such as growth and pregnancy. I would like to draw your attention to the fact that sex hormones have an influence on serum proteins. A considerable increase in alpha$_1$-antitrypsin and a slight increase in alpha$_2$-macroglobulin levels during pregnancy and under long-term treatment with hormonal contraceptives has been observed [1, 2, 3]. However, other conditions, such as tissue injury and inflammation may cause alpha$_1$-antitrypsin increase but do not affect the alpha$_2$-macroglobulin [2, 3, 4].

References

[1] Laurell, C. B., S. Kullander and J. Thorell, Effect of administration of a combined estrogen-progestin contraceptive on the level of individual plasma proteins. Scand. J. Clin. Lab. Invest. **21**, 337 (1968).

[2] Schumacher, G. F. B., unpublished data.

[3] Schumacher, G. F. B. und H. D. Schlumberger, Über Veränderungen der alpha$_1$-Globuline des Serums. Dtsch. Med. Wschr. **88**, 645 (1963).

[4] Schumacher, G. F. B. und H. D. Schlumberger, Haptoglobin und „Nicht-Haptoglobin-Anteile" der alpha$_1$-Globuline bei posttraumatischer Entzündung. Proc. 8th Congr. Europ. Soc. Haemotal., Wien 1962. S. Karger, Basel/New York (1962).

Proteinase Inhibitors in Clinical Studies

Eugen Werle

Institut für Klinische Chemie und Klinische Biochemie der Universität München
(Direktor Prof. Dr. Dr. E. Werle)

It is somewhat difficult to discuss clinical problems in connection with proteinase inhibitors within the program of this conference. The problems which had to be solved by the scientists assembled here were certainly most difficult, but still there is some advantage over clinical problems. Chemical and physical results are demonstrable and measurable in a more exact way and moreover single reactions can be examined apart from others. Within the living body disturbances of the physiological state mostly force us to deal with many interdependent events and the explanation of one reaction only in very few cases throws light on the others and their mutual relations. Freqently this proves to be true when we try to find out something about the participation of proteolytic enzymes in pathological conditions and how they are influenced e. g. by the bovine kallikrein trypsin inhibitor, named Trasylol*.

Because of the limited time it is not possible to discuss all clinical conditions in which Trasylol is said to have a therapeutic effect. Therefore I will confine my discussion to some general problems of proteinase inhibitor therapy.

Favoured by the neighbourhood of the hospital we soon after the discovery of the polyvalent bovine inhibitor tried to use it as a therapeutic tool in the treatment of acute pancreatitis [1].

---

* registered trade mark (Bayer, Wuppertal-Elberfeld)
 = BPTI, basic bovine pancreatic trypsin inhibitor
 = trypsin-kallikrein inhibitor from bovine organs.

At that time it was generally accepted — though it had not been proved experimentally — that prematurely activated trypsin is responsible for the pathological events, mainly for the destruction of the pancreas. This idea was based on the knowledge of the enzymatic equipment of the pancreas and of the possibility of tissue destruction by trypsin. When methods for exact measurement of enzymatic activities in the diseased pancreas were available, the results of experiments done by several research groups at least questioned the trypsin theory (for review see [2]). Today we assume that other enzymes too play an important role and that the inhibitor interferes with several reactions, mainly in the early stages of the pathological events. Therefore the greatest effect is obtained by application of the inhibitor in high amounts as early as possible in the course of the disease.

The proof is still lacking whether Trasylol acts mainly by inhibition of trypsin; but on the other hand trypsin cannot be ruled out as activator of the other preenzymes of the pancreas. Creutzfeldt and coworkers [3] and Wanke and coworkers [4] regard lysolecithin, a split product of lecithin by phospholipase A, which occurs as preenzyme in especially high amounts in human pancreas, as the main factor for cell destruction. However, Wanke and coworkers [4] stress the point that phospholipase A is activated even by minimum amounts of active trypsin. This fact again points to the important role of active trypsin.

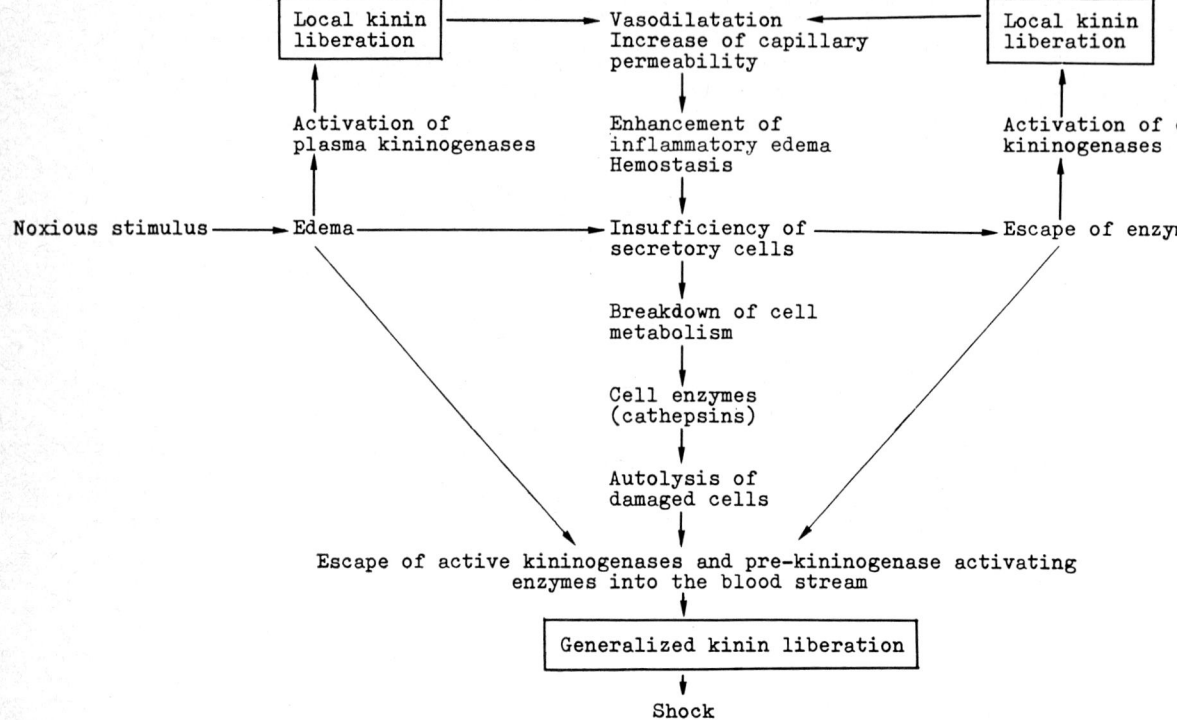

Fig. 1. Participation of plasma kinins in the pathophysiological reactions of acute pancreatitis (TRAUTSCHOLD, WERLE, ZICKGRAF—RÜDEL [2]).

We now regard inhibition of kininogenases which liberate plasma kinins as the important therapeutic mechanism of Trasylol action in pancreatitis (Fig. 1). Locally Trasylol causes an improvement of circulatory disorders which contribute to cell destruction. Systemically a reduction of shock symptoms can be achieved. When the bovine inhibitor had been used first for therapy, we had already discussed this point, but more or less as a side action. Later it proved that a main factor of the therapeutic effect of the inhibitor in several pathological conditions is its capacity of inhibiting kininogenases. So the course of pancreatitis research shows in an impressive way how one assumption, especially if it is problematic in nature, stimulates research work in a wide field.

The following figure (Fig. 2) shows our concept about kinin liberation in several pathological states and the point of action of Trasylol. The upper part of the figure gives an impression of the numerable conditions under which kinin liberation can take place and which at the same time means a chance for a therapeutic effect of Trasylol.

You already can see that it is the unspecificity and the polyvalent efficacy which make the inhibitor a worthy object for clinical studies. This also means an advantage as long as we cannot tell with certainty which enzymes are involved in processes which can be favourably influenced by Trasylol.

In recent years the suitable model of an enzymatic cascade became popular. It was first used for the enzymatic events during blood coagulation. The following scheme, given by EISEN [5] (Fig. 3) shows the enzymatic reactions which can be influenced by Trasylol. EISEN himself

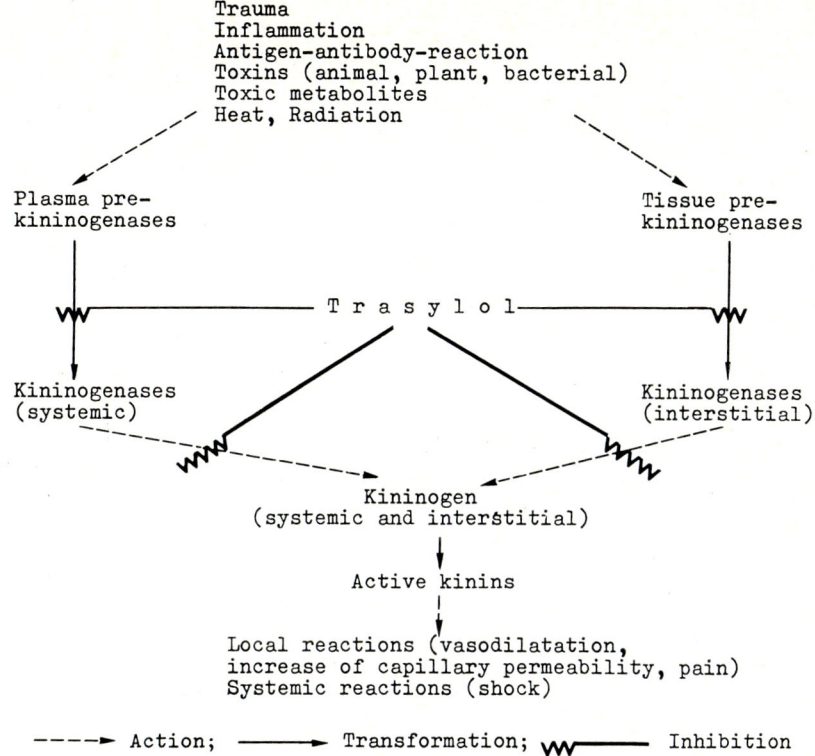

Fig. 2. Inhibition of kinin formation in the organism by the polyvalent bovine inhibitor in various pathological conditions (TRAUTSCHOLD, WERLE, ZICKGRAF-RÜDEL [2]).

called this figure "complicated and confusing" and we agree with him.

This figure now could be completed by combining the diverse enzymatic mechanisms with such different clinical conditions as disorders of blood coagulation and fibrinolysis, inflammation, tissue damage, shock or allergic reactions. In any case, the therapeutic aim is to interrupt the enzymatic cascade at any step possible.

Considering the high levels of proteinase inhibitors which are normally present in plasma and which can hardly be elevated by high doses of Trasylol, it is still a point of controversy how Trasylol acts. Trasylol shows a very short half life time of 30 to 60 minutes after intravenous injection. The inhibitor is accumulated in the liver, in the lung, and to some degree in the cartilage and finally stored in the kidney from where it is excerted (for review see [6] and [7]). There are several points which could be a hint for the understanding of the mechanism of action of Trasylol. Because of its polyvalent inhibitory capacity toward estero proteinases Trasylol can interfere with several enzymatic reactions at the same time and therefore could be superior to the more specific serum inhibitors. On the other hand, there are conditions as generalized intravascular coagulation where a distinct diminution of serum α_1-antitrypsin could be demonstrated [8]. An important point in favour of Trasylol is its small molecular size which enables the inhibitor to gain access to the site of local pathological events far easier than the high-molecular plasma inhibitors. Also a direct action of Trasylol on the vessel wall has been discussed [9], which in the case of edema could contribute to the restitution of the normal permeability of the vessel wall.

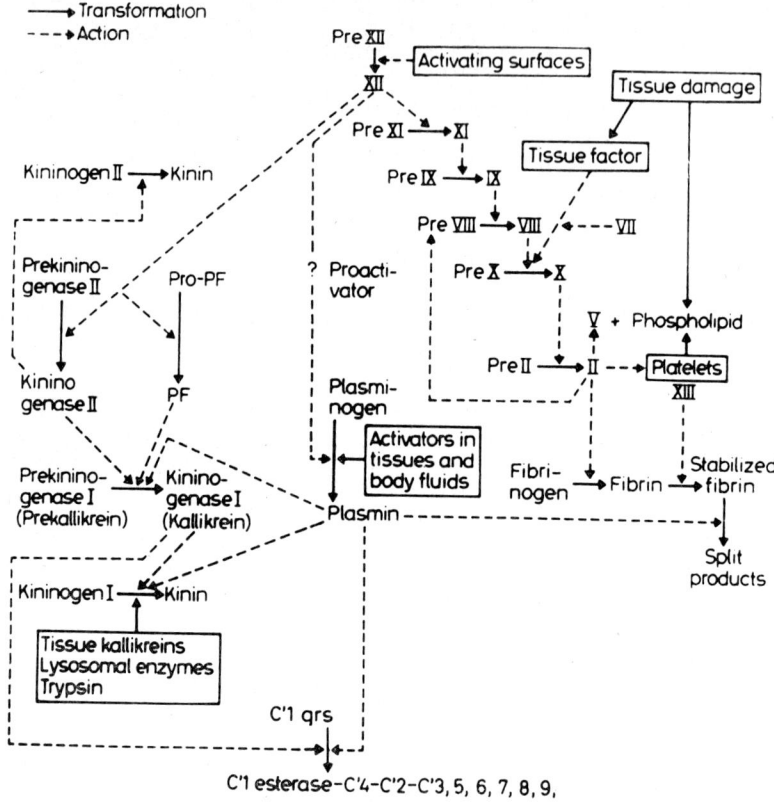

Very important for handling Trasylol in the clinic is its good compatibility. On the other hand allergic and inflammatory reactions were observed in experiments with soybean inhibitor and ovomucoid (for review see [6]). For this reason these inhibitors are, according to our knowledge, no longer used in clinical studies. Furthermore they are lacking the polyvalent inhibitory capacity of the bovine inhibitor and therefore have only minor possibilities of action.

A problem which is inherent in the therapeutic use of Trasylol results from the nature of the pathological conditions in question. The therapeutic effect of the inhibitor is — in most cases — bound to the events in the very beginning of pathological enzymatic reactions. This means e. g. that once a pancreas is destroyed, it cannot be restored by the inhibitor. Once clotting factors are depleted in generalized intravascular coagulation, deficiency of clotting factors cannot possibly be repaired by Trasylol. Shock symptoms are favourably influenced also mainly in the beginning of events where we assume a liberation of kinins and hence the possibility of inhibiting kininogenases. More examples could be given. Furthermore a long-term substitution therapy as it would be desirable in hereditary angioneurotic edema is not possible because of the short half-life time of Trasylol.

Before I come to the end, let me shortly refer to another problem which I already mentioned in the beginning. It is almost impossible to measure all reactions which contribute to a pathological state. For a long time e. g. it was agreed, that the changes in the kininogen level of serum reflect the amount of kinin liberation. Recent results from HABERMANN's [10] and from our laboratory [11] which correlate the kininogen level with other parameters question the validity of several experiments. This is — as I want to emphasize — not a cause to question the effectiveness of the inhibitor, but rather a challenge to examine the sequence and the compounds and products of enzymatic reactions which are involved in more detail in order to find out the mechanism of action of proteinase inhibitors in pathological conditions. This in our research work would mainly refer to demonstration of active kinins. Another aim should be to provide the physician with simple diagnostic methods which would enable him to use inhibitors as a rational therapeutic tool.

It is important not to examine pathological problems apart from physiological ones. If we would exactly know the physiological functions of serum and organ proteinase inhibitors, we certainly could gain access to the solution of the discussed problems. I want to express the hope that all the theoretical and empiric work which is presented in this conference can contribute to the solution of our physiological and clinical problems.

References

[1] WERLE, E., K. TAUBER, W. HARTENBACH and M. M. FORELL, Münchn. med. Wschr. **100**, 1265 (1958).

[2] TRAUTSCHOLD, I., E. WERLE and G. ZICKGRAF-RÜDEL, Biochem. Pharmacol. **16**, 59 (1967).

[3] SCHMIDT, H. and W. CREUTZFELDT, Scand. J. Gastroenterol. **4**, 39 (1969).

[4] WANKE, M., W. NAGEL, M. M. LINDNER and H. SEBENING, Z. Gastroenterologie **6**, 434 (1968).

[5] EISEN, V. and F. E. BRUCKNER, in: Neue Aspekte der Trasylol Therapie, ed. by G. L. HABERLAND, P. HUBER and P. MATIS, F. K. Schattauer Verlag, Stuttgart **4**, 3 (1970).

[6] VOGEL, R., I. TRAUTSCHOLD and E. WERLE, Natural Proteinase Inhibitors, Academic Press, New York-London (1968).

[7] ARNDTS, D., K. O. RÄKER, P. TÖTÖK and E. HABERMANN, Arzneim.-Forsch. **20**, 667 (1970).

[8] DUCK, H.-J., K. EBELING and J. WAGNER, Proc. 7. Internisten Tagung der Gesell. Innere Med. der DDR, Leipzig 1970.

[9] KALLER, H., in: Neue Aspekte der Trasylol Therapie, ed. by R. GROSS and G. KRONEBERG, F. K. Schattauer Verlag, Stuttgart (1966).

[10] HABERMANN, E., in: Neue Aspekte der Trasylol Therapie, ed. by G. L. HABERLAND and P. MATIS, F. K. Schattauer Verlag, Stuttgart **3**, 37 (1969).

[11] WERLE, E. and P. ZACH, Z. klin. Chem. u. klin. Biochem. **8**, 186 (1970).

Specific Isolation and Modification Methods for Proteinase Inhibitors and Proteinases

Hans Fritz, Bruni Brey, Martin Müller and Maria Gebhardt

Institut für Klinische Chemie und Klinische Biochemie der Universität München, D-8 München 15, Germany
(Direktor: Prof. Dr. Dr. E. Werle)

Summary: Water-insoluble trypsin resins were widely used for the isolation of proteinase inhibitors from animal tissues and fluids. Very suitable for isolation procedures are trypsin resins with polyamphoteric supports or cellulose and polyacrylamide carriers.

Limited cleavage of peptide bonds occured with some inhibitors during contact with the enzyme resin. After the resin step further purification is easily achieved by chromatographic methods.

Synthesis of chemically modified inhibitors retaining full inhibitory activity is simplyfied by application of enzyme resins.

Purification of pig pancreas kallikrein yields satisfying results, if an inhibitor resin is employed in which the guanidinated basic pancreatic trypsin inhibitor is linked to cellulose fibers.

Abbreviations: BPTI, basic pancreatic trypsin inhibitor = trypsin-kallikrein inhibitor from bovine organs; EMA, ethylene-maleic acid; IU, inhibitor unit (trypsin inhibition throughout), definition is given in ref. [39].

Introduction

The cristallization of an inhibitor-trypsin complex and its subsequent dissociation by Kunitz and Northrop in 1936 were the guiding steps for the isolation of protein proteinase inhibitors using specific methods [1]. Kassell, Laskowski Sr. and Co-workers modified this procedure to get pure basic pancreatic trypsin inhibitor: They purified firstly the inhibitor-trypsin complex by ion exchange chromatography and subsequently dissociated the complex by dialysis and gel filtration in weakly acidic solutions [2].

Insolubilized Enzymes

Polyanionic and Polyamphoteric Supports

Insolubilized enzymes available since the last decade are by far the most useful tools for inhibitor preparation. In our earlier investigations we used EMA-resins introduced by Katchalski and Co-workers [3] primarily because of their very simple preparation and high binding capacity for the inhibitors we were interested in. In these resins the enzymes are covalently bound to a polyanionic carrier, an ethylene-maleic acid (EMA) copolymer cross-linked with 1,6-dia-

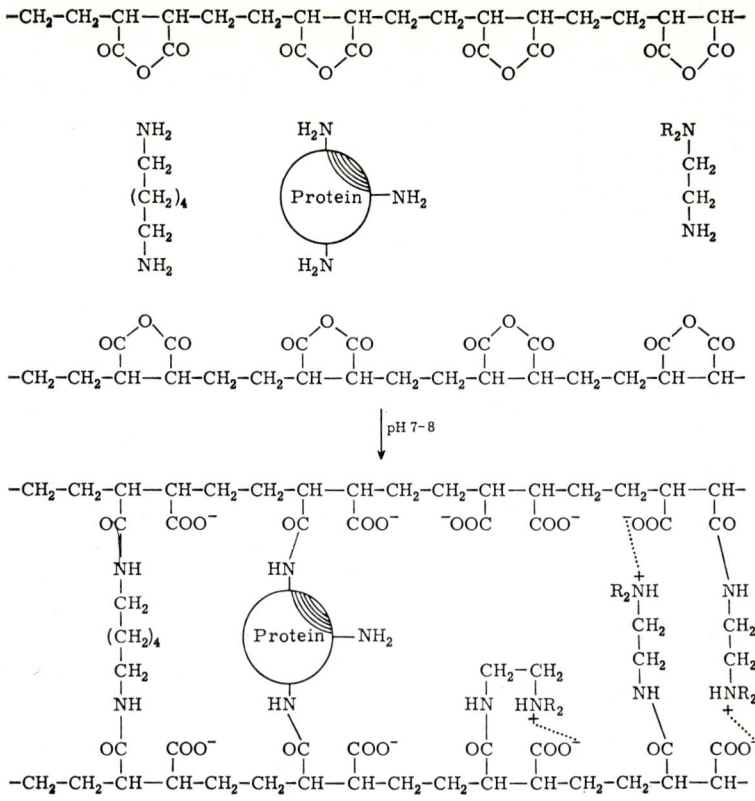

Fig. 1. Preparation of Polyamphoteric Enzyme EMA-Resins. The ethylene-maleic anhydride (EMA) chains are cross-linked by 1,6-diaminohexane. By acylation of ε-amino groups of lysine residues the protein is fixed to the support (3). Incorporation of N,N-dimethylethylenediamine groups yields polyamphoteric resins [5, 10].

minohexane. Our team, HOCHSTRASSER, and TSCHESCHE and Co-workers used these *polyanionic* trypsin resins with considerable advantage for the isolation of pancreatic and other trypsin inhibitors of lower molecular weight from various species [4—8].

The purification itself is based on the selective adsorption of an inhibitor from crude, neutral, or weakly basic extracts, removal of contaminating material by treatment with salt-buffer solutions and dissociation of the insoluble complex under acidic conditions, followed by desalting of the inhibitor solution.

Unfortunately the binding capacity for inhibitors with a relatively high net negative charge is very low — probably the reason why GOLDSTEIN failed in 1964 to isolate the soybean KUNITZ inhibitor with this method [9]. Further drawbacks are:

1. Significant non-specific adsorption due to the polyanionic character of the supporting medium,
2. remarkable solubility of the protein resins,
3. variable swelling of the protein resins in solutions of different ionic strength and acidity, and
4. severe differences in hydrolysis rates of commercial EMA-products which complicate the preparation of the resins [10].

Polyamphoteric trypsin EMA-resins with incorporated N,N-dimethylethylenediamine-groups (Fig. 1) to neutralize electrostatically the

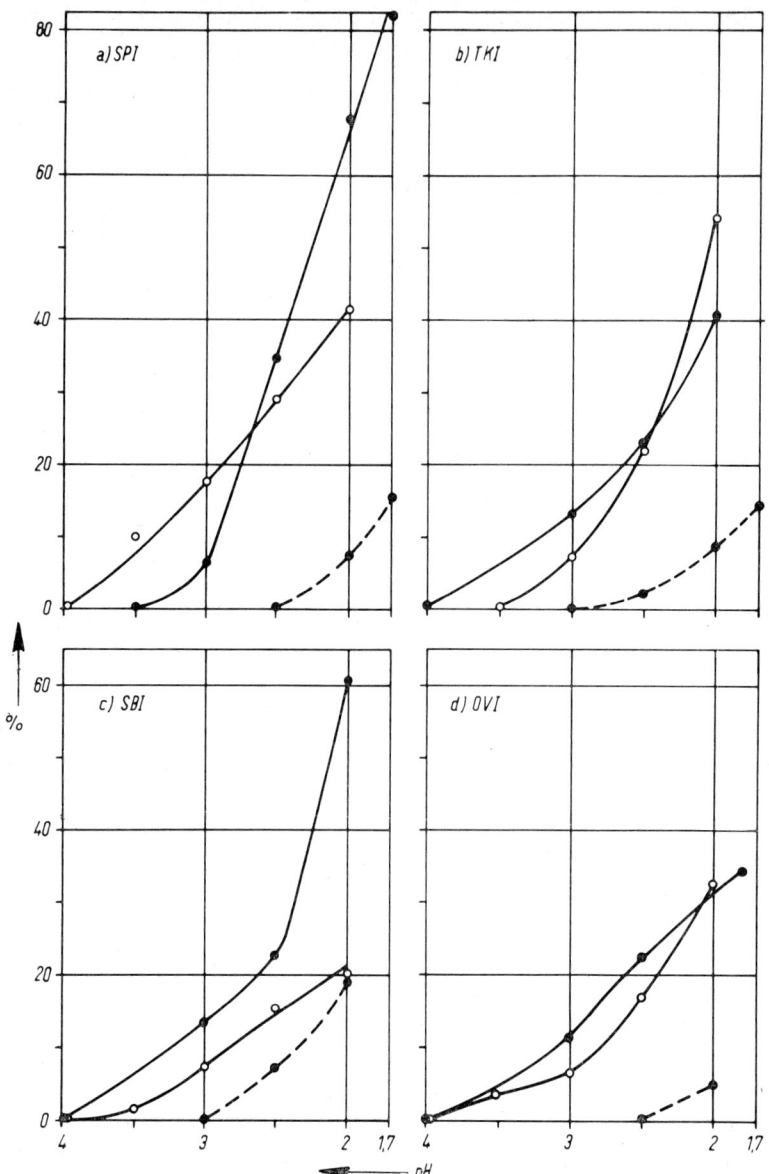

Fig. 2. Dissociation Curves of Inhibitor-Trypsin-Resin Complexes with Polyanionic (broken lines) and Polyamphoteric (full lines) Supports.

SPI: Inhibitor from pig pancreas, TKI: Trypsin-kallikrein inhibitor from bovine organs,
SBI: Soybean Kunitz trypsin inhibitor, OVI: Inhibitor from egg white (ovomucoid).
Ordinates: Percent of inhibitory activity found in solution related to the amount of inhibitor bound to the resin, *Abszissa*: pH value of the suspension. The full lines with closed circles refer to the trypsin resin containing the smaller amount of the polyamphoteric carrier (compared with the lines with open circles). For experimental details see ref. [10].

highly negative charge of the support also bound inhibitors with low isoelectric points in amounts sufficient for isolation purposes; e. g. the soybean Kunitz trypsin inhibitor [10], inhibitors from plant seeds [11], from leeches [12], and from seminal vesicles [13, 14].

The important finding in working with these resins is that in addition to the molecular volume, particulary the electrostatic charge of the inhibitor relative to that of the resin support determines the binding capacities of the resins. An explanation of this phenomenon [15] would be a strong lowering of the concentration of negatively charged proteins in the immediate vicinity of the polyanionic support, whereby the association constants of the inhibitor-enzyme complexes are apparently diminished. Positively charged inhibitors are enriched in the vicinity of the polyanionic carrier, so that the association constants are apparently increased. These effects are reduced by polyamphoteric supports. However the existing method is not suitable to synthesize reproducibly polyamphoteric enzyme resins containing a definite and predetermined amount of alkylated amino groups.

In Figure 2 dissociation curves of some inhibitor-trypsin complexes are shown bound to polyanionic and polyamphoteric carriers respectively. The complexes fixed to the polyamphoteric resin dissociate under more moderate conditions, i. e. at higher pH-values; this is probably due to an increased repulsion between inhibitor and support as a consequence of equidirectional charging.

The dissociation as well as the adsorption of the inhibitor to the enzyme resin should be done preferably continuously in cooled columns rather than by the batch method as the recoveries with the latter are about 20% lower. The salt concentration of the inhibitor solutions should be high enough, approximately 0.3 to 1.0 M, in order to prevent non-specific adsorption of inhibitor to the polyionic EMA-support.

According to our experience, water-insoluble supports for enzyme resins should have the following qualities: The amount of reactive groups binding the enzyme molecules should be definable and determinable just as the number of positively and negatively charged groups in the support, so that the whole range from polyanionic to neutral and polycationic resins is available. It is also important to have defined (big) pore sizes, high abrasion stability, high flow rates and minimal swelling by alterations of ionic strength and pH.

Cellulose and Acrylamide Supports

Trypsin resins which satisfy at least partly these conditions are shown in Table 1. It contains some results we obtained with water-insoluble trypsin resins, based on cellulose [16] and polyacrylamide [17] carriers. The results are repro-

Table 1. Binding Capacities of Trypsin Resins
For experimental details see references [16] and [17]

	Water insoluble carrier		Molecular weight	IP[a]
	Cellulose	Polyacrylamide		
[μMol] Trypsin, PNGB[b]-titrated [18]	10	10		
In [g] trypsin-resin (dry weight)	2.0	1.3		
Amount of bound inhibitor in	[μMol]			
Basic pancreatic trypsin inhibitor	12.3	6.2	6500	> 10
Pig pancreas trypsin inhibitor	11.6		5700	8.5
Bdellins (from leeches)	6.4	4.4	5000	4-6
Inhibitor from peanuts	6.0		17000[c]	
Soybean Kunitz trypsin inhibitor	4.6	5.4	22000	4.5

[a] Isoelectric point, [b] p-Nitrophenyl-p'-guanidinobenzoate, [c] Polymer [11b].

ducible. The amount of active trypsin molecules present was titrated with p-nitrophenyl-p'-guanidinobenzoate according to CHASE and SHAW [18]. It is remarkable that the amount of trypsin molecules able to complex with some inhibitors is near the proportion found from their titration curves. The high amount of basic inhibitors bound by the trypsin-cellulose resin is most likely due to additional non-specific adsorption, as this support contains a considerable amount of negatively charged groups [16]. Some loss of pig pancreatic trypsin inhibitor by enzymatic degradation is also possible [8, 19]. From the similar amounts of bdellins and soybean trypsin inhibitor bound to the resin it can be concluded that the net charge of the inhibitor has a much stronger influence on the binding capacity than the molecular volume.

Isolated Inhibitors

In Table 2 the yields and purities of some trypsin inhibitors are shown isolated in our laboratory. Noticeable are the higher yields with permanent inhibitors which do not show the phenomenon of temporary inhibition. Obviously these inhibitors are also resistant to enzymatic degradation during contact with the enzyme resins. — The inhibitors from lung and liver of sheep and cattle are eluted from the resin column in 2 fractions; the first contains approximately 75% of the bound inhibitor and the second, which appears much later in the effluent, the residual amount up to 100%. We have no explanation of this separation. — Enzymatic degradation mainly causes the loss of the other, temporary inhibitors shown in the Table. In these cases rapid working and extensive cooling during the binding, washing, and elution steps increase the recoveries. Some loss may also be due to solubilization of the enzyme resin when a high amount of inhibitor is bound and thus the protein content of the resin is greatly increased.

Summarizing the observations we can say that the purity of the inhibitors after dissociation (of the complex), elution and gel filtration is influenced by the support used, the ionic strength of the washing buffers, and to some extent the degree of contamination (concentration of the basic inhibitor from sheep lung in the extracts is very low just as the degree of purity, see Table 2) in the crude extracts. Further purification of the inhibitors is relatively easily achieved by ion exchange chromatography or gel filtration.

Enzymatic Cleavage

Most of the inhibitors isolated by the resin method are not homogeneous. The permanent inhibitors isolated from plant seeds by HOCHSTRASSER et al. [6, 11b] consisted mainly of two different forms, separable by ion exchange chromatography, but with the same biochemical characteristics and identical amino acid compositions. The temporary inhibitors from

Table 2. Inhibitors Isolated in Larger Amounts
For experimental details see text and references

Inhibitor from	Ref.	Trypsin-resin support	Yield [%]	Purity[a] [%]
Pancreas (pig, sheep)	10	polyamphoteric EMA[b]	60—75	70—90
Lung (sheep)	21	polyamphoteric EMA	90—100	35—50
Lung, liver (cattle)	22	polyanionic EMA	80—100	85—95
Seminal vesicles (guinea pigs)	10, 13, 14	polyamphoteric EMA	80—95	80
Boar seminal plasma	13	polyacrylamide	63—85	> 90
Leeches (Bdellins)	12	polyamphoteric EMA	65—74	60—80

[a] After gel filtration, [b] Ethylene-maleic acid copolymer.

pancreas glands [20], leeches [12] and seminal vesicles [13, 14] showed three and more different forms with identical or similar amino acid compositions.

The occurence of these different inhibitor forms is at least partly due to the trypsin resin method used. Dr. HOCHSTRASSER in nearly each case obtained a mixture of the native inhibitor and a modified form in which an arg-X or lys-X bond had been broken. He could prove that this C-terminal lysine or arginine residue is part of the reactive site of the inhibitor. Temporary inhibitors from animal organs contained forms with broken bonds at the reactive center, but also with other split bonds. In some cases we isolated active inhibitor forms with missing N- or C-terminal peptides [8, 13, 14, 23, 24]; these peptides are apparently not necessary for the inhibitory reaction.

An interesting example is the trypsin-specific inhibitor from guinea pig seminal vesicles [13, 14]. By the resin method we produced this inhibitor in fully modified form (arg-X bond broken) — this form is inactivated by carboxypeptidase B; however, when isolated by ion exchange chromatography, only the native inhibitor (arg-X bond intact) was obtained. When Dr. FINK treated the native form with the trypsin resin, he obtained the modified one, too. The trypsin-plasmin inhibitors from leeches [12] are also modified to some extent at the reactive center during this isolation procedure. These results do not contradict the elegant modification theories of LASKOWSKI Jr. and Co-workers [25], because the dissociation and the elution of the inhibitors from the trypsin resin complexes take place slowly, so that a thermodynamic equilibrium can be approximated. By kinetically controlled dissociation (sudden drop of pH) of the complexes only non-modified inhibitor forms are to be expected.

The modification reactions during the isolation procedure with trypsin resins are not always disadvantageous; in many cases the groups of HOCHSTRASSER, TSCHESCHE and our laboratory were able to elucidate in this way the reactive sites of inhibitors.

Preparation of Modified Inhibitors

Water-insoluble enzyme resins are also suitable for the preparation of chemically modified inhibitors.

The trypsin-kallikrein inhibitor from bovine organs accumulates in the kidneys shortly after i. v. injection. By contrast the tetramaleyl derivative of this inhibitor is excreted in the urine [26]. In this latter derivative all amino groups, except that of the lysine residue in position 15, are substituted [27]. We synthesized the tetramaleyl inhibitor, which is nearly as active against trypsin, chymotrypsin, plasmin, (and kallikrein) proteinases as the native one, at first according to the method published by CHAUVET and ACHER [28]: We treated the inhibitor-trypsin complex with maleic anhydride and separated the substituted components by gel filtration [27].

Our new method is as follows [17]: The water-insoluble trypsin resin (support: polyacrylamide) is saturated with the inhibitor; unbound inhibitor is washed off. The suspension of the complex is reacted with an excess of maleic anhydride, and afterwards the reaction products are washed out with buffer solutions from the resulting insoluble maleylated complex. The complex is dissociated by the action of weakly acidic salt-buffer solutions. After desalting and lyophilisation, the analytically pure inhibitor is obtained, its structure is shown in Fig. 3. It may be mentioned that the degree of purity of the native stock inhibitor (starting material) was only about 70%.

We tried to synthesize for the first time another inhibitor derivative in a similar manner: The insoluble inhibitor-trypsin complex was treated with pyridoxal 5′-phosphate, and reduced with $NaBH_4$ to link covalently phosphopyridoxyl groups to the inhibitor (and bound enzyme). After dissociation of the complex a mixture of mono-, di- and trisubstituted inhibitor derivatives was obtained [17].

As important as the simplicity of this one-step method is the fact that the modified trypsin resin retains full binding capacity. It can be used repeatedly for the same procedure. The method allows the synthesis of many modified and substituted inhibitors with full inhibitory activity as the reactive center is protected during the modification reaction. The search for the reactive center of "lysine"-inhibitors is facilitated for the same reasons.

Insolubilized Inhibitors

In future probably more importance will be given to insolubilized inhibitors than to enzyme resins. Affinity chromatography using inhibitor resins allows the isolation of enzymes more simply and more rapidly than by present methods.

To obtain further basic data we prepared both polyanionic and polyamphoteric EMA-resins of the basic trypsin-kallikrein inhibitor (BPTI) from bovine organs [29]. The specific binding capacity of these resins (Table 3) is sufficiently high for trypsin, chymotrypsin and plasmin, but it is very low for kallikreins from organs in which we are especially interested. All the other problems and results in using these inhibitor EMA-resins are similar to those already mentioned in connection with the trypsin EMA-resins. Dissociation of the insoluble complexes in slightly acidic salt-buffer solutions yields satisfying results for trypsin, chymotrypsin and plasmin; the kallikrein complex dissociates in weakly acidic guanidine-HCl solutions.

Investigations aimed to obtain inhibitor resins with higher binding capacities for kallikreins

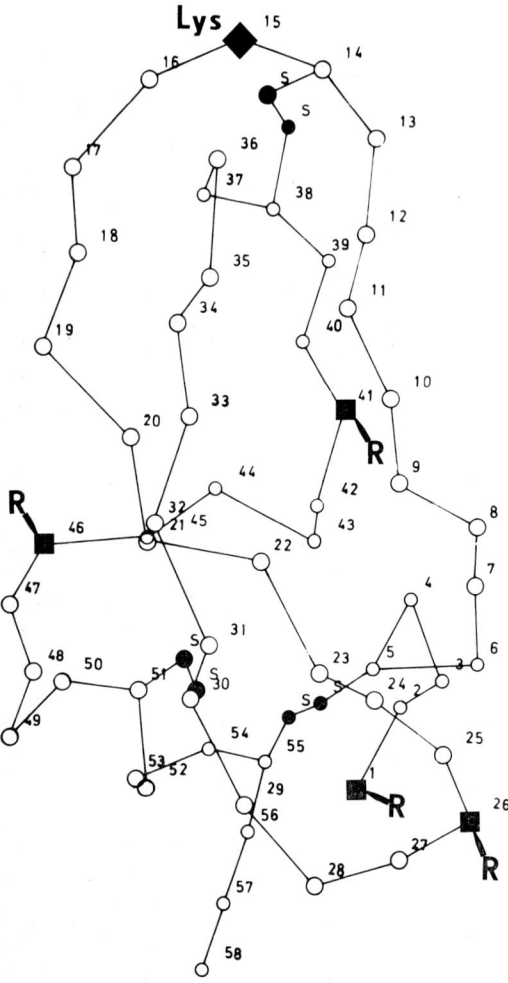

Fig. 3. Tertiary Structure of the Trypsin-Kallikrein Inhibitor from Bovine Organs According to HUBER et al. [32].

R.: Site of substitution by maleyl or 5'-phosphopyridoxyl residues.

Table 3

Polyanionic (A) and Polyamphoteric (N) Inhibitor EMA-Resins

For experimental details see references [29, 5]
The trypsin-kallikrein inhibitor from bovine organs was used

	Weight ratio by preparing the inhibitor resin			
	A-resin		N-resin	
EMA: inhibitor	1:5	3:5	1:5	3:5
Inhibitor amount in [%][a] still able to complex with				
Trypsin	50	18	48	48
Chymotrypsin	28	6	30	37
Plasmin	> 30		> 42	
Kallikrein	< 1		≈ 5	≈ 5

[a] Related to the inhibitor fixed to the resin.

using new carriers (cellulose and polyacrylamide supports) were not successful, perhaps because a large amount of the inhibitor was linked to the resin at or near the lysine residue in position 15 (see Fig. 3), so that complex formation cannot occur. Therefore we tried the following way: The native inhibitor was complexed with active trypsin in which all lysine residues had been guanidinated (L-NH$_2$ → L-NH-C(=NH)-NH$_2$), i.e. converted into homoarginine residues. The rational idea was to dissociate the non-fixable guanidinated trypsin from the complex after the coupling reaction, in which only the free amino groups (R in Fig. 3) of the inhibitor molecule should react with the azide. Surprisingly the guanidinated trypsin, in which no lysine residues could be detected by amino acid analysis, was also bound in high amounts to the resin, probably by aromatic residues.

The guanidinated inhibitor was bound to the resin using the acyl-azide method as well as the unmodified one. The resin containing the guanidinated inhibitor has a sufficiently high binding capacity for kallikreins: About 10% of the overall bound inhibitor are able to complex with kallikrein from pig pancreas (Table 4). This is certainly due to the fact that the guanidinated inhibitor cannot be fixed to the resin in the vicinity of the reactive center; perhaps it is fixed at the non-guanidinated N-terminal α-amino group [30, 31] of the arginine residue (position 1 in Fig. 3) far away from the reactive site, as shown by Dr. HUBER [32], or at non-burried [32, 33] tyrosine residues (position 10, 21, 35 in Fig. 3). The new inhibitor resin works well and yields, after dissociation of the complex with weakly acidic solutions of guanidine salts, reproducable yields from 60 up to 90% related to the overall bound kallikrein (Table 4).

The carbamylated inhibitor has a high affinity only for chymotrypsin, but not for trypsin, kallikrein, nor plasmin [27]. We think, that it is possible to adsorb from crude pancreas extracts specifically chymotrypsin with an insolubilized carbamylated inhibitor.

Table 4

Isolation of Pig Pancreas Kallikrein Using Guanidinated BPTI Fixed to a Cellulose Support

Run No.	Kallikrein			
	Applied To the column [U]	in [m/]	Bound By the resin [U]	Eluted After dissociation [%][a]
2	8365	8	3667	—
3	4110	10.5	2975	30
4	5569	7	2217	65.5
5	4029	5	1953	74
6	4500	17.5	1045	62
7	2140	2.5	1665	64
8	2470	10	1680	70
9	2300	2.5	2100	66
10	2610	2.5	2374	68
11	2300	2.5	2050	72
12	2300	2.5	1940	88

[a] Related to the amount of kallikrein bound to the inhibitor resin.

Procedures: Cellulose hydrazide (300 mg) from Merck AG, Darmstadt, was converted to cellulose azide [34]. 81% from 361 IU for trypsin, or 150 mg guanidinated BPTI [30, 31, 35] were bound to the cellulose support in suspension at pH 8.5 during the reaction with the azide [34].

The inhibitor resin was mixed with 3 times the volume of cellulose powder, filled to a chilled (0° C) column (1.5 × 10 cm) and washed extensively with salt-buffer solutions. The kallikrein samples (specific activity: 14.5 U or 145 KE per mg, for comparison see reference [36]) were solved in 0.3M NaCl, 0.1M triethanolamine buffer, pH 8. The kallikrein-containing solution was slowly applied to the column and excessive unadsorbed kallikrein was washed out with the mentioned salt-buffer solution. Dissociation of the kallikrein-inhibitor resin complex and elution of the kallikrein from the column was done with 0.2—1.0M guanidine-HCl, 0.2M buffer solutions, pH 8 — 4.5.

Results summarized:

1. New prepared resins must be used about three times till constant specific binding capacities are reached. This is caused by solubilization of some material during the first runs.

2. Due to a small degree of dissociation the binding capacity decreases when long washing periods are used (15 — 20 hrs in run 7 and 8; 0.2 — 2 hrs in the other runs).

3. The concentration of the enzyme solution applied to the resin column should be high enough (compare e. g. run 6 with the other runs) to slow down dissociation of the complex.

4. Best yields are obtained if the complex is dissociated in 1.0M guanidine-HCl, 0.2M sodium acetate, pH 6.0, and the kallikrein-containing eluate is collected in deionized water.

5. After ultrafiltration and lyophilisation of the eluted kallikrein fraction a material with a specific activity of 112 U or 1120 KE per mg was obtained which corresponds to an 8 fold increase in activity. The material thus obtained by this one-step method is very pure; for comparison see data in reference [36].

As was demonstrated at this meeting by FEINSTEIN [37] and KASSELL and MARCINISZYN [38] Sepharose-linked enzymes and inhibitors are also employed with considerable advantage for isolation and purification procedures. We expect that much attention will be given in the future to those problems and we expect very interesting results.

Acknowledgements: This research was supported by Sonderforschungsbereich-51 Munich. We wish to thank Farbenfabriken Bayer AG, Wuppertal-Elberfeld, Merck AG, Darmstadt and Novo Industri A/S, Kopenhagen, for gifts of substances.

We are grateful to Prof. Dr. Dr. E. Werle for generously supporting these investigations.

References

[1] KUNITZ, M. and J. H. NORTHROP, J. Gen. Physiol. **19**, 991 (1936); M. KUNITZ, J. Gen. Physiol. **30**, 311 (1947).

[2] KASSELL, B., M. RADICEVIC, S. BERLOW, R. J. PEANASKY and M. LASKOWSKI, J. biol. Chem. **238**, 3274 (1963); KASSELL, B., M. RADICEVIC and M. LASKOWSKI, Fed. Proc. **22**, 529 (1963).

[3] LEVIN, Y., M. PECHT, L. GOLDSTEIN and E. KATCHALSKI, Biochemistry **3**, 1905 (1964).

[4] FRITZ, H., H. SCHULT, M. NEUDECKER and E. WERLE, Angew. Chem. Internat. Ed. **5**, 735 (1966); FRITZ, H., I. TRAUTSCHOLD, H. HAENDLE and E. WERLE, Ann. N. Y. Acad. Sci. **146**, 400 (1968); FRITZ, H., H. SCHULT, M. HUTZEL, M. WIEDEMANN and E. WERLE, Z. physiol. Chem. **348**, 308 (1967); FRITZ, H., M. HUTZEL and E. WERLE, Z. physiol. Chem. **348**, 950 (1967).

[5] FRITZ, H., K. HOCHSTRASSER and E. WERLE, Z. analyt. Chem. **243**, 452 (1968).

[6] HOCHSTRASSER, K., M. MUSS and E. WERLE, Z. physiol. Chem. **348**, 1337 (1967); HOCHSTRASSER, K. and E. WERLE, Z. physiol. Chem. **350**, 249 (1969).

[7] TSCHESCHE, H., Z. physiol. Chem. **348**, 1216 (1967); TSCHESCHE H., Z. physiol. Chem. **348**, 1653 (1967).

[8] TSCHESCHE, H., E. WACHTER and G. KALLUP, Z. physiol. Chem. **350**, 1662 (1969).

[9] GOLDSTEIN, L., personal communication.

[10] FRITZ, H., M. GEBHARDT, E. FINK, W. SCHRAMM and E. WERLE, Z. physiol. Chem. **350**, 129 (1969).

[11] HOCHSTRASSER, K. and E. WERLE, Z. physiol. Chem. **350**, 897 (1969); b) HOCHSTRASSER, K., K. ILLCHMANN and E. WERLE, Z. physiol. Chem. **350**, 929 (1969).

[12] FRITZ, H., M. GEBHARDT, R. MEISTER and E. FINK, this volume, p. 271.

[13] FINK, E., G. KLEIN, F. HAMMER, G. MÜLLER-BARDORFF and H. FRITZ, this volume, p. 225.

[14] FRITZ, H., E. FINK, R. MEISTER and G. KLEIN, Z. physiol. Chem. **351**, 1344 (1970).

[15] GOLDSTEIN, L., Y. LEVIN and E. KATCHALSKI, Biochemistry **3**, 1913 (1964).

[16] FRITZ, H., M. GEBHARDT, R. MEISTER, K. ILLCHMANN and K. HOCHSTRASSER, Z. physiol. Chem. **351**, 571 (1970).

[17] FRITZ, H., M. GEBHARDT, R. MEISTER and H. SCHULT, Z. physiol. Chem. **351**, 1119 (1970).

[18] CHASE, JR., T. and E. SHAW, Biochem. biophys. Res. Commun. **29**, 508 (1967).

[19] TSCHESCHE, H. and H. KLEIN, Z. physiol. Chem. **349**, 1645 (1968).

[20] FRITZ, H., I. HÜLLER, M. WIEDEMANN and E. WERLE, Z. physiol. Chem. **348**, 405 (1967).

[21] FRITZ, H., B. GREIF, W. SCHRAMM, K. HOCHSTRASSER and E. WERLE, Z. physiol. Chem. **351**, 139 (1970).

[22] FRITZ, H., M. HUTZEL and E. WERLE, Z. physiol. Chem. **348**, 950 (1967) and unpublished results.

[23] HOCHSTRASSER, K., W. SCHRAMM, H. FRITZ, S. SCHWARZ and E. WERLE, Z. physiol. Chem. **350**, 893 (1969).

[24] FRITZ, H., W. SCHRAMM, B. GREIF, K. HOCHSTRASSER, E. FINK and E. WERLE, Z. physiol. Chem. **351**, 145 (1970).

[25] LASKOWSKI, Jr., M., R. DURAN, W. R. FINKENSTADT, S. HERBERT, H. F. HIXSON Jr., D. KOWALSKI, J. A. LUTHY, J. A. MATTIS, R. E. MCKEE and C. W. NIEKAMP, this volume, p. 117.

[26] FRITZ, H., K.-H. OPPITZ, D. MECKL, B. KEMKES, H. HAENDLE, H. SCHULT and E. WERLE, Z. physiol. Chem. **350**, 1531 (1969); other references are given there.

[27] FRITZ, H., H. SCHULT, R. MEISTER and E. WERLE, Z. physiol. Chem. **350**, 1531 (1969).

[28] CHAUVET, J. and R. ACHER, J. biol. Chem. **242**, 4274 (1967).

[29] FRITZ, H., B. BREY, A. SCHMAL and E. WERLE, Z. physiol. Chem. **350**, 617 (1969).

[30] KASSELL, B. and R. B. CHOW, Biochemistry **5**, 3449 (1966).

[31] CHAUVET, J. and R. ACHER, Biochem. Biophys. Res. Commun. **27**, 230 (1967).

[32] HUBER, R., D. KUKLA, A. RÜHLMANN and W. STEIGEMANN, this volume, p. 56.

[33] SHERMAN, M. P. and B. KASSELL, Biochemistry **7**, 3634 (1968).

[34] Procedure according to H. D. ORTH and N. HENNRICH, Merck AG, Darmstadt, unpublished data.

[35] FRITZ, H., B. BREY and M. MÜLLER, A somewhat modified procedure of KASSELL & CHOW and CHAUVET & ACHER was used. To be published elsewhere.

[36] FRITZ, H., I. ECKERT and E. WERLE, Z. physiol. Chem. **348**, 1120 (1967).

[37] FEINSTEIN, G., this volume, p. 38.

[38] KASSELL, B. and M. B. MARCINISZYN, this volume, p. 43.

[39] FRITZ, H., E. JAUMANN, R. MEISTER, P. PASQUAY, K. HOCHSTRASSER and E. FINK, this volume, p. 257.

Isolation of Chymotrypsin Inhibitors by Affinity Chromatography through Chymotrypsin—Sepharose

GAD FEINSTEIN

Department of Biochemistry, Tel-Aviv University, Tel-Aviv, Israel

Insoluble trypsin was prepared by LEVIN et al. (1964) by cross-linking of trypsin with EMA*. The insoluble trypsin retained its catalytic properties. FRITZ et al. (1966) have shown that insoluble trypsin retained the capability to form a complex with trypsin inhibitors. They have used this property to isolate trypsin inhibitors by affinity chromatography through insoluble trypsin. Crude extracts from various sources were passed through columns of insoluble trypsin at near neutral pH. The trypsin inhibitors form stable complexes with trypsin at that pH (K_i; 10^{-8}—10^{-11}) and therefore were removed from the extract. After all the extract passed through the columns and the columns were thoroughly washed with the buffer, a new acidic solution was applied to the columns. Due to the unstability of trypsin-inhibitor complexes at low pH values, the trypsin inhibitors were eluted out from the columns. FRITZ et al. (1966) have also prepared other insoluble enzymes, Kallikrein and chymotrypsin, and used them to purify several protein inhibitors. Likewise, they prepared insoluble trasylol and used it to purify trypsin and chymotrypsin. Since then, Prof. WERLE's group has been using this technique to isolate a great number of protein inhibitors from animal and plant sources (VOGEL et al. 1968). One of the adherent difficulties with this method is that the insoluble carrier does also function as an ion-exchanger due to its free carboxylate groups. In order to overcome this difficulty FRITZ et al. (1969) blocked some of the carboxylate groups by cross-linking them using diamine compounds. These polyamphoteric insoluble enzyme preparations have less ion-exchange characteristics and therefore gave better results. AXÉN et al. (1967) have introduced a new insoluble carrier which is devoid of charge. They activated agarose with CNBr and cross-linked proteins through their free amino groups to the activated agarose. PORATH et al. (1967) prepared insoluble chymotrypsin by crosslinking it to Sepharose. The insoluble carrier Sepharose, being a polydextran, is void of charge and therefore there are no problems of ion-exchange. CUATRECASAS et al. (1968) coupled synthetic inhibitors of several enzymes to Sepharose and used the method of affinity chromatography to purify several enzymes by single step of affinity chromatography.

We have tried in this laboratory to use the method of affinity chromatography (FEINSTEIN 1970a, FEINSTEIN, 1970b) with Sepharose as the insoluble carrier. An insoluble soybean trypsin inhibitor was prepared (FEINSTEIN 1970b). STI-Sepharose was packed into a column, equili-

---

* **Abbreviations**: EMA, ethylene maleic anhydride copolymer; STI, soybean trypsin inhibitor; TEA, triethanolamine; BAPA, N-benzoyl-DL-arginine-p-nitroanilide; TOV, turkey ovomucoid; ATEE, N-acetyl-L-tyrosine ethyl ester; TCA, trichloroacetic acid.

brated to pH 8.0 with 0.10M TEA buffer containing 0.02M CaCl$_2$. Twenty mg trypsin in 2.0 ml of TEA buffer were applied to the column and then the column was washed with the same buffer. In the second step of the elution 0.10M KCl, pH 2, solution was used. As can be seen in Fig. 1, a small protein peak emerged with the first step and a big protein peak emerged

volume was slightly retarded on the column. It came out with the impurities and as a result, the specific activity of the first peak varied. The specific activity of the second peak appeared to be higher than the first one and to remain

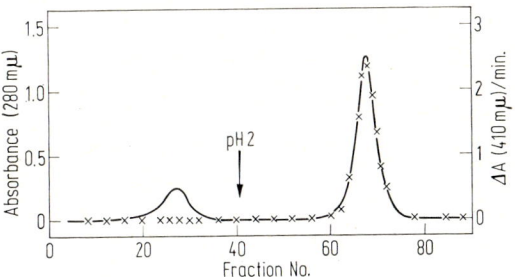

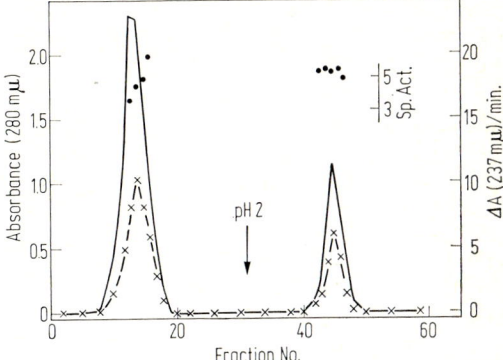

Fig. 1. Chromatography of trypsin on soybean trypsin inhibitor-Sepharose column, 33 cm × 1.8 cm. Twenty mg trypsin were applied to the column and 3.0 ml fractions were collected. Solid line, absorbance at 280 mµ; ××××, tryptic activity in hydrolyzing BAPA. Vertical arrow, elution buffer change from pH 8.0 (TEA buffer) to pH 2.0 (KCl buffer).

Fig. 2. Chromatography of chymotrypsin on turkey ovomucoid-Sepharose column, 22 cm × 1.0 cm. 14.3 mg chymotrypsin were applied to the column and 1.5 ml fractions were collected. Solid line, absorbance at 280 mµ; ××××, chymotrypsin activity in hydrolyzing ATEE; • • • •, specific activity.

at the second step. The tryptic activity of the effluent was determined using BAPA as a substrate (ERLANGER et al. 1961). Only the second peak had tryptic activity. Since commercial crystalline bovine trypsin is known to be about 60% active, it appears that by passing trypsin through the insoluble STI-Sepharose column we were able to get rid of some of the inert impurities and to increase the specific activity of trypsin. Turkey ovomucoid is capable of inhibiting simultaneously trypsin and chymotrypsin. The affinity chromatography of chymotrypsin on TOV-Sepharose is shown in Fig. 2. A great excess of enzyme was applied to the column and so part of the enzyme emerged in the first part of the elution, pH 8.0 TEA buffer. As expected a second protein peak was eluted with the acidic buffer. The chymotryptic activity of fractions were determined using ATEE as a substrate (SCHWERT and TAKENAKA, 1955). It appeared that the excess enzyme that came in the void

constant throughout the peak which would indicate that we got a pure and homogeneous chymotrypsin. Chymotrypsin was also chromatographed on chicken ovomucoid-Sepharose (FEINSTEIN 1970a). The commercial chicken ovomucoid was prepared from egg-white by the method of LINEWEAVER and MURRAY (1947). Chicken ovomucoid is capable of inhibiting trypsin. However, it was found that about 95% of the chymotrypsin was eluted with the void volume and a small protein peak emerged in the second step of elution (Fig. 3). Both peaks had chymotryptic activity. It therefore appeared that the chicken ovomucoid preparation contained also a chymotrypsin inhibitor.

MATSUSHIMA (1958) had found that egg-white contains two trypsin inhibitors, the ovomucoid and additional one that he named ovoinhibitor. He fractionated it from egg white by salt precipitation. RHODES et al. (1960) reported that ovoinhibitor was capable of inhibiting chymotrypsin. FEENEY et al. (1963) established the fact that chicken ovomucoid prepared by the TCA-aceton method of LINEWEAVER and MURRAY (1947) contained ovoinhibitor. TOMIMATSU et al. (1966) fractionated ovoinhibitor from ovomucoid by salt fractionation and characterized it.

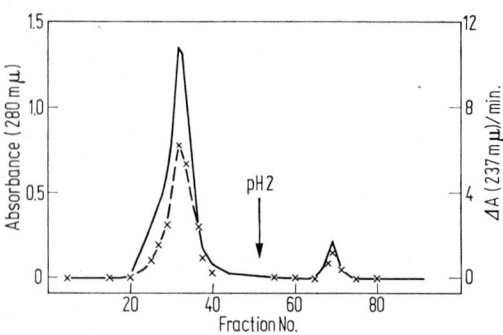

Fig. 3. Chromatography of chymotrypsin on chicken ovomucoid-Sepharose column 35 cm × 1.7 cm. Twenty mg of chymotrypsin were applied to the column and fractions of 2.8 ml were collected. Solid line, absorbance at 280 mμ; X X X X, chymotrypsin activity in hydrolyzing ATEE.

DAVIS et al. (1969) used salt fractionation plus ion-exchange chromatography on DEAE-cellulose to purify ovoinhibitor, and determined its properties.

We have tried in our laboratory to use the method of affinity chromatography through chymotrypsin-Sepharose to isolate chicken ovoinhibitor by single step. Insoluble chymotrypsin cross-linked to Sepharose was prepared according to PORATH et al. (1967). Chicken ovomucoid was prepared by the TCA-acetone method of LINEWEAVER and MURRAY (1947). Chicken ovomucoid was applied to chymotrypsin-Sepharose column equilibrated with pH 8.0, 0.20M, TEA buffer. The column was washed with the buffer until no more protein emerged from the column (Fig. 4). Then, pH 2.0, 0.2M KCl solution was applied to the column. A second protein peak was eluted from the column. When the second protein peak was assayed for its capacity to inhibit proteases it was found that it inhibited trypsin and chymotrypsin. It was found that about one mg of protein inhibited one mg of 3 × crystallized bovine chymotrypsin. In some other preparation 0.7 mg of inhibitor were required for complete inhibition of one mg of chymotrypsin. No chymotrypsin inhibitory activity was found in the first protein peak which emerged at the void

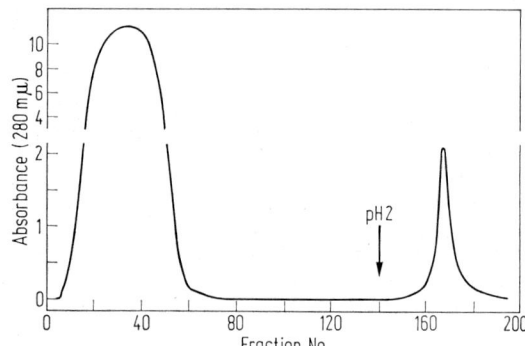

Fig. 4. Chromatography of chicken ovomucoid on chymotrypsin-Sepharose column, 26 cm × 1.6 cm. Three grams of crude chicken ovomucoid were applied to the column and 3.0 ml fractions were collected. Solid line, absorbance at 280 mμ.

volume of the column. Table 1 gives the results of the purification. The yield of inhibitory activity is about 50% and the chymotrypsin inhibitor was 37 fold purified. The yields of activity of different preparations were 40—70%. Acrylamide disc gel electrophoresis had shown that the ovoinhibitor was free of ovomucoid or any other protein contamination. A single broad band was obtained containing several sub-bands. This is in agreement with the results of DAVIS

Table 1. Purification of Ovoinhibitor on Chymotrypsin-Sepharose

	Chymotrypsin I. U.[a]	Yield (%)	Sp. Activity	Purification
Crude	84.3	100	0.028	1
Ovomucoid	0	—	—	—
Ovoinhibitor	39.8	47.2	1.02	36.6

[a] One inhibition unit (I. U.) is an amount that inhibits one mg of chymotrypsin.

et al. (1969) who found that chicken ovoinhibitor is hetrogeneous, apparently due to variation in the content of charged sugars like glucosamine and sialic acid. The amino acid composition of ovoinhibitor was determined after complete hydrolysis on Beckman automatic amino acid analyzer. The results are shown in Table 2. The amino acid composition of ovoinhibitor obtained in this study is very much alike the results that were obtained by TOMIMATSU et al. (1966) and DAVIS et al. (1969).

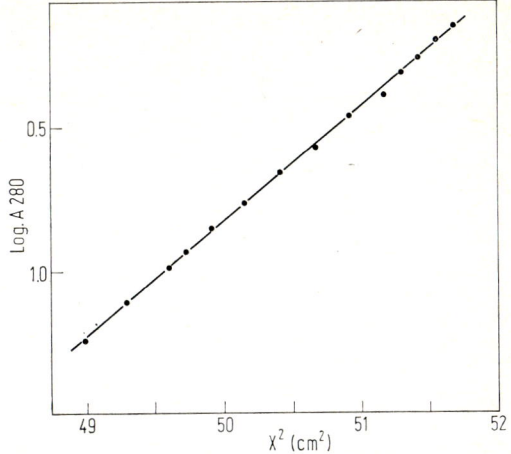

Fig. 5. Sedimentation equilibrium of ovoinhibitor in pH 8.0, 0.10 M TEA buffer.

Table 2. Amino Acid Composition of Chicken Ovoinhibitor

(mole aa/10 000 g)

	A	B	C
Lysine	3.9	4.7	4.4
Histidine	2.3	2.9	2.3
Arginine	3.3	4.3	3.7
Aspartic acid	7.7	9.7	8.5
Threonine	5.2	6.8	5.0
Serine	4.3	5.5	3.6*
Glutamic acid	6.3	8.0	7.0
Proline	2.8	3.7	3.5
Glycine	5.5	6.6	5.5
Alanine	3.4	4.1	3.5
Half-cystine	4.1	7.0	4.1*
Valine	4.3	5.6	4.5
Methionine	0.5	0.8	0.7
Isoleucine	2.9	3.5	2.9
Leucine	3.7	4.5	3.9
Tyrosine	2.2	3.5	3.0
Phenylalanine	1.6	1.2	1.1

A — TOMIMATSU et al. (1966), B — DAVIS et al. (1969), C — This study

* No corrections were made for the destruction of these amino acids during hydrolysis.

Equilibrium sedimentation studies in Beckman Model E analytical ultracentrifuge (17.000 rpm) revealed that as far as size is concerned the ovoinhibitor was homogeneous (Fig. 5). The molecular weight of ovoinhibitor at infinite concentration was found to be 52.400 (Fig. 6). The values of M. W. of ovoinhibitor reported by TOMIMATSU et al. (1966) and DAVIS et al. (1969) were 46.500 and 49.000 respectively.

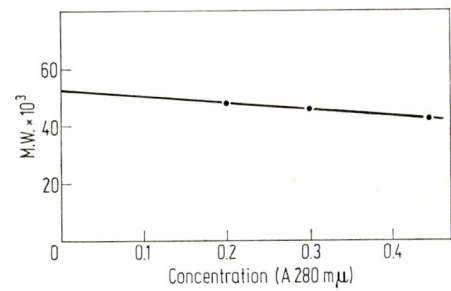

Fig. 6. Molecular weight dependence of chicken ovoinhibitor on concentration. Molecular weight was determined by the sedimentation equilibrium method.

From all the above data it appears that we were able to isolate a pure ovoinhibitor in a single step of affinity chromatography on chymotrypsin-Sepharose.

An attempt was made to use this method for isolation of chymotrypsin inhibitors from soybean. One hundred mg of commercial soybean trypsin inhibitor were applied to chymotrypsin-Sepharose column (Fig. 7). It is well known that soybean contains several proteins which are capable of inhibiting trypsin and chymotrypsin. It is apparent that a portion of the STI did complex with the insoluble chymotrypsin. It was

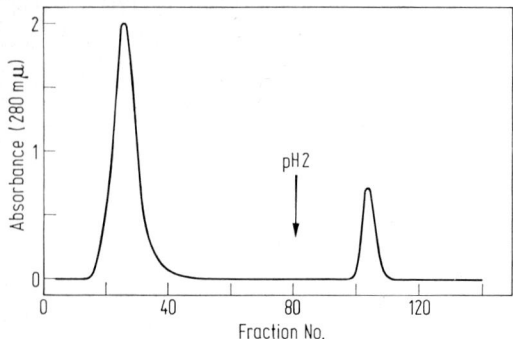

Fig. 7. Chromatography of commercial soybean trypsin inhibitor on chymotrypsin-Sepharose column, 26 cm × 1.6 cm. One hundred mg of commercial STI were applied to the column and 3.0 ml fractions were collected Solid line, absorbance at 280 mμ.

found that the second protein peak was capable of inhibiting chymotrypsin and trypsin. This preparation was compared to pure STI_{AA} (kindly supplied by Dr. A. Gertler of the Hebrew University). Commercial STI, STI_{AA} and chymotrypsin inhibitor prepared by affinity chromatography showed the following specific activities of inhibiting chymotrypsin, 0.70, 3.1 and 4.5 respectively. The chymotrypsin inhibitor was purified about 6 fold by affinity chromatography and its specific activity is near that of pure STI_{AA}. However, acrylamide disc gel electrophoresis revealed that there were several components in the chymotrypsin inhibitor preparation.

We demonstrated in this study the usefulness of using insoluble chymotrypsin cross-linked to the non-charged carrier Sepharose for affinity chromatography of chymotrypsin inhibitors. In a single step of affinity chromatography, we were able to get highly pure preparations of inhibitors. When several chymotrypsin inhibitors were present in a given source, we got all of them removed from the mixture and an additional procedure would be required to separate them from each other. However, the ease and speed in separating proteases' inhibitors from a crude mixture will probably make this method a very useful and common procedure in the near future.

References

[1] Levin, Y., M. Pecht, L. Goldstein and E. Katchalski, Biochemistry **3**, 1905 (1964).
[2] Fritz, H., H. Schult, M. Neudecker and E. Werle, Angew. Chem. internat. Edit. **5**, 735 (1966).
[3] Vogel, R., I. Trautschold and E. Werle, Natural Proteinase Inhibitors, Academic Press (1968).
[4] Fritz, H., M. Gebhardt, E. Fink, W. Schramm and E. Werle, Z. Physiol. Chem. **350**, 129 (1969).
[5] Axén, R., J. Porath and S. Ernback, Nature **214**, 1302 (1967).
[6] Porath, J., R. Axén and S. Ernback, Nature **215**, 1491 (1967).
[7] Cuatrecases, P., M. Wilchek and C. B. Anfinsen, Proc. Nat'l. Acad. Sci. U. S. **61**, 636 (1968).
[8] Feinstein, G., FEBS Letters **7**, 353 (1970a).
[9] Feinstein, G., Biochem. Biophys. Acta **214**, 224 (1970b).
[10] Erlanger, B. F., N. Kokowsky and W. Cohen, Arch. Biochem. Biophys. **95**, 271 (1961).
[11] Schwert, G. W. and Y. Takenaka, Biochem. Biophysp. Acta **16**, 570 (1955).
[12] Lineweaver, H. and C. W. Murray, J. Biol. Chem. **171**, 565 (1947).
[13] Matsushima, K., Science **127**, 1178 (1958).
[14] Rhodes, M. B., N. Bennett and R. E. Feeney, J. Biol. Chem. **235**, 1686 (1960).
[15] Feeney, R. E., F. C. Stevens and D. T. Osuga, J. Biol. Chem. **238**, 1415 (1963).
[16] Tomimatsu, Y., J. J. Clary and J. J. Bartulovich, Arch. Biochem. Biophys. **115**, 536 (1966).
[17] Davis, J. G., J. C. Zahnley and J. W. Donovan, Biochemistry **8**, 2044 (1969).

A Simple Method of Purification of the Basic Trypsin Inhibitor of Bovine Organs*

BEATRICE KASSELL and MEREDITH B. MARCINISZYN

Department of Biochemistry, The Medical College of Wisconsin, 561 North Fifteenth Street, Milwaukee, Wisconsin 53233, U.S.A.

Although a number of methods have been proposed (e. g. 1—9) for the preparation of the basic trypsin inhibitor since the original crystallization by KUNITZ and NORTHROP [10], these methods are quite laborious. The preparation of inhibitors by affinity chromatography on trypsin-resins originated with the Munich group [11, 12]. The present report is concerned with a modification based on the use of trypsin-Sepharose.

The method, starting with frozen lung, involves only four steps: extraction of the lung, trichloroacetic acid precipitation, affinity chromatography on Trypsin-Sepharose 4B, and chromatography on CM-cellulose.

Materials — Sepharose 4B was obtained from Pharmacia Fine Chemicals, Piscataway, N. J. Bovine trypsin was a gift from Novo Industries, Copenhagen. Cyanogen bromide was purchased from Eastman Organic Chemicals, Rochester, N. Y. Frozen lung was obtained from Pel-Freez Biologicals, Rogers, Ark.

Preparation of the affinity column — The attachment of trypsin to the resin is based on the methods used by AXÉN et al. [13] and ANFINSEN et al. [14] for other proteins. The activation of the Sepharose was carried out in a 200 ml wide-mouth bottle containing a pH electrode and placed in a 25° water bath on a magnetic stirrer. Sepharose 4B was washed on a coarse fritted glass filter with water to remove azide and sucked dry. The damp resin (37.5 g) was suspended in 25 ml of water and cyanogen bromide (3.75 g dissolved in 50 ml of water) was added. The pH was rapidly adjusted to 11.0 and maintained at that pH with 5M NaOH (about 10 ml was used). After 8 minutes the rate of addition of NaOH slowed down. The mixture was immediately filtered by suction on a coarse sintered glass funnel (9 × 7 cm) and washed quickly with 2 L of ice water and 2 L of cold 0.05M sodium borate buffer, pH 9.0. Filtration and washing took only about 5 minutes.

A solution of 1.00 g of trypsin was prepared in 0.05M sodium borate buffer, pH 9.0, containing 0.01M $CaCl_2$, just before the Sepharose activation and was kept at 0°. The activated Sepharose was added promptly to this solution and the mixture was rotated gently on an inverting shaker in the cold room overnight. Two identical batches were prepared and combined at this stage.

The resin (equivalent to 2 g of trypsin) was poured into a 3.9 cm column. It was washed with 1.2 L of the borate buffer and 600 ml of 0.5M NaCl, both solutions containing 0.01M

---

* This work was supported by grants from the National Science Foundation (GB-12630) and from the Patrick and Anna M. Cudahy Fund.

CaCl$_2$. A Mariotte flask was used at a liquid level only slightly higher than the top of the resin. The total A$_{280}$ of the first effluent corresponded to 126 mg of trypsin (6.3%). The NaCl wash had almost no absorbance. Finally the resin was transferred to a column of 2 cm diameter and was washed to constant pH and conductivity with 0.1M sodium acetate buffer, pH 4.0, containing 0.3M NaCl and 0.01M CaCl$_2$. Just before each use, the column was washed again with this buffer until the A$_{280}$ of the effluent was below 0.025. After each use, the column was promptly restored to pH 4 and kept at this pH with toluene in the buffer as preservative.

Purification Procedure

1. Preparation of lung extract — Subsequent operations were carried out at 4° except as noted. The lung was partially thawed, cut up and put through a meat grinder. The ground lung (1 kg) was homogenized in a large blender with 900 m*l* of 0.1M triethanolamine buffer, pH 7.8, containing 0.3M NaCl and 0.01M CaCl$_2$ [12]. The homogenate was transferred to a beaker, rinsed with 100 m*l* buffer, stirred gently for 1 hour and then centrifuged for 1 hour at 13.000 g in a refrigerated centrifuge. The sediment was resuspended in an equal volume of buffer, stirred and centrifuged again. The supernatant solutions were combined.

2. Trichloroacetic acid precipitation — At room temperature, trichloroacetic acid solution (50%, w/v) was added under stirring to the combined supernatant solutions to a concentration of 2.5%. After 30 minutes the mixture was centrifuged for 30 minutes in the cold. The sediment was washed with an equal volume of cold 2.5% trichloroacetic acid. The combined supernatant solutions were brought to pH 4,0 with 1M NaOH (about 50 m*l*) and filtered by suction through a pad of Celite. A clear yellow solution (1850 m*l*, containing a minimum of 200 mg of inhibitor) was obtained.

3. Affinity chromatography — The crude inhibitor solution was passed into the column from a Mariotte flask at the rate of 1.5 m*l*/min. The column was washed with the acetate-NaCl-CaCl$_2$ buffer until the absorbance was almost zero and then with unbuffered 0.01M CaCl$_2$ in 0.3M NaCl. The inhibitor was eluted with 0.01M HCl containing 0.3M NaCl and 0.01M CaCl$_2$. The elution pattern is shown in Figure 1.

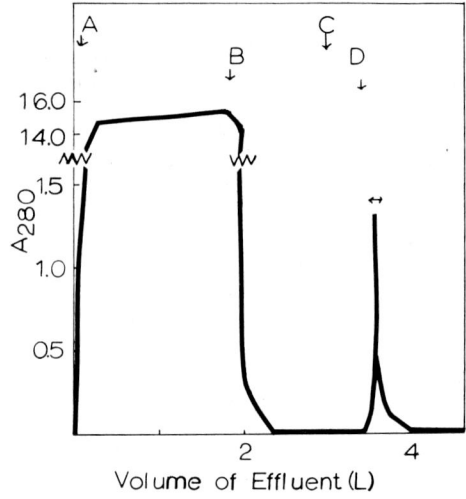

Fig. 1. Chromatography of the crude inhibitor solution on Trypsin-Sepharose 4B. Column 2 × 26 cm. Fractions 15 m*l*. Flow rate 1.5 m*l*/min. A. Sample. B. Buffer: 0.1M sodium acetate, pH 4.0, containing 0.01M CaCl$_2$ and 0.3M NaCl. C. Unbuffered 0.01M CaCl$_2$ in 0.3M NaCl. D. 0.01M HCl containing 0.01M CaCl$_2$ and 0.3M NaCl. The horizontal arrow indicates the inhibitor peak.

Assay of the effluent fractions for inhibiting activity [15] showed that no inhibitor was eluted from the column before the addition of the HCl. The activity in the first few tubes on the ascending side of the inhibitor peak was usually somewhat low. The remainder of the peak was pooled as shown by the arrow in Figure 1. The pooled solution was adjusted to pH 5 with 4M sodium acetate and lyophilized. The salts were removed by passage through a column (2 × 94 cm) of Sephadex G-25, using 0.1M ammonium bicarbonate, pH 7.0, as eluent. The protein peak was lyophilized.

4. *Chromatography on CM-cellulose* — This step was carried out according to the procedure of AVINERI-GOLDMAN et al. [9]. A small scale experiment with the product of step 3 is shown in Figure 2. Details are given in the legend. The second peak contained pure inhibitor as indicated by the criteria below.*

Criteria of purity — The amino acid analysis of the purified inhibitor is shown in Table I. All of the stable amino acids which give good values in a 24 hour hydrolysate are very close to the known values (lysine, arginine, aspartic acid, glutamic acid, glycine, alanine, leucine and phenylalanine). The best check of the purity of this protein is the absence of histidine. Figure 3 shows the amino acid analyzer recording of the short column run, revealing only a minute trace of histidine.

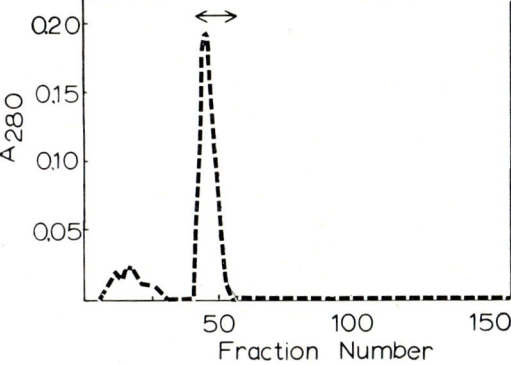

Fig. 2. Chromatography of a sample of the inhibitor peak of Figure 1 on CM-cellulose. Column 0.9 × 20 cm. Load 15 mg. Fractions 2 ml. Flow rate 30 ml/hr. Starting buffer, 250 ml of 0.01M sodium phosphate, pH 6.2, in a closed mixing chamber. Gradient buffer 0.25M NaCl in the same buffer. The arrow indicates the portion of the inhibitor pooled.

Table 1. Amino Acid Analysis of the Inhibitor Peak of Fig. 2

(Residues/mole, uncorrected values of a 24 hr hydrolysate)

Amino Acid	Found	Known	Amino Acid	Found	Known
Lys	4.10	4	Ala	5.97	6
Arg	6.00	6	1/2 Cys	5.35	6
Asp	5.00	5	Val	.90	1
Thr	2.87	3	Met	.82	1
Ser	.88	1	Ileu	1.20	2
Glu	3.00	3	Leu	1.96	2
Pro	4.29	4	Tyr	3.55	4
Gly	5.95	6	Phe	3.84	4

* The protein was desalted as in step 3.

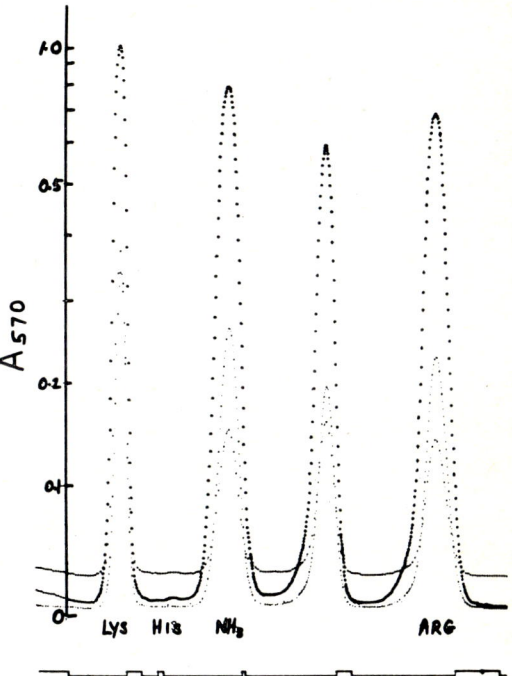

Fig. 3. Amino acid analyzer recording of the short column analysis of the inhibitor peak of Fig. 2, showing the absence of histidine. The unlabeled peak is the internal standard, 1-amino-2-guanido-propionic acid [16].

Discussion

The relatively small amount of labor involved in this method and the high purity of the product recommend this procedure. The Trypsin-Sepharose is easy to prepare and appears to be stable at pH 4. It has a high capacity; the column described, containing 1.9 g of trypsin, retained 500 mg of inhibitor in one experiment.

The inclusion of the trichloroacetic acid precipitation has a two-fold purpose. It completely clarifies the crude extract so that the column does not become clogged and it removes enzymes such as amylases that might attack the Sepharose. A similar method of linking trypsin to a solid support for the preparation of inhibitors has been suggested previously. BEELEY [17] and TOMÁSEK et al. [18] used trypsin linked to Sephadex G-200.

Summary

A new method of preparation of the basic trypsin inhibitor of bovine organs is based on affinity chromatography on a column of Trypsin-Sepharose 4B, followed by chromatography on CM-cellulose. A product of very high purity is obtained.

References

[1] GREEN, N. M. and E. WORK, Biochem. J. **54**, 257 (1953).
[2] KRAUT, H. and R. KÖRBEL-ENKHARDT, Z. Physiol. Chem. **309**, 243 (1957).
[3] KRAUT, H. and N. BHARGAVA, Z. Physiol. Chem. **338**, 231 (1964).
[4] SCEVOLA, M. E., M. FRANCHINI and A. BARBIERI, Boll. Soc. Ital. Biol. Sper. **38**, 1771 (1962).
[5] KASSELL, B., M. RADICEVIC, S. BERLOW, R. J. PEANASKY and M. LASKOWSKI, SR., J. Biol. Chem. **238**, 3274 (1963).
[6] SACH, E., M. THÉLY and J. CHOAY, Compt. Rend. **260**, 3491 (1965).
[7] DLOUHÁ, V., J. NEUWIRTHOVA, B. MELOUN and F. ŠORM, Coll. Czech, Chem. Commun. **30**, 1705 (1965).
[8] SCHULTZ, F., U. S. patents 3.181.997 and 3.308.026.
[9] AVINERI-GOLDMAN, R., I. SNIR, G. BLAUER and M. RIGBI, Arch Biochem. Biophys. **121**, 107 (1967).
[10] KUNITZ, M. and J. H. NORTHROP, J. Gen. Physiol. **19**, 991 (1936).
[11] FRITZ, H., H. SCHULT, M. HUTZEL, M. WIEDEMANN and E. WERLE, Z. Physiol. Chem. **348**, 308 (1967).
[12] FRITZ, H., M. GEBHARDT, E. FINK, W. SCHRAMM and E. WERLE, Z. Physiol. Chem. **350**, 129 (1969).
[13] AXÉN, R., E. HEILBRONN and A. WINTER, Biochim. Biophys. Acta **191**, 478 (1969).
[14] CUATRECASAS, P., M. WILCHEK and C. B. ANFINSEN, Proc. Natl. Acad. Sci. (U. S.) **61**, 636 (1968).
[15] KASSELL, B. and T.-W. WANG, this Volume, p. 89.
[16] WALSH, K. A. and J. R. BROWN, Biochim. Biophys. Acta **58**, 596 (1962).
[17] BEELEY, J. G., Biochem. J. **117**, 70P (1970).
[18] TOMÁŠEK, V., S. A. KUDINOV and F. ŠORM, Abstracts, 5th FEBS meeting, Prague, 1968, p. 259.

The Molecular Architecture of the Serine Proteinases

David Shotton

The Molecular Enzymology Laboratory, Department of Biochemistry, University of Bristol, Bristol BS 8 1 TD, England

Introduction

The extraordinary efficiency and specificity characteristic of enzyme catalysed reactions is in large measure due to the possession by the enzyme of a specialized surface region, the substrate binding site, at which substrate molecules can be specifically bound by selective adsorption, and the possession of unusually reactive groups, usually amino acid side chains, which are positioned adjacent to the bound substrate in precisely the correct spatial configuration for them to participate in and hence catalyse the required reaction. The properties are intimately related to the specific three-dimensional conformations which the polypeptide chains of different enzymes adopt, and hence a study of the detailed three-dimensional structures of a number of related enzymes which possess a common catalytic mechanism but differing substrate specificities would allow a correlation to be made between the enzymic properties exhibited by these enzymes and their genetically determined amino acid sequences. The recent determinations of the three-dimensional structures of three pancreatic serine proteinases have now made such structural comparisons possible.

The Pancreatic Serine Proteinases

Serine proteinases are endopeptidases possessing at their catalytic site a reactive serine residue, which can be covalently inhibited by diisopropylphosphorofluoridate (DFP) (Hartley, 1960). There are two well characterized classes of serine proteinases, a bacterial class possessing a -Thr-Ser-Met- active centre sequence, of which the subtilisins are examples (Smith et al., 1966; Wright et al., 1969) and another class with members in many phyla which has an -Asp-Ser-Gly- active centre sequence. The major mammalian pancreatic endopeptidases all belong to this second class, bovine trypsin, the bovine chymotrypsins -A and -B, and porcine elastase being the best studied members, although other isoenzymes and representatives from other mammalian species are known.

These pancreatic enzymes are secreted as inactive zymogens into the duodenum, and on activation hydrolyse peptide bonds of dietary protein to give smaller peptides, all using the same catalytic mechanism. They differ from one another, however, in their substrate specificities, trypsin hydrolysing peptide bonds formed from the carboxyl groups of the amino acids arginine and lysine which bear basic side chains, the chymotrypsins hydrolysing the peptide bonds on the carboxyl-terminal side of large aromatic amino acid residues such as tyrosine and tryptophan, and elastase preferentially cleaving those of small uncharged amino acids such as alanine.

The first indication that trypsin, chymotrypsin-A and elastase possessed common structural features was given by NAUGHTON, SANGER, HARTLEY and SHAW (1960) who showed that electrophoresis of partial acid hydrolysates of these enzymes after inhibition with ^{32}P-DFP yielded identical patterns of peptides derived from the common -Asp-Ser-Gly- active centre sequence. In the ten years since that discovery the detailed three-dimensional structures of all three enzymes have been determined, by combining the results of chemical investigations of their amino acid sequences and X-ray diffraction studies of the electron density distributions of their crystalline states.

The Structure of α-Chymotrypsin

The structure of bovine chymotrypsin-A was the first to be fully determined, the amino acid sequence being deduced in 1964 by HARTLEY (HARTLEY, 1964; HARTLEY & KAUFFMAN, 1966) and confirmed by KEIL, ŠORM and their colaborators (MELOUN et al., 1966), while the three-dimensional crystal structure of α-chymotrypsin (a form of chymotrypsin-A) was elucidated three years later by BLOW and his collaborators (MATTHEWS et al., 1967; SIGLER et al., 1968).

The α-chymotrypsin molecule consists of 3 polypeptide chains linked by disulphide bridges, derived from the single polypeptide chain of the zymogen, chymotrypsinogen, by tryptic and autolytic cleavages during activation (DESNUELLE, 1960). It has a compact globular shape (Fig. 1), and is largely made up of long stretches of almost fully extended polypeptide chain, which tend to fold back on themselves to form large antiparallel loops. These are hydrogen bonded together into two curved barrel-shaped irregular antiparallel β-pleated sheets which form the main structural elements of the two halves into which the molecule is organized. Two short stretches of α-helix are present, at the C-terminus and in the Methionine Loop, and the N-terminus of the B chain lies buried in an internal cavity.

The two halves of the molecule lie in intimate contact, held together by hydrogen bonds and hydrophobic interactions, the active centre being formed where essential residues from the two halves are brought into close proximity.

Sequence Comparisons

Following the publication of the chymotrypsin-A sequence, the trypsin amino acid sequence was reported from NEURATH's laboratory (WALSH et al., 1964) and independently from ŠORM's laboratory (MIKEŠ et al., 1966), the chymotrypsin-B sequence was determined by SMILLIE and his collaborators (SMILLIE et al., 1968), and the determination of the elastase sequence was completed by SHOTTON and HARTLEY (1970).

When these sequences, each of about 240 amino acid residues, are aligned and compared (Tab. 1), the very close similarity between the isoenzymes, Chymotrypsin-A and -B, is obvious, and a striking degree of sequence identity is apparent between all the enzymes around the active centre serine residue, Ser-195*, the essential histidine residue, His-57, four of the disulphide bridges and in several other regions. These observations led to strong predictions that the three-dimensional structures of these enzymes would all be very similar.

The Structure of Elastase

Direct experimental demonstration of the truth of these predictions recently became possible when the three-dimensional structure of elastase was elucidated (SHOTTON and WATSON, 1970a and 1970b). It was clear from a comparison of the atomic model of elastase with that of α-chymotrypsin that the overall conformation of these two enzymes are very similar (compare Figs. 1 and 2). Where local differences in chain length

---

* The chymotrypsinogen-A numbering scheme (HARTLEY and KAUFFMAN, 1966) has been used throughout to facilitate comparison between corresponding residues of these homologous proteins.

Table 1. Amino acid sequences of porcine elastase and bovine trypsiogen, chymotrypsinogen-A and chymotrypsinogen-B

```
       1   2   3   4   5   6   7   8   9  10  11  12  13  14  15  16  17  18  19  20  21  22  23
E:                                                                 VAL-VAL-GLY-GLY-Thr-GLU-ALA-Gln-
T:                                         Val-Asp-Asp-Asp-Asp-LYS-ILE-VAL-GLY-GLY-Tyr-Thr-Cys-Gly-
CA: Cys-Gly-Val-Pro-Ala-Ile-Gln-Pro-Val-Leu-Ser-Gly-Leu-Ser-ARG-ILE-VAL-Asn-GLY-Glu-GLU-ALA-Val-
CB: Cys-Gly-Val-Pro-Ala-Ile-Gln-Pro-Val-Leu-Ser-Gly-Leu-Ala-ARG-ILE-VAL-Asn-GLY-Glu-ASP-ALA-Val-

      24  25  26  27  28  29  30  31  32  33  34  35  36  36A 36B 36C 37  38  39  40
E:  -Arg-ASN-SER-TRP-PRO-Ser -GLN-ILE-SER-LEU-GLN-Tyr-ARG-Ser-Gly-Ser-SER-Trp-Ala -HIS-
T:  -Ala-ASN-THR-Val -PRO-TYR-GLN-VAL-SER-LEU-ASN —   —   —   —   SER-GLY-TYR-HIS-
CA: -Pro-Gly-SER-TRP-PRO-TRP-GLN-VAL-SER-LEU-GLN-Asp-LYS —   —   —   THR-GLY-PHE-HIS-
CB: -Pro-Gly-SER-TRP-PRO-TRP-GLN-VAL-SER-LEU-GLN-Asp-Ser —   —   —   THR-GLY-PHE-HIS-

      41  42  43  44  45  46  47  48  49  50  51  52  53  54  55  56  57  58  59
E:  -Thr-CYS-GLY-GLY-THR-LEU-ILE-Arg-GLN-ASN-TRP-VAL-Met-THR-ALA-ALA-HIS-CYS-Val-
T:  -PHE-CYS-GLY-GLY-SER-LEU-ILE-ASN-Ser-GLN-TRP-VAL-VAL-SER-ALA-ALA-HIS-CYS-Tyr-
CA: -PHE-CYS-GLY-GLY-SER-LEU-ILE-ASN-GLU-ASN-TRP-VAL-VAL-THR-ALA-ALA-HIS-CYS-Gly-
CB: -PHE-CYS-GLY-GLY-SER-LEU-ILE-Ser-GLU-ASP-TRP-VAL-VAL-THR-ALA-ALA-HIS-CYS-Gly-

      60  61  62  63  64  65  65A 66  67  68  69  70  71  72  73  74  75  76  77  78
E:  -Asp-Arg-Glu-LEU-Thr-Phe-ARG-VAL-VAL-Val-GLY-GLU-His-ASN-LEU-ASN-Gln-Asn-Asn-GLY-
T:  -Lys-SER-Gly-ILE-Gln-VAL-ARG-Leu —   —   GLY-GLN-Asp-ASN-ILE-ASN-Val-Val-Glu-GLY-
CA  -Val-THR-Thr-Ser-Asp-VAL —   VAL-VAL-Ala-GLY-GLU-Phe-ASP-Gln-Gly-Ser-Ser-SER-Glu-
CB  -Val-THR-Thr-Ser-Asp-VAL —   VAL-VAL-Ala-GLY-GLU-Phe-ASP-Gln-Gly-Leu-Glu-THR-Glu-

      79  80  81  82  83  84  85  86  87  88  89  90  91  92  93  94  95  96  97
E:  -Thr-GLU-GLN-TYR-VAL-Gly-VAL-Gln-LYS-ILE-VAL-VAL-HIS-PRO-Tyr-TRP-ASN-THR-ASP-
T:  -ASN-GLN-GLN-PHE-ILE-Ser-Ala-Ser-LYS-Ser-ILE-VAL-HIS-PRO-Ser-TYR-ASN-SER-ASN-
CA: -Lys-Ile-GLN-Lys-LEU-Lys-ILE-Ala-LYS-VAL-Phe-Lys-Asn-Ser-Lys-TYR-ASN-SER-Leu-
CB: -ASP-Thr-GLN-Val-LEU-Lys-ILE-Gly-LYS-VAL-Phe-Lys-Asn-PRO-Lys-PHE-Ser-Ile-Leu-

      98  99 99A 99B 100 101 102 103 104 105 106 107 108 109 110 111 112 113 114
E:  -Asp-VAL-Ala-Ala-Gly-Tyr-ASP-ILE-Ala-LEU-LEU-ARG-LEU-ALA-Gln-Ser-Val-THR-LEU-
T:  -THR-LEU —   —   ASN-ASN-ASP-ILE-Met-LEU-ILE -LYS-LEU-Lys-SER-ALA-ALA-SER-LEU-
CA: -THR-ILE —   —   ASN-ASN-ASP-ILE-Thr-LEU-LEU-LYS-LEU-Ser-THR-ALA-ALA-SER-Phe-
CB: -THR-VAL —   —   Arg-ASN-ASP-ILE-Thr-LEU-LEU-LYS-LEU-ALA-THR-Pro-ALA-Gln-Phe-

     115 116 117 118 119 120 121 122 123 124 125 126 127 128 129 130 131 132 133 134
E:  -ASN-SER-Tyr-VAL-Gln-Leu-Gly-Val-LEU-PRO-Arg-ALA-Gly-Thr-Ile-Leu-ALA-Asn-Asn-SER-
T:  -ASN-SER-Arg-VAL-Ala-Ser-ILE-Ser-LEU-PRO-THR —   SER-Cys-Ala-Ser —   ALA-GLY-THR-
CA: -Ser-GLN-Thr-VAL-Ser-Ala-VAL-Cys-LEU-PRO-SER-ALA-SER-ASP-Asp-Phe-ALA-ALA-GLY-THR-
CB: -Ser-GLU-Thr-VAL-Ser-Ala-VAL-Cys-LEU-PRO-SER-ALA-Asp-GLU-Asp-Phe-Pro-ALA-GLY-Met-

     135 136 137 138 139 140 141 142 143 144 145 146 147 148 149 150 151 152 153 154
E:  -Pro-CYS-Tyr-ILE-THR-GLY-TRP-GLY-LEU-THR-ARG —   THR-ASN-Gly-GLN-Leu-Ala-Gln-Thr-
T:  -Gln-CYS-Leu-ILE-SER -GLY-TRP-GLY-Asn -THR-LYS-Ser-SER-Gly-Thr-Ser-Tyr-PRO-ASP-Val-
CA: -Thr-CYS-Val-Thr-THR-GLY-TRP-GLY-LEU-THR-ARG-Tyr-THR-ASN-Ala-ASN-Thr-PRO-ASP-ARG-
CB: -Leu-CYS-Ala-Thr-THR-GLY-TRP-GLY-Lys -THR-LYS-Tyr-Asn -Ala-Leu-Lys-Thr-PRO-ASP-LYS-
```

	155	156	157	158	159	160	161	162	163	164	165	166	167	168	169	170	170A	170B	171
E:	-LEU	-GLN	-GLN	-ALA	-Tyr	-LEU	-PRO	-Thr	-VAL	-Asp	-Tyr	-Ala	-Ile	-CYS	-Ser	-SER	-Ser	-Ser	-TYR-
T:	-LEU	-Lys	-Cys	-Leu	-Lys	-Ala	-PRO	-ILE	-LEU	-SER	-ASN	-SER	-Ser	-CYS	-LYS	-SER	—	—	Ala -
CA:	-LEU	-GLN	-GLN	-ALA	-SER	-LEU	-PRO	-LEU	-LEU	-SER	-ASN	-THR	-ASN	-CYS	-LYS	-Lys	—	—	TYR-
CB:	-LEU	-GLN	-GLN	-ALA	-THR	-LEU	-PRO	-ILE	-VAL	-SER	-ASN	-THR	-ASP	-CYS	-ARG	-Lys	—	—	TYR-

	172	173	174	175	176	177	178	179	180	181	182	183	184	184A	185	186	187	188	188A
E:	-TRP	-GLY	-SER	-Thr	-VAL	-LYS	-ASN	-Ser	-MET	-VAL	-CYS	-ALA	-GLY	—	Gly	-Asn	-GLY	-VAL	-ARG-
T:	-TYR	-Pro	-Gly	-Gln	-ILE	-THR	-Ser	-Asn	-MET	-Phe	-CYS	-ALA	-GLY	-Tyr	-Leu	-Glu	-GLY	-Gly	-LYS -
CA:	-TRP	-GLY	-THR	-LYS	-ILE	-LYS	-ASP	-Ala	-MET	-ILE	-CYS	-ALA	-GLY	—	Ala	-Ser	-GLY	-VAL	—
CB:	-TRP	-GLY	-SER	-ARG	-VAL	-THR	-ASP	-Val	-MET	-ILE	-CYS	-ALA	-GLY	—	Ala	-Ser	-GLY	-VAL	—

	189	190	191	192	193	194	195	196	197	198	199	200	201	202	203	204	205	206	207	208
E:	-SER	-Gly	-CYS	-GLN	-GLY	-ASP	-SER	-GLY	-GLY	-PRO	-LEU	-His	-CYS	-Leu	-Val	-ASN	-GLY	-Gln	-TYR	-Ala-
T:	-Asp	-SER	-CYS	-GLN	-GLY	-ASP	-SER	-GLY	-GLY	-PRO	-Val	-VAL	-CYS	-Ser	-Gly	-Lys	—	—	—	—
CA:	-SER	-SER	-CYS	-Met	-GLY	-ASP	-SER	-GLY	-GLY	-PRO	-LEU	-VAL	-CYS	-Lys	-Lys	-ASN	-GLY	-Ala	-TRP	-Thr-
CB:	-SER	-SER	-CYS	-Met	-GLY	-ASP	-SER	-GLY	-GLY	-PRO	-LEU	-VAL	-CYS	-Gln	-Lys	-ASN	-GLY	-Ala	-TRP	-Thr-

	209	210	211	212	213	214	215	216	217	217A	218	219	220	221	221A	222	223	224	225	226
E:	-Val	-His	-GLY	-VAL	-Thr	-SER	-PHE	-Val	-SER	-Arg	-Leu	-GLY	-CYS	-Asn	-Val	-THR	-Arg	-LYS	-PRO	-Thr -
T:	-LEU	-Gln	-GLY	-ILE	-VAL	-SER	-TRP	-GLY	-SER	—	—	-GLY	-CYS	-Ala	-Gln	-Lys	-Asn	-LYS	-PRO	-GLY-
CA:	-LEU	-Val	-GLY	-ILE	-VAL	-SER	-TRP	-GLY	-SER	—	Ser	-Thr	-CYS	-Ser	—	THR	-Ser	-Thr	-PRO	-GLY-
CB:	-LEU	-Ala	-GLY	-ILE	-VAL	-SER	-TRP	-GLY	-SER	—	Ser	-Thr	-CYS	-Ser	—	THR	-Ser	-Thr	-PRO	-Ala -

	227	228	229	230	231	232	233	234	235	236	237	238	239	240	241	242	243	244
E:	-VAL	-PHE	-THR	-ARG	-VAL	-SER	-ALA	-TYR	-ILE	-SER	-TRP	-ILE	-ASN	-ASN	-Val	-ILE	-ALA	-SER-
T:	-VAL	-TYR	-THR	-LYS	-VAL	-Cys	-Asn	-TYR	-VAL	-SER	-TRP	-ILE	-Lys	-GLN	-THR	-ILE	-ALA	-SER-
CA:	-VAL	-TYR	-Ala	-ARG	-VAL	-THR	-ALA	-Leu	-VAL	-Asn	-TRP	-VAL	-GLN	-GLN	-THR	-LEU	-ALA	-Ala -
CB:	-VAL	-TYR	-Ala	-ARG	-VAL	-THR	-ALA	-Leu	-Met	-Pro	-TRP	-VAL	-GLN	-GLU	-THR	-LEU	-ALA	-Ala -

	245
E:	-ASN
T:	-ASN
CA:	-ASN
CB:	-ASN

The numbering is that of chymotrypsinogen-A. "Insertions" are numbered 36A, 36B, etc.; "deletions" are indicated thus —. At each sequence position chemically similar residues are shown in capitals and identical residues in semi-bold type, except for identities between chymotrypsin-A and -B. Disulphide bridges are as follows: 1—122, 22—157, 42—58, 128—232, 136—201, 168—182 and 191—220. Activation is brought about by cleavage of the 15—16 peptide bond. The sequence of the activation peptide of the zymogen of elastase is not yet known. See text for acknowledgements.

do occur, these "insertions" are generally found at the ends of external loops, which can expand or contract to accommodate such changes without disturbing the overall conformation of the surrounding regions.

In addition to this overall similarity of structure between elastase and α-chymotrypsin, we found that the orientation and packing of the amino acid side chains in the two enzymes are almost identical, especially in the molecular interior where the polypeptide chains are highly homologous. With the exception of two "buried" negatively-charged aspartic acid residues, numbers 102 and 194, and the α-amino group of residue 16, whose special functions are described below, these internal regions contain no charged groups and are mainly composed of "oily" hydrophobic residues (Fig. 3).

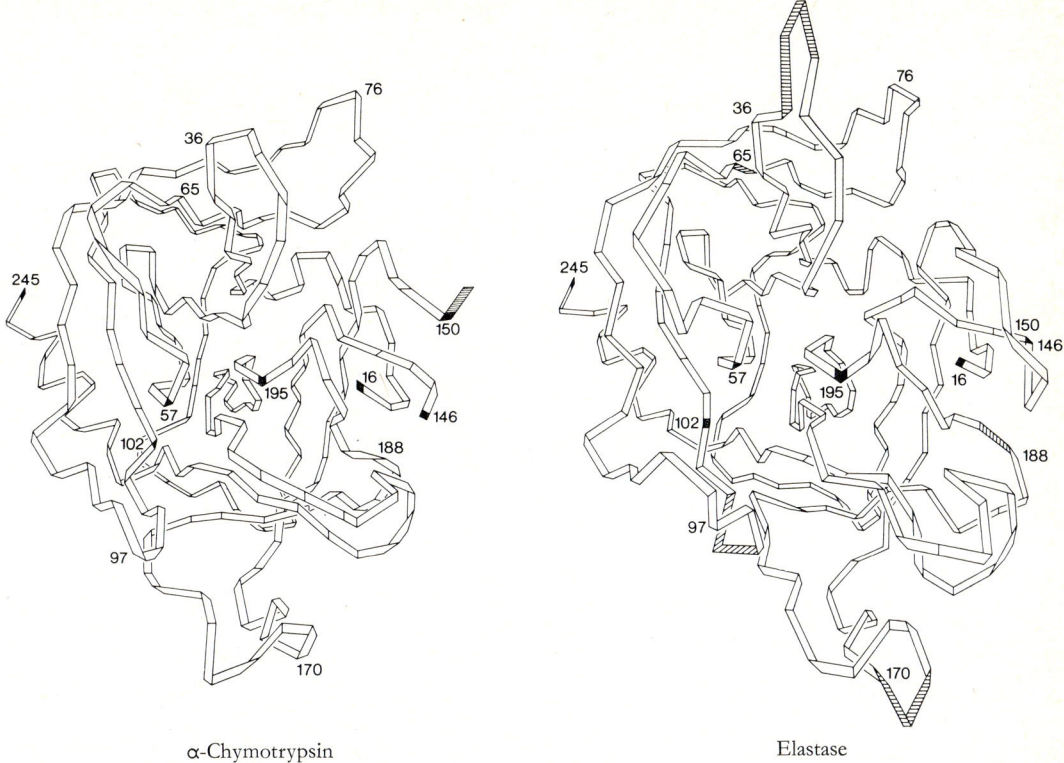

α-Chymotrypsin   Elastase

Fig. 1 and 2. Diagrams of the main chain conformations of α-chymotrypsin (Fig. 1) and elastase (Fig. 2), drawn in similar orientations to illustrate the degree of identity between the two structures. Corresponding residues bear the same numbers. The activation peptide (A-chain, residues 1—15) of α-chymotrypsin has been omitted for clarity.
(Diagrams prepared by Dr. D. M. Blow and Dr. B. S. Hartley from the α-carbon co-ordinates of the two molecules).

A Common Structural Plan

In the atomic models of both elastase and α-chymotrypsin it was observed that the pairs of residues 22 and 157, and 127 and 232, which in trypsin alone are bridged by disulphide bonds, lie close enough together to enable disulphide bonds to be built with little or no alteration in the conformation of the chains, indicating strongly that the structure of trypsin is also very similar.

The picture which thus emerged was that these pancreatic serine proteinases are a family of closely related enzymes built on the same basic plan. The homologous hydrophobic interiors provide the common nuclei of highly stabilized local conformation (represented by the central portions of the two halves of Fig. 3) around which the rest of the molecules can condense to give the similar overall structures which the enzymes possess. By contrast, most of the differences between the enzymes seem to occur in relatively unimportant positions of low homology on the molecular surfaces, which are studded with charged residues and which seem merely to provide insulating layers one polypeptide chain thick serving to shield the hydrophobic interiors from the disruptive hydrating effects of the solvent.

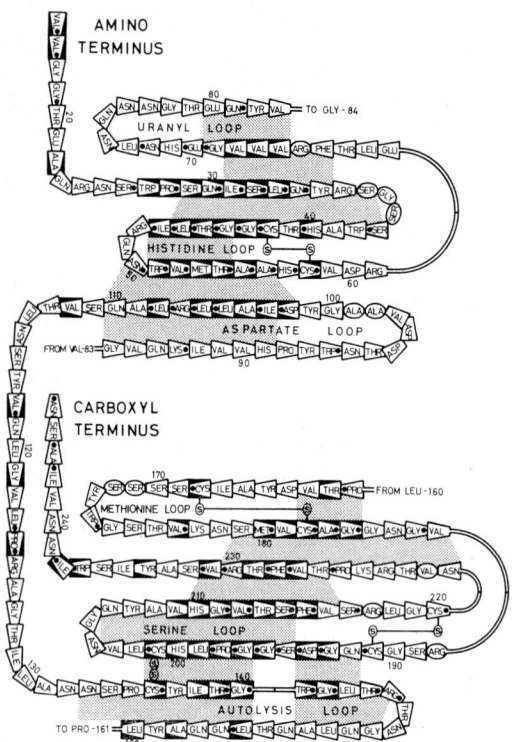

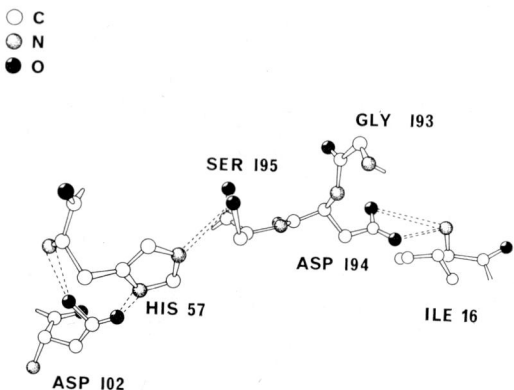

Fig. 4. The conformation of the amino acid side chains which form the "charge relay system" and the internal Ile-16 — Asp-194 ion pair at the catalytic site of α-chymotrypsin. Hydrogen bonds are dashed. Ile-16, Asp-102, the side chain of Asp-194 and one side of the imidazole ring of His-57 are buried within the molecule, inaccessible to water (from BLOW and STEITZ, 1970).

The Common Catalytic Mechanism

BLOW, BIRKTOFT and HARTLEY (1969) discovered that the active centre serine residue, Ser-195, of native α-chymotrypsin is hydrogen bonded to the imidazole side chain of the active centre histidine residue His-57, which is in turn hydrogen-bonded to the carboxylate group of a buried aspartic acid residue, Asp-102. They proposed that this system of hydrogen bonds forms a "charge relay system" (Fig. 4) whereby charge can be transferred from the buried aspartic through a hydrophobic environment to the serine γ-oxygen atom, thereby relieving the "strain" entailed by having an unpaired negative charge in a region of low dielectric constant, and increasing the electronegativity of the oxygen atom in a unique way. By this special mechanism the enzyme is able to make the normally inert hydroxyl group of the serine residue powerfully nucleophilic, imparting it with the ability to initiate a nucleophilic attack on the carbonyl carbon atom of the peptide bond to be hydrolysed. Hydrolysis is followed by the release of the C-terminal half of the substrate as a free amine, the N-terminal half probably forming a transient ester bond with the Ser-195 γ-oxygen,

Fig. 3. A diagrammatic representation of the conformation of the elastase polypeptide chain, showing the complete amino acid sequence and the manner in which the molecule is folded into two similar halves. The looping of the chain brings neighbouring regions into close antiparallel contact, close enough to be stabilized by hydrogen bonds (indicated by shading). Those amino acid residues buried within the molecule are boxed in black. Residues homologous with trypsin, chymotrypsin-A and chymotrypsin-B are indicated by a black dot. Two breaks in the chain have had to be made to represent the three-dimensional conformation on a plane surface. (Originally published in Nature, **225**, 811 (1970) where it is more fully explained).

The recent determination, at high resolution, of the three-dimensional structures of γ-chymotrypsin (another crystal form of chymotrypsin-A) (DAVIES et al., 1970) and of trypsin (STROUD et al., 1970) have fully confirmed these views, γ-chymotrypsin being virtually identical to α-chymotrypsin, and trypsin resembling α-chymotrypsin and elastase as closely as they do one another.

giving an "acyl enzyme" intermediate which is itself then hydrolysed by a reversal of the electron shifts to release the free acid and leave the enzyme ready to attack the next substrate molecule.

The reason why chymotrypsin, trypsin and elastase all exhibit the same catalytic mechanism has been clearly shown by the structural studies to be due simply to the fact that they all possess those residues essential for the catalytic processes, arranged in exactly the same three-dimensional conformations, and hence are able to function similarly.

It is of interest that subtilisin BPN, while having a completely different three-dimensional organization, also possesses at its active centre an Asp-His-Ser hydrogen-bonded system which appears to function similarly to the one found in the mammalian enzymes (WRIGHT et al., 1969), suggesting that this efficient catalytic mechanism has been independently acquired at least twice.

The Substrate Binding Sites

The differences in substrate specificity exhibited by trypsin, the chymotrypsins and elastase are easily understood from a study of their three-

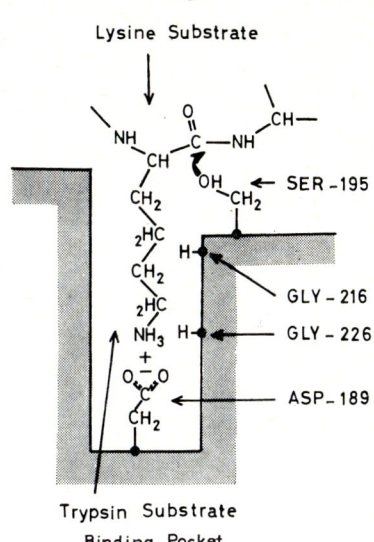

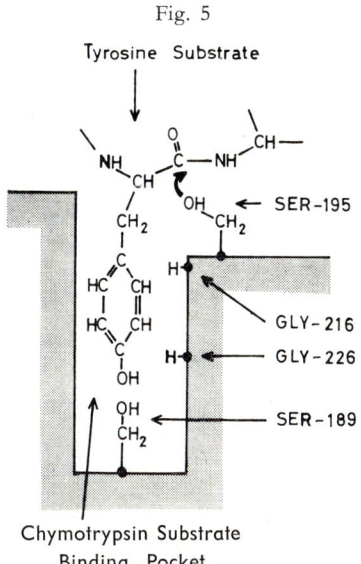

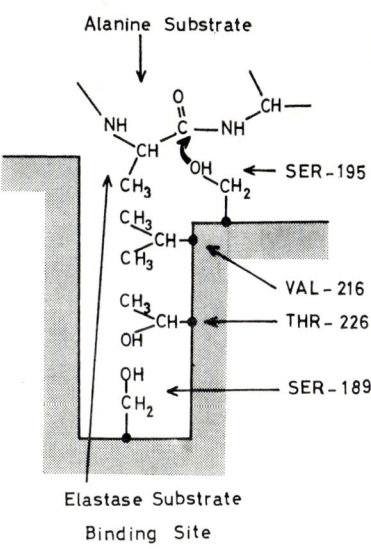

Fig. 5, 6 and 7. A diagrammatic representation of the differences between the substrate binding specificity sites of the chymotrypsin (Fig. 5), trypsin (Fig. 6) and elastase (Fig. 7). The nucleophilic attack of the active centre serine γ-oxygen upon the substrate is indicated by the curly arrow. The "charge relay system" is not indicated.
[First published in New Scientist **45**, 546 (1970)].

dimensional structures. STEITZ, HENDERSON and BLOW (1969) showed that adjacent to the catalytic site of chymotrypsin there is a deep hydrophobic pocket, the substrate binding pocket, perfectly designed to allow the aromatic side chain of a chymotrypsin substrate to bind in the correct orientation for hydrolysis (see Fig. 5).

The only essential difference in the substrate binding pocket of trypsin is the presence of an aspartic acid residue, Asp-189, which lies at the bottom of the pocket, hydrogen bonded to Gln-221A, instead of a serine residue (see Fig. 6), the negative charge on this residue being perfectly placed to form an internal ion-pair with the positively-charged side chain of a basic amino acid residue entering the pocket, explaining the high specificity of trypsin for such substrates.

In elastase, in contrast, Ser-189 is unchanged, but instead Gly-216 which lies at the mouth of the pocket and Gly-226 which lies inside have become changed to a valine and a threonine residue respectively (see Fig. 7). The extra atoms of these side chains so occlude the binding site in elastase that it is sterically impossible for large side chains to be bound, explaining the specifity of elastase for small residues. There is a growing body of evidence (THOMSON & BLOUT, 1970; ATLAS et. al., 1970; SHOTTON & WATSON, unpublished results) that in contrast to trypsin and chymotrypsin, and like papain, the substrate binding site of elastase is extended beyond this one specificity site, being made up of a number of surface subsites which bind other amino acid residues on either side of the bond to be hydrolysed.

The Activation Mechanism

One of the first conclusions to be drawn from the structure of α-chymotrypsin concerned the means whereby the active enzyme is produced from the inert zymogen, chymotrypsinogen. HESS and his collaborators (OPPENHEIMER et al., 1966) had shown that the essential process in activation is the tryptic cleavage of the Arg-15—Ile-16 peptide bond in chymotrypsinogen, which is accompanied by an observable conformational change. Blow and his colleagues observed that in the α-chymotrypsin molecule this newly formed α-amino group of Ile-16 is to be found buried in an internal cavity, where it forms an ion-pair with the acidic side chain of Asp-194, in a position it could not adopt were it still covalently linked to Arg-15. Because Asp-194 is adjacent to the active centre serine residue, Ser-195, it was concluded that the formation of this ion-pair in an internal hydrophobic environment would so alter the conformation of Ser-195 as to bring it into its active orientation.

Recently KRAUT and his colleagues have determined the three-dimensional structure of chymotrypsinogen at high resolution (FREER et al., 1970) This has revealed that the orientation of Asp-194 and certain neighbouring regions, especially at the substrate binding site, indeed are different from those in α-chymotrypsin, but that the positions of Asp-102, His-57 and Ser-195 are virtually unchanged. It is therefore not quite clear at this time how the structural differences between α-chymotrypsin and its zymogen account for the enzymic activity of the one and complete inactivity of the other, even against small non-specific substrates which do not utilize the substrate binding site.

Another apparent paradox related to this phenomenon is the difference in behaviour of α-chymotrypsin and elastase on modification of this ion-pair. In α-chymotrypsin deprotonation or acetylation of the α-amino group of Ile-16 ($pK_a = 8.3$) leads to a loss of enzymic activity (OPPENHEIMER et al., 1966; GHELIS et al., 1967). Elastase shows a similar bell-shaped activity curve against protein substrates, but when assayed against the small specific substrate N-benzoyl-L-alanine methyl ester it shows no decrease in activity on acetylation of the α-amino group of Val-16 (pK_a 9.7; KAPLAN et al., 1970), or on titration up to pH 10.5, at which pH the enzyme starts to unfold (KAPLAN & DUGAS, 1969). It seems likely that the explanation of these observations will be found firstly in the differences in the equilibrium constants of elastase and α-chymotrypsin between the "native" forms of the enzymes in which residue 16, whether protonated or not, occupies an internal conformation, and the alternative external conformations, and secondly in the different sizes of the side chains of the specific substrates of the two enzymes, which may determine whether or not such substrates can bind when, after modification of residue 16 to destroy the ion pair, Asp-194 has moved to an alternative position in which it can be solvated.

Conclusion

Although some work has therefore still to be done to elucidate further the details of the activation, substrate binding and catalytic processes, the studies of the pancreatic serine proteinases described here have for the first time given us a detailed insight into the unity and diversity which exists within one enzyme family, and illustrate the types of relationships which we may expect to find between members of other enzyme families.

Acknowledgement

The autor is grateful to his colleagues at the M. R. C. Laboratory of Molecular Biology, Cambridge, and the Molecular Enzymology Laboratory of the Department of Biochemistry, Bristol, especially Dr. B. S. HARTLEY and Dr. H. C. WATSON, for their contributions to this study, and to the Beit Memorial Medical Research Fellowships for financial support.

References

ATLAS, D., S. LEVIT, I. SCHECHTER and A. BERGER (1970), in press.

BLOW, D. M., J. J. BIRKTOFT and B. S. HARTLEY, Nature **221**, 337 (1969).

BLOW, D. M. and T. A. STEITZ, Annual Review of Biochemistry **39**, 63 (1970).

DAVIES, D., et al. (1970), manuscript in preparation.

DESNUELLE, P., The Enzymes (edit. by P. D. BOYER, H. LARDY and K. MYRBÄCK), **4**, 93 (1960). Academic Press, N. Y.

FREER, S. T., J. KRAUT, J. D. ROBERTUS, H. T. WRIGHT and NG. XUONG, Biochemistry **9**, 1997 (1970).

GHELIS, C., J. LABOUESSE and B. LABOUESSE, Biochem. Biophys. Res. Comm. **29**, 101 (1967).

HARTLEY, B. S., Ann. Rev. Biochem. **29**, 45 (1960).

HARTLEY, B. S., Nature **201**, 1284 (1964).

HARTLEY, B. S. and D. L. KAUFMAN, Biochem. J. **101**, 229 (1966).

KAPLAN, H. and H. DUGAS, Biochem. Biophys. Res. Comm. **34**, 681 (1969).

KAPLAN, H., K. J. STEVENSON and B. S. HARTLEY (1970), in press.

MATTHEWS, B. W., P. B. SIGLER, R. HENDERSON and D. M. BLOW, Nature **214**, 652 (1967).

MELOUN, B., I. KLUH, V. KOSTKA, L. MORÁVEK, Z. PRUŠIK, J. VANĚČEK, B. KEIL and F. ŠORM, Biochim. Biophys. Acta **130**, 543 (1966).

MIKEŠ, O., V. TOMÁŠEK, V. HOLEYŠOVSKÝ and F. ŠORM, Biochim. Biophys. Acta, **117**, 281 (1966).

NAUGHTON, M. A., F. SANGER, B. S. HARTLEY and D. C. SHAW, Biochem. J. **77**, 149 (1960).

OPPENHEIMER, H. L., B. LABOUESSE and G. P. HESS, J. Biol. Chem. **241**, 2720 (1966).

SHOTTON, D. M. and B. S. HARTLEY, Nature, **225**, 802 (1970).

SHOTTON, D. M. and H. C. WATSON, Phil. Trans. Roy. Soc. B **257**, 111 (1970a).

SHOTTON, D. M. and H. C. WATSON, Nature **225**, 811 (1970b).

SIGLER, P. B., D. M. BLOW, B. W. MATTHEWS and R. HENDERSON, J. Mol. Biol. **35**, 143 (1968).

SMILLIE, L. B., A. FURKA, N. NAGABHUSHAN, K. J. STEVENSON and C. O. PARKER, Nature **218**, 343 (1968).

SMITH, E. L., F. S. MARKLAND, C. B. KASPAR, R. J. DELANGE, M. LANDON and W. H. EVANS, J. Biol. Chem. **241**, 5974 (1966).

STEITZ, T. A., R. HENDERSON and D. M. BLOW, J. Mol. Biol. **46**, 337 (1969).

STROUD, G. M., L. M. KAY, A. M. COOPER and R. E. DICKERSON (1970), manuscript in preparation.

THOMPSON, R. C. and E. R. BLOUT (1970), Proc. U. S. Natl. Acad. Sci., in press.

WALSH, K. A., D. L. KAUFFMAN, K. S. V. S. KUMAR and H. NEURATH, Proc. U. S. Natl. Acad. Sci. **51**, 301 (1964).

WRIGHT, C. S., R. A. ALDEN and J. KRAUT, Nature **221**, 235 (1969).

The Atomic Structure of the Basic Trypsin Inhibitor of Bovine Organs. (Kallikrein Inactivator)

R. Huber, D. Kukla, A. Rühlmann and W. Steigemann

Max-Planck-Institut für Eiweiß- und Lederforschung, und Physikalisch-Chemisches Institut der Technischen Universität, München, Germany

Introduction

Several aspects of the crystal structure analysis concerning mainly experimental detail have already been described [Huber, Kukla, Rühlmann, Epp, Formanek (1970)]. A Fourier synthesis at 2.5 Å resolution was reported, phases being derived from five derivatives. Later a sixth derivative was included (hydroxymercuri-toluol-sulfonic acid), which slightly improved the Fourier synthesis.

The molecular boundary can easily be recognized. The polypeptide main chain is clearly defined in regions of high electron density. Approximately 85% of the main chain carbonyl oxygen atoms are visible as protuberances of electron density. This is of great value for defining the orientation of the peptide planes. Many amino acid side chains can easily be recognized from their shape in the map, in particular the cystines, aromatic rings, prolines and glycines. Several amino acid side chains (approximately 15% of the total) at the molecular surface which are obviously not fixed by intra- or intermolecular hydrogen bonds are not visible in the Fourier map beyond their β-carbon atoms. This is due to free rotation of these side chains in the crystal.

A model of the molecule was built using a half silvered mirror as described by Richards (1968). Undefined side chains were built in their most probable conformations. The coordinates were read with a plumb line and refined using the model-building procedure of Diamond (1968). A crude estimate of the error in the atomic coordinates is 0.5 Å.

The main chain conformation

The chemical sequence of the inhibitor is shown in Fig. 1 [Kassell and Laskowski (1965), Chauvet, Nouvel and Acher (1964), Dlouha, Pospisilova, Meloun and Šorm (1965), Anderer and Hörnle (1966)].

Fig. 2 shows the positions of the α-carbon atoms of the 58 amino acids. The molecule appears pear shaped with a length of 29 Å and a maximum diameter of about 19 Å.

The major determinant of the structure appears to be a double stranded antiparallel β-sheet made up of amino acids Ala (16) to Gly (36). This β-sheet has a twist of nearly 180° degrees. Its two strands are connected by the disulfide bridge Cys (14) — Cys (38) at the top of the molecule. The segment 38 to 47 crosses the β-sheet at Phe (45), which is hydrogen bonded to it through its main chain carbonyl oxygen- and nitrogen atoms thus forming a short piece of triple stranded β-structure. Furthermore the main carbonyl oxygen atom of Thr (11) is

Atomic Structure of the Basic Bovine Inhibitor

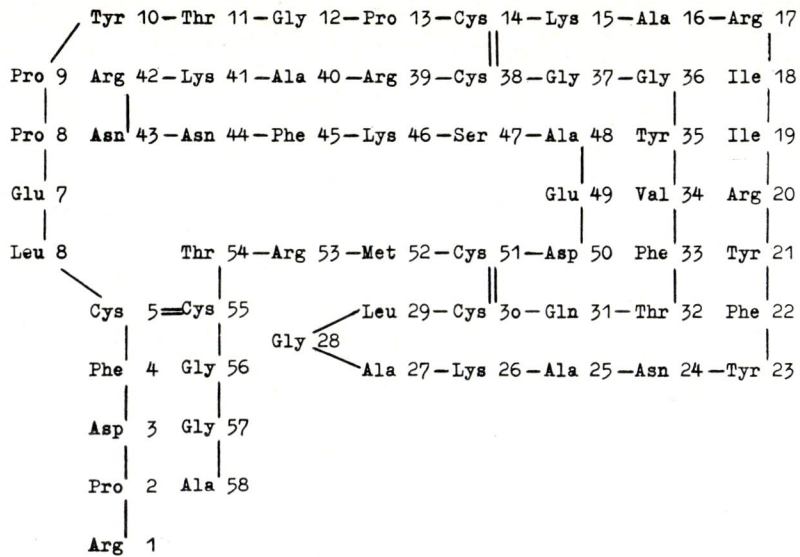

Fig. 1. Amino acid sequence of the inhibitor.

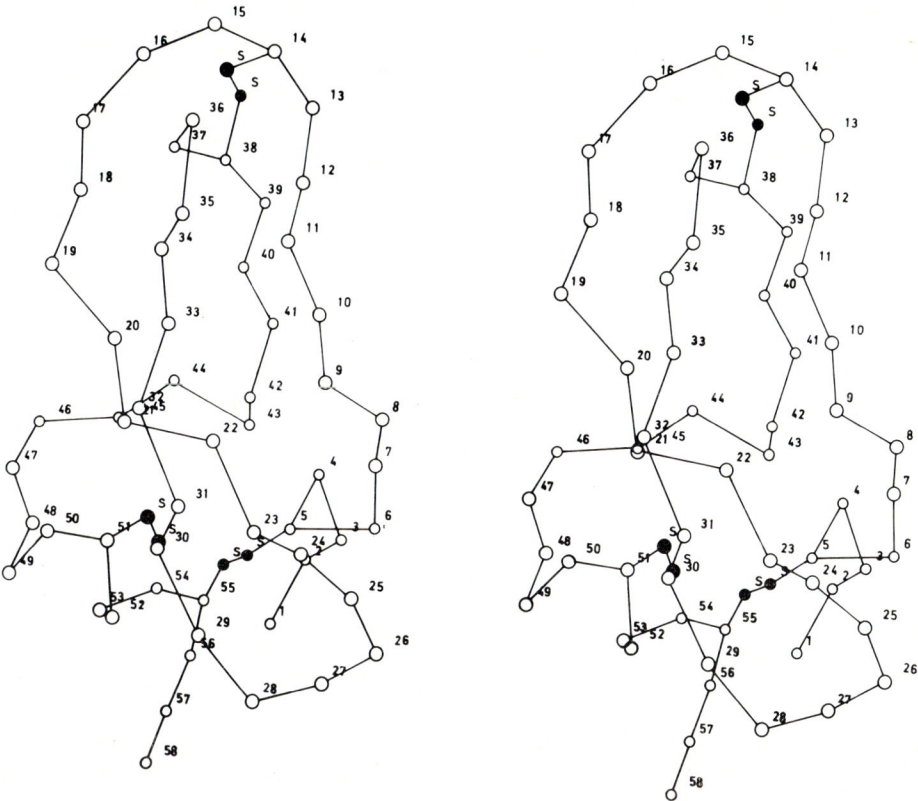

Fig. 2. Stereo pair showing the positions of the α-carbon atoms and the main chain folding.

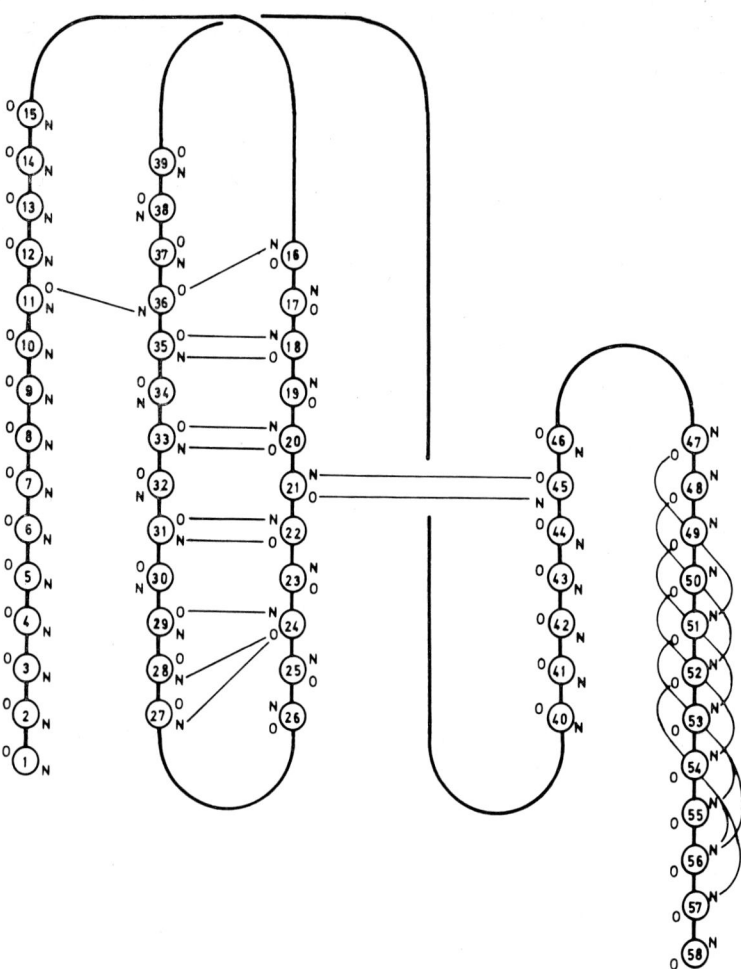

Fig. 3. Diagram of the hydrogen bonds between main chain atoms.

bonded to the NH group of Gly (36) in a parallel β-sheet fashion.

Three turns of α-helix forming the C-terminus of the molecule are attached to the β-sheet. The conformation of this helix is quite regular except that it is opened out at the last turn. Another helical segment near the N-terminus from amino acids Asp (3) to Glu (7) is heavily distorted forming no intra chain hydrogen bonds at all. A reason for this distortion will be put forward later.

Fig. 3 is a schematic representation of all hydrogen bonds formed between main chain atoms. The stereopair 4 shows the positions of all main chain atoms. The hydrogen bonds are dashed.

An interesting feature of the conformation of the molecule is the occurance of two segments forming a poly-proline II helix [COWAN and MCGARWIN (1955) and SASISEKHORAN (1959)]. Glu (7) to Tyr (10) has a nearly ideal polyproline II conformation. The same holds for the chain Arg (39) to Lys (41). The segment Glu (7) to Tyr (10) includes the sequence Pro (8) — Pro (9) which obviously favours the formation of a poly-proline helix.

Fig. 5 is a two-dimensional plot of the dihedral angles φ and ψ for all 58 amino acids. The

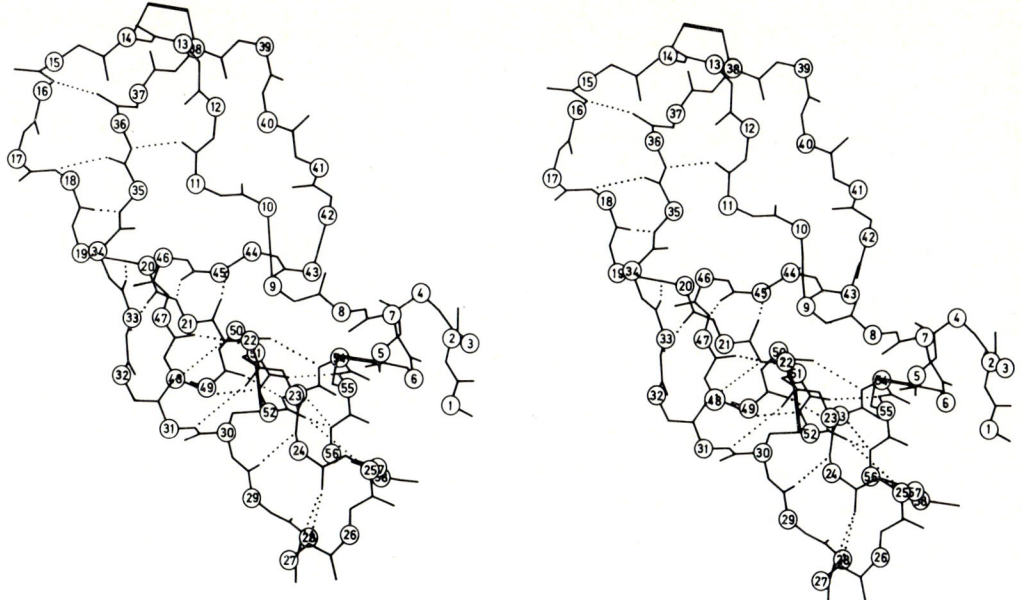

Fig. 4. Stereo pair of all main chain atoms. Hydrogen bonds between main chain atoms are dashed.

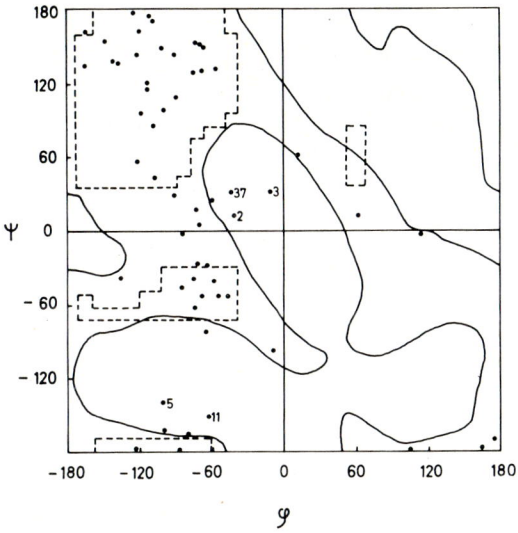

Fig. 5. Plot of the observed dihedral angles φ and ψ. The energetically allowed region is drawn according to PULLMANN, MAIGRET, PERAHIA (1970).*

---

* Nomenclature as recommended by the IUPAC—IUB Commission on Biochemical Nomenclature [Biochemistry **9**, 3471 (1970)].

allowed and forbidden regions are as calculated for alanine residues [PULLMANN, MAIGRET, PERAHIA (1970)]. It can be seen, that most points lie in allowed areas. A considerable deviation from the values of the antiparallel β-sheet occurs [φ — 139°, ψ + 135°, ARNOTT, DOVER, ELLIOT (1967)].

Five residues are in forbidden regions, but can easily be moved into allowed areas if small adjustments of the order of 10 to 20 degrees are made to the conformational angles.

The three disulfide bridges apparently stabilize the compact three-dimensional structure. Cys (14) — Cys (38) links the two loops forming the top of the molecule, Cys (30) — Cys (51) connects the C-terminal helix to the β-sheet and Cys (5) — Cys (55) binds the N-terminal helical segment to the C-terminal helix.

Distribution of animo acids

Although a small protein molecule like the inhibitor has not much interior, the core of the molecule is indeed formed by hydrophobic residues, in particular phenylalanines and cystines. Hydrophobic interactions apparently also play an important role in binding the N- and C-terminal helices to the β-sheet through residues Phe (4), Cys (5), Leu (6) and Ala (48), Cys (51), Met (52), Cys (55) respectively. The hydrophilic sides of the helices are exposed to the solvent. The C-terminal helix begins with Ser (47), a typical helix initiator [HUBER, EPP, STEIGEMANN and FORMANEK (1971)]. Several hydrophobic residues occur in crevices at the molecular surface.

Nearly all hydrophilic residues are at the surface of the molecule and protrude into the surrounding solution. A remarkable exception is an internal asparagine (43) which probably causes the distortion of the N-terminal helical segment. Its NH_2 group forms hydrogen bonds with the main chain carbonyl oxygens Cys (5) and Glu (7), while its carbonyl oxygen is bonded to the main chain NH of Tyr (23). This internal asparagine seems to play an important role in the stabilisation of the molecule by anchoring the segment Cys (38) — Phe (45) to the rest of the molecule. There are several further side chain-main chain hydrogen bonds involving in particular tyrosines. The phenolic hydroxyl group of Tyr (35) is bonded to the main chain carbonyl oxygen of Cys (38), and the hydroxyl group of Tyr (23) interacts with the CO of Gly (56). Both tyrosines are buried within the structure, whereas the tyrosines (10) and (21) appear relatively accessible. This is in accordance with nitration studies [MELOUN, FRIC, SORM (1968)].

Fig. 6 shows the distribution of amino acid side chains, which are charged at neutral pH. It is remarkable that all negative charges are at the base of the molecule, whereas the positive charges appear concentrated at the top of the molecule, which forms the contact area with the proteases in the inhibitor-protease complexes as will be discussed later.

The inhibitor molecule therefore must have a large dipole moment, which might help in finding the correct orientation to the protease molecules, although no complementary dipole moment of the proteases has been reported.

Two of the three disulfide bridges, Cys (5) — Cys (55) and Cys (30) — Cys (51) are completely buried within the molecule. Cys (14) — Cys (38) on the other hand is exposed. This is in accordance with the observation, that only Cys (14) — Cys (38) can readily be reduced and reoxydized without loss of activity (KRESS, LASKOWSKI (1967)].

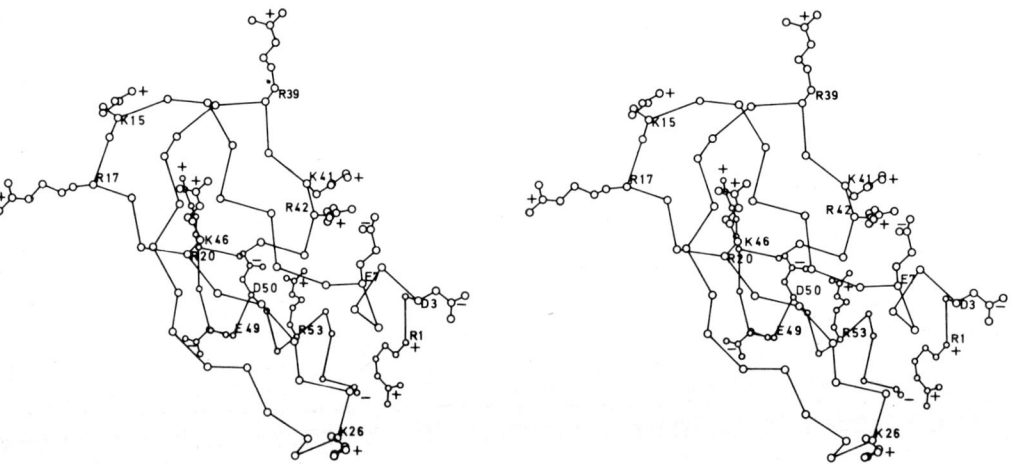

Fig. 6. Stereo pair showing the distribution of charged residues. All negative charges are away from the active center at Lys (15). In this and the following diagrams the one-letter code for amino acids is used. The α-carbon atoms and the atoms of charged residues are drawn.

Structure and mechanism of action

The basic trypsin inhibitor is a poly-valent inhibitor which inactivates trypsin, chymotrypsin, plasmin and kallikrein [FRITZ, SCHULT, MEISTER, WERLE (1969)]. A number of chemical modification studies of the exclusion type has been performed, which show that the bonding to trypsin is through Lys (15). [FRITZ et al. (1969), CHAUVET and ACHER (1967)].

This was further confirmed by KRESS and LASKOWSKI (1967) and LASKOWSKI (1971) for trypsin-binding and surprisingly by RIGBI (1971) for chymotrypsin-binding as well. It was found that the peptide bond between Lys (15) and Ala (16) is cleaved by trypsin and chymotrypsin if the inhibitor has been chemically modified by splitting Cys (14) — Cys (38). Furthermore the peptide bond between Arg (39) and Ala (40) is also cleaved by both proteases. Lys (15) and Arg (39) are at the top of the molecule in an extremely exposed position (fig. 2). Furthermore the confirmation of the polypeptide chains Tyr (10) to Ala (16) and Tyr (35) to Ala (40) respectively is remarkably similar.

Fig. 7 shows the α-carbon positions of Thr (11) to Lys (15) and Tyr (35) to Arg (39), which are approximately related by a twofold axis of symmetry through the disulfide bridge perpendicular to the plane of the paper. We also want to point out, that the sequences near the disulfide bridge Cys (14) — Cys (38) appear to be homologous

Tyr (10) — Thr (11) — Gly (12) — Pro (13) —
Cys (14) — Lys (15) — Ala (16) — Arg (17) —
Tyr (35) —   — Gly (36) — Gly (37) —
Cys (38) — Arg (39) — Ala (40) — Lys (41).

This reminds one of the lima bean inhibitor where two nearly identical sequence regions have been found at the two binding sites of the molecule [STEVENS (1969), STEVENS (1971)].

The experiments of KRESS, LASKOWSKI and RIGBI suggest strongly that the chemically modified inhibitor binds in a substrate-like manner to the proteases. It is tempting to extend this conclusion to the native inhibitor, although no bond cleavage has been observed.

Substrate-like binding has been definitely proved for several other inhibitors. In the presence of catalytic amounts of trypsin a true equilibrium was found to exist between the native inhibitors and their modified forms with a susceptible peptide bond cleaved [LASKOWSKI (1969), NIECAMP, HIXSON, LASKOWSKI (1969), RIGBI, GREENE (1968)]. The equilibrium constant of hydrolysis is of the order of magnitude of unity in sharp contrast to the general expectation. If this scheme is also valid for the basic trypsin inhibitor, its constant of hydrolysis should be extremely small. Such a high stabilization of a peptide bond obviously has to be explained by certain structural features of the active site of

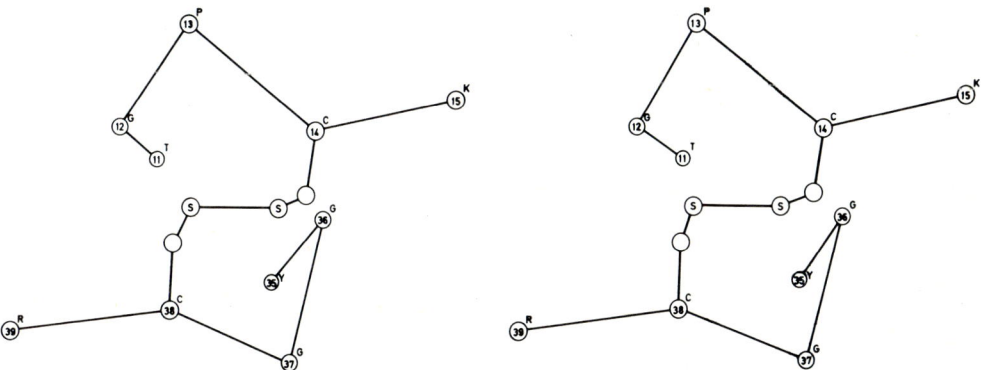

Fig. 7. Stereo diagram of the positions of α-carbon atoms Thr (11) to Lys (15) and Tyr (35) to Arg (39). The approximate two-fold axis relating these atoms can be observed.

the inhibitor. A most obvious explanation would be that the new N- and C-termini are still held to the rest of the structure, so that the potential entropy gain of the peptide bond cleavage could not be realized (LASKOWSKI (1969)].

Fig. 8 shows the structure of the trypsin inhibitor molecule in the vicinity of the active Lys (15). In analogy to other inhibitors the susceptible peptide bond is assumed to be between Lys (15) and Ala (16). The main chain NH group of Ala (16) is hydrogen-bonded to the main chain carbonyl of Gly (36). This is the first hydrogen bond of the β-sheet structure described before, so that cooperative effects might stabilize this bond further. Hydrophobic interactions of the side chains of Ala (16) and Ile (18) may also play a role in tying the peptide chain to the molecule. The Cys (14) — Cys (38) disulfide bond on the other hand rigidly connects the chain on the Lys (15) side with the rest of the molecule. Furthermore the peptide group Cys (14) — Lys (15) is so close to the peptide chain Tyr (35) to Cys (38) that virtually no rotational freedom could be gained upon cleavage of Lys (15) — Ala (16).

A quantitative evaluation of the energy of stabilisation from the structural terms given above seems impossible. However a few kilo calories would be sufficient to obtain a low value of the constant of hydrolysis and appear to be well within the capabilities of the structural features discussed above. This explanation for the stabilization of the Lys (15) — Ala (16) peptide bond is in accordance with the above mentioned observations of bond cleavage in an inhibitor with the disulfide bond Cys (14) — Cys (38) split. Here the new C-terminus is no longer rigidly connected with the rest of the structure and the constant of hydrolysis increases.

The validity of these structural principles for other inhibitors is difficult to evaluate, although it is tempting to generalize. In fact the structural restrictions imposed on the conformation of the active sites of the inhibitors are very severe.

Firstly, in order to act as a substrate, the susceptible bond of the inhibitors has to take up an exactly defined position and orientation relative to the catalytic site of the protease within the complex. [STEITZ, HENDERSON, BLOW (1969), SHOTTON (1971)]. Therefore the contact areas of both molecules must be complementary. Indeed in cooperation with D. BLOW we were able to show by model-building that the basic trypsin inhibitor and chymotrypsin or trypsin fit nicely at the contact areas. Lys (15) and the peptide bond Lys (15) — Ala (16) can be put in the position postulated for a substrate. Several further favourable contacts (hydrogen bonds, hydrophobic interactions) occur between other groups in the contact region of the two molecules (BLOW, KUKLA, RÜHLMANN, HUBER to be published).

Secondly, the susceptible peptide bond has to be stabilized within the structure of the inhibitor molecule. This can be achieved by attaching the peptide group to the rest of the molecule by

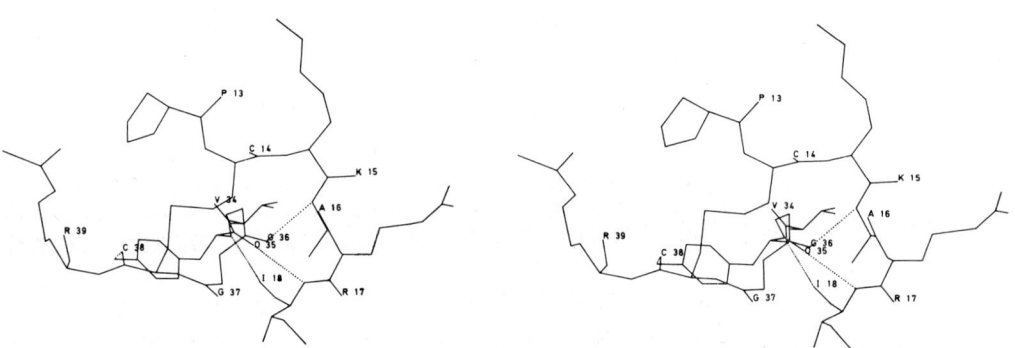

Fig. 8. Stereo diagram of all atoms near Lys (15).

covalent or other bonds. In analogy to the basic trypsin inhibitor a disulfide bond is close to the susceptible bond in several other inhibitors. The β-sheet, on the other hand, which occurs in the basic trypsin inhibitor, appears to be extremely well suited to hold the suceptible peptide group in the proper orientation with respect to the catalytic residues of the protease and to stabilize it simultaneously.

A large number of questions remains to be answered concerning primarily the protease-inhibitor complex.

Probably only a crystal structure determination may provide an ultimate answer and several of these are in progress. The model building procedure mentioned before should be regarded as a preliminary step. It will be described in detail in a later publication after some further refinement has been done.

Arrangement of the molecules in the crystal

The inhibitor molecules in the crystal form endless rods of molecules linked head to tail.

The exposed disulfide bridge Cys (14 — Cys (38) at the top of the molecule is close to methionine (52) and the peptide group in the C-terminal helix of another molecule, thus exhibiting favourable hydrophobic interactions. Furthermore Arg (39) forms a salt bridge with Glu (49).

Side to side packing is much less dense, but also stabilized by several hydrogen bonds. In the highly concentrated phosphate buffer in which the crystals are kept, several phosphate ions and water molecules are bound at the molecular surface. They obviously form hydrogen bonds with several tyrosines, glutamines and asparagines and main chain CO and NH groups.

Heavy atom positions in derivatives

Cysteines and histidines preferentially bind the heavy-metal compounds used in protein crystallography. Methionine and cystine also bind some compounds in particular platinum derivatives.

The inhibitor contains neither cysteine nor histidine and the exposed cystine (14—38) and methionine (52) are buried in the crystal structure and not accessible to reagents. An analysis of the binding sites in this case might therefore be of interest.

The reagents that were found to bind isomorphously to the inhibitor are primarily aromatic mercury containing compounds and seleno-cyanate

p-chloro-mercury-benzoate   PCMB
3-and 5-hydroxy-mercury-salicyclic acid HMSA
3-hydroxy-mercury-5-sulfo-salicyclic acid MSSS
2-hydroxy-mercury-toluol-4-sulfonic acid HMTS
seleno-cyanate   SECN

Each derivative showed multiple substitution. All these sites can roughly be divided into five groups: The highly occupied PCMB sites is near the C-terminal carboxyl group. Several weakly occupied SECN and HMTS sites are also in the close vicinity of the C-terminus. The highly occupied MSSS and HMTS sites and two minor sites of PCMB and HMSA are at the N-terminus. The major sites of HMSA and SECN and some minor sites of HMSA and MSSS are in a cage formed by Asn (24) and Gln (31) of one molecule and Lys (15) of another. Two minor sites of PCMB are in the vicinity of Lys (41) and Tyr (10). A minor site of MSSS is near Tyr (21) and the Arg (19) of another molecule. A further minor site of MSSS is near Lys (46).

It is therefore obvious that in particular amino- and amido groups are able to form complexes with certain mercury containing reagents.

This work was supported by the Deutsche Forschungsgemeinschaft, SFB 51.

References

Anderer, F. A., S. Hörnle, J. Biol Chem. **241**, 1568 (1966).

Arnott, S., D. Dover, A. Elliott, J. Mol. Biol. **30**, 201 (1967).

Chauvet, J., R. Acher, J. Biol. Chem. **242**, 4274 (1967).

Chauvet, J., G. Nouvel and R. Acher, Biochim. Biophys. Acta **92**, 200 (1964).

Cowan, P. M., S. McGarwin, Nature **176**, 501 (1955).

Diamond, R., Acta Cryst. **21**, 253 (1966).

Dlouha, V., P. Pospisilova, B. Meloun and F. Sorm, Coll. Czech. Chem. Comm. **30**, 1311 (1965).

Fritz, H., H. Schult, R. Meister and E. Werle, Hoppe-Seyler's Z. physiol. Chem. **350**, 1531 (1969).

Huber, R., D. Kukla, A. Rühlmann, O. Epp and H. Formanek, Naturwissenschaften **57**, 389 (1970).

Huber, R., O. Epp, W. Steigemann and H. Formanek, Eur. J. Biochem. (1970) **19**, 42 (1971).

Kassell, B., M. Laskowski, Jr., Biochem. Biophys. Res. Comm. **20**, 463 (1965).

Kress, L. F. and M. Laskowski, Sr., J. Biol. Chem. **242**, 4925 (1967).

Laskowski, M. Sr., this issue (1971) p. 66.

Laskowski, M. Jr., in "Structure-Function Relationships of Proteolytic Enzymes" edited by P. Desnuelle, H. Neurath, M. Otteson, Munksgaard, Copenhagen (1969).

Meloun, B., I. Fric and F. Sorm, Europ. J. Biochem. **4**, 112 (1968).

Niecamp, C. W., H. F. Hixson and M. Laskowski, Jr. Biochemistry **8**, 16 (1969).

Pullmann, B., B. Maigret and D. Perahia, Theoret. chim. Acta **18**, 44 (1970).

Richards, F. M., J. Mol. Biol. **37**, 225 (1968).

Rigbi, M. and L. J. Greene, J. Biol. Chem. **243**, 5457 (1968).

Rigbi, M., this issue (1971) p. 74.

Sasisekhoran, V., Acta Cryst. **12**, 897 (1959).

Shotton, D., this issue (1971) p. 47.

Steitz, T. A., R. Henderson and D. M. Blow, J. Mol. Biol. **46**, 337 (1969).

Stevens, F. C., this issue (1971) p. 149.

Stevens, F. C., Proc. Can. Fed. Biol. Soc. **12**, 16 (1969).

Discussion Remarks: **Comments to the Reactive Site**

M. Laskowski, Jr.

I wish to congratulate you, Dr. Huber, on the X-ray crystallographic investigation of the pancreatic inhibitor and on the superb interpretation of these data. I am particularily delighted by your explanation of why pancreatic inhibitor cannot be modified. It appears to me that a combination of X-ray crystallography of inhibitors and of some simple statistical mechanics will allow a rather precise explanation of the value of K_{hyd} of the reactive site peptide bonds of various inhibitors. These values already appear to fall into a pattern. The Kunitz inhibitor with Cys residue immediately preceding the reactive site has $K_{hyd} \cong 0$, the secretory inhibitors with Cys-Pro-reactive site sequence have relatively low K_{hyd} and soybean inhibitor (Kunitz) with Cys preceding the reactive site by many residues has a large value of K_{hyd}. Of course, the simple correlation of K_{hyd} to the position of Cys preceding the reactive site is unlikely to hold up. What is likely to hold up is the relationship between K_{hyd} and the distance between the nearest strong interaction preceding the reactive site and the reactive site. Cys is obviously such a strong interaction, but the interaction can, of course, be located by X-ray crystallography.

I wish to call attention to the fact that another K_{hyd} can be defined in the trypsin inhibitor system, the equilibrium constant for removal of the COOH terminal Arg or Lys from the modified inhibitor. This equilibrium constant is very large and therefore, difficult to measure, however, it is possible to measure the ratio of this constant in the native and denatured protein by comparing the "melting curves" of modified and des Arg or des Lys modified inhibitor. Dr. Carl Niekamp [1] carried out such measurements on soybean trypsin inhibitor (Kunitz) and found the rather startling result; it is easier to remove (by a factor of 2) Arg^{64} from native than from denatured modified inhibitor i. e. that the presence of Arg^{64} destabilizes the native form of the inhibitor. This is contrary to the common expectation and to the experiment of HARTLEY [2] on barnase, who found that the COOH terminal arginine of that protein stabilized it by a factor of 5. We have tentatively explained our observation by assuming that Arg^{64} in the modified soybean inhibitor is not involved in any interaction with the native protein (and thus does not stabilize the native protein) but that its conformational freedom is restricted somewhat by the native structure (by a factor of 2). Therefore, there is a greater entropy gain on denaturing modified rather than des Arg^{64} modified inhibitor or alternately on removing Arg^{64} from the native rather than from denatured modified inhibitor. The strange part of this reasoning was total lack of positive interactions between Arg^{64} and the remainder of the protein. I am, therefore, very gratified by your finding that Lys^{15} of pancreatic inhibitor is not involved in any interactions (its side chain sticks up and neither its NH nor $C = O$ are hydrogen bonded) and by the further possibility from model building studies that this may be a general requirement for trypsin inhibitors.

I would like to point out that in an earlier publication [3] I have misstated Dr. NIEKAMP's preliminary results on this problem by saying that there is no difference between the melting curve of modified and of des Arg^{64} modified inhibitor. Small differences can be seen in the data of that paper and since then the difference was always confirmed and further detected again by an even more sensitive testing technique.

References

[1] NIEKAMP, C. W., M. LASKOWSKI, Jr., (1971, in preparation); NIEKAMP, C. W., PH. D. THESIS, Purdue University (1971).

[2] HARTLEY, R., Biochem. Biophys. Res. Comm. **40**, 263 (1970).

[3] LASKOWSKI, M. Jr., in Structure Function Relationships of Proteolytic Enzymes (P. DESNUELLE, H. NEURATH and M. OTTESEN, eds.). Munksgaard, Copenhagen pp 89—101 (1970).

Naturally Occuring Trypsin Inhibitors: Further Studies on Purification and Temporary Inhibition

M. Laskowski, Sr., Sara L. Schneider, Karl A. Wilson, Lawrence F. Kress, Jan H. Mozejko, Susan R. Martin, Umberto Kucich and Mark Andrews

Laboratory of Enzymology, Roswell Park Memorial Institute, Buffalo, New York 14203

In 1948, Laskowski, Jr., while working in my laboratory, discovered a *trypsin inhibitor in bovine colostrum*. The inhibitor was subsequently purified, crystallized, and some of its properties were reported [1, 2]. The presence of carbohydrate in the inhibitor was, however, overlooked. Even in 1951 when the method of crystallization was first published there were some doubts concerning the homogeneity of the crystalline preparation. As time progressed evidence for the heterogeneity of the inhibitor accumulated. Two alternatives were then considered: first, that the crystalline inhibitor was a mixture of several similar isoinhibitors; second, and less likely that only one component had inhibitory activity, while the others were inactive.

It was decided to further purify both bovine and porcine colostrum inhibitors using modern criteria of homogeneity. Four isoinhibitors were isolated from porcine colostrum, all of which contained carbohydrate [4]. In addition to the isoinhibitors, there was also some material of low specific activity present.

The initial steps of purification [5] resembled previously published procedures [3], and led to a preparation with a specific activity of 5,700 units per mg. This preparation was subjected to gradient chromatography on CM cellulose, and 5 fractions were pooled (Fig. 1). Peak I is obviously heterogeneous, has a specific activity of about 3,000, and differs significantly in composition from the other four peaks. Peaks II to V had activity of a little below 6,000. Each peak was then chromatographed (Fig. 2) under conditions requiring no change of buffer (equi-

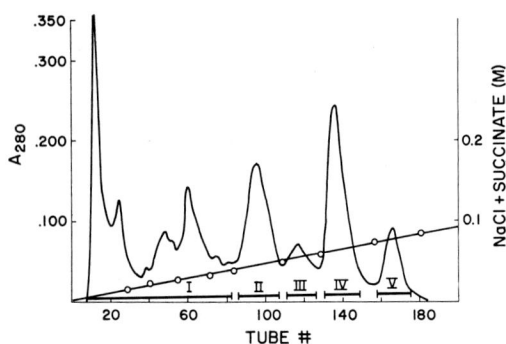

Fig. 1. Gradient chromatography of swine colostrum trypsin inhibitor on microgranular CM-cellulose. Swine colostrum inhibitor (91 A_{280} units) was dissolved in 15 ml of 0.01M sodium succinate, pH 5.5, and dialyzed against the same buffer for 17 h. The solution was placed on a 1.5 × 88 cm column of microgranular CM-cellulose. The protein was eluted with a linear gradient from 0.01 succinate (1.0 1) to 0.01 succinate—0.2M NaCl (1.0 1). Flow rate was 30 ml per h, and 5 ml fractions were collected. The average specific activity of peak I was 3,000 units per A_{280} unit; that of peaks II through V averaged 6,000 units per A_{280} unit in 3 separate chromatographic runs. —, A_{280}; -0-0-0, molarity (succinate+NaCl); ⊢——⊣ tubes pooled.

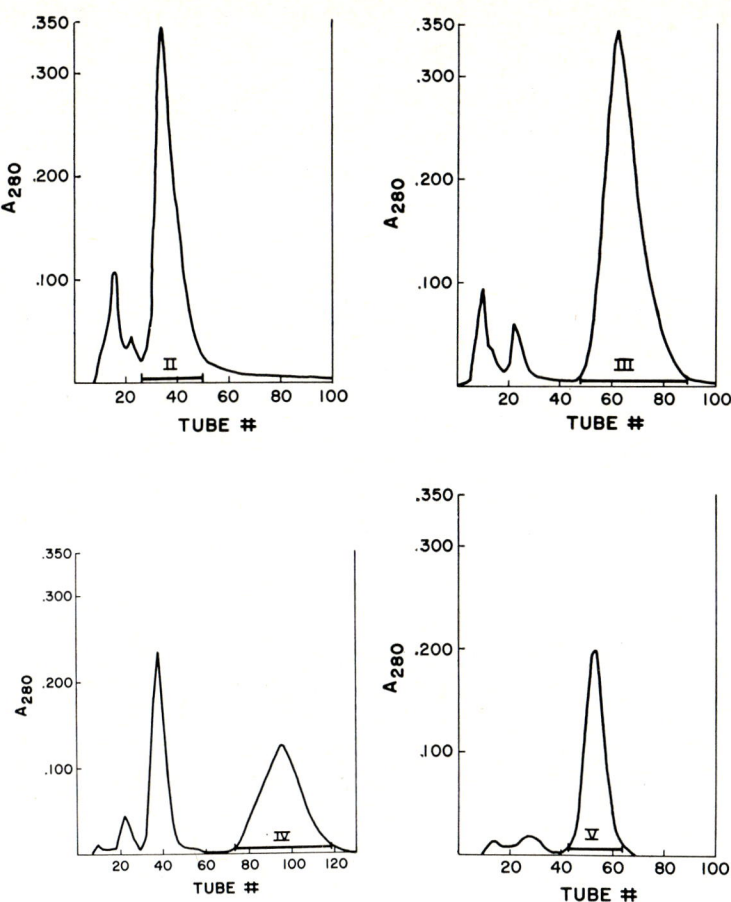

Fig. 2. Equilibrium chromatography of individual swine colostrum trypsin inhibitors on microgranular CM-cellulose. In all the above figures, the Roman numerals correspond to those used in Fig. 1. Each inhibitor was dissolved in 10 ml of the respective buffer to be used and dialyzed against that buffer. The protein solutions were placed on a 1.5 × 88 cm column of microgranular CM-cellulose and eluted with the following buffers: Peak II (22.8 A_{280} units): 0.01M sodium succinate, pH 5.5; Peak III (35.9 A_{280} units): 0.01M succinate—0.015M NaCl; Peak IV (24.2 A_{280} units): 0.01M succinate—0.030M NaCl; Peak V (27.5 A_{280} units): 0.01M succinate—0.045M NaCl. The flow rates were 30 ml per h, and 5 ml fractions were collected. The void volume of the columns occurred at tube 10.

librium chromatography). For each peak a different salt concentration was required. The active major peaks (specific activity about 6,000) were collected and subjected to disc electrophoresis (Fig. 3). Only peak III showed 2 bands; all others gave single bands. Rechromatography of peak III under conditions used for peak II or peak IV gave a single peak emerging at a different position than either peak II or peak IV. Peak III was finally purified by gradient chromatography in 6 M urea. This removed an inactive contaminant (Fig. 4) and resulted in a single band on disc electrophoresis. The composition of each peak was determined and is shown in Table 1. The analytical values and the specific activities of the peaks, except peak I, are very similar. Peak I had a significantly lower activity and a different composition. After subtracting the average composition of peaks II to V, one is left with the following peptide representing

Table 1. Amino Acid and Carbohydrate Composition of Inhibitors Isolated from Swine Colostrum[a]

Component	Peak I	Peak II	Peak III	Peak IV	Peak V
Lysine	1.72	1.23	2.10	2.17	1.97
Arginine	5.01	3.56	3.72	3.68	3.80
Aspartic acid	9.49	5.81	6.08	6.09	6.02
Threonine[b]	3.84	4.27	4.31	4.25	4.23
Serine[b]	5.26	2.21	2.22	2.14	2.08
Glutamic acid	18.95	5.84	6.13	5.88	5.66
Proline	19.74	6.46	6.79	6.71	6.92
Glycine	12.83	4.66	4.85	4.80	4.91
Alanine	5.01	5.05	5.18	5.15	5.29
Cystine[c]	6	6.19	6	6	6
Valine	2.53	1.98	2.20	2.38	2.11
Methionine[d]	0.59	0.71	0.74	0.81	0.82
Isoleucine	(1.00)	(1.00)	(1.00)	(1.00)	(1.00)
Leucine	4.30	4.09	4.28	4.18	3.98
Tyrosine	2.39	2.44	2.54	2.50	2.55
Phenylalanine	2.99	3.76	3.69	3.61	3.28
Glucosamine	1.94	3.70	3.74	3.82	3.29
Galactosamine	0.32	0.39	0.46	0.41	0.35
Sialic acid	0.61	0.80	0.00	0.06	0.01
Fucose	0.23	0.24	0.38	0.42	0.38
Mannose	1.44	1.85	2.22	1.79	2.07
Galactose	1.06	1.06	1.41	1.42	1.19
Glucose	0.07	0.04	trace	0	0.95

[a] Results are expressed as residues per molecule based on a value of Ile = 1.00.
[b] Not corrected for decomposition during hydrolysis.
[c] Determined as cysteic acid in performic acid oxidized peak II. The values for the other peaks were assumed to be the same, since the recoveries of half-cystine were the same for all peaks.
[d] Uncorrected for decomposition; however, the value for methionine sulfone in performic acid oxidized peak II was 0.94 residue.

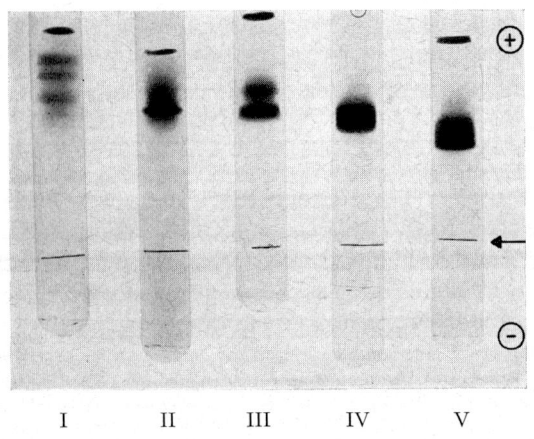

Fig. 3. Polyacrylamide gel electrophoresis of swine colostrum trypsin inhibitors. Approximately 425 μg of the inhibitor fraction purified by equilibrium chromatography were applied to a 15% acrylamide separating gel at pH 4.3. Electrophoresis was performed at 5 ma per gel for 160 min. The arrow marks the position of the tracking dye (bromphenol blue) at the end of the run. The gels were stained in 1% aniline blue-black and de-stained by leaching with 10% acetic acid. From left to right, the gels correspond to the Roman numerals used in Fig. 1.

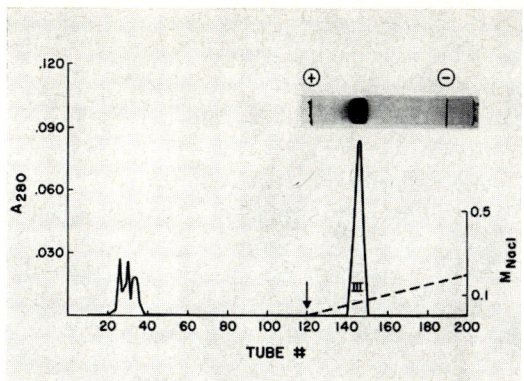

Fig. 4. Purification of Peak III swine colostrum inhibitor on CM-cellulose. Peak III inhibitor (see Fig. 2) was dissolved in 5.5 ml of 0.01M sodium succinate—0.015M NaCl-6M urea, pH 5.15, and placed on a 1.5 × 88 cm column of microgranular CM-cellulose equilibrated with the same buffer. The initial heterogeneous peak lacked inhibitor activity. The arrow indicates the point at which a linear gradient of 0.15M NaCl to 0.5M NaCl in 0.1M succinate-6M urea, pH 5.15, was initiated. The disc electrophoresis pattern of the material which emerged is shown above the peak. The inhibitory material had a specific activity of 5,850 units per A_{280}. —, A_{280}; ----, M NaCl.

the excess of amino acids in peak I: Arg_1 Asp_3 Ser_3 Glu_{13} Pro_{13} Gly_8. The composition is reminiscent of connective tissue proteins. It is not known whether this excess of amino acids represents a separate entity or is a part of a larger molecule connected with the inhibitors.

At this time at least four different active iso-inhibitors have been separated chromatographically from bovine colostrum. The possibility that more than four forms might be present has not as yet been rigorously excluded.

Another venture mostly concerned with purification is a simplified two-step procedure for the preparation of KUNITZ's *pancreatic trypsin inhibitor* [6]. The starting material is fraction E, the mother liquor that remains after trypsin has been crystallized. Fraction E contains about 85% protein with isoelectric point lower than pH 10. The remaining 15% is accounted for by trypsin-trypsin inhibitor complex, some free trypsin, and some autolyzed trypsin. A large Dowex 1—X 2 column (25 lbs. of Dowex) was adjusted to pH 10.1 with 0.1M triethylamine acetate buffer, and fraction E (about 750 gm equivalent to about 200,000 A_{280} units) was dialyzed against the same buffer and passed through the column. About 25,000 A_{280} units passed through, and this eluate was concentrated under vacuum. If the complex was desired, it was crystallized by a standard procedure. If only the free inhibitor was desired the mixture was dialyzed against water, then the pH was adjusted to 1.7 with HCl, and the mixture passed through a 21.5 × 100 cm Pharmacia column containing Sephadex G-50. Upward flow was used (Fig. 5). The first peak contains partially denatured trypsin, and is followed by active trypsin. The inhibitor peak appears after several more liters of eluent have passed through the column. The inhibitor peak is pooled, neutralized with ammonia, and its volume reduced. At this point the inhibitor can be readily crystallized by the standard procedure. If a salt-free preparation is

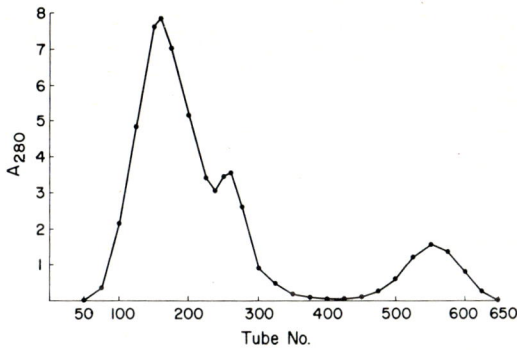

Fig. 5. Separation of free basic pancreatic inhibitor on Sephadex G-50 at pH 1.7. A Pharmacia column 21.5 × 100 cm was equipped with an upward flow adaptor. It was filled with Sephadex G-50 (spherical, medium) and equilibrated with 0.02N HCl. The sample containing 26,850 A_{280} units of protein and obtained from the previous step was dialyzed against water and then adjusted to pH 1.7 with HCl. The total volume was 1 liter. The sample was charged on the column via a peristaltic pump set at 4.2 ml/min, and then eluted at the same rate with 0.02N HCl. The void volume of the column was 10.6 liters, which was discarded, after this fractions were collected. The yield of the inhibitor was 3,487 A_{280} units.

desired, the solution of inhibitor is passed through a column of G-10 (Fig. 6). One passage usually suffices. A cumbersome dialysis in acetylated bags is eliminated.

The rest of the available time I would like to spend discussing *temporary inhibition*. The phenomenon of temporary inhibition was discovered in the early fifties [7]. We used a preparation of secretory pancreatic inhibitor, which was a gift of Dr. KAZAL [8]. The preparation had 3 electrophoretically distinct components. When this inhibitor was mixed with trypsin, tryptic activity first disappeared then gradually reappeared. Paralleling the reappearance of tryptic

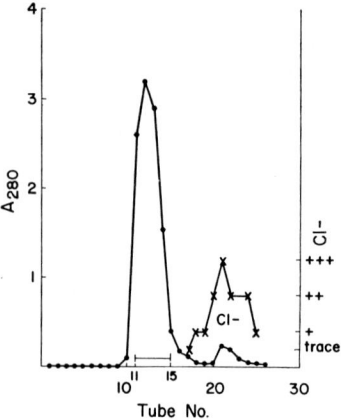

Fig. 6. Desalting of basic pancreatic inhibitor on Sephadex G-10. A Pharmacia column 5×100 cm was equipped with an upward flow adapter. It was filled with Sephadex G-10 (beads form) previously swelled and washed with water. A sample containing 190 A_{280} units of protein in a volume of 10 m*l* was introduced. A peristaltic pump was set at 4.2 m*l*/min and the sample was eluted with water. The first liter of eluate (void volume) was discarded, after which 15 m*l* fractions were collected. The yield of desalted inhibitor was 172 A_{280} units or 206 mg. The ordinate on the right-hand side refers to the appearance of chloride.

activity a loss of inhibitory activity occurred. For historical accuracy, it should be stated that a year earlier GORINI and AUDRAIN [9, 10] had observed a gradual increase in tryptic activity after mixing bovine trypsin with chicken ovomucoid.

Since that time temporary pancreatic inhibitors have been purified, due to the efforts of Burck and coworkers [11,12], Greene and coworkers [13—16], and the Munich group [17, 18]. The primary structures of bovine and porcine secretory inhibitors have been elucidated, and the reactive sites have been established [13—18].

In our original [7] interpretation of the mechanism of temporary inhibition a transient TIT complex was postulated. It was imagined that the active site of the first trypsin molecule reacted with (what is now known to be) the reactive site of the inhibitor, whereas the second trypsin molecule inflicted a cleavage in a position other than the reactive site. TSCHESCHE and KLEIN [19] using porcine secretory inhibitor recently concluded that the two tryptic hits postulated by us occur not simultaneously but in sequence.

We assumed that the mechanism of inhibition of LASKOWSKI, Jr. (see the review [20]) is general and applicable to temporary inhibitors including the porcine secretory inhibitor. To comply with this mechanism the modifying cleavage must preceed the inactivating cleavage. The modifying cleavage, contrary to other tryptic cleavages has an optimum at the acid range (pH from 3 to 4). At this pH range the complex dissociates significantly, and the dissociation of the modified form is favored over the virgin form.

Due to the courtesy of Drs. GRINNAN and BURCK, who provided us with highly purified porcine secretory inhibitor, we were able to study the effect of pH on temporary inhibition.

Fig. 7 shows experiments in which the ratio of trypsin inhibitor to trypsin was kept constant, but the pH was varied. The period of tryptic inactivity was the shortest at pH 3.4, the longest at pH 8.0. The rate of reappearance of tryptic activity is highest at pH 8, lowest at 3.4. TSCHESCHE and KLEIN [19] observed that after exposure of 1 mole of inhibitor to 0.5 mole of trypsin for 24 hrs, only small peptides were found at pH 3.75, whereas at pH 8 a considerable amount of the original inhibitor remained intact. Our experimental findings agree with theirs.

An experiment was then performed in which an excess of the inhibitor and trypsin were preincubated for 2.5 hrs at pH 3.4 until about 20% of the tryptic activity was restored. At that time the sample was divided, and a portion was adjusted to pH 8.0, whereas the other portion was allowed to remain at pH 3.4. The result shows that the return to complete tryptic activity is faster at pH 8.0 than at pH 3.4. The result supports the contention that inactivating clea-

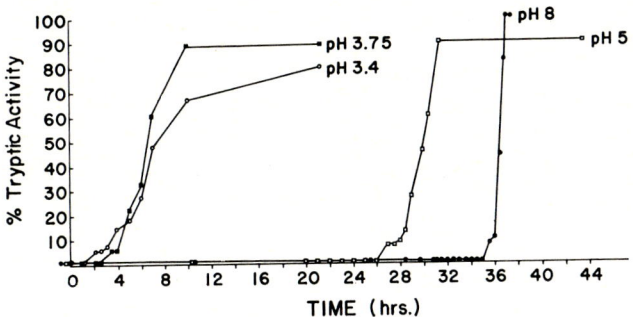

Fig. 7. The effect of pH on the rate of reappearance of tryptic activity. Porcine secretory inhibitor (25 μg) was incubated with porcine trypsin (76 μg) in 0.2 ml of 0.05M sodium citrate and 0.01M $CaCl_2$ at pH 3.4, 3.75, and 5. A 0.05M Tris buffer containing 0.01M $CaCl_2$ was used at pH 8. The amounts of inhibitor and trypsin were calculated from A_{280} readings and optical factors of 2.25 and 0.74 respectively. Aliquots of 10 μl were removed at each time point and assayed for tryptic activity. Results are expressed as the % of the trypsin control at zero time at each pH.

vage occurs after the modifying cleavage, and that the optimum for the inactivating cleavage is around pH 8.0.

Fig. 8. Spectrophotometer tracings at 620 nm of stained polyacrylamide gels containing aliquots from the incubation of porcine secretory inhibitor and trypsin at pH 3.4. Inhibitor (2 mg) and trypsin (5.6 mg) were incubated in 1.2 ml of 0.05M sodium citrate—0.01M $CaCl_2$ buffer at pH 3.4. At 30 min intervals 100 μl aliquots were pipetted into 10 μl of 6N HCl and 10 μl of 25% trichloroacetic acid was added. The samples were frozen and stored. Prior to electrophoresis the incubation aliquots were thawed and centrifuged. The supernatants were applied to polyacrylamide gel columns 0.25 × 10 cm which were 12% in acrylamide and 0.2% in N,N, methylene-bis-acrylamide at pH 3.5. A glycine-acetate buffer, pH 4, was used for electrophoresis from anode to cathode at 5 ma/tube for 3.5 hr.

One would like to speculate that the difference between a "temporary" and a "non-temporary" inhibitor depends on what happens after the modifying cleavage has been inflicted. With the "non-temporary" inhibitor the second tryptic

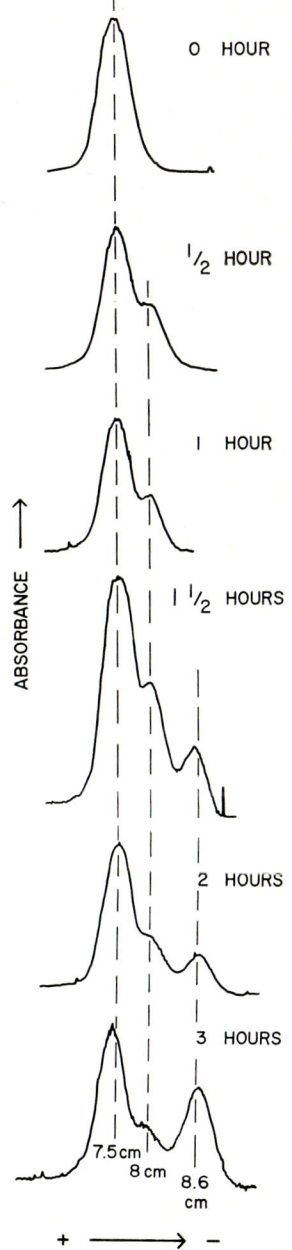

approach favors the reactive site of the inhibitor with the consequence that the original complex is reformed. With the "temporary" inhibitor the second tryptic approach favors a site different from the reactive site. Since the second site behaves as a normal substrate (products dissociate at pH 8.0) the second cleavage inactivates.

To support the conclusions based on measurements of tryptic activity, the reaction was followed by disc electrophoresis [21]. Fig. 8 shows the tracing of stained discs, and is directly comparable to Fig. 7 with respect to timing. It is obvious that the band corresponding to the modified inhibitor is detected much earlier than the reappearance of tryptic activity. Therefore, by inference, the material in the modified disc should represent active inhibitor. Since the appearance of the third disc coincides with the appearance of tryptic activity, by inference, the material in disc 3 should be inactive.

A direct proof that peak 2 is active is supplied by an experiment in which a catalytic amount (1%) of trypsin was used at pH 3.4. The reaction mixture was then tested on disc electrophoresis and showed the presence of only 2 bands corresponding to the intact and the modified

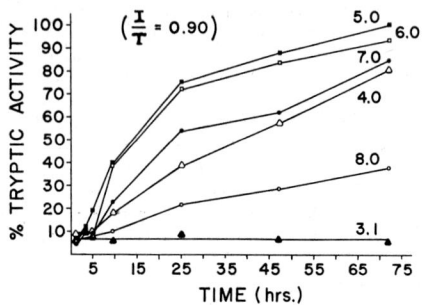

Fig. 10. pH dependence of temporary inhibition. Trypsin, 36.8 μg/ml (sp. act. 41,7 μ/m) and RCAM-PTI*, 10.9 μg/ml (sp. act. 2.6 μ/mg) were incubated in 0.01M CaCl$_2$ and 0.02M Na succinate (pH 3.1, 4.0, 5.0 and 6.0) or 0.02M Tris-HCl (pH 7.0 and 8.0). At the indicated times, 0.2 ml aliquots were removed and added to 0.2 ml of 0.2M Tris-HCl, pH 8.0. After incubation for 5 min at 25°, the samples were assayed for free tryptic activity. The results are expressed in percentage of the activities of corresponding trypsin controls containing no inhibitor.

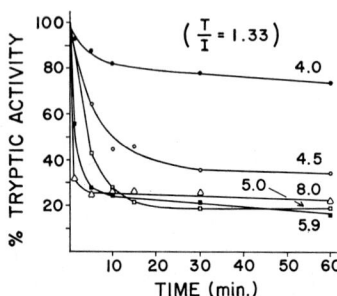

Fig. 9. The dependence of complex formation upon pH. Trypsin, 27.2 μg/ml, and RCAM-PTI*, 6.9 μg/ml, were incubated together in 0.067M Na succinate (pH 4.0, 4.5, 5.0 and 5.9) or 0.067M Tris-HCl, pH 8.0. At the indicated times, 0.3 ml aliquots were withdrawn and assayed for free tryptic activity. Results are expressed in terms of a trypsin control assay at the beginning of the experiment.

---

* RCAM-PTI, reduced carboxamido-methylated Pancreatic Trypsin Inhibitor.

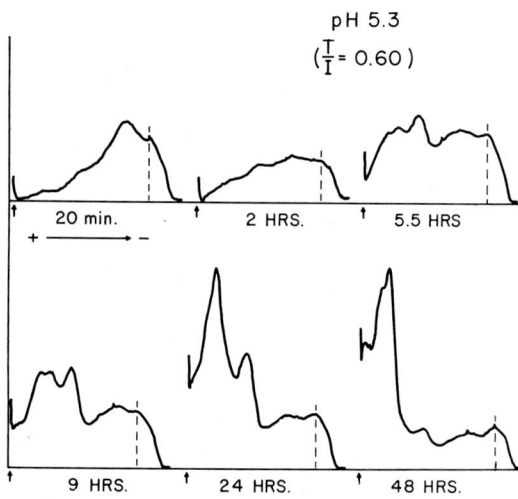

Fig. 11. The reaction of trypsin and RCAM-PTI* at pH 5.3 followed by disc electrophoresis. TPCK-trypsin, 6.37 mg/ml (sp. act. 18 μ/mg) and RCAM-PTI*, 145. mg/ml, were incubated in 0.035M Na succinate plus 0.009M CaCl$_2$, pH 5.3. At the indicated times, 0.1 ml samples were withdrawn and added to 0.1 ml aliquots of 5% TCA. After centrifuging, 0.1 ml samples were run on 10% gels at pH 9.5, 6 ma/tube, with Tris-glycine reservoir buffer, pH 8.3. The gels were stained in amido schwartz, destained electrophoretically, and scanned at 600 nm on a Gilford Spectrophotometer.

inhibitors. The mixture was then chromatographed on CM-cellulose and two active peaks were obtained.

The mechanism established on the naturally occurring temporary inhibitor differs from that seen with an artificial temporary inhibitor [22]. We prepared this artificial inhibitor starting with KUNITZ's bovinep ancreatic inhibitor, in which S—S bond 14—38 has been reduced and substituted with carboxamidomethyl groups. We investigated the effect of pH on this inhibitor's ability to form a complex (Fig. 9). With the artificially prepared temporary inhibitor, the complex starts forming at pH 5 and above. Fig. 10 shows that the rate of reappearance of tryptic activity is also fastest at pH 5.0. We then traced the progress of the reaction by disc electrophoresis [21]. Fig. 11 shows the progressive disappearance of the original form, and the appearance of 3 new peaks representing sequential digestion of the inhibitor. After 24 hrs we accumulated and isolated peak 3. Three peptide bonds were cleaved: Lys 15-Ala 16, Arg 17-Ile 18 and Arg 39—Ala 40. The dipeptide Ala 16—Arg 17 was removed by the cleavages. All 3 cleavages are in the apex of the "pear" as seen by x-ray diffraction [23]. As yet we have not succeeded in establishing the order of cleavages two and three. The inhibitor with 3 bonds cleaved is totally inactive.

Acknowledgements: The work has been generously supported by Grant AM-10481 from the National Institute of Arthritis and Metabolic Diseases. M. A. was supported by the National Science Foundation Research Participation Program in Science, Grant NSF-GY-705.

References

[1] LASKOWSKI, M., Jr. and M. LASKOWSKI, Sr., J. Biol. Chem. **190**, 563 (1951).

[2] LASKOWSKI, M. Jr., P. H. MARS and M. LASKOWSKI, Sr., J. Biol. Chem. **198**, 745 (1952).

[3] LASKOWSKI, M., Sr., B. KASSELL and G. HAGERTY, Biochim. Biophys. Acta **24**, 300 (1957).

[4] KRESS, L. F., S. R. MARTIN and M. LASKOWSKI, Sr., Biochim. Biophys. Acta **229** 836 (1971).

[5] MARTIN, S. R., M. S. Thesis 1968, State University of New York at Buffalo.

[6] MOZEJKO, J. H. and M. LASKOWSKI, Sr., Anal. Biochem. (in press, 1970).

[7] LASKOWSKI, M., Sr. and F. C. WU, J. Biol. Chem. **204**, 797 (1953).

[8] KAZAL, L. A., D. S. SPICER and R. A. BRAHINSKY, J. Am. Chem. Soc. **70**, 3034 (1948).

[9] GORINI, L. and L. AUDRAIN, Biochim. Biophys. Acta **8**, 702 (1952).

[10] GORINI, L. and L. AUDRAIN, Biochim. Biophys. Acta **10**, 570 (1953).

[11] CERWINSKY, E. W., P. J. BURK and E. L. GRINNAN, Biochemistry **6**, 3175 (1967).

[12] BURCK, P. J., R. L. HAMILL, E. W. CERWINSKY and E. L. GRINNAN, Biochemistry **6**, 3180 (1967).

[13] GREENE, L. J., M. RIGBY and D. S. FACKRE, J. Biol. Chem. **241**, 5610 (1966).

[14] GREENE, L. J., J. J. CARLO, A. J. SUSSMAN and D. C. BARTELT, J. Biol. Chem. **243**, 1804 (1968).

[15] GREENE, L. J. and J. S. GIORDANO, Jr., J. Biol. Chem. **244**, 285 (1969).

[16] GREENE, L. J. and D. C. BARTELT, J. Biol. Chem. **244**, 2646 (1969).

[17] TSCHESCHE, H., E. WACHTER, S. KUPFER and K. NIEDERMEIER, Z. Physiol. Chem. **250**, 1247 (1969).

[18] TSCHESCHE, H., E. WACHTER and G. KALLUP, Z. Physiol. Chem. **350**, 1662 (1969).

[19] TSCHESCHE, H. and H. KLEIN, Z. Physiol. Chem. **349**, 1645 (1968).

[20] LASKOWSKI, M., Jr. and R. W. SEALOCK, in: P. L. BOYER, edt. The Enzymes, 3rd edition, Academic Press, New York (1971).

[21] NIEKAMP, C. W., H. F. HIXON and M. LASKOWSKI, Jr., Biochemistry **8**, 16 (1969).

[22] KRESS, L. F., K. A. WILSON and M. LASKOWSKI, Sr., J. Biol. Chem. **243**, 1758 (1968).

[23] HUBER, R., D. KUKLA, A. RÜHLMANN, O. EPP and H. FORMANEK, Naturwissenschaften **57**, 389 (1970).

Studies on the Reactive Site towards Chymotrypsin and Trypsin of the Basic Trypsin Inhibitor of Bovine Pancreas

Meir Rigbi

Department of Biological Chemistry, The Hebrew University of Jerusalem, Jerusalem, Israel

It is well known that many natural trypsin inhibitors also inhibit chymotrypsin. Some of these, such as turkey ovomucoid [1], the lima-bean inhibitor [2, 3] and the Bowman-Birk inhibitor [4] are double-headed, that is, their complex with either enzyme is capable of inhibiting the other. In such cases the reactive sites towards both enzymes are distinct. The sites have sometimes been distinguished by chemical modification. For example, when turkey ovomucoid is acetylated or carbamylated, it becomes inactive towards trypsin, whereas it remains active towards chymotrypsin [5].
Recently, the study of the reactive sites of a large number of trypsin inhibitors has shown that they consist of a lysine-X or an arginine-X sequence, where X is most often an alanine or an isoleucine residue [6—13]. The reactive sites are thus consistent with the specificity requirements of the enzyme. Inhibitor specificity and reactive site sequences are described and discussed in a comprehensive review on proteinase inhibitors by Laskowski, Jr. and Sealock [14].
Much less is known of inhibitor reactive sites towards chymotrypsin. The existence of a chymotrypsin-specific reactive site in the Bowman-Birk inhibitor is indicated by the inactivation of the inhibitor towards chymotrypsin, following incubation with this enzyme at acid pH ([15], see also Birk and Gertler, this volume). In the lima-bean inhibitor the reactive site towards chymotrypsin has recently been identified as the sequence leucine-serine ([16], see also Stevens, this volume). The site is thus specific to the enzyme.
Regarding the basic pancreatic trypsin inhibitor (BPTI), Wu and Laskowski, Sr. [17] preincubated the inhibitor with trypsin and found that the complex is inactive towards α-chymotrypsin. Kraut and Bhargava [18] repeated the experiment of Wu and Laskowski, Sr., and also preincubated the inhibitor with chymotrypsin. They found that the complex formed with either enzyme failed to inhibit the other, and concluded that the same site is involved in the inhibition of both enzymes. More recently it was shown by Chauvet and Acher [19] and by Kress and Laskowski, Sr. [7] that the reactive site of BPTI (Fig. 1) towards trypsin is lysine 15, or the lysine 15-alanine 16 sequence. The latter workers also showed that on reduction of the Cys 14—Cys 38 disulfide bond and carboxamidomethylation of the thiol groups, the inhibitor lost its chymotrypsin inhibitor activity, but retained its trypsin inhibitor activity. A distinction between the two inhibitor activities was thus made. More recently, however, Fritz et al. have shown that pentamaleyl BPTI (four lysines and one N-terminal arginine maleylated) is inactive towards both enzymes [21], whereas the tetramaleyl inhibitor (lysine 15 free) is active towards both [22]. These workers, taking into account

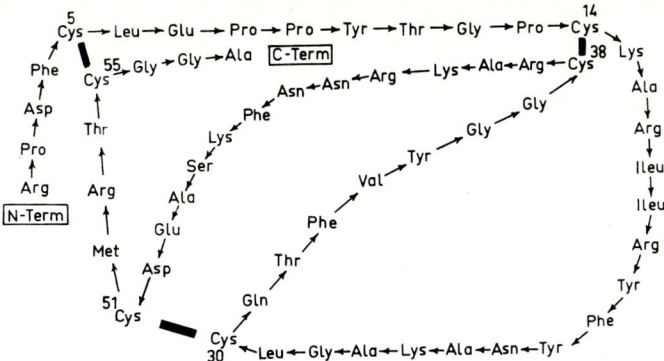

Fig. 1. Structure of the basic pancreatic trypsin inhibitor (KASSELL and LASKOWSKI, Sr. [20]).

the different specificities of the two enzymes, conclude that the reactive site towards chymotrypsin is in the region of the lys 15—ala 16 sequence.

The purpose of the investigation here described was to identify the reactive site of BPTI towards chymotrypsin. This, presumably, would solve on a molecular basis, the problem of whether the dual specificity of the inhibitor is due to distinct, overlapping, or identical sites. In the course of the investigation, some light was also shed on the reactive site towards trypsin.

The investigation consisted of chemically modifying the inhibitor by (a) acetylation and (b) selective dansylation of the active tetramaleyl inhibitor. In order to evaluate the contribution of the lysine 15 side-chain to chymotrypsin and trypsin inhibition, K_I-values were determined and standard free energies of binding were calculated. The interaction between the native or the dansylated inhibitor and either enzyme was also investigated by fluorescence measurements. Finally, (c) the inhibitor was selectively reduced at Cys 14—Cys 38 and subjected to proteolysis by chymotrypsin at acid pH, with or without subsequent action of carboxypeptidase A or B. All modified products were assayed for chymotrypsin and trypsin inhibition.

The results of our investigation indicate that other atoms or amino acid side chains other than lysine 15 are involved in the inhibition of both enzymes. We also show that, as for trypsin, the reactive site towards chymotrypsin includes the Lys 15—Ala 16 sequence. The Arg 39—Ala 40 sequence is another reactive site towards chymotrypsin, and possibly towards trypsin too.

Enzyme and Inhibitor Assay Methods and the Determination of Dissociation Constants

Chymotrypsin and trypsin activities were determined by the rate of hydrolysis of acetyl-L-tyrosine ethyl ester (ATEE) and tosyl-L-arginine methyl ester (TAME), respectively, titrimetrically in the pH-stat or by the change in absorbance in the Gilford recording spectrophotometer. Trypsin activity was also determined with the chromogenic substrate benzoyl-DL-arginine p-nitroanilide (BAPA). Inhibitor activity was assayed by the loss in enzyme activity following a five minute preincubation of enzyme with varying amounts of inhibitor. As will be seen from the following discussion, judicious use of the various assay conditions made it possible to determine a wide range of dissociation constants of native and variously modified BPTI. The experimental conditions of our assays are summarized in Table 1.

At this point it is worth considering the methods used in the determination of dissociation constants. Native inhibitors normally have dissociation constants of the order of 10^{-8} to 10^{-12} M. When these are assayed with trypsin or chymotrypsin, at a concentration of the order of

Table 1. Conditions of Trypsin and Chymotrypsin Assays[a]

	Chymotrypsin Substrate, ATEE		Trypsin Substrate, TAME		Substrate, BAPA
	Titrimetric assay	Spectrophotometric assay	Titrimetric assay	Spectrophotometric assay	Spectrophotometric assay
Enzyme concentration, M	2.1×10^{-7}	2.5×10^{-8}	3.3×10^{-7}	4.2×10^{-8}	1.25×10^{-6}
Substrate concentration(s), M	8×10^{-3}	5.33×10^{-4}	8×10^{-3}	3.30×10^{-4}	0.9×10^{-4}
		10.66×10^{-4}		8.33×10^{-4}	1.5×10^{-4}
Buffer solution:					
Tris, M	1.5×10^{-2}	4×10^{-2}	1.5×10^{-2}	4×10^{-2}	4×10^{-2}
$CaCl_2$, M	2×10^{-2}	—	2×10^{-2}	—	—
KCl, M	0.1	—	0.1	—	—
pH	7.8	8.0	7.8	8.0	8.0
Temperature	25°	25°	25°	25°	15°
Absorbance change	—	$\Delta\varepsilon^{235} = -330$ M^{-1} cm^{-1} [23]	—	$\Delta\varepsilon^{246} = +715$ M^{-1} cm^{-1} [24]	$\Delta\varepsilon^{410} = +8,800$ M^{-1} cm^{-1} [26]
K_m		1.1×10^{-3} [23]		5×10^{-5} [25]	9.39×10^{-4} [26]

[a] Figures in brackets indicate references.

$10^{-7} M$, which is the order of magnitude commonly used in the pH-stat, a mutual depletion system is obtained, that is, one in which the reaction essentially goes to completion and there is a mutual and equivalent reduction of both components. Methods for the determination of K_I in mutual depletion systems have been worked out [27] and indeed, the method of Green and Work [28] makes use of a special case of non-competitive inhibition in mutual depletion systems at only one point, that of inhibitor-enzyme equivalence. In mutual depletion systems, as for other systems, dissociation constants are determined at inhibitor concentrations near or beyond enzyme equivalence. In such systems inhibitor is both free and combined, or mainly free. Enzyme activities are low and it is therefore difficult to obtain sensitive measurements.

Most of our K_I-determinations are based on inhibitor activities at very dilute concentrations of inhibitor, enzyme and substrate such as those found in the spectrophotometric assay systems with ATEE or TAME as substrates. In such systems, dissociation of the inhibitor-enzyme complex is high, and substrate may compete less with inhibitor, which makes for still higher enzyme activity. Inhibitor is mainly free, and therefore the usual equations and methods of graphical analysis may be used. We found it convenient to use Dixon's plot of $1/v$ vs. I at two substrate concentrations [29]. If two straight lines are obtained, their interaction will occur at an abscissa of $-K_I$. If they intersect above the abscissa, the inhibition is competitive. If they meet at a point on the abscissa, it is non-competitive. If hyperbolic curves are obtained, the dissociation constant may be evaluated by further graphical analysis, and if necessary by the use of the general rate equation for enzyme inhibition. One such case will be considered later. The field has been described extensively by Webb [27].

For weak trypsin inhibitors, it was desired to obtain higher trypsin inhibition than is possible in the spectrophotometer with TAME. For this purpose BAPA was employed. In spite of the relatively high concentration of enzyme, inhibitions are higher in this system than with

TAME, due to the low concentrations of DL-BAPA used, and, in particular, to the low affinity of this substrate for enzyme. The D-isomer, included in the substrate preparation, inhibits trypsin, yet the kinetics of the method, which have been worked out by MARES-GUIA and SHAW [30], show that 1/v vs. I plots at two substrate concentrations also yield K_I values as described by DIXON [29].

Acetylation of BPTI*

We reported earlier that acetylation of BPTI with acetic anhydride in the presence of half-saturated sodium acetate results in the loss of activity towards trypsin [32, 33]. Tyrosine groups were not acetylated. In our present study on the reactive site towards chymotrypsin, we sought to acetylate tyrosine groups as well, in view of the known specificity of this enzyme for aromatic amino acid side-chains. BPTI was therefore acetylated (0°, pH 7.7) in the absence of sodium acetate [34]. The total loss of ninhydrin colour indicated that acetylation of amino groups was complete. By the loss in absorbance on acetylation, and its recovery on deacetylation of tyrosines [34], it was calculated that 1.8 tyrosine groups out of 4 were acetylated. Titrimetric assays showed that the inhibitor thus acetylated was active towards chymotrypsin, yet possessed no activity towards trypsin. It was concluded that lysine 15, known to inhibit trypsin, and the unidentified O-acetyl tyrosines, are not required for chymotrypsin inhibition. In this connection it is interesting to note that tyrosine residues 10 and 21 lend themselves to nitration [35], and tyrosines 10, 21 and 35 lend themselves to iodination [36].

Acetylation was next carried out with acetyl-N-hydroxysuccinimide (AcNSI). This reagent preferentially acetylates amino groups [37]. Our work with polyamino acids showed that under the experimental conditions used, tyrosine groups are not acetylated [38]. Acetylation was carried out in a pH-stat at pH 7.8 with a 2.5 molar excess of AcNSI over amino groups. When tritiated (methyl T) AcNSI was used, radioactivity measurements showed that the acetylated inhibitor possessed 5 acetyl groups per BPTI molecule, which correspond to 4 ε-amino groups of lysine and one α-amino group of the N-terminal arginine. Tyrosine groups were not acetylated. By the titrimetric assay, penta-N-acetyl BPTI is inactive towards trypsin (Figure 3).

Figure 2 shows the course of the acetylation, obtained with aliquots acidified at desired time-

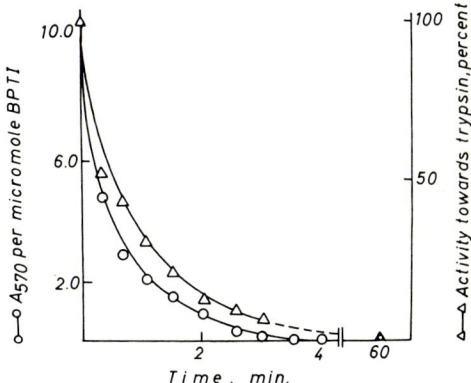

Fig. 2. Time-course of BPTI-acetylation with acetyl-N-hydroxysuccinimide. Trypsin inhibitory activity assayed titrimetrically.

intervals. The activity loss towards trypsin (titrimetric assay) is somewhat slower than the ninhydrin colour loss (570 nm). Both reactions follow second-order kinetics and the constants are within the range of those obtained for model amino acid polymers and copolymers [38].

Though the penta-N-acetyl inhibitor failed to inhibit trypsin with TAME in the pH-stat, some inhibition was observed with TAME spectrophotometrically, and it was still more pronounced with BAPA. Figure 3 shows the results of these experiments. The reasons for the differences have been discussed in the previous section. Figure 4, which shows such a 1/v vs. I plot for trypsin and penta-N-acetyl BPTI,

---

* This research was carried out together with NEHAMA SEVILLA. A preliminary account of some of this work has been published [31].

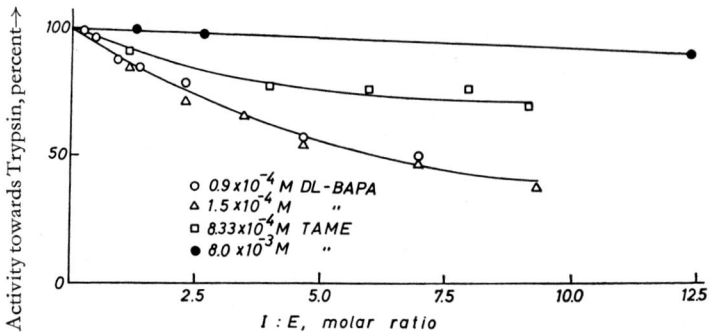

Fig. 3. Trypsin inhibition by penta-N-acetyl BPTI. Full circles, titrimetric assay. All other assays were spectrophotometric.

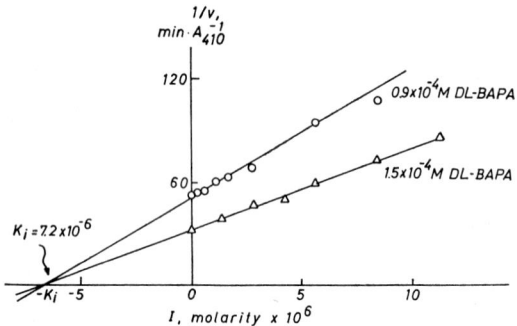

Fig. 4. $1/v$ vs. I plot for inhibition of trypsin by penta-N-acetyl BPTI with BAPA as substrate.

was obtained with BAPA as substrate. The inhibition is non-competitive, and an inhibition constant of 7.2×10^{-6}M is obtained. The inhibition constant of the native inhibitor is smaller than that of the acetylated inhibitor by an order of 10^{-5}. However, when one compares the inhibitors by the standard free energy of association, the native inhibitor is stronger than the acetylated inhibitor only by a factor of 2. These results are included in Table 3. It may be concluded that lysine 15 is an important factor, but not the only one, involved in trypsin inhibition.

We come next to the interaction of penta-N-acetyl BPTI with chymotrypsin. Figure 5 shows the dose-response curves at two substrate levels. The dependence of the inhibition on substrate concentration indicates that the inhibition is not completely non-competitive, and suggests competitive inhibition. A plot of $1/v$ vs. I produced hyperbolic curves (Figure 6). Therefore, the inhibition is not of the completely competitive type.

The following analysis, found to be essential for the accurate determination of K_I, is based largely

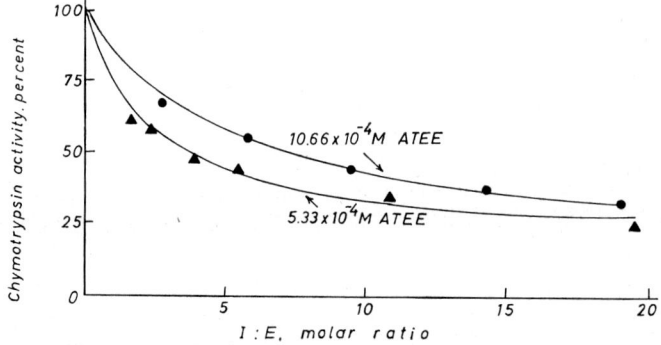

Fig. 5. Chymotrypsin inhibition by penta-N-acetyl BPTI, spectrophotometric assay.

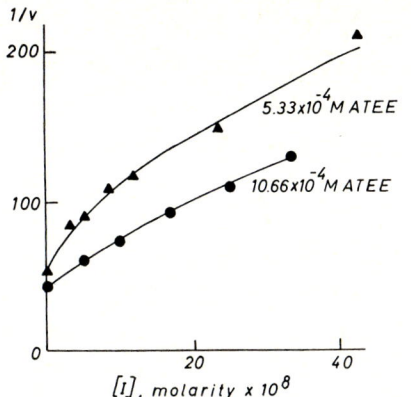

Fig. 6. 1/v vs. I plot for inhibition of chymotrypsin by penta-N-acetyl BPTI.

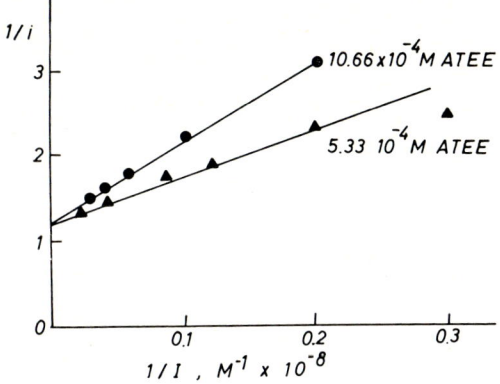

Fig. 7. 1/i vs. 1/I plot for chymotrypsin inhibition by penta-N-acetyl BPTI. i is the fractional inhibition (see text).

on the treatise by WEBB [21]. In order to define the inhibition type, 1/i was plotted against 1/I (Figure 7), where i is the fractional inhibition $\frac{v_0 - v}{v_0}$ (v_0 and v are the velocities of the uninhibited and the inhibited reactions respectively). From the dependence or independence of the slope and intercept of such a plot on substrate concentration, it can be shown that the inhibition is also neither partially competitive, nor partially non-competitive nor completely mixed. By a process of elimination, inhibition must be either partially mixed or coupled. One has to resort to the rate equation for inhibited enzyme reactions in order to derive expressions for the slopes and intercepts of the plots shown in Figure 7.

The general scheme for enzyme-inhibitor-substrate interactions is given by

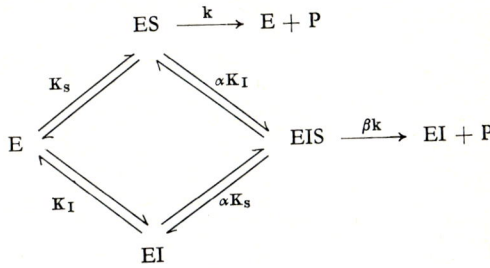

where, apart from the conventional symbols, α is the factor for the equilibrium constants K_I and K_S in the formation of the EIS complex, and β is the factor for the rate constant k in the breakdown of EIS to EI and products. The rate equation is

$$v = V_m \frac{(S)[\alpha K_I + \beta(I)]}{(S)(I) + \alpha[K_I(S) + K_s(I) + K_s K_I]}$$

In the plot of 1/i against 1/I the following relationships may be derived:

$$\text{Intercept} = \frac{(S) + \alpha K_s}{(S)(1 - \beta) + K_s(\alpha - \beta)}$$

$$\text{Slope} = \frac{\alpha K_I(S + K_s)}{(\alpha - \beta) K_s}$$

The values for the slopes and intercepts in Figure 7 were calculated from the data by the least-squares method. From the intercepts at the two concentrations of ATEE, α and β were calculated as 1.56 and 0.27 respectively. These values define the inhibition as partially mixed [$(0 < \beta < 1; \infty > \alpha > 1)$]. When these values were substituted in the equation for the slopes, a K_I value of 3.5×10^{-8} M was obtained for penta-N-acetyl BPTI with chymotrypsin.

In order to compare K_I-values accurately, the K_I-value for native BPTI with chymotrypsin was next determined. Figure 8 shows the dose-response curve. The single curve obtained at the

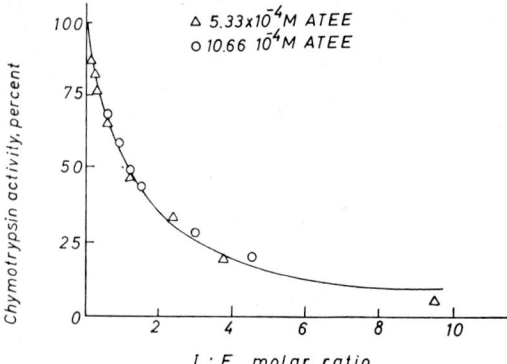

Fig. 8. Chymotrypsin inhibition by native BPTI, spectrophotometric assay.

two substrate levels suggests non-competitive inhibition. A plot of 1/v against I confirms this, and yields a K_I-value of 2.9×10^{-8} (Figure 9). The inhibition constants for native and penta-N-acetyl inhibitor with chymotrypsin are thus 2.9×10^{-8}M and 3.5×10^{-8}M respectively. The difference is negligible.

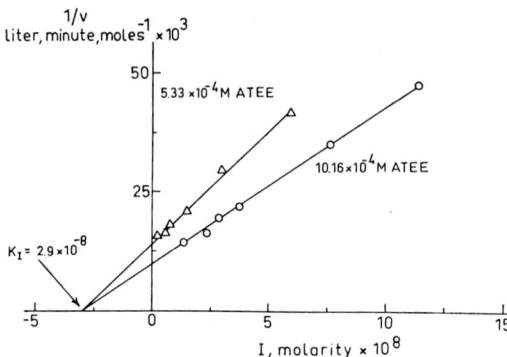

Fig. 9. 1/v vs. I plot for chymotrypsin inhibition by native BPTI.

Full N-acetylation, therefore, changed the inhibition from the completely non-competitive to the partially-mixed type, but did not affect the dissociation constant of the inhibitor-enzyme complex. Partially-mixed inhibition means that there is some dissociation of the EIS-complex into EI and products. It is conceivable that as a result of acetylation, the EI-complex is looser, thus permitting the EIS-complex to yield products. The finding that the dissociation constant of the inhibitor-chymotrypsin complex remains unchanged following acetylation led us to the conclusion that the lysine 15 side chain as well as the other acetylated groups (3 lysines and one N-terminal arginine) were not essential for chymotrypsin inhibition and that other atoms or amino acid side chains other than lysine 15 are involved in chymotrypsin inhibition. As will be shown in the subsequent section, similar conclusions to those arrived at here for both enzymes were obtained by dansylation experiments.

Dansylation of Tetramaleyl BPTI*

As BPTI contains no tryptophan, it was thought that if lysine 15 were dansylated and the product were to inhibit both chymotrypsin and trypsin, excitation of tryptophan on the enzyme moiety of either complex might result in fluorescence of dansyl groups in the inhibitor. Mention should here be made of the work of HAUGLAND and STRYER [40], who prepared inactive mono-anthranoyl chymotrypsin and showed that excitation of tryptophan residues resulted in an enhanced fluorescence of anthranoyl groups. Tryptophan 215 of chymotrypsin and presumably the homologous tryptophan in trypsin, are located in the lining of the pocket at the active site [41] and appear to be particularly well-placed for such radiationless energy transfer.

As the dansylation of native BPTI precipitated the inhibitor, we dansylated tetramaleyl BPTI [22]. This was done in 8M urea at pH 9.3, and the product was purified by dialysis and gel filtration on Sephadex G-15.

The spectrum of dansylated tetramaleyl inhibitor (DTMI) is shown in Figure 10. It possesses an absorption maximum at 325 nm with $\varepsilon^{325} = 11,200$M^{-1}cm^{-1}. If the average molar absorp-

---

* The investigation described in this section was carried out together with Dr. YEHUDIT ELKANA, ALMA GAL, and NEHAMA SEVILLA. A preliminary account of part of the material has been published [39].

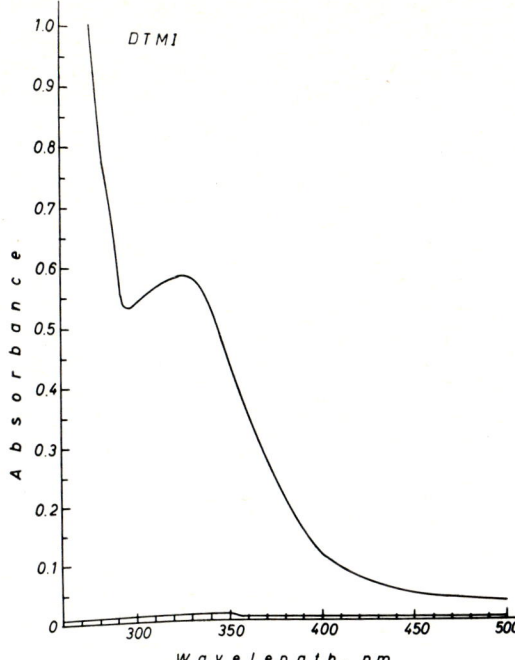

Fig. 10. Absorption spectrum of 0.52×10^{-4}M dansyl tetramaleyl BPTI in 0.01M ammonium bicarbonate.

tivity per protein dansyl group of HARTLEY and MASSEY [42] is used (3,300 $M^{-1}cm^{-1}$), this would correspond to 3.4 dansyl groups per BPTI molecule. However, CHEN [43] found a molar absorptivity, $\varepsilon^{323} = 5,270 M^{-1}cm^{-1}$, for a highly exposed ε-N-dansyl lysine residue in lysozyme, which would correspond to 2.12 dansyl groups per BPTI molecule. Part of the high absorptivity of DTMI is due to adsorbed dansyl sulfonate which can be detected by thin-layer chromatography [44] of DTMI.

When DTMI was hydrolyzed in 6N HCl for only 6.25 hours, to minimize destruction of dansyl groups [44], thin-layer chromatography showed only ε-N-dansyl lysine and dansyl tyrosine*. Amino acid analysis of the hydrolyzate indicates dansylation of two residues — one lysine and one tyrosine (Table 2). It thus appears that DTMI is didansyl tetramaleyl BPTI with one dansyl group blocking the amino group of lysine 15 and the other attached to an exposed tyrosine.

DTMI inhibits both trypsin and chymotrypsin. Figure 11 shows the plot of 1/v against DTMI concentration for trypsin at two TAME concentrations. The inhibition is non-competitive and the inhibition constant is 6×10^{-8}M, one thousand times more than for the native inhibitor. For chymotrypsin, inhibition is competitive and K_I was found to be 3.1×10^{-8}M, practically the same as for the native inhibitor. These results are presented in Table 3. They show, as with acetylation, that the unblocked lysine 15 side chain is an important factor, but not the only one in trypsin inhibition, and is not essential in chymotrypsin inhibition.

Results essentially similar to ours were obtained by FRITZ et al. with tetramaleyl carbamyl (lysine 15) BPTI [22]. Whereas tetramaleyl BPTI is active towards both enzymes, carbamylation of lysine 15 increased the dissociation constant for the interaction with trypsin by a factor of 1.6×10^3, and for the interaction with chymotrypsin by a factor of max. 2.5**.

Table 3 summarizes inhibition constants and standard free energies of binding for native BPTI, variously modified BPTI and some low-molecular weight trypsin inhibitors. A discussion of the data will be presented later.

Fluorimetric titrations were conducted by adding aliquots of a trypsin or chymotrypsin solution to inhibitor at pH 7.8 in an Aminco-Bowman spectrofluorimeter equipped with a mercury-xenon arc. Tryptophan residues were excited in a wavelength range designed not to excite tyrosines. When trypsin was added to DTMI, tryptophan-fluorescence measured at 338 nm decreased in comparison with solutions which contained no inhibitor, the decrease reaching a constant value at about equivalence. Essentially the same phenomenon was observed with the native inhibitor, and has been reported by EDELHOCH and STEINER [49]. With chymo-

---

* The latter compound was not further identified, as the commercial O-dansyl tyrosine gave two spots.

** This figure is by comparison with the native inhibitor.

Table 2. Amino acid analysis of a 6.25 hour acid hydrolyzate of native inhibitor and dansyl tetramaleyl BPTI

Amino acid	Residues per mole			Residues dansylated	
	BPTI		DTMI		
	Correct value [45]	Found	Found	Found	Nearest integer
Lysine	4	3.98	2.90	1.08	1
Arginine	6	6.09	6.02	—	—
Histidine	0	0	0	—	—
Tryptophan	0	0	0	—	—
Aspartic acid	5	4.61	4.86	—	—
Threonine	3	2.75	2.70	—	—
Serine	2	0.87	0.86	—	—
Glutamic acid	3	3.04	2.92	—	—
Proline	4	3.76	3.96	—	—
Glycine	6	6.15	5.86	—	—
Alanine	6	6.15	5.86	—	—
Half-cystine	6	5.42	5.56	—	—
Valine	1	0.79	0.57	—	—
Methionine	1	0.63	0.22	—	—
Isoleucine	2	0.63	0.69	—	—
Leucine	2	1.69	1.68	—	—
Tyrosine	4	3.62	2.64	0.98	1
Phenylalanine	4	3.88	3.65	—	—

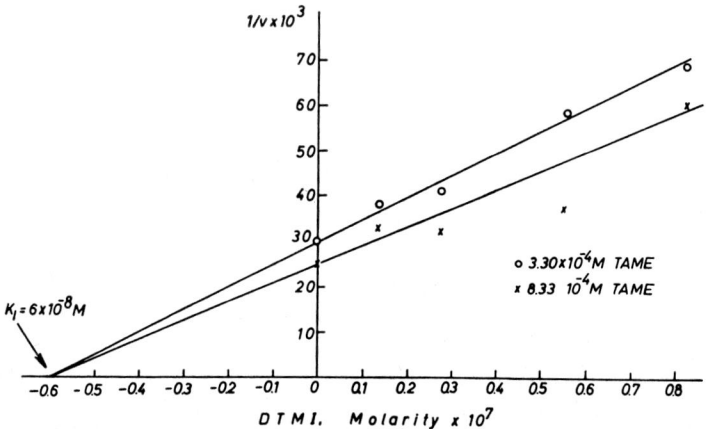

Fig. 11. 1/v vs. I plot for trypsin inhibition by dansyl tetramaleyl BPTI.

trypsin and DTMI, tryptophan fluorescence *increased*, also becoming constant at about equivalence. The same phenomenon was observed with the native inhibitor.

These changes in tryptophan fluorescence may be due to conformational changes in which the position of charged groups with respect to tryptophan residues is altered. Thus, it is known

Table 3. Trypsin and chymotrypsin inhibition by native and modified BPTI and some model compounds[a]

	Substrate	Type of inhibition; comments	K_I M^{-1}	$-\Delta F°$, Kcal/mole	$-\Delta F°$ loss due to modification, Kcal/mole	Ref.
Trypsin Inhibition						
Native BPTI	BAPA	—	6.0×10^{-11}	13.94	—	[22][b]
TM BPTI	BAPA	—	6.0×10^{-11}	13.94	0	[22][b]
TM homoarg BPTI	BAPA	—	6.0×10^{-11}	13.94	0	[22][b]
TM carbamyl BPTI	BAPA	—	1.0×10^{-7}	9.54	4.40	[22][b]
DTMI	TAME	Non-competitive	6.0×10^{-8}	9.84 ± 0.04	4.10	c
Penta-N-acetyl BPTI	BAPA	—	7.2×10^{-6}	7.01 ± 0.04	6.93	c
Butylamine	BAEE	Competitive	1.7×10^{-3}	3.78	—	[46, 47]
N-ALEE	ALEE	As substrate	$K_m = 2.8 \times 10^{-4}$	4.85	—	[48]
Chymotrypsin Inhibition						
Native BPTI	ATEE	Non-competitive	2.9×10^{-8}	10.28 ± 0.03	—	c
TM carbamyl BPTI	SPANA	—	2.0×10^{-7}	9.10	1.27[d]	[22][b]
DTMI	ATEE	Competitive	3.1×10^{-8}	10.25 ± 0.05	0.03	c
Penta-N-acetyl BPTI	ATEE	Partially mixed	3.5×10^{-8}	10.15 ± 0.03	0.13	c
N-ALME	TEE	No inhibition[e]	—	—	—	c

[a] The following abbreviations, which are not found in the text, are used: TM, tetramaleyl; ALEE, acetyl lysine ethyl ester; ALME, acetyl lysine methyl ester; SPANA, succinyl phenylalanine p-nitroanilide; TEE, tyrosine ethyl ester.
[b] K_I values were determined by the method of GREEN and WORK [28] assuming non-competitive inhibition, and are considered to be approximations.
[c] Present communication.
[d] When calculated from the K_I value for native inhibitor of FRITZ et al. [22] the $\Delta F°$ loss is 0.52 Kcal/mole.
[e] Highest concentration of ALME used was 7×10^{-3} M.

that tryptophan fluorescence is exalted by the carboxylate ion and quenched by the $-NH_3^+$ group and the phenolate ion [50, 51].

While the phenomena described are part and parcel of the inhibitor-enzyme interaction, they do not seem to be connected with the role of lysine 15, since they are the same for native and for dansyl tetramaleyl BPTI. Here again is evidence that groups other than lysine 15 are involved in the interaction with chymotrypsin and trypsin.

In the experiments just described, the addition of chymotrypsin or trypsin to DTMI failed to bring about an increase in the fluorescence of dansyl groups as measured at 500 nm. We hereby retract a statement to the contrary in our preliminary note [39]. It therefore appears that these groups failed to enter the specificity pockets of the enzymes (compare, for example, the anthranoyl enzyme [40]), and were at a distance appreciably greater than R_0* from the nearest donor tryptophan group. This again, is evidence for the involvement of groups other than lysine 15 in the interaction with both enzymes.

* R_0 is the distance between donor and acceptor for which radiationless energy transfer efficiency is 0.5. Assuming a fluorescence quantum yield of 0.2, R_0 for the pair tryptophan-dansyl group is 23Å [52].

Limited Proteolysis at Acid pH*

Acetylation and dansylation experiments indicated that lysine 15 is not essential for chymotrypsin inhibition. We now set out to identify the reactive site towards chymotrypsin by incubating BPTI with chymotrypsin at acid pH. This work is an extension by analogy of experiments first described by Ozawa and LASKOWSKI Jr. [6] in which inhibitor is reacted with catalytic amounts of trypsin, with the ensuing cleavage of a specific peptide bond. Chymotrypsin at acid pH was first used to recognize a reactive site towards chymotrypsin in the BOWMAN-BIRK inhibitor [15].

When BPTI is reacted with 2 mole percent chymotrypsin in the pH range 3.7 to 6.5, and this is followed by incubation with carboxypeptidase A at pH 7.6, there is no loss in inhibitor avtivity. There is also no loss in inhibitor activity with 13 mole percent chymotrypsin at pH 5. Similar negative results have been described for trypsin with BPTI at acid pH [9].

We therefore turned to the work of KRESS and LASKOWSKI Sr. [54] who selectively reduced BPTI at the Cys 14—Cys 38 bond (Figure 1). The reduced inhibitor reoxidizes spontaneously at pH 8. When carboxamidomethylated at Cys 14 and 38, the product loses activity towards chymotrypsin, but retains activity towards trypsin [55]. KRESS and LASKOWSKI, Sr. incubated the dicarboxamidomethyl BPTI with trypsin at pH 5.5, and were able to identify lysine 15 as the reactive site towards trypsin [56]. Their result was in agreement with the finding of CHAUVET and ACHER [19] who used a different procedure.

As dicarboxamidomethyl BPTI is unreactive towards chymotrypsin, it is of no use in incubation experiments with this enzyme. We therefore decided to work with the reduced inhibitor. We found that it possesses a weakened activity towards chymotrypsin which, however, is completely recoverable on reoxidation. Fortunately for incubation experiments at acid pH, we found that between pH 3 and 5, spontaneous reoxidation of the reduced inhibitor is slight, and does not exceed 10 percent in 24 hours. The incubation at acid pH would be followed by a second incubation at or near neutrality, during which reoxidation would take place, making the recovery of activity possible.

Reduced inhibitor was incubated with chymotrypsin at acid pH in aconitate buffer containing $CaCl_2$ for 12 hours. This was followed by a second incubation for 24 hours at pH 7.6, with or without carboxypeptidase A or B. Such experiments, compared with controls which contained no chymotrypsin, resulted in marked loss of inhibitor activity. Figure 12 shows the pH-dependence of activity loss, or modification. The modification is optimal at pH 5.1. Figure 13

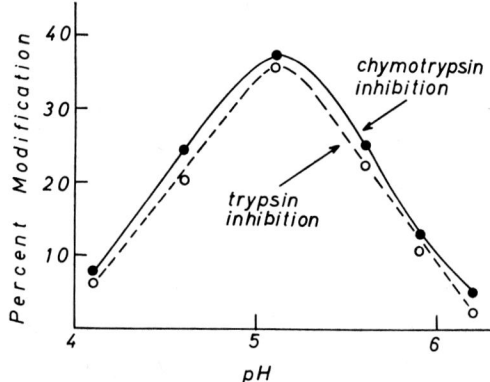

Fig. 12. pH-dependence of modification by chymotrypsin of selectively reduced BPTI.

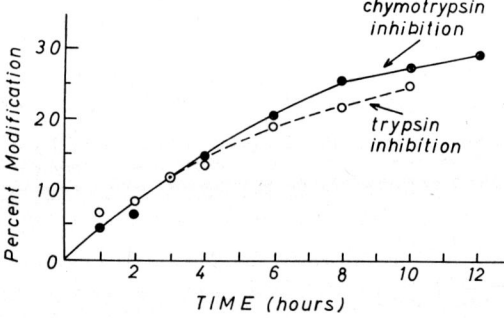

Fig. 13. Time-course of modification by chymotrypsin of selectively reduced BPTI at pH 5.1.

---

* This part of the investigation was carried out in collaboration with SYBIL HERSCHKOPF. A preliminary account of part of the work has been published [53]. In this part of the investigation activities were determined titrimetrically.

shows the time-course of the modification at this pH. The percent modification remained the same whether or not carboxypeptidase A or B was added. It is seen from Figures 12 and 13 that modification by chymotrypsin affects inhibitor activity towards chymotrypsin and trypsin almost equally. A similar activity loss towards both enzymes is obtained when reduced inhibitor is incubated with TLCK-treated chymotrypsin. The modification is therefore not due to contamination by trypsin. Moreover, similar activity losses are obtained when reduced inhibitor is incubated with trypsin at pH 5. The same modification, therefore, almost equally affects both enzymes, and, so it appears, both enzymes bring about the same modification.

In order to identify the site of the modification, while keeping non-selective cleavage at a minimum, the reduced inhibitor was next incubated with 10 mole percent chymotrypsin at pH 5.1 for 7 hours only (Fig. 13). The reaction mixture was lyophilized, the powder was taken up in 0.012N HCl to dissociate inhibitor-chymotrypsin complex, and the inhibitor was separated from chymotrypsin and salts by gel filtration on a Sephadex G-75 column with 0.012N HCl as eluant. The inhibitor fraction was found to contain a mixture of 74.1 percent native, and 25.9 percent modified (inactive) inhibitor. The inhibitor fraction was lyophilized, the powder was taken up in 0.2M ammonium bicarbonate pH 8.5, and aliquots were incubated separately with carboxypeptidase A and B. Carboxypeptidase A yielded no amino acid. Carboxypeptidase B released equivalent amounts of lysine and

Table 4. Basic amino acids released from modified inhibitor by carboxypeptidase B[a]

Amino acid	Micromoles	Amino acid released/total inhibitor molar ratio	Amino acid released/modified inhibitor molar ratio
Lysine	0.060	0.288	1.11
Arginine	0.051	0.245	0.95

[a] Total inhibitor per analyzer column: 0.206 micromoles. Percent modification by the titrimetric chymotrypsin assay: 25.9.

arginine, each of which was equivalent to the percentage modification (Table 4). It appears, therefore, that chymotrypsin splits *two* peptide bonds in the reduced inhibitor, one lysine bond and one arginine bond, for both of which the enzyme is normally not specific. As a result of the cleavages, the activity of the reoxidized inhibitor is lost. It should be emphasized that the incubation mixture of the lyophilized inhibitor fraction with carboxypeptidase A or B contained no chymotrypsin. Thus the cleavage found is due solely to the prior action of chymotrypsin at acid pH.

In order to identify the exposed N-terminal amino acids, the modified inhibitor was dansylated with native BPTI as a control. The products were hydrolyzed in 6N HCl for 4 hours only, to minimize the loss of dansyl groups, and the hydrolyzate was resolved by thin-layer chromatography on silica gel [44]. Commercial dansyl amino acids were used as markers. These included, in addition to ε-dansyl lysine, O-dansyl tyrosine and dansyl arginine (arginine is at the N-terminus of BPTI), all amino acid residues in BPTI with an -NH- group linked to a lysine or arginine carbonyl. It was found that the hydrolyzate of the modified BPTI differed from that of the native inhibitor by the appearance of only one more fluorescent spot, that of dansyl alanine.

One of the two bonds split must therefore be Arg 39—Ala 40, as BPTI contains no other Arg —Ala sequence. As to the lysine-alanine bond, BPTI contains two such sequences, Lys 15—Ala 16, and Lys 26—Ala 27. We believe that the lysine bond split is Lys 15—Ala 16 on the following grounds: (a) tetramaleyl BPTI (all lysines but Lys-15 maleylated) is active towards both chymotrypsin and trypsin, whereas pentamaleyl BPTI is totally inactive towards them. This finding led FRITZ et al. [22] to conclude that the reactive site towards chymotrypsin is in the region of lysine-15; (b) the work of Dr. HUBER (this volume) shows that the lysine-15 residue fits very well into the chymotrypsin pocket.

We are led to the conclusion that Lys 15—Ala 16 *and* Arg 39—Ala 40, symmetrically arranged on either side of the disulfide bridge, are *two* reactive sites towards chymotrypsin. Both Dr. HUBER and Dr. SHOTTON have told me that the two residues cannot be simultaneously accommodated by chymotrypsin. We conclude therefore, that lysine 15, the reactive site towards trypsin, is also the reactive site towards chymotrypsin, and that arginine 39 is another reactive site towards chymotrypsin. This is presumably so for trypsin too. We have no indication as to the order in which these bonds are split.

Discussion and Conclusions

At first sight it seems difficult to reconcile results which show that the lysine 15 side chain is not essential for chymotrypsin inhibition with the conclusion, arrived at from proteolysis experiments, that lysine 15 is indeed at the reactive site towards this enzyme. Just as the enzyme active site consists of "subsites", so the inhibitor reactive site too should be regarded as consisting of several reactive groups or subsites ([6, 57]; for a critical discussion, see ref. [14] section III, I 2). The contributions of these subsites to the free energy of binding are assumed to be additive [57]. Lysine 15 is of prime importance in trypsin inhibition, but the existence of other contributing atoms or groups may be inferred from the standard free energy of binding remaining when lysine 15 is blocked (Table 3). In the case of chymotrypsin, the lysine 15 side chain makes no contribution to the free energy of binding, but it seems reasonable to assume that the susceptible Lys 15—Ala 16 bond, together with other atoms or groups, makes up the total free energy change. As a result of binding at other points, the lysine side chain might be held in the chymotrypsin pocket without contributing to, or detracting from, the free energy.

It is tempting to think that the interaction of BPTI with chymotrypsin involves the same atoms or groups as with trypsin, with the difference that the lysine 15 side chain makes no contribution to the free energy of binding. Deduct $-\Delta F°$ for trypsin with butylamine (the prototype of the lysine side chain), from $-\Delta F°$ for trypsin with BPTI (Table 3), and you have a $-\Delta F°$ of 10.16 Kcal/mole, which is practically $-\Delta F°$ for the interaction of chymotrypsin with BPTI (10.28 Kcal/mole). When lysine 15 is blocked by carbamylation, $-\Delta F°$ with trypsin (9.54 Kcal/mole) is close to the value of BPTI with chymotrypsin. The larger $-\Delta F°$ loss which is obtained when BPTI is acetylated may be due to the poorer fit of the ε-N-acetyl lysyl group in the trypsin pocket as compared with the chymotrypsin pocket.

Nonetheless, there is need for much caution in interpreting the $-\Delta F°$ values. Fluorescence studies with dansyl tetramaleyl BPTI indicate that in the complex with both enzymes, the ε-N-dansyl (lysine 15) group is neither in the specificity pocket nor near it, and must therefore lie elsewhere. Yet DTMI inhibits both enzymes.

Arginine 39, located symmetrically opposite lysine 15 across the disulfide bridge (Fig. 1), is another reactive site towards chymotrypsin. It may be another reactive site towards trypsin as well. The existence of several reactive sites in an apparently single polypeptide chain has been described for chicken ovoinhibitor [58, 58a].

It seems surprising that chymotrypsin *selectively cleaves* reduced BPTI at residues to which it normally has little or no affinity. Chymotrypsin splits some proteins at lysine residues [59], but not selectively. The cleavage by chymotrypsin at arginine residues in proteins has not been reported [59], yet chymotrypsin slowly hydrolyzes benzoyl arginine ethyl ester [60] and polyhomoarginine, a trypsin and chymotrypsin inhibitor [61, 62].

In our work, lysine 15 has been shown to be a reactive subsite towards chymotrypsin, as it is towards trypsin. The reactive sites towards the two enzymes are therefore not distinct. Whether they overlap or are identical remains a moot point with a bias in favour of the latter alternative.

Discussion remarks (added from memory by M. RIGBI):

D. SHOTTON: I am puzzled by the problem of the dansyl group. You say the dansylated tetramaleyl inhibitor is active. I cannot see any way of accommodating such a large group either in or near the pocket, so how could the interaction take place?

N. HUBER: Possibly the reaction could then take place with arginine 39. This could have the ε-N-dansyl (lysyl 15) group on the outside of the molecule, and there would be no problem of accommodating it.

M. RIGBI: With lysine 15 blocked by a large group which cannot be accommodated, such as the dansyl group, arginine 39 would cease to be *another* reactive site, cleaved in succession either before or after the Lys 15—Ala 16 bond, and become the *alternative* site. I would like to ask you Dr. SHOTTON more about the groove you mentioned in chymotrypsin, near the "specificity" pocket. Could it by any chance accommodate arginine 39 while the "pocket" contains the lysine side chain?

D. SHOTTON: The groove would just about contain the Cys 14—Cys 38 disulfide bridge.

References

[1] RHODES, M. B., N. BENNET and R. E. FEENEY, J. Biol. Chem. **235**, 1686 (1960).
[2] RYAN, C. A. and J. J. CLARY, Arch. Biochem. Biophys. **108**, 169 (1964).
[3] HAYNES, R. and R. E. FEENEY, J. Biol. Chem. **242**, 5378 (1967).
[4] BIRK, Y., Ann. N. Y. Acad. Sci. **146**, 388 (1968).
[5] STEVENS, F. C. and R. E. FEENEY, Biochemistry **2**, 1346 (1963).
[6] OZAWA, K. and M. LASKOWSKI, Jr., J. Biol. Chem. **241**, 3955 (1966).
[7] KRESS, L. F. and M. LASKOWSKI, Sr., J. Biol. Chem. **243**, 3548 (1968).
[8] ČECHOVA, D., V. SVESTKOVA, B. KEIL and F. ŠORM, FEBS Letters **4**, 155 (1969).
[9] RIGBI, M. and L. J. GREENE, J. Biol. Chem. **243**, 5457 (1968).
[10] TSCHESCHE, H., Z. physiol. Chem. **348**, 1216 (1967).
[11] HOCHSTRASSER, K., M. MUSS and E. WERLE, Z. physiol. Chem. **348**, 1337 (1967).
[12] HOCHSTRASSER, K. and E. WERLE, Z. physiol. Chem. **350**, 249 (1969).
[13] HOCHSTRASSER, K., K. ILLCHMANN and E. WERLE, Z. physiol. Chem. **350**, 929 (1969).
[14] LASKOWSKI, M., Jr. and R. W. SEALOCK, in The Enzymes, Third Edition, Vol. II, Academic Press, New York (to be published).
[15] BIRK, Y., A. GERTLER and S. KHALEF, Biochim, Biophys. Acta **147**, 402 (1967).
[16] KRAHN, J. and F. C. STEVENS, Biochemistry **9**, 2646 (1970).
[17] WU, F. C. and M. LASKOWSKI, Sr., J. Biol. Chem. **213**, 609 (1955).
[18] KRAUT, H. and N. BHARGAVA, Z. physiol. Chem. **348**, 1500 (1967).
[19] CHAUVET, J. and R. ACHER, J. Biol. Chem. **242**, 4274 (1967).
[20] KASSELL, B. and M. LASKOWSKI, Sr., Biochem. Biophys. Res. Commun. **20**, 463 (1965).
[21] FRITZ, H., E. FINK, M. GEBHARDT, K. HOCHSTRASSER and E. WERLE, Z. physiol. Chem. **350**, 933 (1969).
[22] FRITZ, H., H. SCHULT, R. MEISTER and E. WERLE, Z. physiol. Chem. **350**, 1531 (1969).
[23] SEVILLA, N. and M. RIGBI (to be published).
[24] HUMMEL, B. C. W., Can. J. Biochem. and Physiol. **37**, 1393 (1959).
[25] CUNNINGHAM, L., in M. FLORKIN and E. H. STOTZ (Editors), Comprehensive Biochemistry, Vol. 16, Elsevier Publishing Company, Amsterdam, 1965, p. 98.
[26] ERLANGER, B. F., N. KOKOWSKY and W. COHEN, Arch. Biochem. Biophys. **95**, 271 (1961).
[27] WEBB, J. L., Enzyme and Metabolic Inhibitors, Vol. I, Academic Press, New York and London, 1963, mutual depletion systems, pp. 66—78; graphical methods, pp. 149—173; general kinetics, pp. 55—60.
[28] GREEN, N. M., and E. WORK, Biochem. J. **54**, 347 (1953).
[29] DIXON, M., Biochem. J. **55**, 170 (1953).
[30] MARES-GUIA, M. and E. SHAW, J. Biol. Chem. **240**, 1579 (1965).

[31] SEVILLA, N. and M. RIGBI, Israel J. Chem. 7, 130p (1969).
[32] AVINERI, R., G. BLAUER and M. RIGBI, Israel J. Chem. 1, 199 (1963).
[33] AVINERI-GOLDMAN, R., I. SNIR, G. BLAUER and M. RIGBI, Arch. Biochem. Biophys. 121, 107 (1967).
[34] RIORDAN, J. F. and B. L. VALLEE in C. H. W. HIRS (Editor) Methods in Enzymology, Vol. XI, Academic Press, New York and London, (1967); effect of sodium acetate, p. 571; deacetylation, p. 574.
[35] MELOUN, B., I. FRIČ and F. ŠORM, Eur. J. Biochem. 4, 112 (1968).
[36] SHERMAN, M. P. and B. KASSELL, Biochemistry 7, 3634 (1968).
[37] LAPIDOT, Y., S. RAPPOPORT and Y. WOLMAN, J. Lipid Res. 8, 142 (1967).
[38] RIGBI, M. and M. LOVE, unpublished results.
[39] GAL, A., Y. ELKANA and M. RIGBI, Israel J. Chem. 8, 177p (1970).
[40] HAUGLAND, R. P. and L. STRYER, in G. N. Ramachandran (Editor) Conformation of Biopolymers, Academic Press, New York and London, (1967), p. 321.
[41] STEITZ, T. A., R. HENDERSON and D. M. BLOW, J. Mol. Biol. 46, 337 (1969).
[42] HARTLEY, B. S. and V. MASSEY, Biochim. Biophys. Acta 21, 58 (1956).
[43] CHEN, R. F., Biochem. Biophys. Res. Commun. 40, 1117 (1970).
[44] GROS, G. and B. LABOUESSE, Eur. J. Biochem. 7, 463 (1969).
[45] KASSELL, B., M. RADICEVIC, S. BERLOW, R. J. PEANASKY and M. LASKOWSKI, Sr., J. Biol. Chem. 238, 3274 (1963).
[46] INAGAMI, T. and T. MURACHI, J. Biol. Chem. 238, PC 1905 (1963).
[47] HEIDBERG, J., E. HOLLER and H. HARTMANN, Ber. Bunsenges. physik. Chem. 71, 19 (1967).
[48] GORECKI, M. and Y. SHALITIN, Biochem. Biophys. Res. Commun. 29, 189 (1967).
[49] EDELHOCH, H. and R. F. STEINER, J. Biol. Chem. 240, 2877 (1965).
[50] EDELHOCH, H. and R. F. STEINER, in Electronic Aspects of Biochemistry, Academic Press, New York (1964) p. 7.
[51] EDELHOCH, H., L. BRAND and M. WILCHEK, Biochemistry 6, 547 (1967).
[52] CONRAD, R. H. and L. BRAND, Biochemistry 7, 777 (1968).
[53] HERSCHKOPF, S. and M. RIGBI, Israel J. Chem. 8, 178p (1970).
[54] KRESS, L. F. and M. LASKOWSKI, Sr., J. Biol. Chem. 242, 4925 (1967).
[55] KRESS, L. F., K. A. WILSON and M. LASKOWSKI, Sr., J. Biol. Chem. 243, 1758 (1968).
[56] KRESS, L. F., and M. LASKOWSKI, Sr., J. Biol. Chem. 243, 3548 (1968).
[57] BERGER, A. and I. SCHECHTER, Phil. Trans. Roy. Soc. Lond. B 257, 249 (1967).
[58] TOMAMITSU, Y., J. J. CLARY and J. J. BARTULOVICH, Arch. Biochem. Biophys. 115, 536 (1966).
[58a] DAVIS, J. G., J. C. ZAHNLEY and J. W. DONOVAN, Biochemistry 8, 2044 (1969).
[59] HILL, R. L., Advan. Protein Chem. 20, 37 (1965).
[60] INAGAMI, T. and J. M. STURTEVANT, J. Biol. Chem. 235, 1019 (1960).
[61] RIGBI, M. and L. ELBAZ, Abstracts 6th FEBS Meeting, Madrid, (1969) No. 730.
[62] RIGBI, M. and L. SCHWARTZ, Israel J. Chem. (to be published).

Discussion remarks: **Modification of Soybean Trypsin Inhibitor by α-Chymotrypsin**

M. LASKOWSKI, Jr.

It might be of interest to point out that Miss Ursula DeVonis in our laboratory has now concluded that bovine α-chymotrypsin is inhibited by soybean trypsin inhibitor (KUNITZ) at the same reactive site (Arg^{64}-Ile) as trypsin. Upon incubation of the inhibitor with catalytic quantities of α-chymotrypsin (pretreated with tosyl-lysine-chloroketone to reduce trypsin contamination) the Arg^{64}-Ile bond is hydrolyzed. The kinetics and pH dependence of this hydrolysis differ from hydrolysis by catalytic quantities of trypsin. α-Chymotrypsin is inhibited by both the virgin soybean trypsin inhibitor and by the modified one (the modified inhibitor was prepared with trypsin). However, the inhibition by modified inhibitor is very slow. There is little or no inhibition by des Arg^{64} modified inhibitor. Kinetic control dissociation experiments, designed to prove that the complexes formed from α-chymotrypsin and virgin inhibitor and from α-chymotrypsin and modified inhibitor are the same substance, are now in progress.

The Action of Thermolysin on the Basic Trypsin Inhibitor of Bovine Organs*

BEATRICE KASSELL and TSUN-WEN WANG

Department of Biochemistry, The Medical College of Wisconsin, 561 North Fifteenth Street, Milwaukee, Wisconsin 53233, U.S.A.

The remarkable resistance of the basic pancreatic inhibitor** to enzymic digestion is well known [1, 2]. The large number of enzymes that do not destroy the inhibitory activity are shown in Table 1.

Table 1. Enzymes that do not inactivate the basic pancreatic inhibitor

Pepsin[a,b]	Elastase[b]
Trypsin[a,b]	Collagenase[b]
Chymotrypsin[a,b,c] A & B	Pronase[a,b]
Carboxypeptidase[b,c] A & B	Ficin[b]
Kallikrein[b]	Papain[b]
Plasmin[b]	Bromelain[b]
Leucostoma venom peptidase A[c,d]	
Russell's viper venom protease[b]	
Aspergillus oryzae protease[b]	
Bacillus subtilis proteases[b] A & B	
Streptomyces griseus proteases[c] VI, VII, VIII	
Penicillium notatum proteinase[c,e]	
Silk worm digestive juice alkaline protease[c,f]	

[a] KASSELL and LASKOWSKI, reference [1], [b] KRAUT and BHARGAVA, reference [2], [c] WANG and KASSELL, unpublished, [d] a gift from Dr. J. M. PRESCOTT, Texas A. & M. University, [e] a gift from Dr. WILLIAM E. MARSHALL, General Foods Corporation, [f] a gift from Dr. JUN-ICHIRO MUKAI, Kyushu University.

* Supported by a grant from the National Science Foundation (GB-12630).
** Also known as Kunitz inhibitor, basic pancreatic inhibitor, polyvalent inhibitor or kallikrein inhibitor of bovine organs. Subsequently called "the inhibitor".

It therefore excited our interest when we found that thermolysin was able to digest the inhibitor both in the free form and as the inhibitor-trypsin complex [3]. The present report reviews the digestion of the free inhibitor and describes the identification and inhibitory activity of some of the products obtained. Their positions have been located in the known sequence [4, 5] of the inhibitor and in the 3-dimensional structure based on x-ray crystallography [6, 7].

Materials, Methods and Controls — The inhibitor was the same as previously described [8]. Thermolysin, an enzyme from Bacillus Thermoproteolyticus ROKKO [9, 10], was purchased from Calbiochem, Los Angeles, Calif. Bovine trypsin was a gift from Novo Industries, Copenhagen. N-Benzoyl-DL-arginine-p-nitroanilide (BAPA) was purchased from Nutritional Biochemicals Corp., Cleveland, Ohio.

Trypsin inhibiting activity was measured at 37.6° by the method of ERLANGER et al. [11] with preincubation of the inhibitor and trypsin for 5 minutes before the addition of the BAPA substrate. Control experiments showed that thermolysin did not interfere with this assay.

Peptide bond hydrolysis was measured by increase in ninhydrin-positive material, using the quantitative ninhydrin method of ROSEN [12] and by paper electrophoretic demonstration of newly formed ninhydrin-positive spots.

Changes in the activity of the thermolysin preparation under the experimental conditions were measured by its ability to digest casein [10]. In tests of the stability of thermolysin at 80°, we have confirmed MATSUBARA's results [10] which showed retention of 50% of activity after 2 hours. A 72-fold molar excess of the inhibitor did not affect the activity of thermolysin.

The digestion products were separated on Sephadex G-25. Pooled portions of the peaks were further purified in some cases on Sephadex G-50 or G-10.

Amino acid analysis was carried out as described by Moore and Stein [13]. Hydrolysis for 48 hours was used to be certain of differentiating between one or two moles of isoleucine in the peptides.

Results

Demonstration of digestion — Figures 1 and 2 show the correspondence between loss of inhibiting activity and increase in ninhydrin-positive digestion products. Exposure to the high temperature alone caused little change in either measurement. In the presence of thermolysin, in 30 mi-

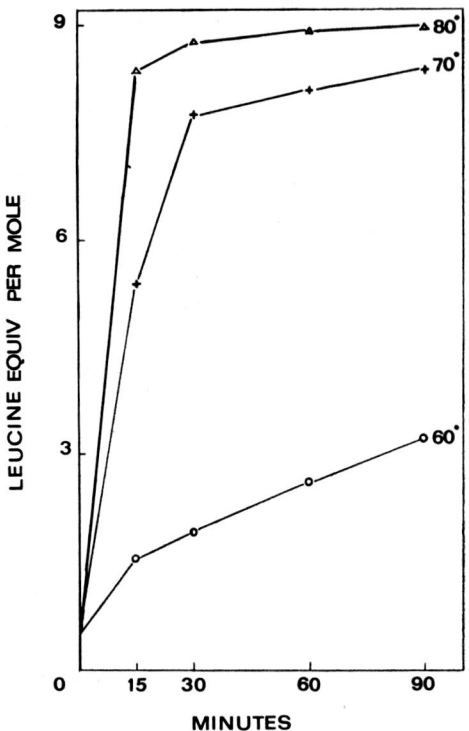

Fig. 2. Increase in ninhydrin color during incubation of inhibitor with thermolysin. Digest: 120 μg of inhibitor and 40 μg of thermolysin in 2 m*l* of 50 mM borate buffer, pH 8.0. The values are corrected for the color due to thermolysin alone and inhibitor alone determined at each time. There was little change in these blanks with time.

nutes at 80°, there was almost complete loss of activity accompanied by formation of new amino groups corresponding to 9 equivalents of leucine per mole. Cessation of digestion at 30 minutes was not due to destruction of thermolysin (see above). At the lower temperatures, both methods show that digestion still occurred, but was slower. Electrophoresis of the digest demonstrated products not corresponding either to the inhibitor or to thermolysin (Fig. 3).

Thermal transition of the inhibitor — To determine whether the digestion was due to unfolding of the inhibitor at these high temperatures, we studied the thermal transition. The transition was determined at pH 8 with a Gilford spectrophotometer equipped with thermospacers and an adjustable water bath. A separate portion of

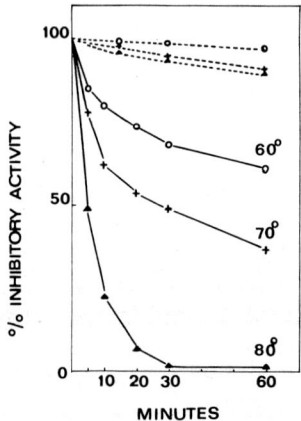

Fig. 1. Loss of trypsin-inhibiting activity during incubation of inhibitor with thermolysin. Digests (solid lines): 6 μg of inhibitor and 1 μg of thermolysin in 1 m*l* of 50 mM borate buffer, pH 8.0. Controls (dotted lines): 6 μg of inhibitor in the same buffer. After incubation for the times shown, the tubes were placed in an ice bath until assayed.

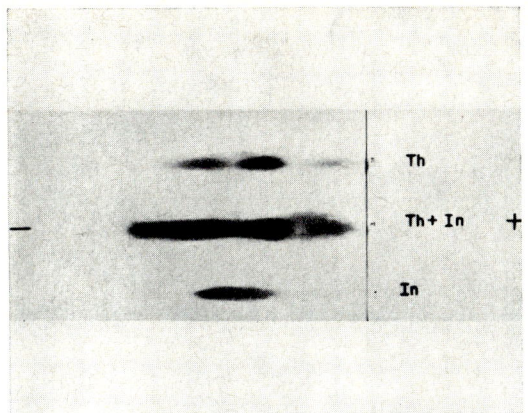

Fig. 3. Paper electrophoresis of the thermolysin digest of the inhibitor. Th: 72 µg of thermolysin. Th + In: 72 µg of thermolysin and 223 µg of inhibitor. In: 223 µg of inhibitor. Digestion: 70°, 1 hour. Whatman 3 MM paper, 0.5M pyridine acetate buffer, pH 6.4, 2 hours at about 75 volts/cm. Color was developed with ninhydrin.

the solution was incubated for each temperature and readings were made at 20 and 30 minutes at 293 nm (the wave length giving maximum change in absorbance). The control was a cold portion of the same solution placed in the cuvette holder only momentarily for the reading. After the readings, the solutions were rapidly cooled and kept at 4° overnight to test reversal. Figure 4 shows that there was a thermal change with a midpoint at 56°. The greater part of the change was reversible, even at 80°, in agreement with retention of most of the inhibiting activity in the control solutions of Figure 1. The increase in absorbance was very small, indicating a minimal conformational change.

Identification of the digestion products — In order to understand how thermolysin first attacks the inhibitor, mild conditions for digestion were chosen, namely 60° for one hour. A solution of thermolysin (400 µg in 10 ml of 50 mM sodium borate buffer containing 2 mM CaCl$_2$) was added at 60° to 50 mg of solid inhibitor. After one hour, the digest was frozen and lyophilized. The residue, dissolved in 2 ml of water, was applied to a column of Sephadex G-25. The elution pattern shown in Figure 5 was obtained.

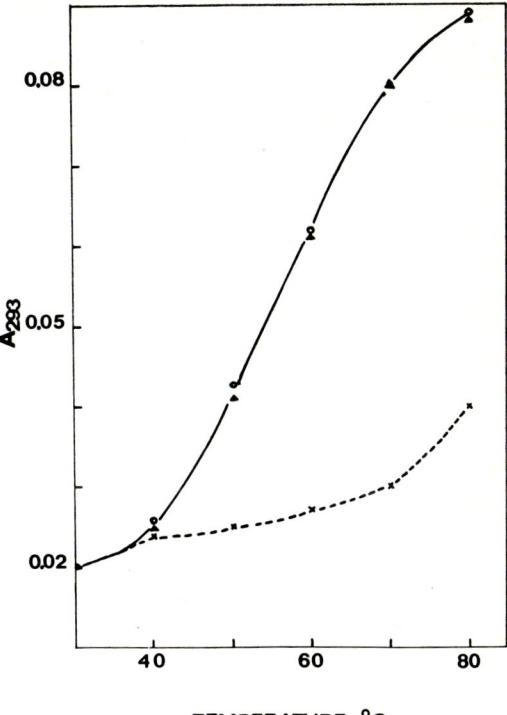

Fig. 4. Change in absorbance at 293 nm with temperature. Inhibitor: 0.2 mM solution in 50 mM borate buffer, pH 8.0. For procedure, see text. Solid line: incubation for 20 (△) and 30 (o) minutes at each temperature. Dotted line: after reversal overnight at 4° (X).

The fractions, pooled as indicated in Figure 5, were subjected to amino acid analysis. Fractions 1 and 2, comprising the ascending and descending portions of the main peak based on absorption at 280 nm, and accounting for about 90% of the protein, did not differ from the starting inhibitor in composition. Fraction 3 is the tail portion of the large peak and comprises an area in which the ratio of A_{220}/A_{280} was higher. The composition of fraction 3 is shown in Table 2 in comparison to the total composition of the inhibitor. Inspection of the sequence shown in Figure 6 locates the 8 missing amino acids in two tetrapeptides: Ileu-Arg-Tyr-Phe (19—22); Ala-Lys-Arg-Asn (40—43).

Other peptides identified are shown in Table 3. The two peptides of peak 6 contain the amino acids of the tetrapeptides missing from peptide 3,

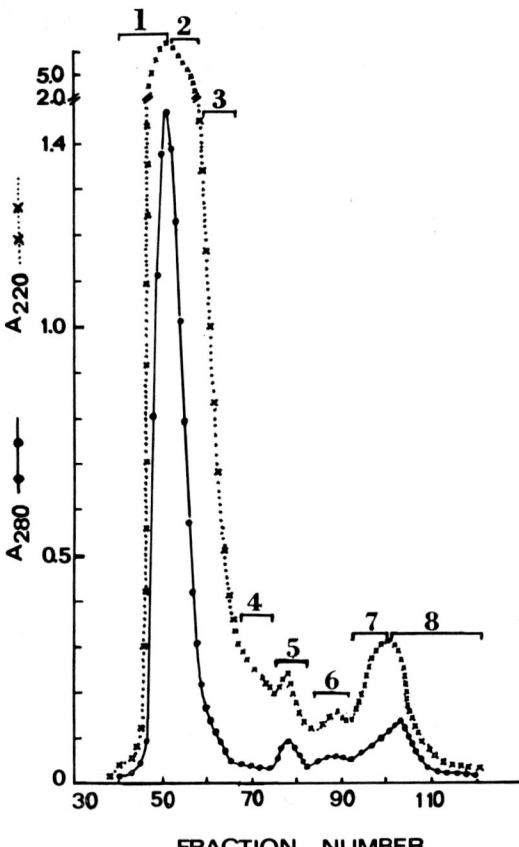

Fig. 5. Separation of the products after digestion of the inhibitor (50 mg) with thermolysin (400 μg) at 60° for 1 hour. Column: Sephadex G-25, medium, 2 × 94 cm. Eluent: 0.1M NH$_4$HCO$_3$, pH 7.0. Fractions 3 ml, flow rate 12 ml/hr.

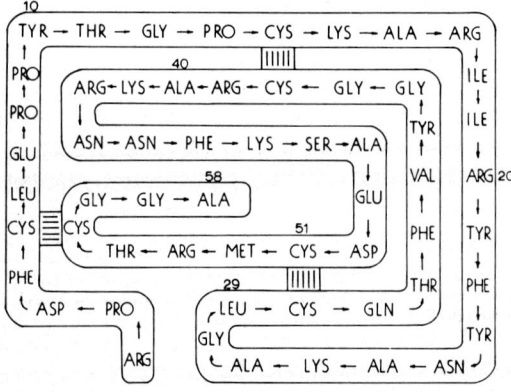

Fig. 6. Primary structure of the inhibitor (from references [4] and [5]).

Table 2. Comparison of the Amino Acid Composition of Fraction 3 (Fig. 5) to the known Composition of the Inhibitor

Amino Acid	Residues/mole		
	Inhibitor	Fraction 3[a]	Missing[c]
Lys	4	3.05	0.95
Arg	6	4.35	1.65
Asp	5	4.01	0.99
Thr	3	2.80	
Ser	1	0.64	
Glu	3	2.93	
Pro	4	3.82	
Gly	6	6.00	
Ala	6	4.99	1.01
1/2 Cys	6	4.62[b]	
Val	1	1.12	
Met	1	0.64	
Ileu	2	1.01	0.99
Leu	2	2.08	
Tyr	4	3.13	0.87
Phe	4	2.84	1.16

[a] Uncorrected values of a 48 hour hydrolysate, [b] The loss of cystine is attributed to the prolonged hydrolysis, [c] The 8 missing residues correspond to positions 19—22 (Ile-Arg-Tyr-Phe) and 40—43 (Ala-Lys-Arg-Asn).

except that asparagine 43 has apparently been removed. Peak 7 contains an overlap peptide of peptide 3, showing that not all of the initial digestion proceeded through peptide 3.

Peptide 3 (Tab. 2), the largest of the degradation products isolated, had four bonds split and therefore does not define the position at which thermolysin first attacked the inhibitor. We investigated the possibility that there were cleavages in the main peak of Figure 5, with the pieces being held together by disulfide linkages. Fraction 2 of Figure 5 was used for this experiment and was first further purified by passage through Sephadex G-50. A small amount of impurity separated, peak B (Fig. 7). Performic acid oxidation [14] was used to break the disulfide bonds and oxidized peak A was applied to a column of Sephadex G-25. The pattern of Figure 8 was obtained. All but a trace of the absorbance at 280 nm was in peak A. Amino

Table 3. Structure of the Additional Peptides derived from Fig. 5

Peak[a]	Structure and location in the sequence
5	12　　　14　　　　17 Gly-Pro-Cys-Lys-Ala-Arg 　　　　　　　｜ 33　　　　　　｜38　　　　44 Phe-Val-Tyr-Gly-Gly-Cys-Arg-Ala-Lys-Arg-Asn-Asn
6a	19　　　　22 Ile-Arg-Tyr-Phe
6b	40　　42　　15　　17 Ala-Lys-Arg (or Lys-Ala-Arg)[b]
7	16　　　　　　　　　　24 Ala-Arg-Ile-Ile-Arg-Tyr-Phe-Tyr-Asn
8	22　　24 Phe-Tyr-Asn

[a] Peak 4 could not be identified. It contained a complex mixture in very low yield.
[b] The second peptide (15—17) is less likely; thermolysin is not known to hydrolyze on the amino side of a lysine residue.

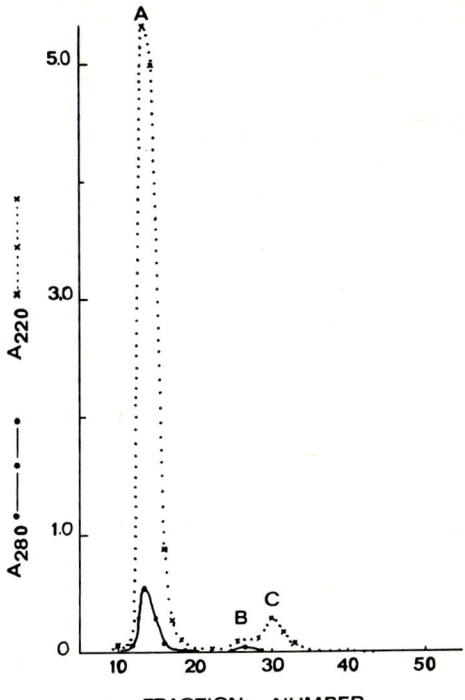

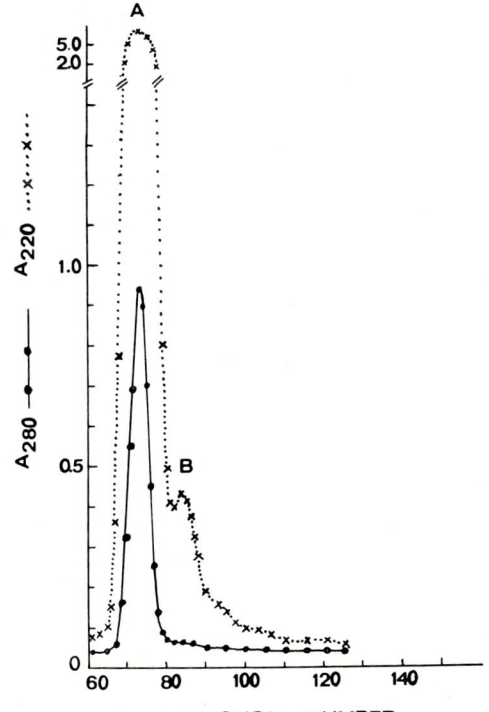

Fig. 7. Purification of fraction 2 of Fig. 5 on Sephadex G-50, fine. Colum 2 × 97 cm. Other conditions as in Fig. 5.

Fig. 8. Separation of performic acid-oxidized peak A of Fig. 7 on Sephadex G-25. Same column and conditions as in Fig. 5.

acid analysis of this peak gave the correct total composition for *undigested* oxidized inhibitor; no residues were missing. Analysis of peaks B and C of Figure 8 yielded only very small quantities of amino acids not corresponding to any portion of the inhibitor; they were probably derived from some impurity still present in the material that was used for oxidation. We concluded that the original digest did not contain any digestion product between the inhibitor itself and peptide 3 with four bonds split.

Inhibiting activity of the digestion products — The larger peptides of Figure 5 were tested for activity against trypsin and chymotrypsin. The main peak (1 and 2) showed slightly less activity than native inhibitor; this was expected from the unchanged composition and the fact that the solution had been heated at 60° for one hour. Peptide 3 (Tab. 2), an inhibitor derivative modified by the removal of eight amino acids, still inhibited both trypsin and chymotrypsin. Peptide 5 (Tab. 3) showed little or no activity against both enzymes.

Table 4. Bonds hydrolyzed during Digestion of the Inhibitor by Thermolysin at 60°

Bond	Location	Bond	Location
Thr-Gly	11—12	Asn-Ala	24—25
Lys-Ala	15—16	Thr-Phe	32—33
Arg-Ile	17—18	Arg-Ala	39—40*
Ile-Ile	18—19*	Arg-Asn	42—43
Tyr-Phe	21—22	Asn-Asn	43—44*
Phe-Tyr	22—23*	Asn-Phe	44—45

* Bonds hydrolyzed to form peptide 3 (Table 2), the largest degradation product isolated.

Discussion

The process of digestion of the basic trypsin inhibitor by thermolysin has been partially elucidated. The bonds hydrolyzed are shown in Table 4. All points of digestion are in agreement with the known specificity of thermolysin based on the studies of MATSUBARA et al. [15, 16] and of AMBLER and MEADWAY [17].

We cannot say at this time which single bond is the first point of attack by thermolysin. Although the conditions of digestion were so mild that more than 90% of the inhibitor remained undigested (the first peak of Figure 5), the largest degraded product, peptide 3, already had four bonds hydrolyzed; these are marked with asteriks in Table 4. In the 3-dimensional structure of the inhibitor, determined by x-ray crystallography [6, 7], three of these bonds are located in exposed positions in the loops forming the upper section of the pearshaped-molecule (see this volume, p. 56). The fourth bond, Phe-Tyr 22—23, is buried [6, 18].

Thermolysin digestion produced a modified inhibitor (peptide 3) that retained activity against trypsin and chymotrypsin. Thus the eight amino acids removed from this derivative (Table 2) are not essential for inhibitory activity. Peptide 5 (Table 3) showed little or no activity against both enzymes, although it contains lysine 15 and the adjacent disulfide bond. The available amounts of these peptides were not sufficient to characterize fully their inhibitory properties and further work is required.

Summary

The basic trypsin inhibitor of bovine organs is extensively digested by thermolysin at 70—80°. At 60° less digestion occurs and large and small digestion products of the inhibitor have been isolated and located in the known structure. One of these products lacking two tetrapeptides, Ile-Arg-Tyr-Phe (19—22) and Ala-Lys-Arg-Asn (40—43) is an inhibitor of trypsin and chymotrypsin.

References

[1] KASSELL, B. and M. LASKOWSKI, Sr., J. Biol. Chem. **219**, 203 (1956); Fed. Proc. **24**, 593 (1965).
[2] KRAUT, H. and N. BHARGAVA, Z. Physiol. Chem. **334**, 236 (1963); **348**, 1948 (1967).
[3] WANG, T.-W. and B. KASSELL, Biochem. Biophys. Res. Commun. **40**, 1039 (1970).
[4] KASSELL, B. and M. LASKOWSKI, Sr., Biochem. Biophys. Res. Commun. **20**, 463 (1965).
[5] ANDERER, F. A. and S. HÖRNLE, J. Biol. Chem. **241**, 1568 (1966).
[6] HUBER, R., D. KUKLA, A. RÜHLMANN, O. EPP and H. FORMANEK, Naturwissenschaften **57**, 389 (1970).
[7] HUBER, R., D. KUKLA, A. RÜHLMANN and W. STEIGEMANN, this volume, p. 56.
[8] KASSELL, B., M. RADICEVIC, S. BERLOW, R. J. PEANASKY and M. LASKOWSKI, Sr., J. Biol. Chem. **238**, 3274 (1963).
[9] ENDO, S., J. Fermentation Tech. **40**, 346 (1962).
[10] MATSUBARA, H., Publ. Amer. Assoc. Advan. Sci., No. 84 (1967) p. 283.
[11] ERLANGER, B. F., N. KOKOWSKY and W. COHEN, Arch. Biochem. Biophys. **95**, 271 (1961).
[12] ROSEN, H., Arch. Biochem. Biophys. **67**, 10 (1957).
[13] MOORE, S. and W. H. STEIN, Methods in Enzymology **6**, 819 (1963).
[14] HIRS, C. H. W., Methods in Enzymology **11**, 197 (1967).
[15] MATSUBARA, H., A. SINGER, R. SASAKI and T. H. JUKES, Biochem. Biophys. Res. Commun. **21**, 242 (1965).
[16] MATSUBARA, H., R. SASAKI, A. SINGER and T. H. JUKES, Arch. Biochem. Biophys. **115**, 324 (1966).
[17] AMBLER, R. P. and R. J. MEADWAY, Biochem. J. **108**, 893 (1968).
[18] SHERMAN, M. P. and B. KASSELL, Biochemistry **7**, 3634 (1968).

On the Mechanism of the Interaction between Basic Pancreatic Trypsin Inhibitor and Trypsin

V. Keil-Dlouhá, J.-M. Imhoff and B. Keil

Institut de Chimie des Substances Naturelles, C.N.R.S., Gif-sur-Yvette, France

Basic pancreatic trypsin inhibitor (BPTI)* was the first of the large family of naturally occurring protease inhibitors the primary structure of which was entirely elucidated in the laboratories of Acher (Chauvet et al., [1]), Kassel [2] and in our laboratory [3].

An intense study has been undertaken by means of chemical modifications to elucidate the relative importance of individual groupings within the inhibitor molecule for complex formation with trypsin. Our recent interest was concentrated on the trypsin moiety in the enzyme-inhibitor complex.

Trypsin possesses a very narrow specificity. It is generally assumed that this narrow specificity is structurally due to an "anionic site", to which substrates or small inhibitors, like benzamidine, bind electrostatically through their positive charge, and a hydrophobic binding site, a crevice which is responsible for the binding of the carbon side chain of the substrate molecule; the two sites are sterically close.

The purification of *β-trypsin and its derivatives*, as well as the elucidation of their structural relationships [4, 5], created a new impetus for studies of their active centers. Fig. 1 represents schematically the polypeptide chain of β-trypsin (II) and of its active derivatives. Cleavage in β-trypsin of the bond Lys 131 — Ser 132 leads to the formation of α-trypsin (III). This new form differs in the enzymatic properties from intact β-trypsin (e. g. the activity towards N-α-benzoyl arginine p-nitroanilide is in the case of α-trypsin 50% lower). The differences of the esterase activity in TLCK-derivatives of α- and β-trypsin and in TLCK-TPCK-derivatives of β-trypsin indicate the presence of an independent "chymotryptic" site in the active center of β-trypsin. This "chymotryptic" site seems to be absent in α-trypsin [6].

A further degradation of α-trypsin at the bond Lys 176 — Asp 177 yields another still active derivative, ψ-trypsin (IV). This differs markedly from α- and β-trypsin. The opening of the bond Lys 176 — Asp 177 generates an additional free amino group on the residue Asp 177 which in both α- and β-trypsin binds positively charged substrates. This structural change in ψ-trypsin results in a complete loss of the amidase activity. Similarly ψ-trypsin has lost, in contrast to the α- and β-forms, the ability to be specifically alkylated by TLCK at the residue His 46. The reaction of ψ-trypsin with the active center titrant, p-nitrophenyl-p-guanidino benzoate (NPGB), is slow, but after 20 minutes the extent

---

* **Abbreviations:** BPTI, basic pancreatic trypsin inhibitor; TLCK, tosyl-lysine chloromethyl ketone; TPCK, tosyl-phenyl-alanine chlormethyl ketone; NPGB, p-nitrophenyl-p-guanidino benzoate; Ac-Tyr-OEt, acetyl tyrosine ethyl ester; Bz-Arg-OEt, benzoyl arginine ethyl ester; TCA, trichloro acetic acid; TR, trypsin.

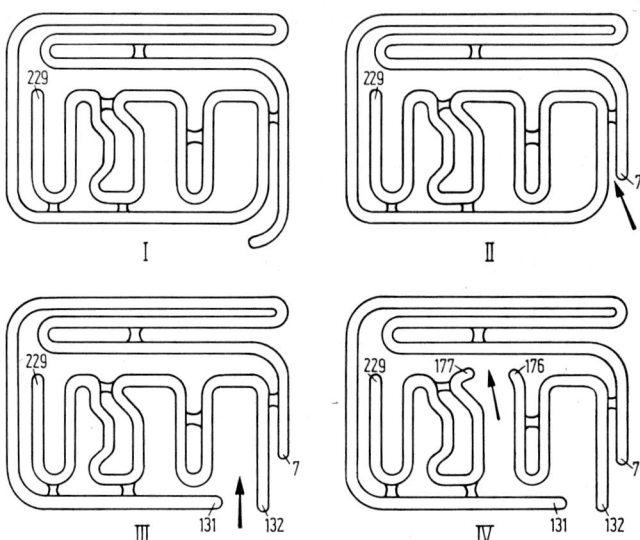

Fig. 1. Conversion of trypsinogen(I) to β-trypsin(II), α-trypsin(III) and ψ-trypsin (pseudotrypsin) (IV).

of acyl complex formation is equal to that of α-trypsin [5].

Our studies of the enzyme-inhibitor interactions were undertaken with purified α-, β- and ψ-derivatives of trypsin [4, 5]. The content of active centers found by NPGB titration was 90% for the α-form and 82% for the β-form.

First we reinvestigated the *values of the dissociation constants* for α- and β-trypsin using NPGB as active center titrant. When NPGB was added to α- and β-trypsin, the initial burst of p-nitrophenol released was reduced in proportion to the concentration of BPTI. The extrapolated ratio for 100% inhibition is 1.07 moles of BPTI/mole of active β-trypsin. The dissociation constant for BPTI-β-trypsin was too low to be evaluated under our conditions (see also Fig. 2), the value

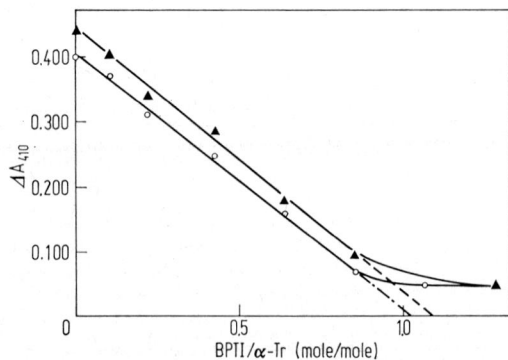

Fig. 2. Inhibition of the β-trypsin-NPGB reaction by BPTI in 0.05M Tris 0.02M CaCl$_2$ buffer pH 8.3 (0—0) in 0.05M Tris 0.02M CaCl$_2$ buffer pH 8.3, 40% ethanol (▲—▲).

Fig. 3. Inhibition of the α-trypsin-NPGB reaction by BPTI. Conditions were as in Fig. 2.

calculated for BPTI-α-trypsin complex from the data presented in Fig. 3 is 5.25×10^{-7}M.

The value of the dissociation constant for the complex of α-trypsin, which was calculated on the basis of active enzyme concentration, is clearly higher than that of β-trypsin as well as that reported for crystalline trypsin (10^{-10}—10^{-12}M). This difference could be explained by the heterogenous character of crystalline trypsin which contains several components interacting differently with BPTI.

During the interaction with trypsin, a slow replacement of the pancreatic inhibitor by NPGB takes place. NPGB forms a very stable covalent acyl-complex with the hydrolytic site of trypsin, and it is simultaneously bound by an electrostatic bond to the residue Asp 177, which is the substrate binding site. The difference in the dissociation constants obtained with NPGB is related to the different stability of bonds between the inhibitor and the active sites of the α- and β-forms of trypsin. Although the obtained difference in the dissociation constants can be considered for several reasons only as an approximation, it nevertheless indicates that the cleavage of one bond between Lys 131 and Ser 132 in the trypsin molecule slightly decreases the stability of the complex.

For studies of *hydrophobic interactions*, a medium composed of 40% ethanol and 0.05M Tris buffer pH 8.3 containing 0.02M $CaCl_2$ was used. In the cases of α- and β-trypsin ethanol was practically without effect (Figs. 2 and 3). In the case of ψ-trypsin, the NPGB-reaction cannot be exploited to the same extent as for α- and β-forms. Whereas both α- and β-trypsin are rapidly acylated by small excess of NPGB, simultaneously releasing an equivalent amount of p-nitrophenol, the reaction of ψ-trypsin with NPGB proceeds slowly. An immediate "burst" of p-nitrophenol will not be observed [5]. From the shape of the curve, which we have obtained as a result of the reaction between our sample of ψ-trypsin and NPGB, no quantitative conclusion can be drawn for the active site (Fig. 4-II). Nevertheless a clear difference was observed when BPTI was present in the system. No p-

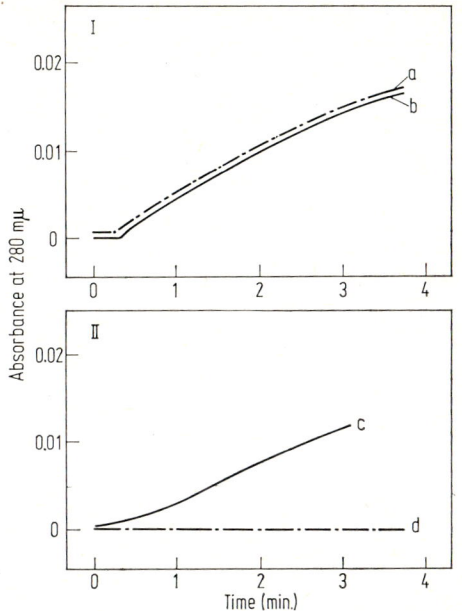

Fig. 4. ψ-trypsin — NPGB reaction with (—·—·) and without (———) BPTI
in: I) 0.05M Tris 0.02M $CaCl_2$ buffer pH 8.3, 40% ethanol,
II) 0.05M Tris 0.02M $CaCl_2$ buffer pH 8.3.

nitrophenol was released at all, which seems to witness a specific interaction between ψ-trypsin and BPTI. The study of the same reaction in the presence of 40% ethanol (Fig. 4-I) was invalidated by the denaturing effect of ethanol on ψ-trypsin. The raise of absorbance due to progressive precipitation was observed even in the absence of NPGB.

Results summarized in Table 1 show that in aqueous Tris buffer solution α- and β-trypsin are equally active towards benzoyl arginine ethyl ester (Bz-Arg-OEt). On the other hand, β-trypsin possesses higher activity towards acetyl tyrosine ethyl ester (Ac-Tyr-OEt) than the α-derivative, in which the secondary "chymotryptic" site is lost.

The affinity of ψ-trypsin for Bz-Arg-OEt and Ac-Tyr-OEt is 1.5—3% of that shown by β-trypsin. Complete inhibition of the hydrolytic activity towards both ester substrates in the presence of BPTI in Tris buffer was found for α-, β- and ψ-trypsin.

Table 1. Influence of BPTI on esterase activity of trypsin and its derivatives

Trypsin derivatives	Molar ratio BPTI/Trypsin	TRIS		TRIS-Ethanol 40%	
		Bz-Ar-OEt	Ac-Tyr-OEt	Bz-Arg-OEt	Ac-Tyr-OEt
		Specific activity in μeq/min/mg			
β—	0	74	42.5	134	7.0
	4	0	0	0	0
α—	0	72	29.2	116	5.8
	4	0	0	0	0
ψ—	0	2.3	0.6	—	—
	4	0	0	—	—

Substantial difference was observed using the system 40% ethanol — 0.1M Tris solution. As is well known [7], the free energy of hydrophobic interaction is significantly reduced in 40% ethanol. The results of Table 1 show that in the presence of ethanol in the reaction medium, enhancement of β-trypsin activity towards Bz-Arg-OEt is doubled to that in aqueous Tris solution. On the other hand, the cleavage of a tyrosine ester substrate is reduced. Inhibitor blocks entirely the activity towards both substrates.

In the case of α-trypsin, the activity towards both ester substrates is changed by the presence of ethanol analogically. In the presence of BPTI, however, even this reduced esterase activity was completely lost.

To corroborate the conclusions drawn from the activity measurements a different approach using chromatographic separation was used to prove the interaction between the different forms of trypsin and BPTI. On the columns of Sephadex G 75 or G 100 the complex or the free trypsin moiety will be readily separated from the inhibitor. After trichloroacetic acid treatment of the high molecular fraction the inhibitor which has been bound to the trypsin moiety can be assayed by usual methods. This technique has been used in our laboratory previously to study the interaction between BPTI and trypsinogen. A 1:1 ratio complex formation has been proved at that time.

Gel filtration of a mixture of BPTI with ψ-trypsin (in a molar ratio of 2:1) in Tris buffer (Fig. 5a) shows two peaks of inhibitor activity. The first one corresponds to BPTI bound to ψ-trypsin, while the second represents the excess of BPTI. Gel filtration of the same mixture in ethanol-Tris buffer shows (Fig. 5b) that in this case all the inhibitory activity is concentrated in the second peak. The susceptibility of ψ-trypsin to denaturate by ethanol was mentioned. Accordingly no interaction with BPTI took place. On the other hand, a complex formation of β-trypsin and the inhibitor could be observed under similar conditions (Fig. 6), which is in agreement with the results presented in Table 1. From the complexing ability in aqueous solutions of ψ-trypsin which is devoid of the anionic binding site and on the other hand from the same ability of β-trypsin under conditions which labilize hydrophobic bonds we can deduce that at least two independent types of bonds are important for the specific complex formation.

CHAUVET and ACHER have shown [8] that Lys 15 of BPTI is responsible for the inhibition of trypsin activity. In the earlier work of Green [9] there were certain indications that the mechanism of interaction of trypsin with BPTI could be explained by formation of ionic bonds.

From all these data we can assume with confidence that electrostatic bond is involved in the formation of the complex with BPTI. This bond is between a basic group of BPTI, Lys 15, and Asp 177 of trypsin which is the anionic binding site of the enzyme. The results available support

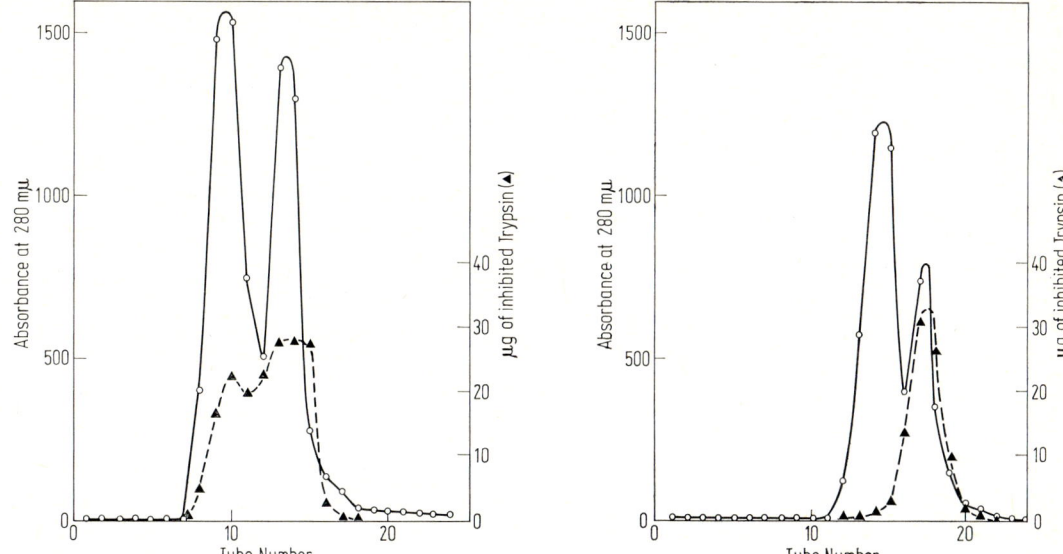

Fig. 5a. Gel filtration of BPTI-ψ-trypsin complex on Sephadex G-75 (column 1.0×45 cm) in 0.05M Tris 0.02M CaCl$_2$ buffer pH 8.3 at room temperature. The flow rate was 1.5 m*l* per hour; 0.5 m*l* fractions were collected. The inhibitory activity was determined as the difference between the tryptic activity of a standard and the tryptic activity of the assayed fraction, and then calculated per μg of inhibited trypsin with the aid of a standard curve.

Fig. 5b. Gel filtration of BPTI-ψ-trypsin on Sephadex G-100 (column 1.0×45 cm) in 0.05M Tris 0.02M CaCl$_2$ buffer pH 8.3, 40% ethanol at room temperature. Other conditions were as in Fig. 5a.

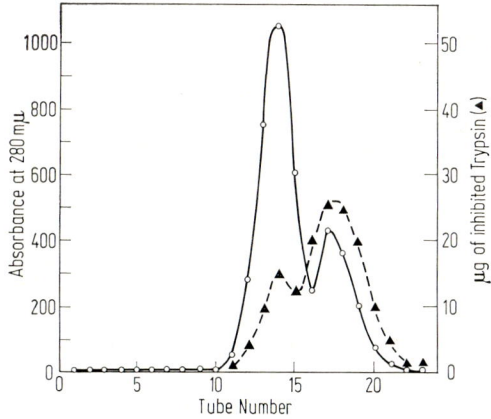

Fig. 6. Gel filtration of BPTI-β-trypsin complex on Sephadex G-100 (column 1.0×45 cm) in 0.05M Tris 0.02M CaCl$_2$ buffer pH 8.3, 40% ethanol at room temperature. Remaining conditions as in Fig. 5a.

the view that also the hydrophobic interactions play an important role in complex formation.

An independent series of experiments, undertaken with the pancreatic inhibitor and specifically alkylated derivatives of α- and β-trypsin has led us to the conclusion that *a "chymotryptic" site exists in the active center of β-trypsin* [6].

Fig. 7 represents the hypothetical relative positions of certain residues in the area of the active center of trypsin. This scheme was derived from the analogy in the primary structures of trypsin and chymotrypsin and from the three-dimensional structure of chymotrypsin [10]. According to this working hypothesis, His 29 occurs in trypsin in the area of its active center; Ser 183 is between His 46 and His 29; Trp 127 (fourth residue from Lys 131) is in close proximity to His 29.

From the data already available and from our new experimental results [6], the following conclusions can be drawn:

1. Cleavage of the bond Lys 131 — Ser 132 in α-trypsin diminishes the ability of the enzyme

for the interaction with BPTI, compared to β-trypsin.
2. Specific alkylation of His 46 by TLCK inhibits the "primary" active site, which is composed of His 46, Ser 183 and Asp 177.

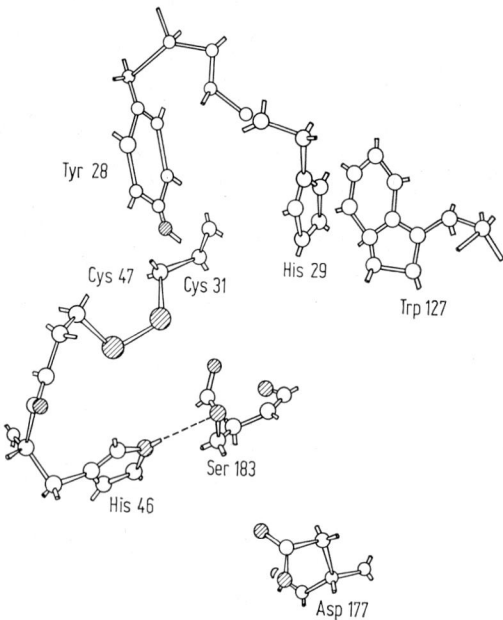

Fig. 7. The relative positions of certain residues in the area of the trypsin active center.

3. Additional alkylation by TPCK probably of a second histidine (His 29) inhibits the "secondary", "chymotryptic" site of β-trypsin.
4. Blocking of both histidine residues (46, and probably 29) destroys the esterase activity completely.
5. Alkylation of His 46 in α-trypsin is followed by complete loss of esterase activity which indicates that in α-trypsin the "secondary", "chymotryptic" site does not exist.

The premise of a "chymotryptic" site in the trypsin molecule helps to explain the participation of hydrophobic forces in the interaction between trypsin and the pancreatic inhibitor. Comparison of the profiles presented in Figures 8a and 8b shows that TLCK-β-trypsin forms a complex with BPTI in Tris buffer as well as in Tris-40% ethanol, i. e. alkylation of His 46 with TLCK does not prevent the interaction of β-trypsin with pancreatic inhibitor. The results obtained by chromatography seem to clearly exclude the idea, that the interaction with the inhibitor moiety is due to the presence of traces of unsubstituted active β-trypsin.

In contrast to β-trypsin, the substitution of α-trypsin with TLCK lead to a complete loss of esterase activity (Tab. 2). The gel filtration of a mixture of TLCK-α-trypsin and 4 fold molar excess of BPTI (Fig. 9a and 9b) showed that no complex formation takes place between TLCK-α-trypsin and the pancreatic inhibitor. All inhibitory activity is concentrated in the second peak, which corresponds to the excess of free inhibitor.

As is well known, α-trypsin differs from the β-form in the cleavage of one bond between Lys 131 and Ser 132. From all the column separations we assume that this particular cleavage destroys the "chymotryptic" active site of the trypsin molecule. It has been shown also that only one electrostatic bond exists between Asp 177 of trypsin and Lys 15 of the inhibitor and that the other interactions are due to hydrophobic forces. Alkylation of His 46 in α- and β-trypsin with TLCK leads to the same chemical modification resulting in the inactivation of the specific tryptic active site which is resposible for cleavage of positively charged substrates. On the other hand, the difference between α- and β-forms consists in their "chymotryptic" active site. The complex formation with pancreatic inhibitor was observed only in the case of TLCK-β-trypsin, the "chymotryptic" active site of which is not affected. In contrast to the β-form, TLCK-α-trypsin has completely lost its affinity for BPTI as a result of the disarrangement of its "chymotryptic" active site.

The substitution of TLCK-β-trypsin with TPCK brought about the same change of enzymatic properties as in the case of TLCK-α-trypsin. As can be seen from Table 2, TLCK-TPCK-β-trypsin lost the residual esterase activity. The results presented in Fig. 10a and

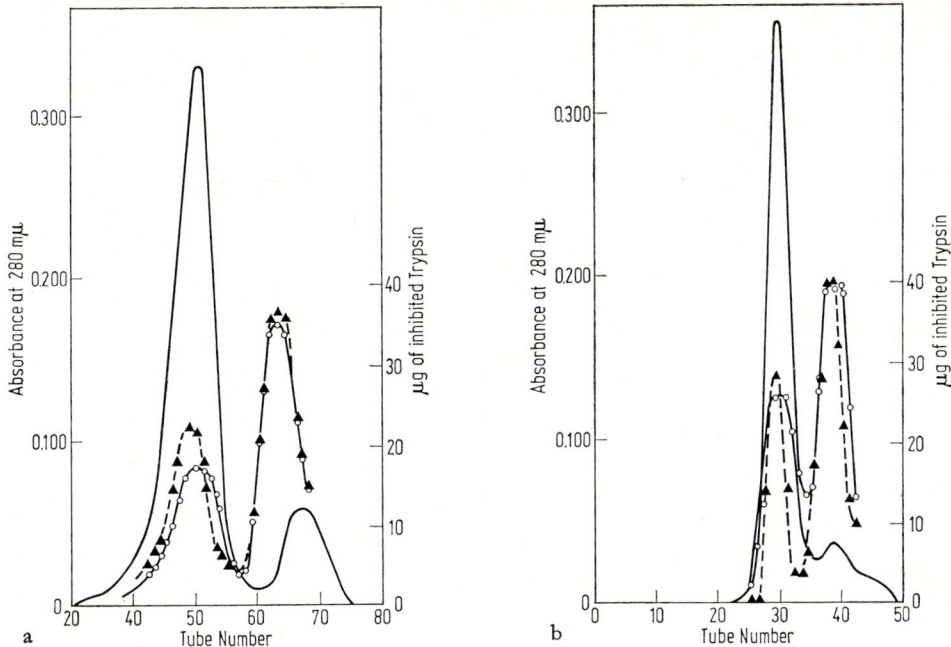

Fig. 8a. Gel filtration of BPTI-(TLCK) β-trypsin complex on Sephadex G-75 (column 1.0×45 cm) in 0.05M Tris 0.02M CaCl$_2$ buffer pH 8.3 at room temperature. Flow rate: 1.5 ml per hour; 0.5 ml fractions were collected. 0—0 Inhibition of β-trypsin.

▲—▲ Inhibition of β-trypsin after treatment of the samples with 5% TCA.

Fig. 8b. Gel filtration of BPTI and TLCK -β-trypsin complex on Sephadex G-100 (column 1.0×45 cm) in 0.05M Tris 0.02M CaCl$_2$ buffer pH 8.3, 40% ethanol at room temperature. Flow rate: 1.5 ml per hour; 0.5 ml fractions were collected. 0—0: inhibition of β-trypsin.

▲—▲: inhibition of β-trypsin after treatment of the samples with 5% TCA.

Table 2. Influence of BPTI on esterase activity of trypsin and its derivatives

Trypsin derivatives	Molar ratio BPTI/Trypsin	TRIS		TRIS-Ethanol 40%	
		Bz-Arg-OEt	Ac-Tyr-OEt	Bz-Arg-OEt	Ac-Tyr-OEt
		Specific activity in µeq/min/mg.			
β-TR	0	74	42.5	134	7.0
	4	0	0	0	0
α-TR	0	72	29.2	116	5.8
	4	0	0	0	0
TLCK-β-TR	0	0.7	0.2	—	—
	4	0	0	—	—
TPCK-β-TR	0	74	4.6	—	—
	4	0	0	—	—
TLCK-TPCK-β-TR	0	0	0	0	0
	4	—	—	—	—
TLCK-α-TR	0	0	0	0	0
	4	—	—	—	—

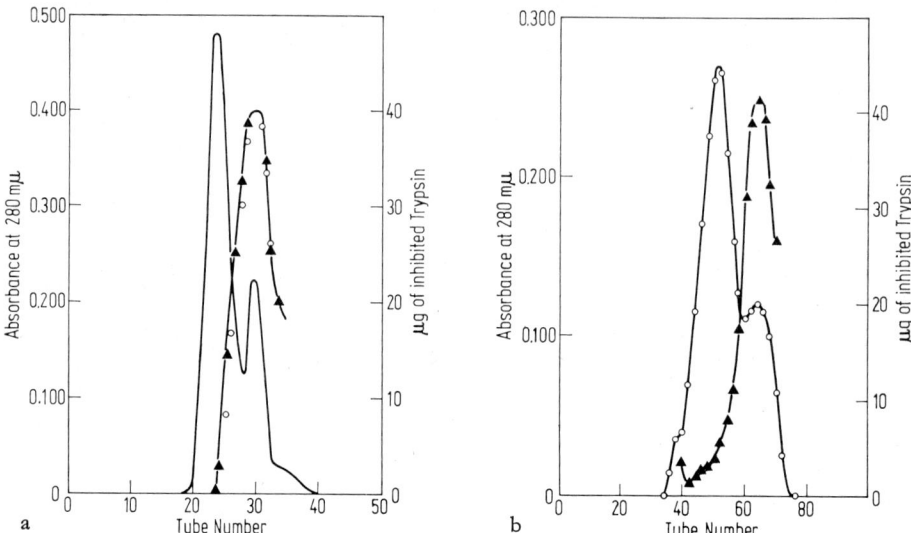

Fig. 9a. Gel filtration of BPTI and TLCK-α-trypsin mixture on Sephadex G-75 (column 1.0 × 45 cm) in 0.05M Tris 0.02M CaCl$_2$ buffer pH 8.3. Conditions were as in Fig. 8a.

Fig. 9b. Gel filtration of BPTI and TLCK α-trypsin mixture on Sephadex G-100 (column 1.0 × 45 cm) in 0.05M Tris 0.02M CaCl$_2$ buffer pH 8.3, 40% ethanol at room temperature. Flow rate: 1.8 ml per hour, 0.6 ml fractions were collected. ▲——▲: inhibition of β-trypsin after treatment of the samples with 5% TCA.

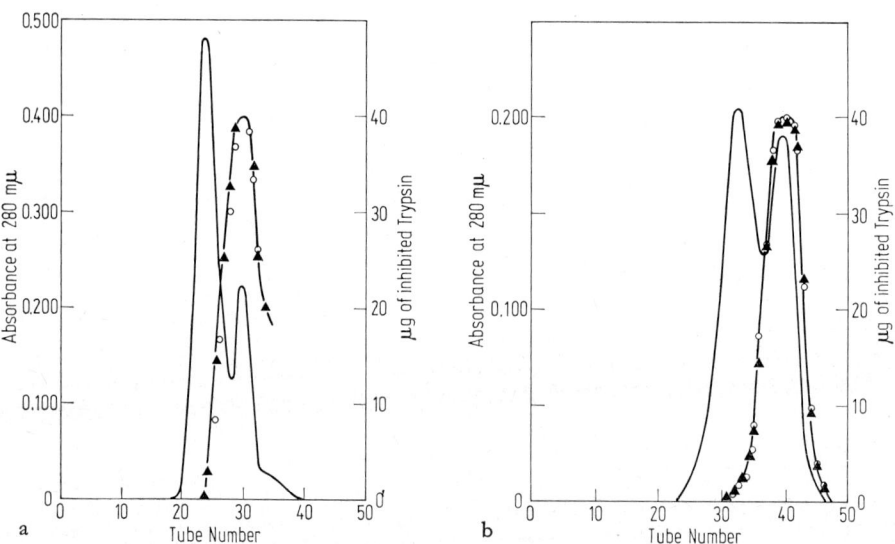

Fig. 10a. Gel filtration of BPTI and TLCK-TPCK-β-trypsin mixture on Sephadex G-75 (column 1.0 × 45 cm) in 0.05M Tris 0.02M CaCl$_2$ buffer pH 8.3. Conditions were as in Fig. 8a.

Fig. 10b. Gel filtration of BPTI and TLCK-TPCK-β-trypsin mixture on Sephadex G-100 (column 1.0 × 45 cm) in 0.05M Tris 0.02M CaCl$_2$ buffer pH 8.3, 40% ethanol. Conditions were as in Fig. 8b.

10 b show that it also lost the affinity for complex formation with pancreatic inhibitor.

On summarizing the results we are inclined to think, that an alternative interaction between trypsin and pancreatic inhibitor occurs in the "chymotryptic" active site of trypsin, an interaction probably of hydrophobic character. Which residue of the trypsin moiety can be involved? All of the evidence supported earlier by fluorescence and spectroscopic studies is clearly consistent with the involvement of one tryptophyl residue in the contact area between trypsin and the inhibitor [11]. Considering that there is no tryptophan in the pancreatic inhibitor we must conclude that it belongs to the trypsin moiety of the complex.

The elucidation of the three-dimensional structure of BPTI [12] helps enormously to explain the *specific functional features* of this inhibitor. This structure is represented in Figure 11.

The positions of the hydrophobic residues are marked by black circles. There is a certain "hydrophobic" region in the front of the molecule. Hydrophobic residues (including cystine) represent 40% of the total. High hydrophobicity of BPTI should serve as a hydrophobic microenvironment in the region of active site of trypsin in the enzyme-inhibitor complex. This provides an explanation for the low dissociation constant of the complex. In such a microenvironment, the electrostatic bond between the carboxyl group of Asp 177 in trypsin and the NH_3-group of Lys 15 in BPTI is extremely strong. Aside from one electrostatic bond there are also regions of hydrophobic interaction. One of them is between the inhibitor and the "chymotryptic" binding site; another (if not identical), is between the carbon chain of Lys 15 of BPTI and the hydrophobic crevice of trypsin which binds the carbon chains of the charged substrates. This crevice is, according to Mares-Guia and Shaw, on a straight line between Asp 177 and the hydrolytic site [13]. The peptide bond between Lys 15 and Ala 16 of BPTI comes consequently in the close neighborhood of the catalytic site. Only the presence of the disulfide bond Cys 14-Cys 38 in native BPTI and the very low dissociation constant of the complex protect the bond Lys 15 — Ala 16 from being split.

It is well known that in a polypeptide chain the rate of cleavage of a peptide bond next to a basic residue is decreased by an adjacent cystine and

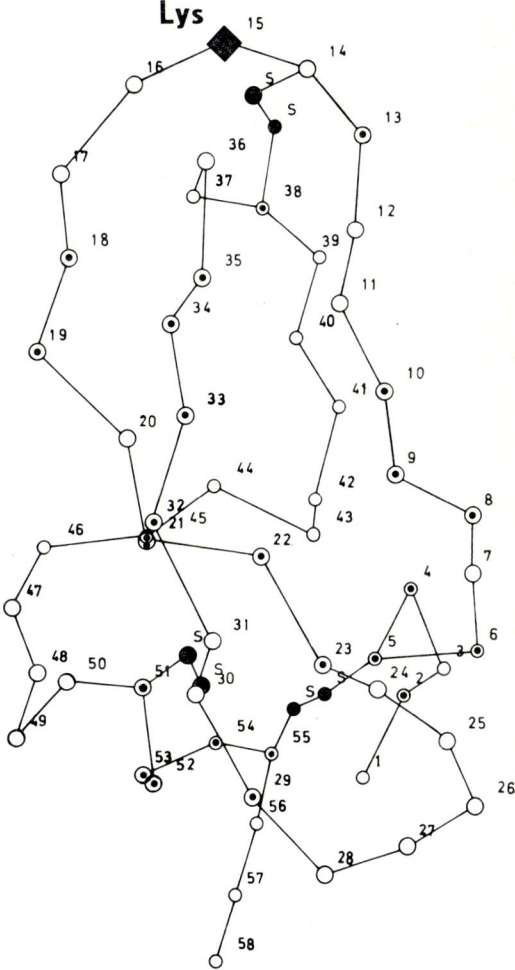

Fig. 11. Three-dimensional structure of BPTI [12].

practically stopped in the case of an adjacent proline; the latter case exists in BPTI. The sequence around Lys 15 is Pro-Cys-Lys-Ala, which is common for many other inhibitors of mammalian origin [14]. The splitting of a bond during complex formation between native BPTI and trypsin has never been proved; on the contrary, in an independent study, we have presented evidence that BPTI remains intact in the complex [15]. Only after partial reduction of the disulphide bridge Cys 14—Cys 38 was some splitting of the bond Lys 15 — Ala 16 observed. Although a mechanism of bond splitting associated with the inhibitor-trypsin interaction has been demonstrated in many other cases, we cannot consider this hydrolysis of a covalent bond in the inhibitor as a generally indispensable prerequisite for the interaction with trypsin. Rather, such a mechanism is a coincidental phenomenon which depends on the primary structure of the inhibitor and on the dissociation constant for the interaction between its basic group and Asp 177 of trypsin.

All the results presented above contribute to the *following conclusions*:

1. An independent active site with "chymotryptic" specificity exists in the molecule of β-trypsin.
2. The mechanism of complex formation between trypsin and pancreatic trypsin inhibitor involves the inhibition of both tryptic and chymotryptic binding sites.
3. Inhibition of the tryptic binding site occurs through the formation of an electrostatic bond between Asp 177 of trypsin and Lys 15 of the inhibitor.
4. Inhibition of the chymotryptic binding site takes place through a hydrophobic interaction between an amino acid residue of trypsin which represents the binding site for chymotryptic substrates and hydrophobic groups of the inhibitor.
5. BPTI forms complexes with all three forms of trypsin and with TLCK-β-trypsin.

References

[1] CHAUVET, J., J. NOUVEL and R. ACHER, Biochim. Biophys. Acta **115**, 121, 130 (1966).
[2] KASSEL, B., M. RADICEVIC, M. J. Ansfield and M. LASKOWSKI, Sr., Biochem. Biophys. Res. Commun. **18**, 225 (1965).
[3] DLOUHÁ, V., D. POSPÍŠILOVÁ, B. MELOUN and F. ŠORM, Collection Czechoslov. Chem. Commun. **30**, 1311 (1965).
[4] SCHROEDER, D. D. and E. SHAW, J. Biol. Chem. **243**, 2943 (1968).
[5] SMITH, R. L. and E. SHAW, J. Biol. Chem. **244**, 4704 (1968).
[6] IMHOFF, J. M. and V. KEIL-DLOUHÁ, FEBS Letters, **12**, 345 (1971).
[7] TANFORD, C., J. Am. Chem. Soc. **84**, 4240 (1962).
[8] CHAUVET, J. and R. ACHER, J. Biol. Chem. **242**, 4274 (1967).
[9] GREEN, N. M. and E. WORK, Biochem. J. **54**, 347 (1953).
[10] BIRKTOFT, J. J., B. W. MATTHEWS and D. M. BLOW, Biochem. Biophys. Res. Commun. **36**, 131 (1969).
[11] EDELHOCH, H. and R. F. STEINER, J. Biol. Chem. **240**, 2877 (1965).
[12] HUBER, R., D. KUKLA, A. RÜHLMANN, O. EPP and H. FORMANEK, Naturwissenschaften **57**, 389 (1970).
[13] MARES-GUIA, M. and E. SHAW, J. Biol. Chem. **240**, 1579 (1965).
[14] LASKOWSKI, M., Jr. and R. W. SEALOCK, "Protein Proteinase Inhibitors" in "The Enzymes", in press.
[15] DLOUHÁ, V., B. KEIL and F. ŠORM, Biochem. Biophys. Res. Commun. **31**, 66 (1968).

Amino Acid Sequence of Trypsin Inhibitor from Cow Colostrum

D. Čechová, V. Jonáková and F. Šorm

Institute of Organic Chemistry and Biochemistry, Czechoslovak Academy of Sciences, Prague, Czechoslovakia

The trypsin inhibitor from cow colostrum can be resolved by ion-exchange chromatography on CM-Sephadex at pH 3.4 into three main types, A, B, and C. The distribution of these peaks differs with individual batches of the starting material, peak B, however, is always dominant. The rechromatography of all three peaks on DEAE-cellulose has shown that neither of these components are homogeneous but rather represent certain inhibitor families [1]. The analysis of tryptic digests of S-sulfonated proteins in peaks A, B, and C by the technique of peptide maps has demonstrated a high degree of structural similarity of these proteins. The difference in their behaviour can be in most cases accounted for by differences in the content of the sugar moiety bound firmly to the inhibitors. We have not found as yet whether these differences exist in the colostrum or whether they arise from the liberation of a part of the sugar moiety during the treatment. It has been shown, however, that one of the components of peak C differs also in amino acid sequence.

Our sequential study was carried out with the inhibitor from group B which we had obtained from cow colostrum in a dominant quantity. The results of amino acid analysis and end-group analysis, together with the absence of free SH-groups in the material indicated that the molecule of the inhibitor consists of one single polypeptide chain of 67 amino acids, cross-linked by three disulfide bonds. Histidine and tryptophan residues are absent. The inhibitor thus resembles the bovine basic trypsin inhibitor and the acidic inhibitor from pancreatic juice.

The sequential study was based on the analysis of tryptic fragments of the S-sulfonated protein. These fragments, which accounted for the whole molecule of the inhibitor, were isolated by high-voltage electrophoresis and paper chromatography. Larger peptides were cleaved enzymatically to smaller fragments. The amino acid sequence of individual peptides was determined for the most part by Edman degradation, sometimes in combination with the dansyl technique. The order of tryptic peptides was determined by the analysis of the chymotryptic digest of the inhibitor. Difficulties were encountered with peptide T-4 containing 33 residues. We have not been able to distinguish between aspartic acid and asparagine in positions 27 and 46. We have observed that these two residues bind the sugar moiety which complicates the sequential analysis. However, from the comparison of the site in question with homologous positions in the pancreatic inhibitor it would appear that in both cases the positions are occupied by asparagine.

The comparison of the amino acid sequence of the cow colostrum trypsin inhibitor with the known primary structure of the basic inhibitor from bovine pancreas indicates that the genes for both inhibitors are of the same phylogenetic origin. The homology in these structures, which exceeds 40%, is especially obvious in the amino acid sequence around the active sites of both inhibitors (lysine 18 in colostrum inhibitor [2]), in the position of half-cystine residues, and in the positions of most aromatic amino acids. Since both inhibitors inactivate to a lesser degree also chymotrypsin, the active site responsible for the inhibition of chymotrypsin will obviously lie in one of the homologous regions of the molecule.

References

[1] ČECHOVÁ, D., V. JONÁKOVÁ-ŠVESTKOVÁ and F. ŠORM, Collection Czechoslov. Chem. Commun **35**, 3085 (1970).

[2] ČECHOVÁ, D. and G. MUSZYNSKÁ, FEBS Letters **8**, 84 (1970).

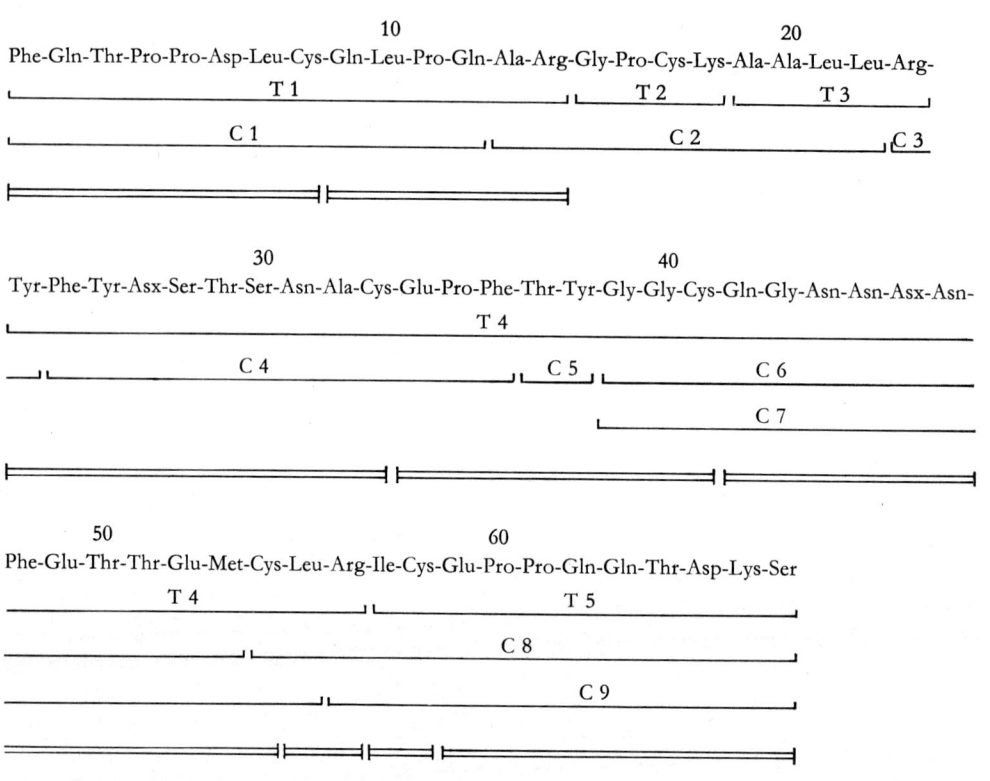

T 1—5 peptides obtained by tryptic digestion of the S-sulfonated inhibitor.
C 1—9 peptides obtained in dominant quantity by chymotryptic digestion of the oxidized inhibitor.
⊢──⊣ peptides obtained by tryptic cleavage of aminoethylated tryptic peptides.

Figure 1

Amino Acid Sequence of Cow Colostrum Inhibitor 107

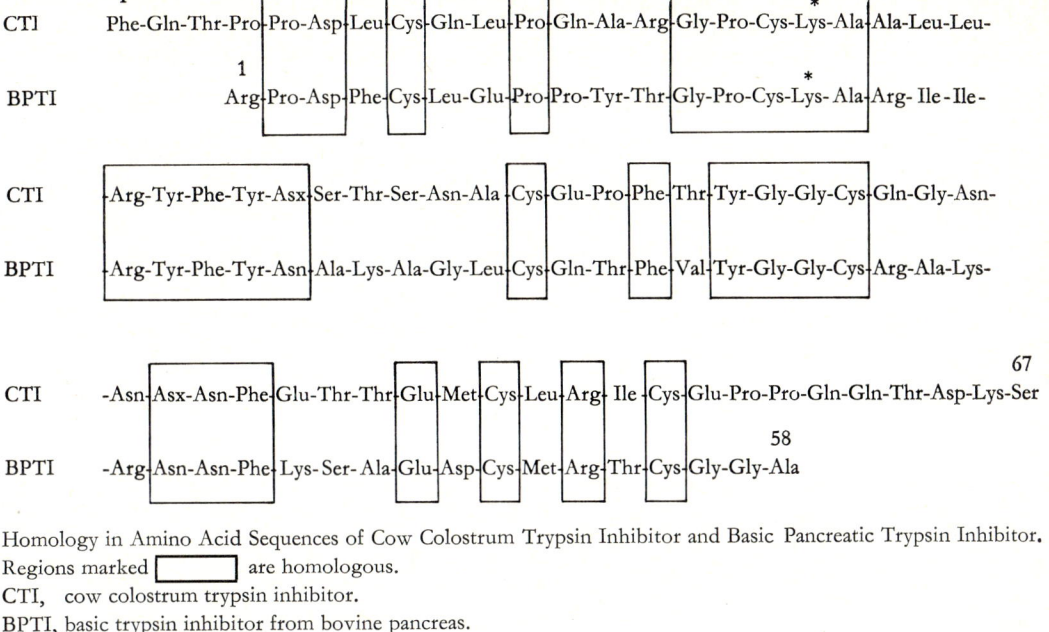

Homology in Amino Acid Sequences of Cow Colostrum Trypsin Inhibitor and Basic Pancreatic Trypsin Inhibitor.
Regions marked ☐ are homologous.
CTI, cow colostrum trypsin inhibitor.
BPTI, basic trypsin inhibitor from bovine pancreas.
* active sites of both inhibitors.

Figure 2

Structure and Chemical Modification of Kunitz Soybean Trypsin Inhibitor

Tokuji Ikenaka*, Takehiko Koide* and Shoji Odani

Department of Chemistry, Osaka University College of Science, Toyonaka

Interest in the study of the inhibition mechanism of proteinase inhibitors has been increased greatly by the investigation of the reactive site of Kunitz soybean trypsin inhibitor (Kunitz STI) by Laskowski, Jr. and his co-workers [1, 2, 3]. For the complete understanding of the inhibition mechanism of proteinase inhibitors, the elucidation of the amino acid sequence of these proteins is very important. The amino acid sequences of several trypsin inhibitors, bovine pancreatic basic [4, 5, 6], bovine pancreatic Kazal's [7] and ascaris [8] inhibitors have been reported. It is very interesting to compare the amino acid sequence of Kunitz STI with those of other proteinase inhibitors, therefore, we studied the primary structure of Kunitz STI.

During the isolation and purification of two kinds of soybean trypsin inhibitors, namely Kunitz STI and Bowman-Birk STI**, from soybean flakes by the method of Yamamoto and Ikenaka [9], we obtained an accidentally modified Kunitz STI in which one tryptophan, one tyrosine and two methionine residues were chlorinated or oxidized, though this modified STI had 80 percent inhibitory activity of the original native STI. In this paper, we would like to present the partial amino acid sequence of Kunitz STI and also discuss the relationship between the amino acid residues in the protein and inhibitory activity based on the results of the elucidation of the location of the amino acid residues modified during the purification procedure.

1. Structure of Kunitz soybean trypsin inhibitor

For the elucidation of the amino acid sequence of this inhibitor, the protein was first divided into 4 fragments as shown in Figure 1.
The native STI was partially hydrolyzed with trypsin at pH 3.75 by the method of Ozawa and Laskowski, Jr. [2].
Reduction and carboxymethylation, followed by gel filtration of the tryptic hydrolysate gave Fragments A and BCD. Chemical cleavage of two methionyl bonds in the protein with cyanogen bromide was carried out in 70% formic acid by the method of Steers et al. [10]. As shown in Figure 1, Fragment D was separated from Fragment ABC by gel filtration on a Sephadex G-100 column. After reduction and carboxymethylation of Fragment ABC, Fragment C was separated from Fragment AB on a Sephadex G-50 column. The amino acid compositions of the Fragments A, B, C and D are listed in Table 1. The amino acid composition of Fragment B was

---

* Present address: Department of Biochemistry, Niigata University School of Medicine, Niigata.
** In our paper [9] this inhibitor was named as 1.9S inhibitor.

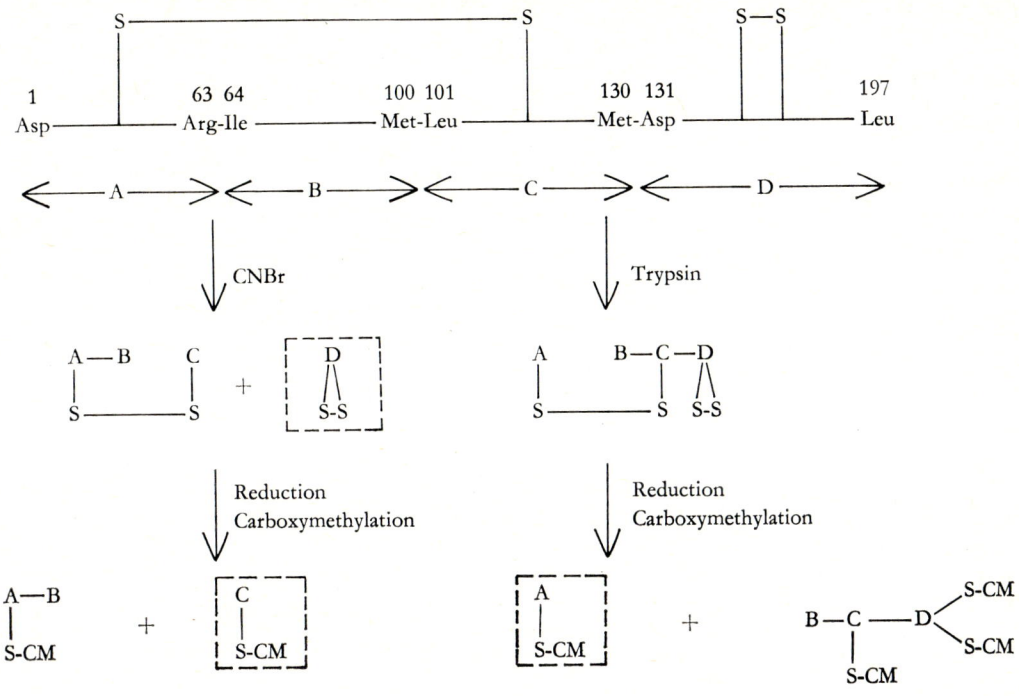

Figure 1. Fragmentation of KUNITZ soybean trypsin inhibitor

```
                  1             5                  10                 15                 20                 25
       Asp-Phe-Val-Leu-Asp-Asn-Glu-Gly-Asn-Pro-Leu-Glu-Asn-Gly-Gly-Thr-Tyr-Tyr-Ile-Leu-Ser-Asp-Ile-Thr-Ala-Phe-

                        30                 35          I   40                 45                 50
       Gly-Gly-Ile-Arg-Ala-Ala-Pro-Thr-Gly-Asn-Glu-Arg-Cys-Pro-Leu-Thr-Val-Val-Gln-Ser-Arg-Asn-Glu-Leu-Asp-Lys-

                 55                 60                 65              70  71
       Gly-Ile-Gly-Thr-Ile-Ile-Ser-Pro-Ser-Tyr-Arg-Ile-Arg-Phe-Ile-Ala-Glu-Gly-Val,Phe-Ile-His-Ala-Glu(Pro,Gly)Leu,Ser-

                                                                              100       II    105
       Leu-Lys,Asx,Asx,Asx,Asx,Thr,Ser,Ser,Glx,Pro,Gly,Ala,Val,Ile,Leu,Phe,Lys,Arg,Met-Leu-Cys-Val-Gly-Ile-Pro-Thr-

                 110                115                120                125                130
       Glu-Trp-Ser-Val-Val-Glu-Asp-Leu-Pro-Glu-Gly-Pro-Ala-Val-Lys-Ile-Gly-Glu-Asn-Lys-Asp-Ala-Met-Asp-Gly-Trp-

                 135                140                145       148           III
       Phe-Arg-Leu-Glu-Arg-Val-Ser-Asp-Asp-Glu-Phe-Asn-Asn-Tyr-Lys,Leu-Val-Phe-Cys-Pro-Gln-Gln-Ala-Glu-Asp-

            IV
       Asp-Lys-Cys-Gly-Asp-Ile-Gly(Ile,Ile,Asx,Asx,Asx,Thr,Ser,Gly,His)Arg,Arg,Leu-Val-Val-Ser-Lys-Asn-Lys-Pro-Leu-

                                      195       197
       Val-Val-Gln-Phe-Glu-Lys,Leu-Asp-Lys-Glu-Ser-Leu
```

Figure 2. Amino Acid Sequence of KUNITZ Soybean Trypsin Inhibitor

One S—S linkage is between Cys **I** (39)—Cys **II** (102), the other is Cys **III** (152)—Cys **IV** (161)

Table 1. Amino acid composition of STI and its fragments

Amino acids	STI	Fragments			
		A	B	C	D
Aspartic acid	29	9	4	3	13
Threonine	8	5	1	1	1
Serine	12	4	3	1	4
Glutamic acid	20	5	3	4	8
Proline	11	4	2	3	2
Glycine	18	8	3	3	4
Alanine	9	3	3	2	1
Half-cystine	4	1	0	1	2
Valine	15	3	2	4	6
Methionine	2	0	1	1	0
Isoleucine	15	6	4	2	3
Leucine	15 (16)	5	3	2	5 (6)
Tyrosine	4	3	0	0	1
Phenylalanine	9	2	3	0	4
Tryptophan	2	0	0	1	1
Lysine	11	1	2	2	6
Histidine	2	0	1	0	1
Arginine	10	4	2	0	4
Total	196 (197)	63	37	30	66 (67)

calculated by subtracting the result of Fragment A from that of Fragment AB.

The amino acid sequences of the fragments were determined by the usual methods. By this time, complete sequences of Fragments A [11] and C, and tryptic peptides of Fragment D had been determined as shown in Figure 2. The amino terminal sequence of Fragment B had been established to be Ile-(Arg)-Phe-Ile-Ala-Glu-Gly-Val- by Edman's direct method [12] using Fragment BCD [11].

There was no special amino acid sequence in this protein or around the reactive site, Arg 63. However, it might be pointed out that there are many Ile, Thr and Pro residues in the sequence between Ile 54 and Ile 67. Since these amino acids are known to form non-helical structures in proteins, this part might have a random structure, and be attacked very easily by the enzyme.

In the tryptic hydrolysate of Fragment D, two peptides, Leu-Asp-Lys-Glu-Ser and Leu-Asp-Lys-Glu-Ser-Leu, were found in a ratio of about 3:1. These peptides seem to be derived from the C-terminal end of Fragment D because of the lack of lysine and arginine at the C-terminal of these peptides. When the native STI was digested with carboxypeptidase A, the release of about 0.3 mole of leucine per mole of protein was observed. On the other hand, the hydrazinolysis method revealed the presence of 0.8 mole of serine and 0.2 mole of leucine as the C-terminal amino acids. These results are in very good agreement with the isolation of the two C-terminal peptides. Therefore, it might be concluded that some of the inhibitor molecules seem to lack leucine at the C-terminus.

2. Isolation and characterization of accidentally modified Kunitz soybean trypsin inhibitors (M-STI)

For the chemical and physicochemical studies of STI, we isolated the inhibitors from soybean flakes [9] many times in the last few years. One of such STI preparations (prepared during

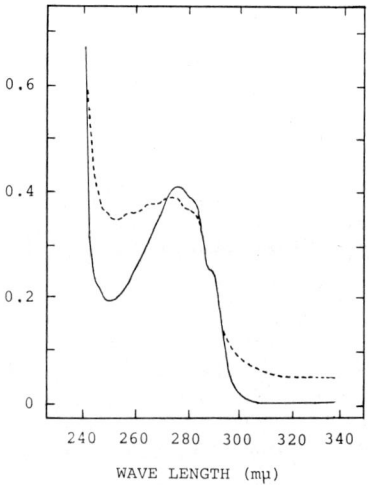

Fig. 3. Ultraviolet spectra of native and modified STI. ———— Native STI, -----M-STI.

summer time) was found to show a different ultraviolet absorption spectrum from that of native STI as shown in Figure 3, though this protein (M-STI) had trypsin inhibitory activity. These findings led us to study the mode of modification and the locations of these modified amino acid residues in the protein in connection with inhibitory activity.

Through a Sephadex G-100 column (3 × 100 cm), M-STI was separated into three fractions, namely fractions I, II and III in the order of elution (Fig. 4). These three fractions closely resembled each other in their amino acid compositions, though tryptophan, tyrosine and lysine contents of these fractions seemed to be less than those of native STI (Table 2). The elution position and the amino acid composition suggested that fractions II and III were dimer and monomer of M-STI, respectively, whereas, fraction I was trimer or higher polymer of the protein. After dialysis and lyophilization, fraction III was again separated into three fractions as those of Figure 4.

When fraction III was gel-filtrated without dialysis and lyophilization, this fraction gave a single peak in the position of monomer. Fraction II after dialysis and lyophilization was also separated on the column into fractions I, II and small amounts of III. These results suggest that the polymerization of M-STI might be caused by dialysis and/or lyophilization. In case of native STI, these phenomena were not observed. Fractions II and III had 40% and 83% trypsin inhibitory activity of native STI, respectively, while fraction I was completely inactive (Tab. 2).

When M-STI was subjected to cyanogen bromide oxidation to cleave methionyl bonds in a similar manner to the fragmentation of native STI, no formation of homoserine or homoserinelactone was observed. This result seemed to suggest the oxidation of methionine residues to methionine

Table 2. Amino acid composition and inhibitory activity of modified and native STI

Amino acids	M-STI Fractions			Native STI	
	I	II	III	Found	Reported value
Aspartic acid	27.1	28.1	29.3	28.8	29
Threonine	7.4	7.9	7.7	7.4	8
Serine	11.5	11.6	11.8	11.7	12
Glutamic acid	20.2	20.3	20.7	20.7	20
Proline	11.1	11.0	11.3	11.4	11
Glycine	17.8	18.5	17.5	17.7	18
Alanine	9.1	9.3	9.1	9.0	9
Half-cystine	3.9	3.9	3.7	4.1	4
Valine	14.3	14.8	14.1	14.6	15
Methionine	1.7	1.6	1.8	1.7	2
Isoleucine	15.1	14.8	14.8	14.6	15
Leucine	15.2	15.2	15.2	15.2	15 (16)
Tyrosine	3.0	2.9	3.1	4.0	4
Phenylalanine	9.4	10.2	9.6	9.6	9
Tryptophan	1.0	1.0	1.0	2.0	2
Lysine	9.0	8.8	8.9	10.6	11
Histidine	1.9	2.1	1.9	2.1	2
Arginine	9.3	10.0	10.3	10.1	10
Inhibitory activity	0	40	83	100	

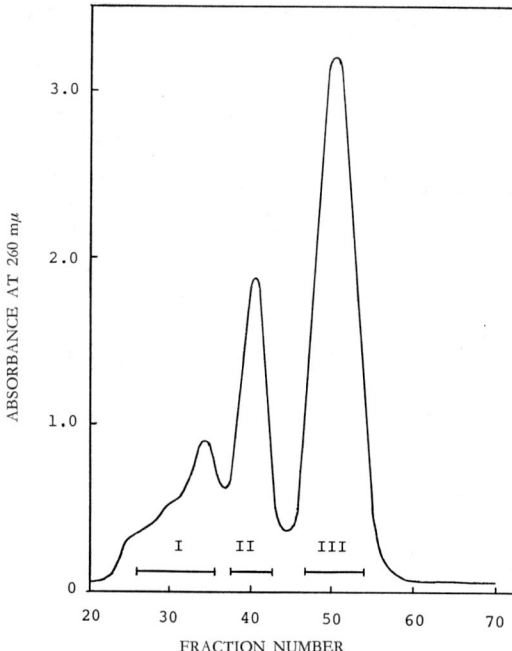

Fig. 4. Gel filtration of M-STI on Sephadex G-100. The sample (3 ml) was applied to a column (3 × 100 cm) of Sephadex G-100 (eluant: 0.1M sodium acetate buffer, pH 6.0) and 10 ml fractions were collected.

sulfoxide, which was converted to methionine during acid hydrolysis and was not susceptible to the cyanogen bromide oxidation. From the data of amino acid analysis and the cyanogen bromide oxidation of fraction III, one mole each of tryptophan and tyrosine and two moles of methionine in M-STI (fraction III) seemed to be modified.

3. Inspection of the modified amino acid residues in the modified Kunitz soybean trypsin inhibitor

The states and locations in M-STI of the modified amino acid residues were analyzed.
Tryptophan residue — There are two tryptophan residues in STI. One, 109, is located in Fragment C and the other, 133, in Fragment D as shown in Figures 2 and 5. In order to know the location of the modified tryptophan, M-STI was divided into 3 fragments as shown in Figure 5. The sulfide bonds of M-STI were first reduced and carboxymethylated, and then methionine sulfoxide residues were reduced to methionine with 5M thioglycolic acid. After removal of the reagents by passing through a Sephadex column, the reduced CM-M-STI was oxidized with cyanogen bromide. Gel-filtration of the oxidized product on a Sephadex G-75 column showed two peaks, one consisted of Fragments AB + D and the other of Fragment C. The amino acid compositions of Fragments AB + D and C thus obtained were compared with those of the similar fragments isolated from native STI (Tab. 3). The compositions in Table 3 show the lack of tryptophan in Fragments AB + D obtained from M-STI, and the presence of tryptophan in Fragment C obtained from both native and M-STIs and also in Fragments AB + D obtained from native STI. These results might conclude the modification of tryptophan 133 in Fragment D of M-STI.

The modified state of tryptophan 133 was not confirmed at the present time, though an unknown peak X was detected on an amino acid chromatogram of a hydrolysate of Fragments AB + D obtained from M-STI with hydrochloric acid containing 4 per cent of thioglycolic acid [13]. Assuming the ninhydrin color value of X equals to the average value of the other amino acids, its content was calculated to be 0.7 mole per mole of the peptide.

STEINER [14] reported that only one of three tryptophan residues* was oxidized with N-bromosuccinimide with a minor loss of activity, whereas two tryptophans were oxidized by hydrogen peroxide with complete inactivation. It is probable that the same tryptophan residue

---

* STEINER reported the presence of three tryptophan residues reacted with N-bromosuccinimide in denatured state.

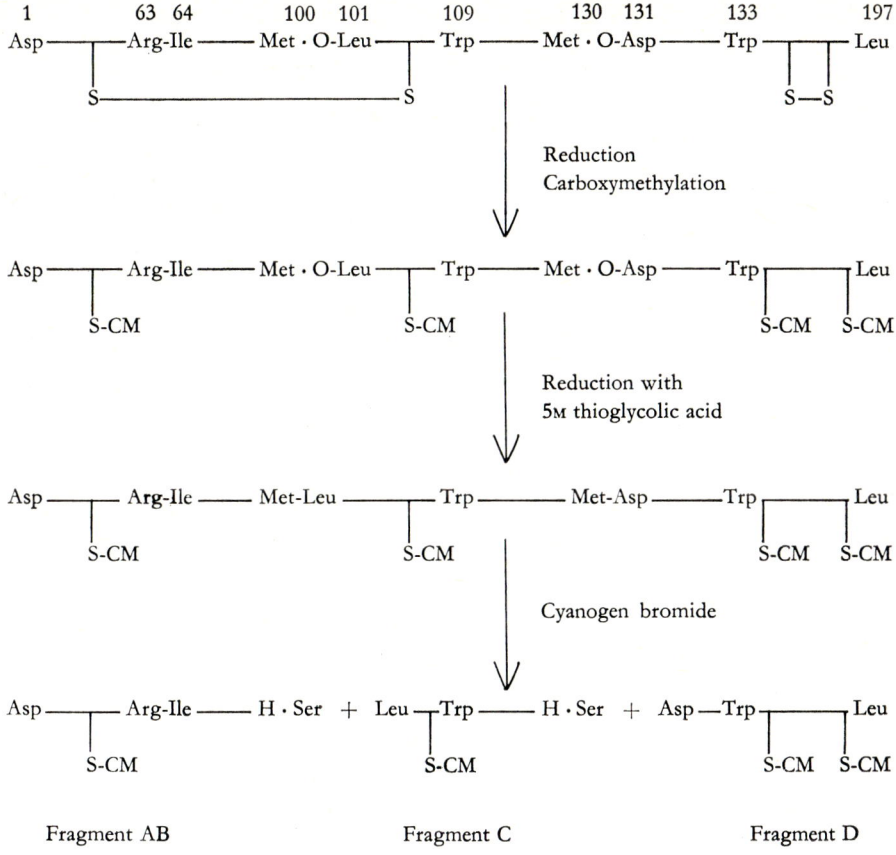

Figure 5. Cyanogen bromide degradation of M-STI after reduction of methionine sulfoxide.
CM: carboxymethyl group,　H·Ser: homoserine,　Met·O: methionine sulfoxide

133 is modified in our case and in case of the oxidation by N-bromosuccinimide, and this tryptophan is not essential for the inhibitory activity of STI, while, tryptophan 109 might play an important part in the interaction with trypsin.

Tyrosine residue — There are four tyrosine residues in STI, three residues 17, 18 and 62 are in Fragment A and the fourth is in Fragment D. One of these residues in M-STI was not detected by the amino acid analysis (Tab. 2). To know the location of this modified tyrosine residue, the fragmentation of M-STI into Fragments A and BCD was carried out by the procedure of Ozawa et al. [2] (Fig. 1). The amino acid analyses of Fragments A and BCD show that Fragment A has only two tyrosine residues, indicating the modification of one tyrosine residue in this fragment. On the other hand, Fragment BCD contains one tyrosine, showing no modification of the tyrosine residue in the fragment.

Tyrosine 62 adjacent to the reactive site arginine, residue 63, was expected to have been modified in M-STI, since the reactive site must be exposed to the surface of the molecule.

By the digestion of Fragment A with carboxypeptidases A and B, arginine 63 and tyrosine 62 were liberated in an amount of 1 mole/mole of peptide. This result revealed that tyrosine 62

Table 3. Amino acid composition of Fragments AB + D and C isolated from modified and native STI after CNBr degradation

Amino acids	Fragments AB + D			Fragment C		
	Found		Expected[a]	Found		Expected[a]
	Modified	Native		Modified	Native	
Aspartic acid	27.1	26.4	26	3.2	3.5	3
Threonine	6.1	6.4	7	0.9	1.0	1
Serine	10.0	9.7	11	1.0	1.2	1
Glutamic acid	16.0	16.6	16	4.0	4.2	4
Proline	8.5	8.9	8	3.1	2.9	3
Glycine	14.7	14.0	15	2.8	3.0	3
Alanine	7.6	6.9	7	1.9	2.0	2
Half-cystine[b]	—	—	3	—	—	1
Valine	10.8	10.0	11	3.1	2.8	4
Methionine[b]	—	—	1	—	—	1
Isoleucine	13.3	12.7	13	2.0	2.1	2
Leucine	13.0	13.0	13	2.0	2.0	2
Tyrosine	3.1	3.9	4	0.0	0.0	0
Phenylalanine	8.5	9.1	9	0.2	0.3	0
X[c]	0.7	0.0	0	0.0	0.0	0
Tryptophan[d]	0.0	1.1	1	1.0	0.9	1
Lysine	7.8	8.9	9	1.8	1.7	2
Histidine	1.8	2.1	2	0.0	0.0	0
Arginine	10.0	9.6	10	+	+	0

Values are shown in molar ratio with respect to leucine taken as 13 (Fragments AB + D) or 2 (Fragment C) residues.
[a] Calculated from Table 1. [b] Cystine (as carboxymethyl cysteine) and methionine (as homoserine and homoserine lactone) were not determined. [c] A substance thought to be the modification product of tryptophan. [d] Determined by the method of Matsubara and Sasaki [13].

is intact. Fragment A of M-STI was digested with trypsin (the specimen contains a small amount of chymotrypsin) and the tryptic peptides were separated on a Dowex 50 × 2 column by the similar procedure used for the elucidation of the amino acid sequence of Fragment A of native STI. Two peaks corresponding to the heptadecapeptide (Asp 1 — Tyr 17) and the octadecapeptide (Asp 1 — Tyr 18) were analyzed for their amino acid compositions. Table 4 shows the amino acid compositions of these two peptides, indicating the presence of one tyrosine in each peptide. These observations suggest that tyrosine 17 might be intact and tyrosine 18 modified. On the amino acid chromatogram of the acid hydrolysate of the octadecapeptide, the presence of unnatural amino acid was detected. The elution position of this amino acid corresponded to monochlorotyrosine prepared by the method of Thompson [15]. Therefore, a part of tyrosine 18 might be converted to monochloro-derivative. However, the recovery of monochlorotyrosine was only 22% of the lost tyrosine. It is of interest that only tyrosine 18 of two neighboring tyrosines 17 and 18 has been modified. Baba et al. have reported the acetylation of one tyrosine residue of STI by N-acetyl imidazole without loss of the activity [16]. This acetylated tyrosine seems to be also tyrosine 18. The reactivity of two tyrosine residues in STI for acetylation with N-acetyl imidazole [17], iodination [14], modi-

Table 4. Amino acid composition of the peptides obtained by tryptic digestion of Fragment A of M-STI

Amino acids	heptadecapeptide		octadecapeptide	
	M-STI	Expected[a]	M-STI	Expected[a]
Aspartic acid	3.75	5	3.73	5
Threonine	1.00	1	1.01	1
Glutamic acid	2.34	2	2.58	2
Proline	0.97	1	1.13	1
Glycine	3.22	3	4.05	3
Valine	0.88	1	0.92	1
Leucine	2.00	2	2.00	2
Tyrosine	1.21	1	1.29	2
Phenylalanine	0.98	1	0.91	1
Y[b]	0.00	0	0.22	0

Values are shown with respect to leucine taken as two residues.
[a] Taken from Fig. 2.   [b] A substance thought to be monochlorotyrosine.

fication with cyanuric fluoride [17] and oxidation with tyrosinase [18] with a minor loss of inhibitory activity has been reported. Taking our results of the modification of tyrosine 18 into consideration, these two reactive tyrosine residues seem to be tyrosines 17 and 18.

Methionine residue — Two methionine residues in STI have been oxidized to methionine sulfoxide as shown above. The location of these residues was at 100 and 130.

It might be concluded from these results that tyrosine 18, methionines 100 and 130 and tryptophan 133 are exposed to the surface of the protein molecule, and are easily modified with chlorine in tap water, though these residues are not essential for trypsin inhibitory activity.

The possibility of the modification of these amino acid residues in the protein during the isolation and purification procedure should be taken into account. The other preparations of STI from the same lot of soybean flakes by the similar procedure* were found to be intact in any respects. This fact excluded the possibility of the modification of STI in the starting soybean flakes and during the extraction with 0.25N sulfuric acid at room temperature (26° C). Moreover, the latter possibility was denied by the high stability of native STI during incubation with 0.25N sulfuric acid at 37° for 48 hours. The high possibility for the oxidation or chlorination of these amino acid residues in the protein lies in the modification during dialysis against running tap water in summer time. To survey this possibility, native STI was incubated in chlorine water (0.5 to 2.0 mM) and the change of ultraviolet absorption spectrum was examined. The spectrum of the protein incubated in 1.0 mM chlorine-water was found to be very similar to that of M-STI.

The knowledge of the chemically reactive residues obtained with accidentally modified materials might afford suggestion for advanced experiments in the chemical modification of STI, and caution should be taken against oxidation and chlorination of the reactive amino acid residues in proteins during dialysis against tap water.

* An exception was dialysis. When M-STI was obtained, the dialysis was run against running tap water in summer time.

References

[1] FINKENSTADT, W. R. and M. LASKOWSKI, Jr., J. Biol. Chem. **240**, PC 962 (1965).
[2] OZAWA, K. and M. LASKOWSKI, Jr., J. Biol. Chem. **241**, 3955 (1966).
[3] SEALOCK, R. W. and M. LASKOWSKI, Jr., Biochemistry **8**, 3703 (1969).
[4] KASSELL, B. and M. LASKOWSKI, Sr., Biochem. Biophys. Res. Commun. **20**, 463 (1965).
[5] CHAUVET, J. and R. ACHER, Bull. soc. chim. biol. **49**, 985 (1967).
[6] DLOUHÁ, V., D. POSPÍŠILOVÁ, B. MELOUN and F. ŠORM, Coll. Czech. Chem. Commun. **33**, 1363 (1968).
[7] GREENE, L. J. and D. C. BARTELT, J. Biol. Chem. **244**, 2646 (1969).
[8] FRAEFEL, W. and R. ACHER, Biochim. Biophys. Acta **154**, 615 (1968).
[9] YAMAMOTO, M. and T. IKENAKA, J. Biochem. **62**, 141 (1967).
[10] STEERS, E., Jr., G. R. CRAVEN, C. B. ANFINSEN and J. L. BETHUNE, J. Biol. Chem. **240**, 2478 (1965).
[11] IKENAKA, T., S. TSUNASAWA and T. KOIDE, J. Biochem. **69**, 251 (1971).
[12] IWANAGA, S., P. WALLEN, N. J. GROENDAHL, A. HENSCHEN and B. BLOMBAECK, European J. Biochem. **8**, 189 (1960).
[13] MATSUBARA, H., R. M. SASAKI, Biochem. Biophys. Res. Commun. **35**, 175 (1969).
[14] STEINER, R. F., Arch. Biochem. Biophys. **115**, 257 (1966).
[15] THOMPSON, E. O. P., Biochim. Biophys. Acta **15**, 440 (1954).
[16] BABA, M., K. HAMAGUCHI and T. IKENAKA, J. Biochem. **65**, 113 (1969).
[17] GORBUNOFF, M. J., Biochemistry **7**, 2547 (1968).
[18] CORY, J. T. and E. FRIEDEN, Biochemistry **6**, 121 (1967).

Kinetics and Thermodynamics of Interaction Between Soybean Trypsin Inhibitor (Kunitz) and Bovine β Trypsin*

Michael Laskowski, Jr., Ruth W. Duran[a], William R. Finkenstadt, Sarah Herbert[b], Harry F. Hixson, Jr.[c], David Kowalski, James A. Luthy, Jeffrey A. Mattis[d], Raymond E. McKee and Carl W. Niekamp[e]

Department of Chemistry Purdue University, Lafayette, Indiana 47907 U.S.A.

Previous symposium presentations from our laboratory concerning soybean trypsin inhibitor (Kunitz) (e. g. [1]) were focused primarily on the unusual chemistry of its reactive site. These studies stressed the ability to resynthesize the cleaved peptide bond by processes involving both thermodynamic [2] and kinetic [3, 4] control and to restore the arginyl residue in the desarginyl[64] modified inhibitor [5]. They were culminated by an enzymatic mutation — replacement of arginyl[64] by lysyl[64] in the intact peptide chain [5].

The objective of this presentation is to review the former studies and to introduce new, thus far unpublished, kinetic and thermodynamic data in order to show what can be learned about the mechanism of the trypsin-inhibitor association and the chemistry of the complex. This knowledge appears to be particularly relevant since we are now on the advent of having available from X-ray crystallography detailed 3-dimensional structures of several enzyme-protein inhibitor complexes. A structure of one inhibitor is already at hand [6, 7]. Once the structures of complexes are available, what should the X-ray crystallographers look for? What unusual features do we hope for them to explain?

---

* Supported by Grants GM 10831 and GM 11812 from the National Institutes of General Medical Sciences, National Institutes of Health, U. S. Public Health Service.
[a] Present address Photo Products Division, DuPont de Nemours Company, Parlin, New Jersey 08859.
[b] National Institutes of Health Postdoctoral Fellow.
[c] National Institutes of Health Predoctoral Fellow. Present address, Xerox Research Laboratories, Rochester, New York 14580.
[d] Predoctoral Trainee, National Science Foundation.
[e] Predoctoral Trainee, National Institutes of Health Grant GM 01195. Present address Department of Chemistry, Yale University, New Haven, Connecticut 06520.

The Chemistry of the Reactive Site

As a result of a laborious, indirect chain of reasoning [8, 9] we have surmised and later clearly demonstrated [10] that upon incubation of soybean trypsin inhibitor (Kunitz) with catalytic quantities of trypsin a single peptide bond in the inhibitor Arg[64]—Ile[65] is hydrolyzed (Fig. 1). The phenomenon occurs at all pH values tested, but the rate of hydrolysis dramatically depends on pH. We have called the in-

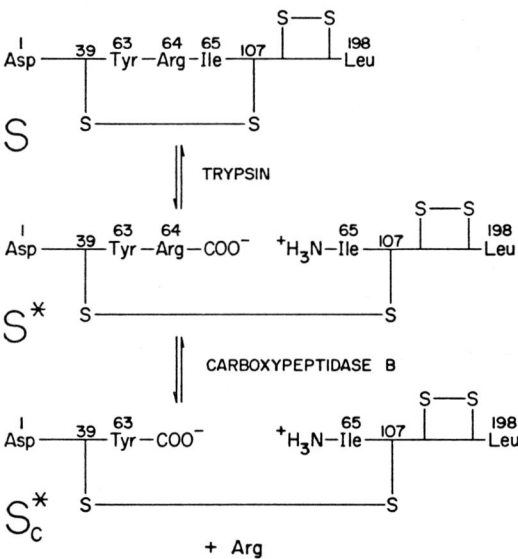

Fig. 1. Chemical events occuring on conversion of virgin to modified and of modified to des arg[64] modified soybean trypsin inhibitor (KUNITZ). The assignments of position are based on [10] and references contained therein and on partial sequences of IKENAKA et al. [42, 43] and of BROWN et al. [44]. For the sake of consistency with one former work we refer to the reactive site sequence as Tyr[63] Arg[64] Ile[65] even though IKENAKA et al. [42, 43] show Tyr[62]Arg[63] Ile[64]. All other assignments follow the work of IKENAKA.

hibitor with its entire peptide chain intact virgin inhibitor and the one with the peptide bond cleaved modified. Both the virgin and the modified inhibitors form 1:1 inactive complexes with trypsin with comparable and (at neutral pH) very high association constants. However, virgin inhibitor forms the inactive complex much faster than the modified one does — for soybean inhibitor of KUNITZ about 30—40 times faster. Treatment of modified inhibitor with carboxypeptidase B leads to the release of precisely 1 mole/mole of arginine and produces desarginine[64] modified inhibitor which, however, is called an inhibitor only because of the source from which it was produced. It does not inhibit trypsin.

HAYNES and FEENEY [11] have shown first that specific substitution of the NH_2 terminal isoleucyl in the modified soybean inhibitor (KU-NITZ) by trinitrobenzenesulfonate leads to complete inactivation. We have extended this observation to specific maleylation, citroconylation and carbamylation of the NH_2 terminus. Further, we have shown that specific citroconylation or carbamylation of the newly formed NH_2 terminus of modified chicken ovomucoid leads to a total loss of activity. It is our present opinion that any blockage of the newly formed NH_2 terminus of any modified inhibitor will lead to a loss of activity [12].

Identity of Complexes made from Virgin and Modified Inhibitors

Combination of trypsin with either virgin or modified inhibitor ultimately leads to a stable trypsin-inhibitor complex. The question to be asked now is: Are the two complexes the same? It becomes apparent that this is not a trivial question when we realize that complexes made of virgin inhibitor and β trypsin (intact single chain) and of virgin inhibitor and α trypsin (two chains, Lys^{131}—Ser bond hydrolyzed [13]) differ. They can be separated by disc gel electrophoresis at pH 9 on the basis of their expected difference in charge, and they dissociate at different rates at low pH. Clearly in the β trypsin-inhibitor complex the Lys^{131}—Ser bond is intact; in the α trypsin-inhibitor complex it is hydrolyzed.

We assert that the complexes made from virgin and from modified inhibitors are the same, and proof is clearly needed. Since both modified and virgin inhibitors are available, complexes were prepared from each and compared. They were found indistinguishable on the basis of disc gel electrophoresis and of rates of dissociation at a variety of pH values [14, 15]. The most convincing proof is based on kinetic control dissociation [3, 4]. If trypsin-inhibitor complex is suddenly transferred from conditions where it is stable to conditions where it is highly unstable (very low pH, 6M guanidine HCl etc.), dissociation takes place. The products may be either virgin inhibitor and trypsin or modified inhibitor and trypsin. If the conditions of dissociation are so harsh that after an initial dissociation recombination of products is highly improbable, kinetic control dissociation will be observed, i. e. the distribution of products will be governed by the relative values of the dissociation rate constants and not by the equilibrium constant between modified and virgin inhibitor. When complexes made from either virgin or modified inhibitor are subjected to kinetic control dissociation under a variety of conditions, predominantly virgin ($\geq 95\%$) inhibitor is obtained, as shown by disc gel electrophoresis, ΔpH overshoot on combination with trypsin and presence of a single chain after reduction of the disulfide bridges [3, 4]. This result must be due to kinetic control dissociation since equilibrium favors modified inhibitor under all conditions [2] and, therefore, *proves* that the complexes made from both forms of the inhibitor are the same chemical species.

This result, coupled with previous ones, allows us to state a large number of conclusions. First, the simplest possible mechanism for the interaction is

$$T + I \underset{k_{-a}}{\overset{k_a}{\rightleftarrows}} C \underset{k_{-b}}{\overset{k_b}{\rightleftarrows}} T + I^* \qquad (1)$$

where T is trypsin, I and I* are virgin and modified inhibitors respectively and C is the stable complex. Second, we know that $k_a \gg k_{-b}$ (direct measurement) and $k_{-a} \gg k_b$ (kinetic control dissociation), so the reactions on the left hand side are fast, while those on the right hand side are slow. Third, since the stable complex does not know whether it was made from virgin or from modified inhibitor, enzymatic action of trypsin must be involved in either the k_b steps (the reactive site peptide bond is intact in the complex and is not involved in any covalent interaction with trypsin) or in the k_a steps (the reactive site peptide bond is hydrolyzed in the complex) or in both (the reactive site peptide bond is in some intermediate state in the complex, e. g. an acyl or a tetrahedral intermediate involving a covalent bond between inhibitor and enzyme).

Of these three possibilities two have strong advocates. We have chosen [3] the intermediate state for the reactive site peptide bond in the complex, e. g. acyl enzyme; many of our critics choose the peptide bond fully intact and no covalent bond. To the best of our knowledge the remaining option has no admirers.

It must be very clear that at first glance the evidence presented so far favors our critics. The simplest way to account for the almost exclusive production of virgin inhibitor upon kinetic control dissociation is to assume that it is already in this form in the complex. The reactions on the left hand side of equation (1) are fast — they could be thought of as simple association-dissociation steps coupled, if need be, with conformational changes. The reactions on the right hand side are slower — these could be viewed as a complex sequence of steps leading to the production of virgin inhibitor and only then to its combination with the enzyme to form a stable complex.

In rebutting these arguments we are fortunate in being able to cite an analogy to a system where similar phenomena are involved. TOBIAS et al. [16] introduced lactones I and II in the study of chymotrypsin mechanism. These lactones are highly useful since it can be readily shown that

I II

upon their combination with chymotrypsin, acyl enzymes are formed. The enzyme-substrate complexes can be isolated by Sephadex chromatography. Spectroscopy of the isolated complexes shows that the *p*-nitrophenol portion of the substrate is in the *p*-nitrophenol form at low pH and in *p*-nitrophenoxide form at high pH, i. e. that the lactone bond is cleaved and phenolic-

-OH or -O⁻ is free. These acyl enzymes can be subjected to kinetic control dissociation; in the case of lactone II-complex the expected hydroxyacid is the predominant product, but in the case of lactone I-complex the predominant product is the lactone even though equilibrium considerations also favor the hydroxyacid. Thus, it is seen that kinetic control dissociation yielding the original substrate does not necessarily imply that the substrate is in the unaltered form in the complex. The nature of the product formed appears to depend on the restriction of the motion of the hydroxy group in the acyl enzyme. When that group is rigidly constrained (as in lactone I) it effectively competes with water and yields a lactone on dissociation; when the group is free to move (as in lactone II) attack by water on the acyl enzyme is favored, and the hydroxyacid is the predominant product. For some time we have been suggesting that in modified trypsin inhibitors the newly formed amino terminal amino acid is rigidly constrained [1, 4, 17]. This supposition gains additional support from the work of HUBER [7] presented at this meeting showing that in bovine pancreatic trypsin inhibitor (KUNITZ) the NH of Ala[16] is hydrogen bonded to the C=O of Gly[36] (Lys[15]—Ala[16] is the reactive site peptide bond of this inhibitor).

Therefore, the analogy is certainly consistent with our postulate. The work of TOBIAS et al. was extended by HEIDEMA and KAISER [18] to the sultone below.

Again, the sultone forms an acyl enzyme with chymotrypsin. Measurement of the rates of decomposition of the acyl enzyme to the sultone and to the hydroxy sulphonic acid shows that the sultone predominates in the decomposition products by a factor of 100. What is of additional great interest to us is that the pH dependence of the decomposition to either product was almost precisely the same.

Stopped Flow Studies on Trypsin-Inhibitor Association Kinetics [15]

Analogies, however useful, are not convincing enough. To strengthen our point we embarked long ago on a program of detailed kinetic characterization of the trypsin-inhibitor interaction in the hope that unravelling of the mechanism would give us additional clues about the nature of the complex.

In order to obtain the least ambiguous data, freshly purified β trypsin was used. This material was on the average 90% active, as judged by the all-or-none p-nitrophenyl-p'-guanidobenzoate (NPGB) assay at pH 8 [19]. The active fraction was made up of roughly 85—90% of a major component, and 10—15% of a minor component as measured by either disc gel electrophoresis of complexes with soybean inhibitor or by NPGB kinetic assay at pH 4 [20].

Since the association of either virgin or modified soybean trypsin inhibitor with trypsin is quite rapid, stopped flow measurements are the obvious technique for study. Such a study, however, requires a decision about the signal that will be observed. The published method [21] is to use proflavin as an indicator for free (uncomplexed) trypsin. Such a method has an advantage in producing large and conveniently followed signals in the visible, but it suffers from the fact that it is an indirect technique. On the other hand, EDELHOCH and STEINER [22] and BENMOUYAL and TROWBRIDGE [23] have shown that upon complex formation there is a small but measurable perturbation of ultraviolet absorption spectra of trypsin and/or of inhibitor. Difference spectra between complex and equivalent amounts of free trypsin and free inhibitor are shown in Figure 2. We [15] chose to base our stopped flow measurements on the signal at 260 nm since it is very broad and the monochromator could be used essentially wide open. Our choice has the great advantage of direct measurement of complex; its obvious disadvantage is that the signal is small and occurs in a range where the proteins themselves have a large absorbance. These drawbacks made the signals

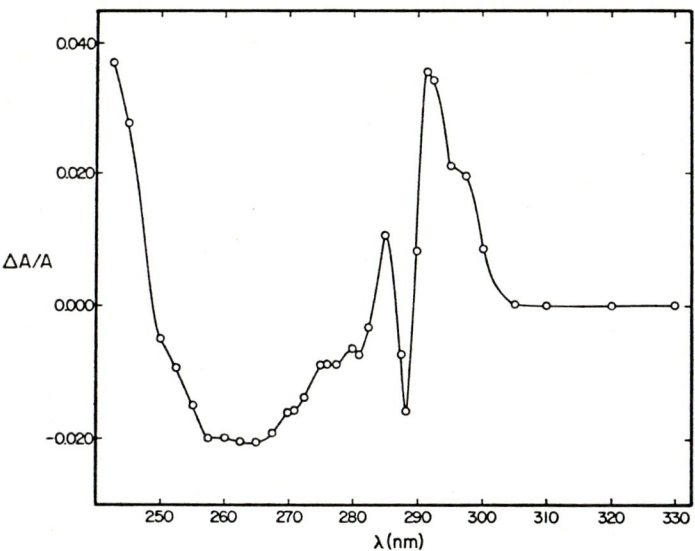

Fig. 2. Ultraviolet difference spectrum between soybean trypsin inhibitor-trypsin complex (sample cell) and free inhibitor and free trypsin (reference cell). Equimolar solutions of trypsin and of inhibitor were placed in separate compartments of the reference tandem double cell [45]. Equal volumes of the two solutions were mixed to form complex and placed in each of the two compartments of the sample tandem double cell. The results are expressed as $\frac{\Delta A}{A}$, i. e. the absorbance difference at each wave length is divided by total absorbance of the reference cell at the same wave length [15].

obtained difficult to read with very high accuracy and further restricted us to mixing trypsin and inhibitor only in 1 : 1 mole ratio in order to maximize the signal.

Equation (1) suggests that the trypsin-inhibitor association should be second order, and indeed, most workers studying such systems in the past (generally at very low concentrations) found it to be so. When our measurements were conducted at very low concentrations (long path length cell) the traces obtained were second order with respect to time and showed the expected inverse proportionality of half-time to starting concentration. As the concentrations were appreciably raised (short path length cell) the time dependence of the traces approximated first order rather than second order, and the dependence of half-time on reactant concentration diminished and almost disappeared as the concentrations were raised to the highest measurable levels. These results are summarized in Figure 3. Such results clearly suggest that the order switches from second to first as the concentration is raised. The simplest interpretation is to invoke a rapidly equilibrating intermediate, which then in a first order process yields the stable complex, i. e.

$$T + I \rightleftharpoons L \longrightarrow C \qquad (2)$$

and

$$T + I^* \rightleftharpoons L^* \longrightarrow C \qquad (3)$$

Such mechanisms are commonly involved for either noncovalent protein-ligand interaction where the second step is then a conformation change or for covalent protein modification where the second step is then a chemical reaction. HAYNES and FEENEY [24] have already suggested that the trypsin-inhibitor reaction is not diffusion controlled. Therefore, the involvement of an intermediate such as L is highly likely. What is of considerable interest to us here is the symmetry in the behavior of the T+I and T+I*

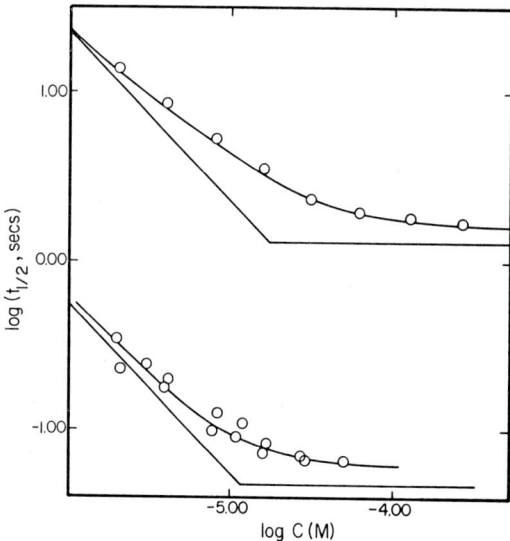

Fig. 3. The logarithm of the half life of absorbance changes observed at 260 nm in the Gibson-Durrum stopped flow apparatus on mixing of equimolar solutions of β trypsin and of virgin (lower curve) or of modified (upper curve) inhibitor as a function of the logarithm of the concentration of these solutions. At lower concentrations a 20 mm path length and at higher concentrations a 2 mm path length were employed, pH 5.6, 23°C [15].

systems (equations (2) and (3) and Fig. 3). On these bases the mechanism of equation (1) must be revised to yield

$$T + I \underset{k_{-1}}{\overset{k_1}{\rightleftarrows}} L \underset{k_{-2}}{\overset{k_2}{\rightleftarrows}} C \underset{k_{-3}}{\overset{k_3}{\rightleftarrows}} L^* \underset{k_{-4}}{\overset{k_4}{\rightleftarrows}} T + I^* \quad (4)$$

Data such as those of Figure 3 should be interpretable to yield $K_L = \dfrac{k_{-1}}{k_1}$ and k_2 from the $T+I$ association and $K_L^* = \dfrac{k_4}{k_{-4}}$ and k_{-3} from the $T+I^*$ association. However to resolve these parameters a decision must be made whether the spectral signal arises from the $T+I \rightarrow L$ step or from the $L \rightarrow C$ step. Since a large time dependent signal persists even at high (T) and (I) concentrations, where L is formed essentially instantaneously on our time scale, we have assumed that the latter was correct[1].

With this assumption in hand data such as of Figure 3 have been obtained at a variety of pH values. They were resolved into the two component parameters by an elaborate computer procedure. The resolution itself is rather inaccurate, and changes in computing procedure or new data may still give rise to sizeable revisions. However, it appears that such revisions have only a very small effect on the second order rate constants $\dfrac{k_2}{K_L}$ and $\dfrac{k_{-3}}{K_L^*}$ and since in all subsequent calculations numbers are used in that form, the comparisons with measured equilibrium constants are not likely to be significantly affected by later revisions.

Figures 4 and 5 show, respectively, the pH dependence of k_2 and k_{-3} and of K_L and K_L^*. One is struck by the obvious symmetry in the pH dependence of these numbers, except for the deviation in the behavior of pK_L^* in the 6—8 pH region. Whether this is real or a flaw in our resolution procedure is still an open question.

If we assume that the complex, C, is a central intermediate in the conversion of I to I* and probably an acyl enzyme then the values of k_2 and K_L and their pH dependence should be compared with k_{cat} and K_m values for hydrolysis of amide and peptide substrates, which are normally assumed to be acylation steps. The normal values of k_{cat} are about 1/sec and of K_m about 10^{-3} M^{-1}, and, therefore, the values seen in Figure 4 appear too large.

However, ABITA et al. [25] have shown that the k_{cat} and K_m values may vary by orders of magnitude away from the normal values depending upon the sequence of residues surrounding the hydrolyzed Lys-X or Arg-X peptide bond. In particular, presence of neighboring negatively charged Asp and Glu residues markedly decreases the parameters, while their absence

---

[1] Obviously both steps could produce a signal. The data are not precise enough to say anything more than that $L \rightarrow C$ step is responsible for most of the signal.

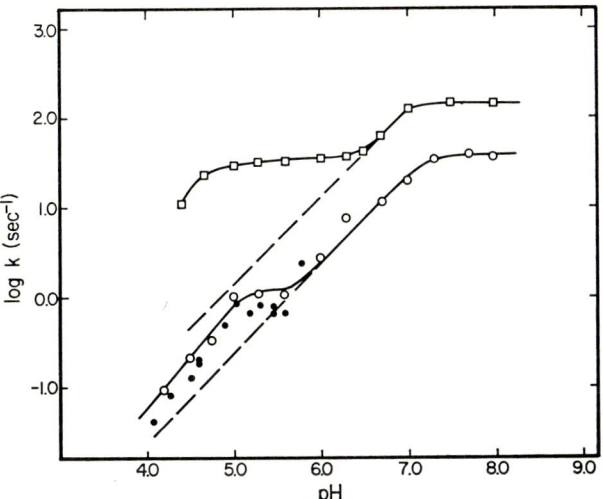

Fig. 4. The logarithm of the rate constants k_2 (L $\longrightarrow$ C in equation (4)), □, and of k_{-3} (L* $\longrightarrow$ C in equation (4)), ○, as a function of pH. These data are based on analysis of results similar to Fig. 3 at each pH value. β trypsin was used throughout, 23°C. [15].

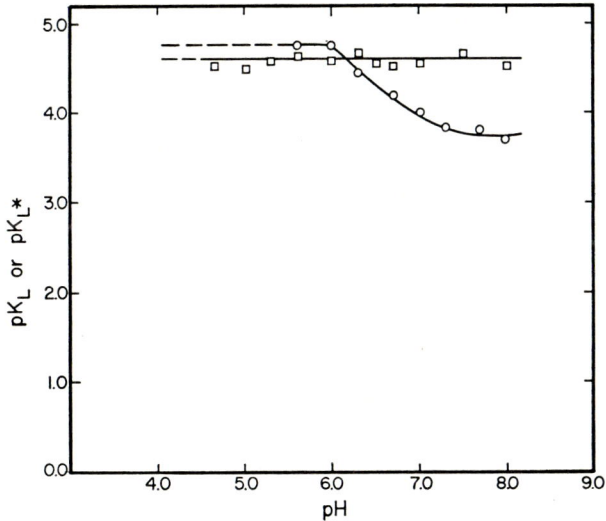

Fig. 5. The negative logarithm (pK_L or pK_L^*) of dissociation constants K_L and K_L^* of loose, non-covalent complexes L and L* (equation (4)) as a function of pH. In calculating these values essentially instantaneous equilibration of $T + I \rightleftharpoons L$ and $T + I^* \rightleftharpoons L^*$ was assumed. β trypsin was used throughout, 23°C. [15].

over long stretches of sequence makes the catalysis and binding much more favorable than "normal" values. In this connection it is worth noting that **1)** in the sequence surrounding the Arg[64]-Ile reactive site of soybean inhibitor (KUNITZ) the nearest negatively charged residue is Glu[70]; **2)** in the sequences surrounding the reactive sites of all trypsin and chymotrypsin inhibitors whose sequences were determined, negatively charged residues are notably absent

([26] and the various papers at this meeting); and 3) in the X-ray crystallographic structure [6] of pancreatic trypsin inhibitor (Kunitz) all negatively charged residues are quite far from the Lys[15]-Ala[16] reactive site and therefore, presumably far from the contact area with trypsin.

The pH dependence of k_2 shows that the formation of the stable complex critically depends on ionization of a group with pK 7 — in close analogy with known acylation reaction, where this pK is usually assigned to the loss of a proton from His of trypsin (or more properly to the Asp-His-Ser charge relay). The plateau at pH 5—6 is unexpected. Most workers anticipate that the logarithm of acylation rate should decline linearly with declining pH (dotted line in Fig. 4), however, no definite information is at hand. The linear decline over a broad pH range has been shown for ester hydrolysis [27] but this is deacylation limited. In the work on amide and peptide bond hydrolysis data collection was usually terminated at pH 6 or higher since the rates become inconveniently slow for study at lower pH. The plateau problem is intriguing and deserves detailed investigation. It is of interest that similar plateau occurs in the data of Pütter [28] on the second order association constant of pancreatic trypsin inhibitor (Kunitz) with trypsin.

The almost negligible pH dependence of K_L and rather small one of K_L* are consistent with the generally small pH dependences of K_m's for trypsin and chymotrypsin substrates.

The impression that emerges from the examination of the k_2 and k_{-3} data and of K_L and K_L* data is that they are fully consistent with the acylation rate constant for truly excellent substrates for trypsin and that the pH dependences are very similar for the reaction of $T + I$ and of $T + I*$.

In examining stopped flow data there is always a nagging doubt that even though the data are correct the rates are assigned to the wrong reaction. In the case of $T + I*$ reaction we are fortunate in being able to test the assignment by measuring the $L* \rightarrow C$ rate in much more direct experiments. Solutions of T and I* were mixed, and after various short time intervals the pH was suddenly lowered to affect kinetic control dissociation. In such an experiment the product of dissociation of L* will be I*, while the product of dissociation of C will be I. Thus, we were able to determine the rate of conversion of I* to I as a function of the time interval between mixing with trypsin and kinetic control dissociation [4]. This rate is obviously $L* \xrightarrow{k_{-3}} C$ since at the high concentrations of T and I* employed the formation of L* is essentially instantaneous and complete, and the time course of the reactions is essentially first order. The data shown as black points in Figure 4 agree rather well with the stopped flow data.

Measurement of Rate Constants for Complex Dissociation

After we make the assumption that the complexes L and L* equilibrate rapidly, complete characterization of the mechanism of equation (4) requires 6 parameters: K_L, K_L*, k_2, k_{-2}, k_3, k_{-3}. Four of them were determined by stopped flow association experiments; the two dissociation rate constants of the stable complex k_{-2} (to virgin inhibitor) and k_3 (to modified inhibitor) remain to be determined. From kinetic control dissociation we know only that $k_{-2} \gg k_3$ but neither the magnitude nor the pH dependence.

The rate constant, k_3, was determined by analyzing the steady state conversion of virgin to modified inhibitor by catalytic quantities of trypsin [2,29]. The distribution of the two inhibitor species as a function of time was monitored by disc gel electrophoresis and produced data similar to the bottom curve in Figure 6 at each pH value. It turns out that not only is the conversion of virgin to modified inhibitor easier to follow experimentally, but also on mathematical analysis it provided more useful kinetic information than the conversion of modified to virgin inhibitor. Curves such as the bottom curve of Figure 6 readily yield the value

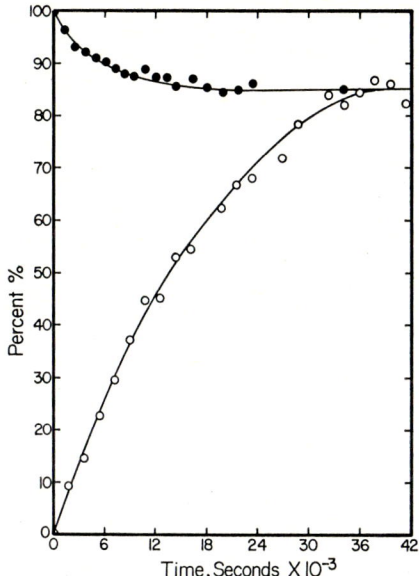

Fig. 6. Tryptic conversion of both pure virgin (I) and pure modified (I*) soybean trypsin inhibitor into the equilibrium mixture as monitored by disc gel electrophoresis. For the I → I* conversion 2 mole percent and for the I* → I conversion 0.5 mole percent of commercial trypsin were employed, pH 4.00, 20°C [2].

of K_{hyd}, the equilibrium constant between modified and virgin inhibitor, and k_{cat}, the catalytic rate constant for formation of I*. (The latter is simply obtained provided that the starting inhibitor concentration $(I)_0$ is much greater than K_m, which is certainly the case above pH 4.) Steady state analysis [30] of the mechanism of equation (4) yields for k_{cat}

$$k_{cat} = \frac{k_2 k_3}{k_2 + k_{-2} + k_3} \quad (5)$$

Since from kinetic control experiments we know that $k_{-2} \gg k_3$ this simplifies to

$$k_{cat} = \frac{k_2 k_3}{k_2 + k_{-2}} = k_3 \times \frac{\frac{k_2}{k_{-2}}}{1 + \frac{k_2}{k_{-2}}} \quad (6)$$

The quantity $\frac{k_2}{k_{-2}}$ is the equilibrium constant between L and C in mechanism of eq. (4). The fraction $\frac{\frac{k_2}{k_{-2}}}{1+\frac{k_2}{k_{-2}}}$ is just the fraction of C in an equilibrium mixture of L and C. At pH values above 4, $\frac{k_2}{k_{-2}}$ is very large (stable complex predominates over loose complex) and the fraction is essentially unity. Under such circumstances

$$k_{cat} = k_3 \quad (7)$$

At lower pH values $\frac{k_2}{k_{-2}}$ is low enough to yield

$$k_{cat} = \frac{k_2}{k_{-2}} k_3 \quad (8)$$

Since k_3 and k_{-2} have closely similar pH dependences (as will be shown) the pH dependence of $\frac{k_3}{k_{-2}}$ is very small and the pH dependence of k_{cat} will be governed by that of k_2, which as we have already shown rises with rising pH. Therefore, the frequently observed low pH maximum for conversion of virgin to modified inhibitors arises because at low pH k_{cat} has the pH dependence of k_2, at higher pH values the pH dependence is that of k_3.

Above pH 4, k_{cat} is k_3 (Fig. 7). If the mechanism of equation (4) is to be compared to a simple enzyme-substrate mechanism and C to an acyl enzyme then k_3 is analogous to the deacylation rate constant. We note that its value at neutral pH (ca 10^{-6}/sec) is about 8 orders of magnitude lower than the common value of about 100/sec for deacylation of good substrates. Further, its pH dependence is precisely opposite of that normally expected. It is, of course, clear that proteinase inhibitors are inhibitors just because this is true.

We were thus far unable to obtain reliable k_{-2} data from steady state catalytic conversion of modified to virgin inhibitor. However, since from kinetic control dissociation data we know that $k_{-2} \gg k_3$ and since the rate of dissociation of the stable complex, $k_D = k_{-2} + k_3$ (see equation (4)) it is simply sufficient to measure the rate of dissociation of the complex and assume that $k_D = k_{-2}$.

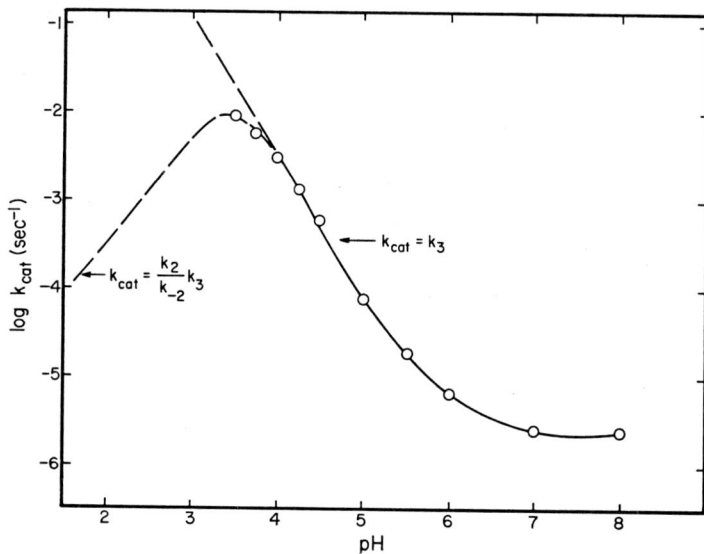

Fig. 7. The logarithm of the rate constant, k_{cat}, for enzymatic conversion of virgin to modified inhibitor by catalytic amounts of β trypsin as a function of pH. These data are essentially based on analysis of initial slopes of curves such as the bottom curve of Fig. 6. It is well known that at very low pH the rate of conversion becomes very slow. However, in this pH range enzyme-substrate binding is also quite weak and extensive work needs to be done to resolve K_m and k_{cat}. This has not yet been completed. 20°C. [29].

At very low pH the dissociation of complex can be measured by stopped flow simply by mixing unbuffered solutions of complex near neutral pH with very strongly buffered acidic solutions. The sudden pH drop causes the complex to dissociate and the dissociation can be monitored by following the rise in optical density at 260 nm (see Fig. 2, this experiment is simply the reverse of the association experiment). The reactions are expected to be and are found to be strictly first order. The top points in Figure 8, are obtained in this manner. Unfortunately rates at pH values lower than 2 could not be obtained since the dissociation rate constant is 100 sec^{-1} at this pH (half life 6 msec), and coupled with the weak signal used to

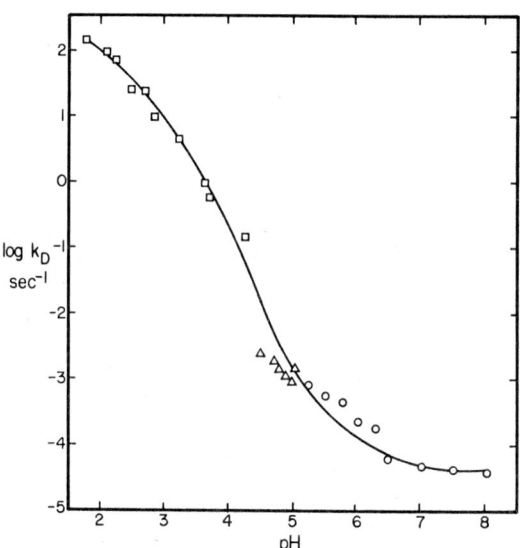

Fig. 8. The logarithm of the rate of dissociation of trypsin-inhibitor complex, k_D, as a function of pH. The data are essentially used as k_{-2} data of equation (4) (see text). □, Obtained by mixing equal volumes of complex and of acidic buffer and monitoring the 260 nm absorbance change in stopped flow [15], △, obtained by adding a small amount of complex to a TAME solution and observing the increase in rate of hydrolysis in a pH stat, i. e. GREEN's [31] Method I [14], ○, obtained by adding small amounts of trypsin to a solution containing TAME and soybean trypsin inhibitor and observing the decrease in rate of hydrolysis — GREEN's [31] Method II [32]. 25°C.

monitor this is at our experimental limit. On the higher pH end the stopped flow data are restricted by the fact that simple pH drop is insufficient to cause complete dissociation of the complex (at the lowest concentrations acceptable in the experiment) above pH 3.5. Thus, the data are complicated by the necessity of including partial reversibility corrections. At only slightly higher pH no appreciable dissociation is observed.

An alternate method of studying dissociation is to rely on the enzymatic activity of trypsin as a criterion of complex dissociation. Since such enzymatic methods are sensitive to very low concentrations of free trypsin, very much lower concentrations of complex can be employed, thus assuring greater dissociation. Far more important is the fact that concentrated solutions of substrates with very low K_m value, such as TAME and BAEE, powerfully compete with the inhibitor for free trypsin and thus insure at least partial dissociation. Two methods of taking advantage of these conditions have been suggested in qualitative form by N. MICHAEL GREEN [31] and elaborated for quantitative application by DURAN [14] and HIXSON [32]. In the first, a small quantity of complex is added to a concentrated substrate solution — at first no tryptic activity is observed, but later, as the complex dissociates, the substrate keeps free trypsin from recombining with the inhibitor, and ultimately the maximal possible tryptic activity is observed. The rate of appearance of this maximal trypsin activity is the rate of dissociation of complex. This method is reasonably direct but suffers greatly when the rate of dissociation of complex is very small (half time greater than 30 min) because it imposes very rigid requirements on instrument stability. The second method due to GREEN [31] is quite versatile but much less direct. A small quantity of free trypsin is added to concentrated solution of substrate containing some dissolved inhibitor. At first the rate of substrate hydrolysis is high since all of the added trypsin interacts with substrate, but slowly some of the trypsin combines with inhibitor and an equilibrium rate of hydrolysis (slower than initial) is attained. From the fraction of trypsin ultimately inhibited, the rate of attainment of equilibrium, the inhibitor and substrate concentration and K_m for substrate, the dissociation rate constant can be calculated. It is clear, however, that both methods used to measure k_D suffer from complications due to introduction of substances other than trypsin and inhibitor and from variety of assumptions made in calculations. The data shown at the bottom of Figure 8 are clearly the least satisfactory set in this study — they suffer further since most of them have been obtained with commercial trypsin rather than purified β trypsin. We are still searching for more satisfactory method of measuring k_{-2}, either by interpreting the steady state rates of modified to virgin inhibitor conversion or by monitoring the rate of complex dissociation.

Figure 8 shows that the rate of dissociation of the stable trypsin inhibitor complex is very small at neutral pH, thus accounting for the stability of the complex. This slow dissociation of complexes may be important physiologically since even when an enzyme-inhibitor complex is transferred to a solution where it must dissociate because it is, say, very dilute or because a great deal of substances which compete for either the free enzyme or for free inhibitor are present, the enzyme activity will still appear only after a sizeable delay.

The second observation is that $k_{-2} > k_3$ as expected from the kinetic control dissociation experiments. An even more important finding is the inverse pH dependence of the two rate constants k_{-2} and k_3 for breakdown of the central complex. By contrast, the reactions of most proteolytic enzymes with good specific substrates are characterized by increasingly more rapid breakdown of the central (acyl) complex (i. e. higher turnover of substrate) as pH is increased toward neutrality. As was pointed out earlier the kinetic parameters for formation of the stable complex of trypsin with I or I* (k_2 and k_{-3}) are in accord with those for normal enzyme-substrate reactions. It is therefore obvious that the extraordinary stability of trypsin-inhibitor complex is due to the inverse pH dependences of k_{-2} and of k_3. This fact is reflected in the very high pH dependence of the trypsin-STI association equilibrium constant.

The strange pH dependence of k_{-2} and k_3 becomes even more intriguing when we realize that it implies that complex must become protonated in order to dissociate rapidly to either virgin or to modified inhibitors. The pK's of the relevant groups in the complex can, in principle, be determined by analysis of the k_{-2} and k_3 data as a function of pH, but the present data are too incomplete to warrant such an analysis. Particularly striking is the fact that the rate of dissociation continues to increase significantly all the way down to pH 2. This implies that pK of one of the critical groups in the complex is less than 2.

As pointed out before, the great speed of dissociation of the soybean complex prohibits extending the data to lower pH. However, the complex of pancreatic trypsin inhibitor (KUNITZ) and trypsin dissociates far more slowly at comparable pH values than the soybean complex. We have carried out preliminary stopped flow measurements on dissociation of this complex at pH 2, 1 and 0, and we have found a factor of 10 increase in the dissociation rate constant on each pH unit decrease. This result suggests an almost incredibly low pK of less than 0 for the group which is critical to dissociation. An alternate possibility is the consideration of a slow protonation step. The experiments at pH 1 and 0 are an amusing application of stopped flow technique since trypsin is unstable at these pH values. However, to obtain our data the solution had to be monitored at the low pH values only for a few seconds or less, and, therefore, valid data were obtained.

Comparison of Directly Measured Equilibrium Constants with those Calculated from Rate Constants

Having measured all 6 parameters of the kinetic mechanism of equation (4), we are ready to draw some general conclusions about the nature of the complex. However, before we do so we must answer two important questions 1) is the formulated mechanism correct? and 2) are the parameters correct, i. e. have we properly assigned the various rate constants to the outcomes of our various experiments?

Kinetic analysis cannot prove a mechanism, it can only show it to be consistent. The same consistency test can be applied to the correctness of rate assignments. The 6 parameters we measured fully describe the system. The system is also characterized by two pH dependent equilibrium constants, which can be measured quite independently of all of the kinetic measurements carried out here. We prefer to define the following two constants. One is the equilibrium constant between modified inhibitor and virgin inhibitor, K_{hyd}. It is independent of the enzyme used. It can be measured by achieving a steady state in trypsin catalyzed virgin to modified inhibitor conversion or better by approaching the equilibrium from both directions (Fig. 6). In terms of kinetic constants K_{hyd} is given by

$$K_{hyd} = \frac{[I^*]}{[I]} = \frac{k_1 k_2 k_3 k_4}{k_{-1} k_{-2} k_{-3} k_{-4}} \quad (9)$$

In the discussion of the stopped flow results we have already assumed that T and I, and T and I* equilibrate almost instantaneously to yield respectively L and L* and that the dissociation constants for these complexes are given by $K_L = \frac{k_{-1}}{k_1}$ and $K_L{}^* = \frac{k_4}{k_{-4}}$. Introducing these equalities into (9) we obtain

$$K_{hyd} = \frac{K_L{}^*}{K_L} \times \frac{k_2 k_3}{k_{-2} k_{-3}} \quad (10)$$

The values of K_{hyd} have been measured directly by allowing solutions of virgin or of modified inhibitor to incubate with catalytic quantities of trypsin for long periods of time at various pH values. The composition of these solutions (i. e. the virgin to modified inhibitor ratio) was monitored occasionally by disc gel electrophoresis, and when no further change with time was noted the compositions were assumed to be equilibrium compositions and used in calculating K_{hyd}. The data are shown in Figure 9. It should be pointed out that these are preliminary data and not as reliable as the value of $K_{hyd} = 6 \pm 1$

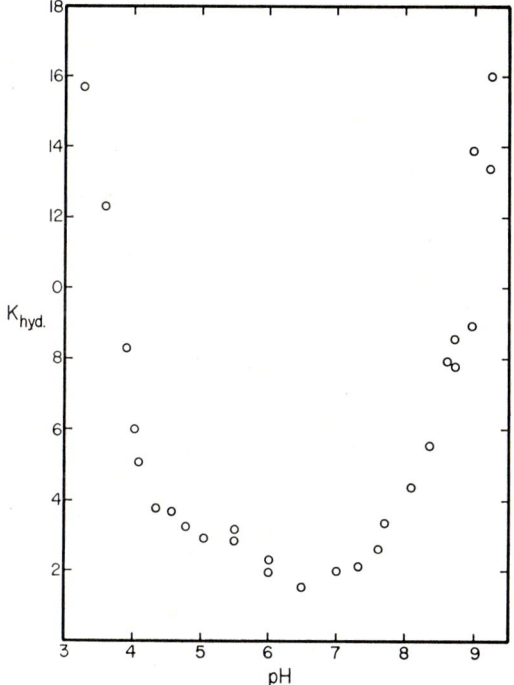

Fig. 9. pH dependence of K_{hyd}, the equilibrium constant for hydrolysis of virgin to modified inhibitor [31]. These are preliminary data subject to revision.

at pH 4.00 which is based on detailed analysis of approach to equilibrium from both directions (Fig. 6).

Since all of the parameters of equation (10) are now known we can also calculate the value of K_{hyd} from kinetic parameters. For example at pH 6.0

$$K_{hyd} = \frac{2 \times 10^{-5} M}{3 \times 10^{-5} M} \times \frac{30/\text{sec}}{1.5 \times 10^{-4}/\text{sec}} \times \frac{7 \times 10^{-6}/\text{sec}}{2/\text{sec}} = 0.5$$

The example is given to show that all of the 6 parameters are critically important in obtaining K_{hyd} and that numbers of greatly different orders of magnitude are involved in the calculation. The observed value at pH 6 obtained by direct measurement is about 2.5 ± 1 (Fig. 9). Similar calculations have been repeated as a function of pH (every 0.2 pH unit) from pH 4.4—8.0. In all cases K_{hyd} calculated from kinetic data agreed with the directly observed one within a factor of 3—5 (kinetic calculation generally yields low values). The pH dependences of the calculated and of the directly determined data are closely similar. In view of the probable experimental errors (both random and systematic) we regard the agreement as very satisfactory.

The second equilibrium constant is much more commonly employed. It is the association constant for the enzyme-inhibitor complex. However, in the terminology of mechanism of equation (4) there exist three forms of complex L, C, and L* and two forms of inhibitor I and I*. When these are included we obtain

$$K_{app} = \frac{(L^*) + (C) + (L)}{(T)[(I) + (I^*)]} \quad (11)$$

In terms of kinetic parameters it can be written as[1]

$$K_{app} = \frac{1 + \dfrac{k_{-3}}{k_3} + \dfrac{k_{-3}}{k_3} \times \dfrac{k_{-2}}{k_2}}{K_{L^*}(1 + 1/K_{hyd})} \quad (12)$$

Note that at pH values above 4, the first and the third term in the numerator (corresponding to (L*) and (L)) are negligible and the term $(1 + 1/K_{hyd})$ in the denominator makes but a small contribution. All terms have been used in the calculation, but it is clear that in fact the relation is

$$K_{app} = \frac{k_{-3}/k_3}{K_{L^*}} \quad (13)$$

and we are coupling only data of Figures 4, 5 and 7.

---

[1] Numerous equivalent methods of writing this are available. We have chosen to write it this way to stress the parameters dealing with the modified inhibitor-trypsin interaction.

Equilibrium constants for trypsin-soybean trypsin inhibitor association have been carefully measured by a potentiometric procedure of LEBOWITZ and LASKOWSKI [8]. In this procedure the average number of protons, $\bar{q}$, released upon forming a molecule of complex is determined at each pH value of interest

$$TH_{t-\bar{h}_t} + IH_{i-\bar{h}_i} = CH_{t+i-\bar{h}_c} + \bar{q}H^+ \quad (14)$$

$$\bar{q} = \bar{h}_c - \bar{h}_t - \bar{h}_i \quad (15)$$

where t and i are the maximal numbers of proton binding sites on T and I respectively, and $\bar{h}_t$, $\bar{h}_i$ and $\bar{h}_c$ are the numbers of protons which have been titrated from the maximally protonated form of T, I, and C respectively at the pH of measurement. The values of $\bar{q}$ are then used by inserting them into equation

$$\frac{d \log K_{app}}{dpH} = \bar{q} \quad (16)$$

$$(\log K_{app})_{pH_2} = (\log K_{app})_{pH_1} + \int_{pH_1}^{pH_2} \bar{q}\, dpH \quad (17)$$

Application of this equation requires one reference value of $(\log K_{app})\, pH_1$. In the LEBOWITZ-LASKOWSKI procedure [8] this is obtained from the curvature of the plot of the number of protons released per molecule of inhibitor added when the inhibitor is added to a constant amount of trypsin, i. e. from the deviation from stoichiometric complex formation. Such an experiment can be done only at relatively low pH where K_{app} is of right order of magnitude. The data of LEBOWITZ and LASKOWSKI [8] have, in fact, been confirmed by measurements of WINZOR and NICHOL [33] (as reinterpreted by GILBERT [34]) by gel exclusion chromatography.

However, in retrospect, it seems clear that three systematic errors have been made by LEBOWITZ and LASKOWSKI [8] and that these are sufficiently serious to make us question the data. These are **1)** unawareness of the true explanation of the "overshoot" and thus of the involvement of both virgin and modified inhibitor. While Lebowitz mixed virgin inhibitor with trypsin, it is clear that instead, in order to get data that conform to the definition of K_{app} given in equation (11), *an equilibrium mixture of I and I* at the pH of interest* must be used instead of virgin I. It turns out, however, that this objection, while fundamentally important, introduces only very small errors into the data; **2)** the optical factor and molecular weight of soybean inhibitor used by LEBOWITZ were incorrect. Further, there was no independent criterion for measurement of active trypsin (it had to be based on the assumption that the inhibitor is 100% active). It turns out, however, that fortunate cancellations lead to this cause introducing almost no error; **3)** α and β trypsin combine with the inhibitor with different equilibrium constants and produce somewhat different $\bar{q}$ values. LEBOWITZ used commercial trypsin, whose active component is usually 50% α and 50% β. Thus, the $\bar{q}$ values were averages of α and β values. More importantly the measured reference equilibrium constant, $(K_{app})\, pH_1$ (see equation (17)) was evidently a complicated average of the constants appropriate to α and β trypsin.

Therefore, the LEBOWITZ and LASKOWSKI experiments were repeated [35, 36] avoiding the three difficulties listed above. The $\bar{q}$ values as a function of pH for α, β and commercial trypsin (LEBOWITZ) are shown in Figure 10. It is noteworthy that both the α and β curves show sharp maxima in $\bar{q}$ while no such maximum is apparent in the data with commercial trypsin. This is largely due to averaging. It is thus clear that a very important feature which contributes a good deal of potential information content is lost by the use of commercial trypsin.

The $\bar{q}$ values as a function of pH are a differential titration curve between complex and the two uncomplexed components. Such a differential titration curve should be of great value in determining molecular details of trypsin-inhibitor interactions when the pK values of the responsible groups are determined before and after complexation and when these pK's are assigned to specific ionizable groups in the trypsin and/or inhibitor molecule. Such an analysis is now going on but is far from complete. It is, however, clear from a most casual inspection of Figure 10 that

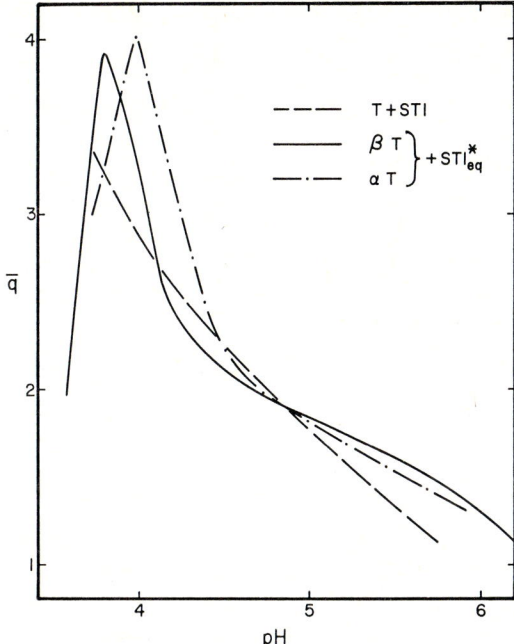

Fig. 10. The average number of hydrogen ions released per molecule of trypsin-inhibitor complex formed as a function of pH.

————— β trypsin and equilibrium mixture of virgin and modified inhibitor prepared at the pH of the experiment [35, 36],

—·—·— α trypsin and equilibrium mixture above [35, 36],

— — — commercial trypsin and virgin inhibitor [8].

several groups (at least 4) ionizing in the acidic range (pK ≤ 7) undergo large downward shifts in pK as a result of complex formation. From an analysis of these and other data on several trypsin-inhibitor systems we have concluded elsewhere [26] that most or all of these groups are on trypsin rather than on the inhibitor molecule.

Application of the data of Figure 10 and of the newly determined reference values $(K_{app})_{pH_1}$ in equation (17) yields K_{app} as a function of pH (Fig. 11). It is seen that (K_{app}) for β trypsin is about 10 times greater than for α trypsin over most of the pH range and that the commercial trypsin results are much closer to those for α trypsin than those for β trypsin. We are now ready to compare the β trypsin equilibrium results obtained by potentiometry with those calculated from kinetic parameters by the use of equation (12). As can be seen in Figure 11 the agreement is again remarkably satisfactory. It should be noted that Figure 11 is a log-log

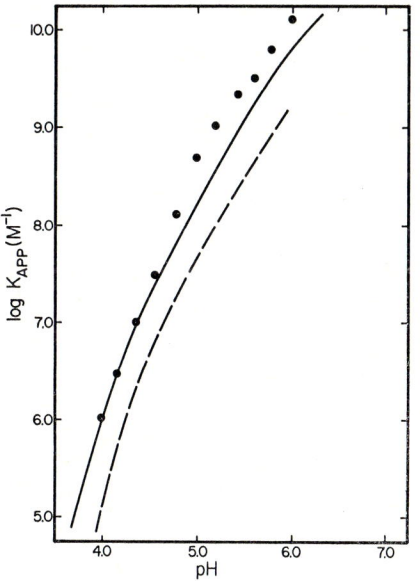

Fig. 11. The logarithm of trypsin-inhibitor association constants, K_{app}, as defined by equation (11), as a function of pH. ————— β trypsin and equilibrium mixture of inhibitors. These values were obtained potentiometrically. Data of Figure. 9 and appropriate reference equilibrium constant were used in equation (17). — — — α trypsin and equilibrium mixture, the data for commercial trypsin [8] are almost identical to these. ● values of K_{app} for β trypsin calculated by using kinetic data of Figs. 4, 5, 7 and 8 in equation (12).

plot. The ratio of the calculated to observed data is also within a factor of 3 to 5, closely similar to the results on K_{hyd}.

Evidence that the Complex is either an Acyl Enzyme or a Tetrahedral Intermediate

Thus, the 6 parameters of the general mechanism (equation (4)) survived what we regard as highly rigid consistencial test, by agreeing within the experimental error to both the K_{hyd} and K_{app} values. It should be noted that these tests were conducted over a relatively broad pH range over which both the equilibrium constant (K_{app}) and the parameters (k_2, k_{-2}, k_3, k_{-3}) varied over several orders of magnitude. We thus gain considerable confidence in the mechanism being reasonably correct.

The mechanism of equation (4) places the stable complex, C, in an exactly symmetric central position between I and I*. Furthermore, the pH dependences of symmetric pairs of rate constants (k_2 and k_{-3}, Fig. 4), (k_{-2} and k_3 Figs. 7 and 8) and quasi-equilibrium constants K_L and K_L^* (Fig. 5) are closely similar. Note that identical behavior within all pairs is not possible since this would require K_{hyd} to be pH independent (eq. 10) which it is not (Fig. 9). Finally, the rates of formation of the stable complex are similar in their magnitude and pH dependence to the rates of formation of "central intermediates" — acyl enzymes from excellent trypsin substrates.

On the basis of all of these considerations we feel that we have *proved* (as well as anything can be *proved* by kinetic analysis) that the stable trypsin-inhibitor complex is the central intermediate in the conversion of I to I* and that the reactive site peptide bond of the inhibitor in the complex must be in an intermediate position between I and I*.

What, then, is the exact chemical nature of the complex. The widely held belief of a majority of scientists in this field is that all serine protease catalyzed hydrolyses must go through an acyl enzyme intermediate and that this acyl enzyme is the "central intermediate" for all reversible hydrolysis reactions [37]. If this general belief is correct the problem is solved, and soybean trypsin inhibitor (KUNITZ) and trypsin are linked in the complex by an ester bond between the carboxyl of Arg[64] of the inhibitor and the hydroxyl of Ser[183] of trypsin. On the other hand, several scientists regard the general proof for mandatory acyl intermediate in the hydrolysis of peptides as far from compelling [38—40] although it is compelling for esters. They suggest instead that tetrahedral intermediates may directly decompose to products (in our case L*) without the mandatory intermediacy of the acyl enzyme.

The possibility that C is a tetrahedral intermediate (peptide bond intact) is attractive since it clearly explains the need for the unblocked NH_2 terminal of Ile[65]. On the other hand, blockage of the amino terminus of modified inhibitors could simply interfere with proper acyl enzyme formation by steric hindrance.

Let us now return to the question of what do we hope the X-ray crystallographers of enzyme-inhibitor complexes will explain. First, they could settle the acyl-enzyme tetrahedral intermediate problem. Second, they can probably account for the fact that inhibitors in the association step act as such excellent substrates by showing that the reactive site sequence of the inhibitors is locked in the very favorable substrate like conformation. This was already accomplished by HUBER et al. [7] in the model fitting of his pancreatic trypsin inhibitor (KUNITZ) to a model for trypsin. It appears that it will be considerably harder to show why the complexes dissociate so slowly and especially hard to show why the protonation of some ionizable groups of trypsin (in the complex) so greatly speeds up the rate of dissociation of the complex either to starting materials (virgin inhibitor) or to products (modified inhibitor). Clearly this stabilization is the major unsolved problem of enzyme-inhibitor research.

The Large pK Shifts on Complex Formation

Very large pK shifts on complex formation of several ionizable groups must be invoked to explain the $\bar{q}$ data (Fig. 10) and especially the pH dependence of complex dissociation at very low pH (Fig. 8 and pancreatic trypsin inhibitor results in the text). Such shifts may well seem paradoxical to many protein chemists. Fortunately we were able to show quite directly in a model experiment that very large pK shifts on

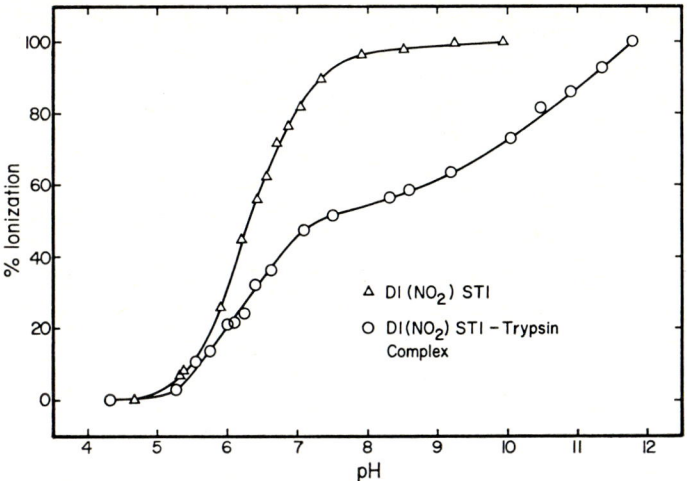

Fig. 12. Spectrophotometric titration curve, followed at 420 nm of △ soybean trypsin inhibitor with two of its tyrosyl residues nitrated ○ an equimolar mixture of that inhibitor and trypsin. Additional experimental evidence shows that trypsin and dinitro inhibitor are fully associated up to roughly pH 9 — above that pH appreciable dissociation takes place [41].

complex formation do indeed take place. Soybean trypsin inhibitor was nitrated (KUNITZ) and a monomeric dinitro derivative that forms a complex with trypsin was isolated from the products. Of the four tyrosyls in soybean trypsin inhibitor, two were nitrated in the derivative — Tyr[63], which is next to the reactive site (see Fig. 1), and either Tyr[17] or Tyr[18]. The derivative shows normal titration behavior with the two nitrotyrosyls having roughly identical pK's of 6.3 (Fig. 12). Titration of the complex, however, shows that only one of the groups ionizes normally with its original pK of 6.3. The other group shows an abnormally shaped titration curve with a midpoint at pH 10 to 11 [41]. We presume that the latter is nitro Tyr[63]. Other data, up to pH 12, indicate that the odd titration behavior is most likely due to the dissociation of the complex and is not due to the ionization of nitro Tyr [63] *in the complex*. Therefore, the pK of the nitro Tyr [63] *in the complex* is even higher than pH 10—11. The important point is that we have shown by direct experiments that pK shifts of at least 4.5 pK units (probably more) may accompany complex formation.

All of this talk has been confined to a detailed analysis of the interaction between bovine β trypsin and soybean trypsin inhibitor (KUNITZ). Yet the analysis of this situation serves well to explain the interactions of many serine proteinases with many protein proteinase inhibitors. The fact that most conclusions described here are relatively general is partially documented in [26] and even more so by several other papers at this meeting.

References

[1] LASKOWSKI, M., Jr. in: Structure Function Relationships of Proteolytic Enzymes, (P. DESNUELLE, H. NEURATH and M. OTTESEN eds.) Munksgaard, Copenhagen (1970) pp. 89—101.

[2] NIEKAMP, C. W., H. F. HIXSON, Jr. and M. LASKOWSKI, Jr., Biochemistry **8**, 16 (1969).

[3] FINKENSTADT, W. R. and M. LASKOWSKI, Jr., J. Biol. Chem. **242**, 771 (1967).

[4] HIXSON, H. F., Jr. and M. LASKOWSKI, Jr., J. Biol. Chem. **245**, 2027 (1970).
[5] SEALOCK, R. W. and M. LASKOWSKI, Jr., Biochemistry **8**, 3703 (1969).
[6] HUBER, R., D. KUKLA, A. RÜHLMANN, O. EPP and H. FORMANEK, Naturwissenschaften **57**, 389 (1970).
[7] HUBER, R., D. KUKLA, A. RÜHLMANN and W. STEIGEMANN, First International Conference on Proteinase Inhibitors, Munich (1970), this volume, p.
[8] LEBOWITZ, J. and M. LASKOWSKI, Jr., Biochemistry **1**, 1044 (1962).
[9] FINKENSTADT, W. R. and M. LASKOWSKI, Jr., J. Biol. Chem. **240**, PC 962 (1965).
[10] OZAWA, K. and M. LASKOWSKI, Jr., J. Biol. Chem. **241**, 3955 (1966).
[11] HAYNES, R. and R. E. FEENEY, Biochim. Biophys. Acta **159**, 209 (1968).
[12] KOWALSKI, D., unpublished results from this laboratory.
[13] SCHROEDER, D. D. and E. SHAW, J. Biol. Chem. **243**, 2943 (1968).
[14] DURAN, R. W., M. S. Thesis, Purdue University 1965.
[15] LUTHY, J. A., M. PRAISSMAN and M. LASKOWSKI, Jr., Abstracts 158th National Meeting of the American Chemical Society, New York, N. Y. 1969, BIOL No 321; full manuscript in preparation.
[16] TOBIAS, P., J. H. HEIDEMA, K. W. LO, E. T. KAISER and F. J. KEZDY, J. Am. Chem. Soc. **91**, 202 (1969).
[17] NIEKAMP, C. W., Ph. D. Thesis, Purdue University 1971.
[18] HEIDEMA, J. H. and E. T. KAISER, J. Am. Chem. Soc. **92**, 6050 (1970).
[19] CHASE, T., Jr. and E. SHAW, Biochem. Biophys. Res. Comm. **29**, 508 (1967).
[20] HRUSKA, J. F., J. H. LAW and F. J. KEZDY, Biochem. Biophys. Res. Comm. **36**, 272 (1969).
[21] FEINSTEIN, G. and R. E. FEENEY, Biochemistry **6**, 749 (1966).
[22] EDELHOCH, H. and R. F. STEINER, J. Biol. Chem. **240**, 2877 (1965).
[23] BENMOUYAL, P. and C. G. TROWBRIDGE, Arch. Biochem. Biophys. **115**, 67 (1966).
[24] HAYNES, R. and R. E. FEENEY, Biochemistry **7**, 2879 (1968).
[25] ABITA, J. P., M. DELAAGE, M. LAZDUNSKI and J. SAVRDA, Eur. J. Biochem. **8**, 314 (1969).
[26] LASKOWSKI, M., Jr. and R. W. SEALOCK, in: The Enzymes, 3rd edition (Boyer, P. D. ed) vol. III, Academic Press, New York 1971 (in press).
[27] KEZDY, F. J., G. E. CLEMENT and M. L. BENDER, J. Am. Chem. Soc. **86**, 3690 (1964).
[28] PÜTTER, J., Z. physiol. Chem. **348**, 1197 (1967).
[29] MATTIS, J. A., unpublished results from this laboratory.
[30] PELLER, L. and R. A. ALBERTY, J. Am. Chem. Soc. **81**, 5907 (1959).
[31] GREEN, N. M., J. Biol. Chem. **205**, 535 (1953).
[32] HIXSON, H. F., Jr., Ph. D. Thesis, Purdue University, 1970.
[33] NICHOL, L. W. and D. WINZOR, J. Biochim. Biophys. Acta **94**, 591 (1965).
[34] GILBERT, G. A., Nature **210**, 299 (1966).
[35] FINKENSTADT, W. R., unpublished results from this laboratory.
[36] LASKOWSKI, M., Jr. and W. R. FINKENSTADT, in: Methods of Enzymology vol. II (C. H. W. HIRS and S. N. TIMASHEFF eds.) Academic Press, New York 1971 (in press).
[37] BENDER, M. L. and F. KEZDY, J. Ann. Rev. Biochem. **34**, 49 (1965).
[38] EPAND, R. M., Biochem. Biophys. Res. Comm. **37**, 313 (1969).
[39] WANG, J. H., Proc. Nat. Acad. Sci. U. S. **66**, 874 (1970).
[40] KAPLAN, H., V. B. SYMONDS and D. R. WHITAKER, Can. J. Biochem. **48**, 649 (1970).
[41] MCKEE, R. E. and C. W. NIEKAMP, unpublished results from this laboratory.
[42] IKENAKA, T., T. KOIDE and S. TSUNAZAWA, Symposium on Protein Structure, Tokyo (1968).
[43] IKENAKA, T., S. TSUNAZAWA and T. KOIDE, J. Biochem. (Tokyo) **69**, (1971).
[44] BROWN, J. R., M. LERMAN and Z. BOHAK, Biochem. Biophys. Res. Comm. **23**, 561 (1966).
[45] HERSKOVITS, T. T. and M. LASKOWSKI, Jr., J. Biol. Chem. **237**, 2481 (1962).

Mass Spectral Determination of Peptide Bond Hydrolysis Equilibria in Protein Proteinase-Inhibitors

HARALD TSCHESCHE and RAINER OBERMEIER*

*Organisch-Chemisches Laboratorium der Technischen Universität München, Germany,
Lehrstuhl für Organische Chemie und Biochemie*

All of the evidence known so far suggests that a basic amino acid residue, either lysine or arginine, is required as an essential residue within the reactive site of protein-like trypsin inhibitors [1, 2]. A great many inhibitors, up to twenty from quite different sources, have been investigated with respect to these residues within the reactive site and all have been found to belong to either class of lysine- or arginine-inhibitors. It is obvious that the bonds formed from these residues within the reactive sites of trypsin inhibitors [(Arg) Lys — Ile (Leu, Ala)] are of the same nature as those found to be sensitive to tryptic hydrolysis and that are cleaved during activation of a variety of zymogens.

It has been observed in a number of trypsin inhibitors that this enzyme-susceptible bond in the reactive site was cleaved during interaction with the enzyme trypsin generating a modified inhibitor from virgin one [1, 4—7]. It has been shown by LASKOWSKI, Jr. and his coworkers that the cleavage of the reactive site peptide bond in soybean trypsin inhibitor is a reversible process leading to a steady state interconversion of virgin and modified inhibitors with true thermodynamic equilibrium. The equilibrium constant ($K_{Hydr.}$) was shown to be a function of pH with a broad minimum in the pH range of 5—8 and a sharp increase at both high and low pH values [8]. These unusually low pH values (pH 2.00—4.50) for tryptic hydrolysis of the reactive site peptide bond are pecularities of trypsin inhibitors since they differ strikingly from common substrates. Their rates of hydrolysis sharply decline with rising pH, while that of other substrates rise [1, 3, 9].

The Reactive Site of Porcine Pancreatic Secretory Trypsin Inhibitor

We found that the porcine pancreatic secretory trypsin inhibitor I (PSTI) is modified at acid pH by a specific, limited proteolysis reaction with catalytic amounts of trypsin [5, 9]. The lysine-isoleucine bond (Residues 18—19) has been identified as the reactive site on the basis of the limited enzymatic proteolysis at acid pH. In order to show that the lysine is really at the reactive site, we tried to remove this residue by carboxypeptidase B treatment, an experiment kindly suggested to us already in 1967 by LASKOWSKI, Jr. Unfortunately we were not able to inactivate the modified inhibitor because of the unfavourable sequence Pro-Lys which enabled this residue to withstand carboxypeptidase B treatment. Therefore determination of the rate

---

* Present address Protein Research Laboratory, University of Pittsburgh, Pittsburgh, Pennsylvania 15213, U.S.A.

of modification by inactivation experiments with carboxypeptidase B — a method used with great success by Ozawa and Laskowski, Jr. [1] and Rigbi and Greene [7] — could not be applied. The site of cleavage within the polypeptide chain (Residues 18—19) was determined by Edman degradation of the equilibrium mixture with p-bromophenyl-isothiocyanate. [10] This reagent was chosen because it facilitated the mass spectral identification of the cleaved p-bromophenyl-thiohydantoines [11] and even permitted their semi quantitative assay. Two sequences could be deduced unequivocally, the amino terminal sequence which had been determined before by degradation of virgin inhibitor I (Thr-Ser-Pro-) (97—100% yield) and the inner partial sequence from the point of cleavage (Ile-Tyr-Asn-) (2—3% yield) in minor amounts [11], Figure 1. Therefore, the site of cleavage within the chain could be located without further separation and investigation of modified and virgin inhibitors as having occurred between residues 18 and 19.

No other points of cleavage could be detected within the polypeptide chain. This result was in agreement with the positioning of the reactive site in the bovine pancreatic secretory inhibitor determined by Rigbi and Greene [7].

Assay of Peptide Bond Hydrolysis Equilibria

It was obvious, that adaptation of this method to a quantitative determination of the cleaved phenyl-thiohydantoins after the first cycle of Edman degradation performed on the equilibrated virgin and modified inhibitor mixture would permit easy assay of the equilibrium constant at a given pH.

The mass spectral determination of the phenyl-thiohydantoins of the aliphatic amino acids is uncomplicated with respect to stability and volatility in the mass spectrometer. These amino acids do not generate problems during Edman

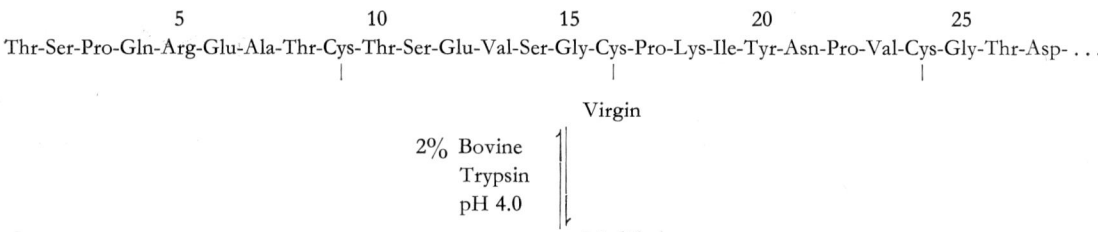

Cycle	1	2	3	Yield
a. of Virgin inhibitor	Thr	Ser	Pro	100—97%
b. of Modified inhibitor	Thr (Ile)	Ser (Tyr)	Pro (Asn)	100—97% (3—1.5%)

Fig. 1. Identification of the reactive site peptide bond within the sequence of porcine pancreatic secretory trypsin inhibitor I by three cycles of quantitative Edman degradation on inhibitor previously modified by a limited acid partial proteolysis reaction at pH 4.0 in 1% solution of 0.05M ammonium formate by 2% of bovine trypsin (Merck AG, 2 U/mg) at 28° for 16 hours. No other amino acids were detected within the mass spectra (Fig. 4) of the PTH indicating no further splits within the polypeptide chain. For the procedure of degradation and quantitative assay of the cleaved residues see ref. [11] and legend of Tab. 1. Amino acid sequence deduced from ref. [11].

degradation as has been observed with glutamine or serine residues. Quantitation of the determination of alanine, leucine or isoleucine — the amino acids encountered as carboxyterminal residues in the reactive sites of the trypsin inhibitors investigated so far — could be achieved by an isotope dilution step and a derivative ratio assay using conventional mass spectrometry. The corresponding penta-deutero-phenylthiohydantoines or the para-fluorophenylthiohydantoines as substitutes were used as internal standards [12, 13]. A similar procedure using ^{15}N-Methylderivatives has been developed by RICHARDS [14, 15]. The internal standards were added to the previously equilibrated mixture of virgin and modified inhibitors after the first cycle of EDMAN degradation but prior to extraction of the cleaved phenylthiohydantoins. In order to eliminate the difficulties encountered with quantitative extraction, the cleaved phenylthiohydantoins and the internal standards were extracted together from the reaction mixture and subjected to mass spectral assay.

To obtain quantitative information from the mass spectral data, the ratios (Ile-PTH : Ile-D$_5$-PTH and Ala-PTH : Ala-F-PTH) in the height of the molecular ion peaks (M^+) present in the mixture were accurately measured by hand from the recorded spetrum. As is demonstrated in Figure 2, these heights are not differing much for different amino acids within the same class, i. e. the aliphatic amino acids, when equal amounts are subjected to mass spectral assay. Nearly identical peak heights within experimental error were found when different derivatives of the same amino acid were investigated in our mass spectrometer (Varian CH-4, Varian MAT, Bremen, Germany). This is obvious from the spectra of the proline derivatives shown in Figure 3.

The accurately measured ratios of the peak heights together with the initial concentration of each of the added standards permitted determination of the exact quantity of each phenylthiohydantoin (Ala, Ile or Leu) formed during Edman degradation. Small differences in the volatility of the derivatives were determined independently and taken into consideration during calculations (Ile-PTH : Ile-D$_5$-PTH = 1.00 : 1.03). Any contribution from other ions was substracted whenever this was necessary.

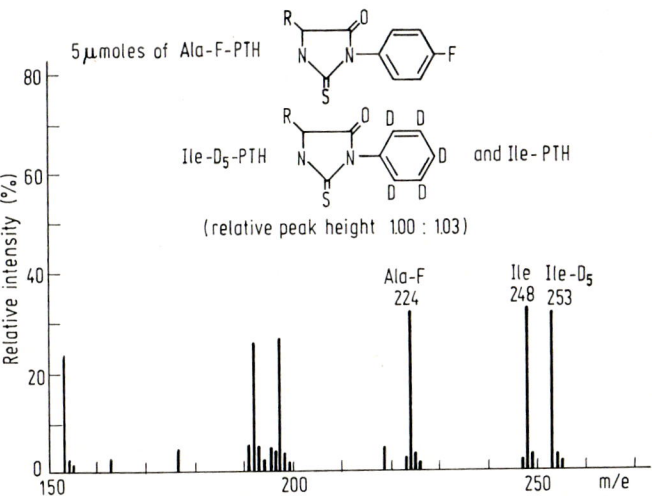

Fig. 2. Mass spectrum of a mixture of 5 μ-mol each of Ala-F-PTH (M^+ 224), Ile-PTH (M^+ 248) and Ile-D$_5$-PTH (M^+ 253) obtained from a Varian CH-4 mass spectrometer (Varian MAT, Bremen, Germany). The PTH-derivatives were adsorbed to coal spectro grade which then was inserted directly into the ion source (type TO 4). The ratios of the peak heights remained constant within $\pm$ 5% over a period of more than 30 min. Ionizing voltage 45 eV.

We have investigated several trypsin inhibitors with respect to their mode and extent of modification by catalytic amounts of trypsin at

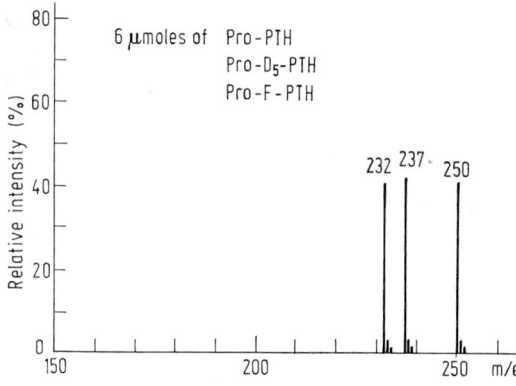

Fig. 3. Mass spectrum of 6μmole each of Pro-PTH (M+ 232)Pro-D_5-PTH(M+ 237) and Pro-F-PTH(M+ 250). Experimental conditions as in Fig. 2.

acidic pH. In this procedure the inhibitors (4.95 or 2.80 mg dry weight in 1% solution) were equilibrated for a period of 20 hours with 2% of trypsin at pH 4.0 (PSTI for 16h) and 28°. The question of reversibility was not examined for the modified inhibitors, nor was checked by longer incubation periods whether the steady state equilibrium indeed was reached within the 20 hours because of short supply of the various inhibitors.

Table 1 summarizes the results obtained with several different trypsin inhibitors. Under the experimental conditions described (see legend to Table 1) we found a 3% modification of porcine pancreatic secretory trypsin inhibitor I. The spectrum used for calculation is shown in Figure 4. The homologous inhibitor from the same class of proteins, the ovine pancreatic secretory trypsin inhibitor, exhibited a modifica-

Table 1. The extent of modification was determined by one cycle of EDMAN degradation. The reaction was performed in 50 per cent aqueous pyridin at 40° for 1 hour with a 50M excess of phenylisothiocyanate under a barrier of nitrogen. The mixture was dried under vacuum at 40°, three times extracted with 2 ml of acetic acid ethylester using centrifugation and then dried again under vacuum. The entire coupling reaction was repeated one time. Cyclisation was achieved with 0.5 ml trifluoroacetic acid for 45 min at 40°. After evaporation of the excess acid the residue was dried under vacuum. For each of the isotope or fluoro-derivatives 1 ml of a solution of 0.5 μmole in acetic acid ethylester was added and the mixture dried. The residue containing the internal standard and the cleaved amino acid derivative was then extracted three times with 2 ml of acetic acid ethylester. The extracts were combined and subject to quantitative mass spectral assay.

Table 1. Modification of Trypsin Inhibitors

Inhibitor	Reactive Site	Modification pH 4.0 %	Method Ref.
Porcine (KAZAL)	Lys—Ile	2.8	Mass spectral
Ovine (KAZAL)	Arg—Ile	20	Mass spectral
Bovine (KAZAL)	Arg—Ile	19	Cpase B Inactiv.[a]
Guinea Pig seminal	Arg—Ile	21	Mass spectral
Dog Gl. submandibul.	Arg—Ala	6	Mass spectral*
Peanut	Arg—Ala	5	Mass spectral
Soybean (KUNITZ)	Arg—Ile	86	Gel electroph.[b]
Bovine (KUNITZ)	Lys—Ala	0 (< 0.2%)	Mass spectal

[a] RIGBI and GREENE 1968.
[b] FINKENSTADT and LASKOWSKI, Jr. 1967; NIEKAMP, HIXSON and LASKOWSKI, Jr. 1969.
* Value not corrected for purity of starting material.

tion of 20% under identical experimental conditions. This extent is in good agreement with the one determined by RIGBI and GREENE [7] on the bovine inhibitor by inactivation experiments with carboxypeptidase B. From our sequence studies [16] we know that both proteins — ovine and porcine — are completely alike within the region of the reactive site

period by 2% of bovine trypsin, since incubation with 2% of porcine trypsin (porcine trypsin, Merck AG, Darmstadt, Germany, 3.5 U/mg) resulted in a higher extent of 6.5% of modification. After prolonged periods of incubation of about 16 hours an increased degradation of the pancreatic secretory trypsin inhibitor by trypsin could be observed due to its temporary inhibitory character. Because conversion took place so slowly it was doubted that true equilibrium of virgin and modified inhibitor could be reached in the temporary porcine pancreatic secretory (lysine-) inhibitor prior to excessive degradation by trypsin at pH 4.0. The equilibrium constant may be as high as in the bovine and ovine (argine-) inhibitors. At pH 3.5 the extent of modification by 2% of porcine trypsin (16 hours, 28°) was 21% indicating the pH dependence of the modification reaction [9], s. Table 2.

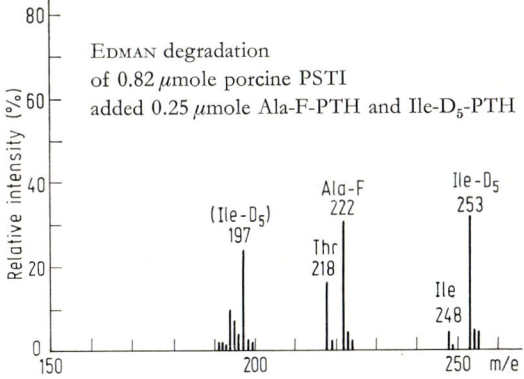

Fig. 4. Mass spectrum of the PTH-derivatives obtained after the first cycle of EDMAN degradation on 0.82 μmole of porcine pancreatic secretory trypsin inhibitor I, previously modified in 1% solution at pH 4.0 by 2% of bovine trypsin (Merck AG, 2 U/mg) for 16 hours at 28°. The peak heights were used for calculation of the extent of modification (m/e 197 is a major fragment of Ile-D_5-PTH). Experimental conditions as in Fig. 2.

sequence (Residues 15 through 28). That means, that exchange of lysine for arginine within the same reactive site sequence drastically changed the rate of modification by a factor of about seven, when the same trypsin preparation and concentration (2% bovine trypsin, Merck AG, Darmstadt, Germany, 2.0 U/mg) was used under identical conditions of incubation. This result is in accordance with the findings of SEALOCK and LASKOWSKI, Jr., that free [64-lysine]-inhibitor obtained by enzymatic replacements from virgin soybean trypsin inhibitor is modified much more slowly than the authentic [64-arginine]-inhibitor [17]. It seemed obvious that a true equilibrium of virgin and modified porcine inhibitor had not been reached within the 16-hour incubation

Table 2. Extent of modification of porcine pancreatic secretory trypsin inhibitor (2.80 mg) by 2% of trypsin, either bovine (Merck AG, 2 U/mg) or porcine (Merck AG, 3.5 U/mg), in 1% solution at pH 4.0 (3.5) for 16 hours at 28°. Determination by mass spectral assay after one cycle of EDMAN degradation via quantitative determination of the cleaved PTH amino acid derivatives.

Table 2. Modification (%) of Porcine Inhibitor (KAZAL)

Trypsin species	pH 4.0	pH 3.5
Bovine Trypsin (2 mole %)	2.5	
Porcine Trypsin (2 mole %)	6.5	21

Another inhibitor tested originated from guinea pig seminal vescicles I [18]. The reactive site was determined to be an arginine-isoleucine bond modified to 21% by 2% trypsin at pH 4.0. The inhibitor from the submandibular gland of the dog [19] was found to contain an arginine-alanine bond that was cleaved to about 5% under our conditions. One plant inhibitor from peanut (Arachis hypogaea) [20] was investigated containing an arginine-alanine bond hydrolyzed to an extent of 4%. No modification at all could be detected in the basic trypsin inhibitor from bovine organs (WERLE's kallikrein inactivator

[21]; Kunitz inhibitor [22]. This result was not changed after prolonged periods of incubation (48 hours) and at 37° and with the trypsin concentration increased to 3%.

The procedure described here allowed easy identification of the amino acid residue participating on the reactive site peptide bond cleaved during the partial proteolysis reaction. Without further steps of separation and purification the cleaved peptide bond may be located within the sequence by consecutive EDMAN degradation. The quantitative assay permitted easy determination of the extent of modification and therefore of the equilibrium constant at a given pH. The accuracy of the determined values is within a $\pm 5\%$ range and may easily be improved. The method may be generally applied whenever modification occurs. It is still feasable for unfavourable sequences, for instance porcine secretory trypsin inhibitor, when the extent of modification cannot be determined by inactivation with carboxypeptidase B. One of its major advantages is, however, the high sensitivity at low levels of modification, since this is an inherent quality of mass spectral assay. The lowest limit of detection in our assay procedure indicates modification at an 0.2 mole % level.

Acknowledgements: We wish to thank N. HUMS and I. SCHMID for the registration of the mass spestra and we are indebted to Dr. FRITZ, Dr. HOCHSTRASSER and E. FINK for generous gifts of inhibitors from guinea pig, peanut and seminal vesicles.

References

[1] OZAWA, K. and M. LASKOWSKI, Jr., J. Biol. Chem. **241**, 3955 (1966).
[2] HAYNES, R. and R. E. FEENEY, Biochemistry **7**, 2879 (1968).
[3] LASKOWSKI, M., Jr., in: Structure Function Relationships of Proteolytic Enzymes, (P. DESNUELLES, H. NEURATH and M. OTTESEN, eds.) Munksgaard, Copenhagen (1970).
[4] FINKENSTADT, W. R. and M. LASKOWSKI, Jr., J. Biol. Chem. **240**, PC 962 (1965).
[5] TSCHESCHE, H., Z. Physiol. Chem. **348**, 1216 (1967).
[6] HAYNES, R. and R. FEENEY, Biochem. Biophys. Acta **159**, 209 (1968).
[7] RIGBI, M. and L. J. GREENE, J. Biol. Chem. **243**, 547 (1968).
[8] NIEKAMP, C. W., H. F. HIXSON, Jr. and M. LASKOWSKI, Jr., Biochemistry **8**, 16 (1969).
[9] TSCHESCHE, H. and H. KLEIN, Z. Physiol. Chem. **349**, 1645 (1968).
[10] TSCHESCHE, H. and E. WACHTER, unpublished results from this laboratory.
[11] TSCHESCHE, H. and E. WACHTER, Eur. J. Biochem. **16**, 187 (1970).
[12] WEYGAND, F. and R. OBERMEIER, Z. Naturforsch. **23b**, 1390 (1968).
[13] OBERMEIER, R., Ph. D. Thesis, Techn. University Munich 1969.
[14] RICHARDS, F. F., W. T. BARNES, R. E. LOVINS, R. SALOMONE and M. D. WATERFIELD, Nature **221**, 1241 (1969).
[15] FAIRWELL, TH., W. T. BARNES, F. F. RICHARDS and R. E. LOVINS, Biochemistry **9**, 2260 (1970).
[16] TSCHESCHE, H., R. OBERMEIER and K. HOCHSTRASSER, this volume, p. 207.
[17] SEALOCK, W. and M. LASKOWSKI, Jr., Biochemistry **8**, 3730 (1969).
[18] FRITZ, H., M. GEBHARDT, E. FINK, W. SCHRAMM and E. WERLE, Z. Physiol. Chem. **350**, 129 (1969).
[19] HAENDLE, H., H. FRITZ, J. TRAUTSCHOLD and E. WERLE, Z. Physiol. Chem. **343**, 185 (1965).
[20] HOCHSTRASSER, K., K. ILCHMANN and E. WERLE. Z. Physiol. Chem. **350**, 929 (1969).
[21] KRAUT, H., E. FREY and E. WERLE, Z. Physiol. Chem. **192**, 1 (1930).
[22] KUNITZ, M. and J. H. NORTHROP, J. Gen. Physiol. **19**, 991 (1936).

Discussion remarks: **Comments to the Modification Reaction**

M. LASKOWSKI, Jr.

In your data there is a much smaller conversion of virgin to modified porcine secretory inhibitor than it is the case for bovine secretory inhibitor. I prefer to think of this result as a kinetic rather than thermodynamic phenomenon. Some time ago SEALOCK (Biochemistry **8**, 3703 [1969]) noted that the virgin to modified conversion of the enzymatic Lys^{64} soybean trypsin inhibitor (KUNITZ) is much slower than of the natural Arg^{64} inhibitor. Since porcine and bovine inhibitors are homologous but differ in the reactive site residue (Lys in porcine, Arg in bovine) it is likely that the difference in the conversion rates of the porcine and bovine inhibitors is a reflection of a more general phenomenon — with equivalent reactive sites Lys inhibitors are converted to modified form more slowly than Arg inhibitors.

The results become even more interesting when we recall that RIGBI and GREENE (J. Biol. Chem. **243**, 5457 [1969]) have noted that pocine trypsin converts virgin bovine secretory inhibitor to modified form much more rapidly than bovine trypsin. Recently, Dr. FINKENSTADT in our laboratory noted that porcine trypsin converts virgin soybean inhibitor (KUNITZ) also very much more rapidly than bovine trypsin does. It seems, therefore, that porcine trypsin is generally more efficient in this reaction than bovine. The single molecular basis (if there is only one) of this difference is unknown. It is, however, striking that porcine inhibitor (Lys) is slow to convert and porcine trypsin is fast and that bovine inhibitor (Arg) is fast and bovine trypsin is slow. It appears that compensating mutations in the two molecules may have been made in order to keep the kinetic parameters of the interaction closely similar in both systems. Mr. SEALOCK is now studying this problem in detail.

The observation of compensating mutations is much more dramatically illustrated by considerations of human trypsin and human secretory inhibitor, discussed in other talks at this meeting, since human trypsin is not inhibited by porcine and bovine secretory inhibitors but is inhibited by the human secretory inhibitor, which is homologous to porcine and bovine ones. However, in this case the changes cannot yet be ascribed to any simple, single amino acid change such as Lys ↔ Arg reactive site mutation.

Chemistry and Biology of Proteinase Inhibitors from Soybeans and Groundnuts

by Yehudith Birk and Arieh Gertler

Faculty of Agriculture, The Hebrew University of Jerusalem Rehovot, Israel

The interest in structure-activity relationship of soybean trypsin inhibitors originates in the concern with nutritional properties of soybean proteins. Soybeans are a rich and important source of plant proteins for foods and feeds. It is a well known fact that they cannot be utilized efficiently by the living organism unless properly heated. The role and possible involvement of soybean trypsin inhibitors in the lower than expected nutritional values of raw soybean proteins and in pancreatic hypertrophy was studied by numerous investigators. Soybeans contain at least three different proteinase inhibitors: 1. The soybean trypsin inhibitor SBTI was isolated, crystallized and characterized by Kunitz [1, 2] and studied extensively ever since, serving also as the reactive site model in the studies of Laskowski Jr. and co-workers [3]. 2. The acetone insoluble trypsin inhibitor, reported by Bowman [4], was purified and further studied in our laboratory [5, 6, 7]. This inhibitor, designated AA, is also a strong chymotrypsin inhibitor. 3. The *Tribolium* larval proteinase inhibitor [8] was separated from the accompanying trypsin and chymotrypsin inhibitors by column chromatography on Ca-phosphate (hydroxylapatite) [9]. It was found to be responsible for the inhibition of *Tribolium* larval midgut proteinases *in vitro* and for the impairment of larval growth when added to their diet. In view of the presence of *Tribolium* proteinase inhibitors in soybeans [8] and in wheat [10] and of trypsin-like enzymes in stored product pests such as *Tenebrio molitor* [11], plant proteinase inhibitors may be considered as possible specific defense mechanisms of seeds evolved against insects.

Trypsin-Chymotrypsin Inhibitor AA from Soybeans

Following is a review of the properties of trypsin and chymotrypsin inhibitor AA [12]. It can be easily separated from SBTI upon extraction from the soybean meal with 60% ethanol in which SBTI is insoluble. Purification of the inhibitor is achieved by chromatography of the dialysed acetone-precipitate on CM-cellulose, yielding ∼ 200 mg inhibitor from 100 g soybean meal. AA has an extinction coefficient of 4.8, and an isoelectric pH of 4.2; it contains no carbohydrates, no free SH groups, no tryptophan but an abundance (∼ 20%) of cystine. It does not lose activity when treated with pepsin or pronase and differs in this respect from SBTI. The resistance to pepsin is of nutritional significance since ingested AA is likely to pass the stomach unaffected whereas SBTI is. AA inhibits the trypsin-like enzyme from pronase [13] but does not inhibit elastase [14]. The complex trypsin-AA or chymotrypsin-AA can be isolated by paper electrophoresis at pH 6.9 or by gel filtration. The complexes dissociate at pH 4.9. The fact that the trypsin-AA complex could inhibit chymotrypsin and the chymotrypsin-AA complex could inhibit trypsin suggested the presence in AA of two different and considerably remote active sites — one against trypsin and the other against chymotrypsin. This was further substantiated when AA was submitted to the reactive site studies according to LASKOWSKI [3]. Upon treatment with trypsin at pH 3.75 AA retained its activity against trypsin and chymotrypsin but it lost its activity towards trypsin when further digested with carboxypeptidase B whereas the activity against chymotrypsin was not affected. Similar treatment of AA with chymotrypsin at pH 3.75 resulted in loss of activity against chymotrypsin but the trypsin-inhibiting activity was not affected [15]. It has been concluded that AA possess two different, independent active sites, the one against chymotrypsin abiding probably in a considerably larger loop than that against trypsin. The amino acid composition of the reactive sites of AA against trypsin and chymotrypsin and the amino and carboxy terminals of the inhibitor are given in Table 1. The similarity to the recent findings on the lima bean inhibitor [16] is striking. It is of interest to point out that tyrosine does not form part of the bond opened by chymotrypsin, although it is in its vicinity. This may perhaps account for the fact that treatment with tetranitromethane affects, but only by 30%, the chymotrypsin inhibiting activity of AA. The finding that the trypsin

Table 1. End Groups and Reactive Sites of Trypsin and Chymotrypsin Inhibitor AA

NH$_2$-terminal amino acid	Asp	1 μmole/μmole AA, MW 8000†	By DNP and cyanate methods	
COOH — terminal amino acid	Gln	1 μmole/μmole AA, MW 8000	By CPA	Inhibitory activity not affected by removal
COOH — terminal sequence	LysGluGln ↓		By CPA followed by CPB	Inhibitory activity not affected by removal
Trypsin-reactive site sequence*	...Lys-Ser		By CPB and DNP, respectively	
Chymotrypsin-reactive site sequence**	...(IleAspValAla PheTyr) (GlnLeu) ↓ Leu-Ser...		By CPA and DNP, respectively	
Treatment with tetranitromethane	Amino acid analysis reveals disappearance of the two tyrosines and appearance of one nitro-tyrosine			Trypsin-inhibiting activity not affected; Chymotrypsin-inhibiting activity decreased by 30%
Treatment with p-azobenzenearsonate	Amino acid analysis reveals disappearance of 1—2 tyrosines			Trypsin-inhibiting activity slightly affected; chymotrypsin-inhibiting activity decreased by 70%

* Identified after cleavage with trypsin at pH 3.75.
** Identified after cleavage with chymotrypsin at pH 3.75.
† MILLAR, WILLICK, STEINER and FRATTALI, J. Biol. Chem. 244, 281 (1969).

DNP, dinitrophenyl method of SANGER
CPA, carboxypeptidase A
CPB, carboxypeptidase B

inhibiting activity of AA is not influenced at all by tetranitromethane further substantiates the independence of the two active sites.

Attempts to modify AA by elastase were carried out in the pH range of 4—9 showing optimal modifying condition at pH 6 to 8. No effect on the trypsin-inhibiting activity was noted but the activity against chymotrypsin was decreased by 70% when assayed on casein and by 35% when assayed on acetyl-tyrosine-ethyl ester. Reaction of the inhibitor modified by elastase with carboxypeptidase A for 24 hours resulted in release of 1 mole leucine and 1 mole aspartic acid per mole of AA.

Trypsin-Chymotrypsin Inhibitor GTCI from Groundnuts

Being interested in the nutritional significance of proteinase inhibitors and in their possible role as built-in insecticides we attempted a comparative study of inhibitors of common legume seeds. Although the presence of a trypsin inhibitor in groundnuts was reported already in 1947 [17], very little information has been added since. While our work on the groundnut trypsin and chymotrypsin inhibitor (GTCI) was in progress, two reports have been published [18, 19]. The different data clearly indicate the presence of more than one inhibitor in groundnuts. GTCI has been isolated by extraction of defatted groundnut meal at pH 5, followed by precipitation with ammonium sulfate (70% saturation) and successive column chromatography on DEAE-cellulose (pH 7.0), Ca-phosphate, hydroxylapatite, (pH 6.8) and CM-cellulose (pH 5.5) [20]. Unlike the other legume seed trypsin-inhibitors GTCI is a basic protein with an isoelectric pH of 8—9 arising probably from amides. No free SH groups or carbohydrates could be detected. It has a molecular weight of about 7700 as derived from sedimentation-diffusion and from short-column sedimentation equilibrium analyses as well as from gel filtration on Sephadex G-50. Unlike inhibitor AA from soybeans the complex that GTCI forms with trypsin is unable to inhibit chymotrypsin and GTCI-chymotrypsin complex does not inhibit trypsin. This suggests the presence of either one active site for both inhibitory activities or two very close sites. Reactive-site studies were performed on GTCI by incubation with trypsin at pH 3.75, followed or not by incubation with carboxypeptidase B(CPB) at pH 8.0. The results of these experiments, summarized in Table 2, show that the

Table 2. Effect of incubation of GTCI with trypsin at pH 3.75 and with carboxypeptidase B at pH 8.0 on the inhibitory activity of GTCI against trypsin and chymotrypsin when assayed on casein

Incubation of GTCI		Residual inhibitory activity against		Arginine released (mole/mole)
		Trypsin	Chymo-trypsin	
		(in %)		
At pH 3.75	for: 24 h	100	100	0.0
With trypsin at pH 3.75	for: 24 h	82	4	0.0
With trypsin at pH 3.75 for 24 h followed by CPB at pH 8	for: 2 h	71	3	0.7
	4 h	55	3	0.8
	16 h	38	2	1.3
	26 h	26	2	1.6
	42 h	26	2	1.6
With CPB at pH 8	for: 24 h	101	100	0.0

active site against trypsin is not identical with that against chymotrypsin although they are probably very close to each other. Mere incubation with trypsin at pH 3.75 inactivates the site against chymotrypsin but does not affect the site against trypsin. Treatment of the modified inhibitor with carboxypeptidase B for 26 hours led to a loss of ~75% of activity and a release of 1.6 mole arginine per mole GTCI, indicating that the trypsin-reactive site comprises of an Arg-X bond. Incubation of GTCI with either chymotrypsin at pH 3.75 or with carboxypeptidase B at pH 8.0 did not affect the trypsin-inhibiting or chymotrypsin-inhibiting activity of GTCI.

The amino acid composition of GTCI as compared to that of other legume seed trypsin inhibitors is given in Table 3. The similarity of

Table 3. Amino acid composition [a] of trypsin and chymotrypsin inhibitors from legume seeds

Amino acid	from soybeans		from lima beans				from groundnuts				from mung beans[g]	from kidney beans[h]	from navy beans[i]
	CSBTI[b]	AA	No. 1[c]	No. 2[c]	No. 3[c]	No. 4[c]	GTCI[d]	Tixier[e]	P_I[f]	P_{II}[f]			
Aspartic acid	29	11	12	14	13	13	8	14	24	24	10	10	30
Glutamic acid	21	8	6	5	7	7	6	16	16	12	7	5	17
Glycine	18	0	1	0	1	1	4	12	8	8	2	2	5
Alanine	9	4	3	3	4	3	3	7	8	8	3	3	8
Valine	12	1	1	1	1	1	5	9	10	12	1	1	2
Leucine	16	2	3	3	3	3	1	4	6	8	2	2	6
Isoleucine	14	2	4	4	5	4	0	1	2	0	2	3	9
Serine	13	8	12	12	15	13	5	10	16	12	10	10	35
Threonine	8	2	4	3	5	5	7	14	12	16	3	5	14
Half-cystine	4	14	12	14	16	14	14	22	12	12	8	10	30
Methionine	3	1	0	0	0	0	0	1	0	0	2	1	1
Proline	10	7	6	6	7	7	7	13	14	18	6	5	16
Phenylalanine	9	2	1	1	2	2	2	4	6	4	1	1	4
Tyrosine	4	2	1	1	2	1	1	3	2	2	1	1	4
Histidine	2	1	5	3	6	6	2	3	2	4	4	3	10
Lysine	11	5	4	4	4	4	2	7	4	4	6	3	11
Arginine	9	2	2	2	2	2	7	14	14	20	4	2	7
Tryptophan	2	0	0	0	0	0	0	0	?	?	0	0—1	0
Amide-NH_3	(14)	(6)	(4)	(5)	(5)	(5)	(12)	(?)	(?)	(?)	(5)	(4)	(17)
Total	194	72	77	76	93	86	74	154	156	164	72	67—78	209
Molecular weight	21700	7995	8404	8291	9892	9243	8089	16812	17124	18146	8032	8000	23030

[a] Expressed as number of amino acid residues per molecule.
[b] Wu and Scheraga, Biochemistry **1**, 698 (1962).
[c] Jones, Moore and Stein, Biochemistry **2**, 66 (1962).
[d] Tur-Sinai, Birk, Gertler and Rigbi, Israel J. Chem. **8**, 178p (1970).
[e] Tixier, C. R. Acad. Sc. Paris **266**, 2498 (1968).
[f] Hochstrasser, Illchmann and Werle, Hoppe-Seyler's Z. Physiol Chem. **350**, 929 (1969).
[g] Hsien-ming, Chu, Shan-Shan, Meih-suan, Cheng-wu and Tien-chin, Scientia sinica **14**, 1454 (1965).
[h] Pusztai, Biochem. J. **101**, 379 (1966).
[i] Wagner and Riehm, Arch. Biochem. Biophys. **121**, 672 (1967).

GTCI to AA and to the lima bean inhibitors is obvious. The abundance of S-S bonds is responsible for the stability of these inhibitors to proteolytic digestion and to elevated temperatures. It is quite possible that the inhibitor isolated by Tixier is a dimer of GTCI but the latter seems to differ markedly from the two inhibitors isolated by Hochstrasser et. al. Varietal differences of the seeds as well as state of maturity and aging should be considered as possible causes of difference, in addition to the genuine existence of more than one inhibitor in groundnuts, as is the case for soybeans and lima beans.

Nutritional Significance of Soybean Trypsin Inhibitors

Finally, we would like to discuss the nutritional significance of soybean trypsin inhibitors. The growth impairment and pancreatic hypertrophy of birds and mammals, resulting from the presence of raw soybean meal (RSBM) as the protein source in their diet, was often attributed to the trypsin inhibitors. Numerous studies were carried out with inhibitor preparations purified to different extents and the results with respect to growth impairment were controversial. Studies performed in our laboratory on the

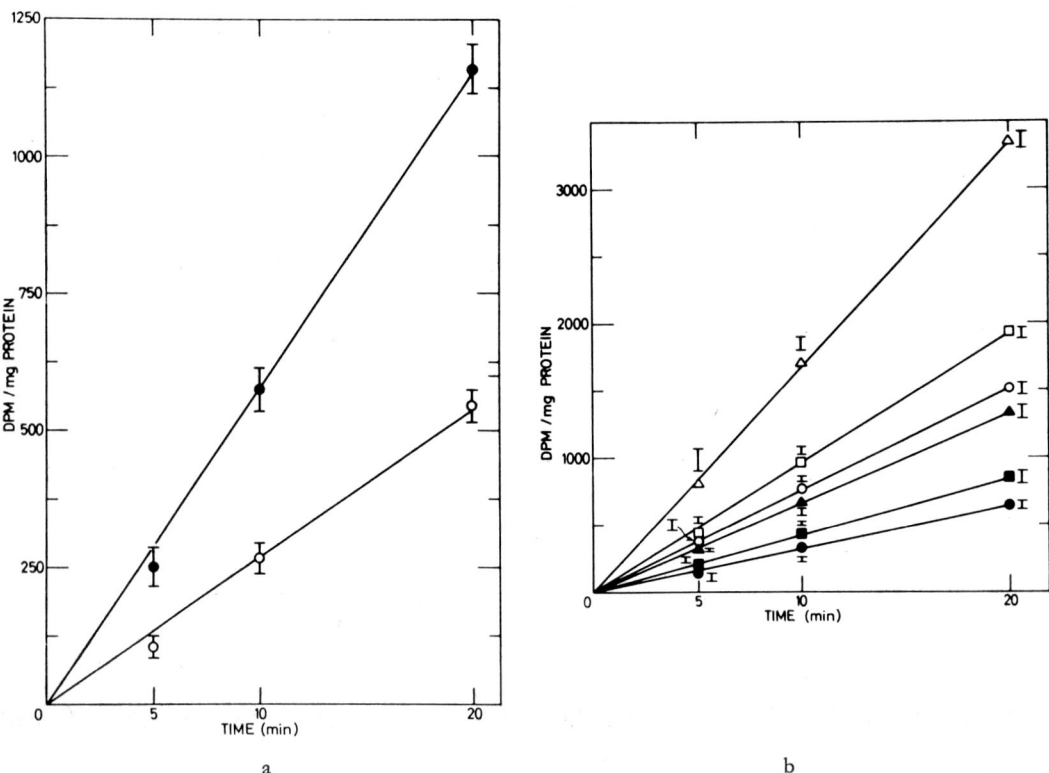

Fig. 1a. Rate of incorporation of L-valine-U-^{14}C into pancreatic amylase of rats adapted to either RSBM* (○) or HSBM* (●).

Fig. 1b. Rate of incorporation of L-valine-U^{14}C into pancreatic trypsinogens (○, □) and chymotrypsinogen (△) of rats adapted to either RSBM (empty symbols) or HSBM (filled symbols).
[Konijn, Birk & Guggenheim, Am. J. Physiol. **218**, 1113 (1970)].

* RSBM, raw soybean meal; HSBM, heated soybean meal.

effect of pure trypsin inhibitors on growth rate, pancreatic hypertrophy and intestinal proteolysis of chicks and rats [21, 22, 23] led to the following conclusions: The inhibitors are fully responsible for the pancreatic hypertrophy but have only a small effect on growth. The intestinal proteolytic activity is decreased by addition of AA to the diet but not by SBTI; this difference arises probably from the small affinity of SBTI for chymotrypsin as well as from it's inactivation by peptic digestion.

The pancreatic hypertrophy caused by ingestion of inhibitors (or of RSBM) is concomitant with an adaptation of the exocrine pancreatic enzymes and is expressed in preferential enhancement of proteolytic enzymes, mainly of chymotrypsin [22, 23]. Similar results have been obtained by DESNUELLE and co-workers [24] who have studied the adaptation of pancreatic enzymes to protein rich diets. One may assume that a RSBM- or trypsin inhibitor-supplemented diet simulates the conditions of a protein rich diet, regarding the reaction of pancreatic enzymes. This similar behavior results from the combination of the inhibitors in RSBM with the intestinal trypsin and chymotrypsin so that a higher amount of enzymes is needed for performing proteolysis, as is the case following ingestion of high levels of protein. The adaptation of the exocrine pancreatic enzymes to RSBM diets could also be demonstrated by *in vitro* incorporation of L-valine-U-^{14}C into amylase, trypsinogens and chymotrypsinogen of rats adapted to either RSBM or to properly heated soybean meal HSBM (Fig. 1a, b) [25]. These data show that pancreases of rats adapted to RSBM incorporated less labelled valine into amylase and more into the proteolytic enzymes than those adapted to HSBM. The mode of action of the inhibitors on the pancreas has not yet been established but one of the current assumptions is that they do not act directly on the gland but through the mucosa of the small intestine, stimulating it to release a humoral factor, such as pancreozymin, which affects the pancreas.

Acknowledgements: The studies performed in our laboratory were carried out with the assistance and collaboration of B. ALONI, N. BEN-SHALOM, S. KHALEF, G. MOGENSEN and A. TUR-SINAI. The authors are grateful to the Fund for Research in Oil Industry, Israel Ministry of Commerce and Industry, for financial support.

References

[1] KUNITZ, M., Science **101**, 668 (1945).
[2] KUNITZ, M., J. Gen. Physiol. **30**, 291 (1947).
[3] FINKENSTADT, W. R., and M. LASKOWSKI, Jr., J. Biol. Chem. **240**, PC 962 (1965).
[4] BOWMAN, D. E., Proc. Soc. Exptl. Biol. Med. **63**, 547 (1946).
[5] BIRK, Y., Biochim. Biophys. Acta **54**, 378 (1961).
[6] BIRK, Y., A. GERTLER and S. KHALEF, Biochem. J. **87**, 281 (1963).
[7] BIRK, Y. and A. GERTLER, Biochem. Prep. **12**, 25 (1968).
[8] LIPKE, H., G. S. FRAENKEL and I. E. LIENER, J. Agric. Food Chem. **2**, 410 (1954).
[9] BIRK, Y., A. GERTLER and S. KHALEF, Biochim. Biophys. Acta **67**, 326 (1963).
[10] APPLEBAUM, S. W. and A. M. KONIJN, J. Insect Physiol. **12**, 665 (1966).
[11] APPLEBAUM, S. W., Y. BIRK, I. HARPAZ and A. BONDI, Comp. Biochem. Physiol. **11**, 85 (1964).
[12] BIRK, Y., Ann. N. Y. Acad. Sci. **146**, 394 (1968).
[13] TROP, M. and Y. BIRK, Biochem. J. **116**, 19 (1970).
[14] GERTLER, A. and Y. BIRK, Eur. J. Biochem. **2**, 170 (1970).
[15] BIRK, Y., A. GERTLER and S. KHALEF, Biochim. Biophys. Acta **147**, 402 (1967).
[16] KRAHN, J., and F. C. STEVENS, Biochemistry **9**, 2646 (1970).
[17] BORCHERS, R., C. W. ACKERSON and L. KIMMETT, Arch. Biochem. **13**, 291 (1947).

[18] Tixier, M. R., C. R. Acad. Sci. Paris **266**, 2498 (1968).

[19] Hochstrasser, V. K., K. Illchmann and E. Werle, Hoppe-Seyler's Z. Physiol. Chem. **350**, 929 (1969).

[20] Tur-Sinai, A., Y. Birk, A. Gertler and M. Rigbi, Israel J. Chem. **8**, 178p. (1970).

[21] Gertler, A., Y. Birk and A. Bondi, J. Nutrition **91**, 358 (1967).

[22] Konijn, A. M., Y. Birk and K. Guggenheim, J. Nutrition **100**, 361 (1970).

[23] Gertler, A. and Z. Nitzan, Brit. J. Nutr. **24** 893 (1970).

[24] Reboud, J. P., G. Marchis-Mouren, L. Pasĕro, A. Cozzone and P. Desnuelle, Biochim. Biophys. Acta **117**, 351 (1966).

[25] Konijn, A. M., Y. Birk and K. Guggenheim, Am. J. Physiol. **218**, 113 (1970).

Lima Bean Protease Inhibitor: Amino Acid Sequence and Active Sites Against Trypsin and Chymotrypsin

Frits C. Stevens

Department of Biochemistry, Faculty of Medicine, University of Manitoba, Winnipeg 3, Canada

Introduction

Lima bean protease inhibitor (LBI) was first extensively purified and studied by Fraenkel-Conrat et al. (1952). Later it was shown that LBI could be further fractionated, by ion exchange chromatography, into at least four biologically active variants (Jones et al., 1963; Haynes and Feeney, 1967). These variants all have identical biological activity and are very similar, but not identical, in their molecular weight (~ 9000) and their amino acid composition. LBI is characterized by the absence of tryptophane and methionine and by a very high content of half cystine residues, all in the form of 6—7 disulfide bridges (Jones et al., 1963). LBI is a good inhibitor of both trypsin and chymotrypsin (Feeney et al., 1969). Recent chemical modification and gel filtration studies (Haynes and Feeney, 1967; Fritz et al., 1969) indicate that LBI is a "double headed" inhibitor with independent, non-overlapping sites for the binding of trypsin and chymotrypsin. Therefore, it was of interest to us to further investigate the interaction of this inhibitor with both trypsin and chymotrypsin and to determine its active sites for these two enzymes. Further, as a first step in elucidating the structure-function relationships in LBI, we decided to determine the amino acid sequence of one of the LBI variants.

The "Double Headed" Nature of LBI

The experiments described here were performed with the mixture of LBI variants. In our assay system (Krahn and Stevens, 1970), using synthetic substrates for trypsin and for chymotrypsin, 1 mg of LBI inhibits approximately 2.5 mg of trypsin and 1.0—1.5 mg of chymotrypsin. Figure 1 summarizes the results of several gel filtration experiments designed to demonstrate complexes between LBI and trypsin and/or chymotrypsin (Krahn and Stevens, 1970). On a column of G-75 LBI, trypsin, and chymotrypsin have elution volumes of 275, 265, and 250 m*l*, respectively (Fig. 1a). The material in peak II (Fig. 1b) was pooled as indicated and freeze-dried. Its properties (elution volume 228 m*l*, no trypsin inhibitory activity, high chymotrypsin inhibitory activity) are those expected for a trypsin-LBI complex if the sites for trypsin and chymotrypsin are independent. The other peaks in Figure 1b are due to a small excess of LBI and to autolysis products. In Figure 1c peak I (elution volume 195 m*l*, no inhibitory activities) has the properties of a LBI-trypsin-chymotrypsin complex and peak II represents excess trypsin-LBI complex. In a similar fashion we also first made the LBI-chymotrypsin complex, from which the ternary complex can then be produced by addition of

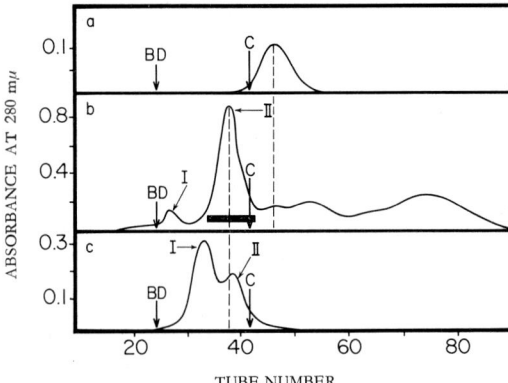

Fig. 1. Gel filtration of LBI and its complexes on Sephadex G-75. Samples (1 ml) were run on a column (2,5 × 90 cm) of Sephadex G-75 previously equilibrated with 0.1M ammonium bicarbonate. The eluent used was also 0.1M ammonium bicarbonate, the flow rate 40 ml/hr and 6 ml fractions were collected. The elution volumes of blue dextran (BD) and chymotrypsin (C) are indicated by arrows. (a) 20 mg of LBI, (b) 20 mg of LBI to which 50 mg of trypsin have been added, and (c) 13.5 mg of peak II (part b) to which 3 mg of chymotrypsin has been added.

trypsin. It should be pointed out that, when assayed with synthetic substrates, the LBI-chymotrypsin complex exhibits residual chymotryptic activity. This is presumably due to the displacement of the inhibitor from the enzyme by the substrate. These results offer substantial support to the idea that LBI has independent sites for the binding of trypsin and chymotrypsin. That the binding of these two enzymes is stoichiometric is evident from the results in Table 1 which compares the theoretical values of the amino acid composition of 1:1 molar complexes between the enzymes and LBI with the experimental values obtained by actual amino acid analysis of the complexes isolated by gel filtration as described above. The good agreement, within experimental error, between theoretical and experimental values establishes beyond doubt that LBI can form 1:1 molar complexes with trypsin or chymotrypsin and can also form a ternary complex with one mole of each of these two enzymes.

Table 1. Amino acid compositions of the complexes of lima bean inhibitor with trypsin and chymotrypsin

Amino acids	LBI-Trypsin		LBI-Chymotrypsin		Chym-LBI-Tryp	
	Theory[a]	Experiment[b]	Theory[a]	Experiment[b]	Theory[a]	Experiment[b]
Lysine	18.42	17.10	18.52	19.16	32.77	34.40
Histidine	8.79	9.09	7.32	6.52	10.74	11.05
Arginine	3.87	3.82	5.05	5.05	6.97	8.08
Aspartic acid	36.05	37.90	35.70	36.20	58.95	59.30
Threonine	16.59	16.19	28.22	29.42	39.72	39.62
Serine	46.53	47.50	38.28	39.50	73.03	73.99
Glutamic acid	21.70	22.20	22.30	23.50	37.20	37.00
Proline	15.05	15.25	15.90	15.04	25.15	23.70
Glycine	29.33	28.35	26.23	27.50	53.73	52.49
Alanine	18.00	17.10	26.38	28.50	41.20	41.60
Valine	15.90	15.74	20.85	22.80	35.45	34.40
Methionine	1.75	1.82	1.92	1.85	3.67	2.19
Isoleucine	17.21	18.00	11.16	11.15	24.40	25.90
Leucine	16.86	17.91	23.61	25.60	37.61	35.48
Tyrosine	12.29	11.00	5.69	5.37	16.69	14.39
Phenylalanine	3.91	4.73	7.79	9.16	10.21	10.95

[a] The theoretical values were obtained by adding the amino acid composition of enzyme and inhibitor, assuming the formation of 1:1 molar complexes.
[b] The experimental values were obtained by actual amino acid analysis of the complexes isolated by gel filtration.

The "Active Sites" of LBI

Ozawa and Laskowski, Jr. (1966) first showed that incubation of Kunitz soybean trypsin inhibitor with catalytic amounts of trypsin at acid pH resulted in an equilibrium mixture of native inhibitor and modified inhibitor in which a single arginyl-isoleucyl peptide bond had been cleaved. The modified inhibitor reacted more slowly with trypsin and subsequent removal of the new carboxyterminal arginine residue resulted in loss of trypsin inhibitory activity. Based on these and other experiments (Ozawa and Laskowski, Jr., 1966; Laskowski, Jr., 1970) Laskowski and coworkers proposed that all naturally occurring trypsin inhibitors contain either an Arg-X or a Lys-X peptide bond in their active site.

With LBI we carried out similar experiments using both trypsin and chymotrypsin for partial proteolysis at acid pH. The complete experimental details of these studies have been previously published (Krahn and Stevens, 1970). Table 2 summarizes the results obtained by partial proteolysis of LBI with trypsin. It can be seen that the trypsin modified inhibitor (which presumably makes up $\sim 30\%$ of the reaction mixture obtained by partial proteolysis) reacts with trypsin more slowly than does native inhibitor; a 15 minute preincubation of the reaction mixture with near molar amounts of trypsin at neutral pH restores full inhibitory activity. Treatment of the reaction mixture with carboxypeptidase B results in the release of ~ 0.3 moles/mole of lysine with a concommittant permanent loss of 30% of the original trypsin inhibitory activity. The chymotrypsin inhibitory activity of LBI is not affected by the above treatments. These results would indicate that partial proteolysis of LBI with trypsin, under the conditions used, results in a reaction mixture containing 30% modified inhibitor with a Lys-X peptide bond cleaved. Removal of this new carboxyterminal lysine results in loss of trypsin inhibitory activity. We concluded that the anti-trypsin inhibitory site of LBI therefore probably contains a Lys-X peptide bond. This is in agreement with the results of chemical modification studies (Haynes and Feeney, 1967, 1968; Fritz et al., 1969) which showed that modification of lysine residues in LBI resulted in loss of trypsin inhibitory activity without any effect on its chymotrypsin inhibitory activity.

Incubation of LBI with catalytic amounts of chymotrypsin at acid pH results in loss of $\sim 70\%$ of the original chymotrypsin inhibitory activity without affecting the trypsin inhibitory activity. Full chymotrypsin inhibitory activity could only be restored after prolonged incubation (45—48 hrs) of the reaction mixture with near molar amounts of chymotrypsin at neutral pH. Incubation of the reaction mixture with carboxypeptidase A resulted in the fast release of 0.7—0.9 moles/mole of new carboxyterminal leucine, indicating a leucyl-X-peptide bond as the site of chymotryptic cleavage and as part of the anti-chymotrypsin active site of LBI.

It is of interest to point out that very similar results were obtained by other workers (Birk et al., 1967; Frattali and Steiner, 1969) upon partial proteolysis of Bowman-Birk soybean inhibitor with both trypsin and chymotrypsin. The striking similarity between LBI and the Bowman-Birk soybean inhibitor has been previously pointed out (Frattali, 1969; Steiner and Frattali, 1969).

Table 2

Inhibitory activities of trypsin-modified LBI

Sample[b]	% Residual Inhibitory Act.[a] vs.			
	Trypsin Preincubation		Chymotrypsin Preincubation	
	None	15 min	None	15 min
Native LBI	80	100	100	100
Trypsin-modified	65	100	100	100
Trypsin-modified and COB treated	70	70	100	100

[a] The inhibition obtained by native LBI in the 15 min preincubation assay was taken as 100%.
[b] The trypsin-modified sample was treated with carboxypeptidase B (COB) for 3 hr.

In an attempt to locate the anti-chymotrypsin site of LBI within the primary sequence of the molecule, the partial proteolysis of LBI by chymotrypsin was carried out on a preparative scale using purified LBI variant III (in the terminology of Jones et al., 1963). We were able to show (Krahn and Stevens, 1970) that partial proteolysis with chymotrypsin resulted in the cleavage of a leucyl-seryl peptide bond located 29 residues in from the carboxyterminal end of the molecule in the sequence -Thr-Leu-Ser-Ile-Pro-. Similar studies on LBI variant IV, of which we have determined the amino acid sequence, show that the same 29 residue fragment is released from it by partial proteolysis with chymotrypsin. Because of this and because of the similarity in peptide maps obtained after trypsin or thermolysin hydrolysis we feel confident in extrapolating the results obtained on variant III to variant IV.

Amino Acid Sequence of LBI (Variant IV)

Our sequence studies were carried out on LBI variant IV purified as described by Jones et al. (1963). This variant contains 84 amino acid residues including 2 arginine and 4 lysine residues. The details of sequence analysis have been published elsewhere (Tan and Stevens, 1971). Figure 2 diagrammatically indicates how the peptides used for sequence determination and for obtaining overlaps were prepared. One of the peptides from the tryptic digest of reduced carboxymethylated LBI was lost during purification because of its poor solubility. Therefore the tryptic digestion was repeated using reduced carboxamidomethylated LBI and with the initial fractionation of the digest on Bio-gel P-6. The tryptic digest of guanidinated LBI, in which the lysine residues have been converted to homoarginine residues, was carried out in the hope of obtaining overlapping peptides by limiting tryptic cleavage to arginine residues.

Seven tryptic peptides were obtained in good yield, as expected from the presence of 2 Arg and 4 Lys. However, examination of these peptides showed that there was one peptide containing two lysine residues and three peptides ending in arginine. During the sequence studies it became obvious that two of the arginine containing tryptic peptides (T-3 and T-3a) were

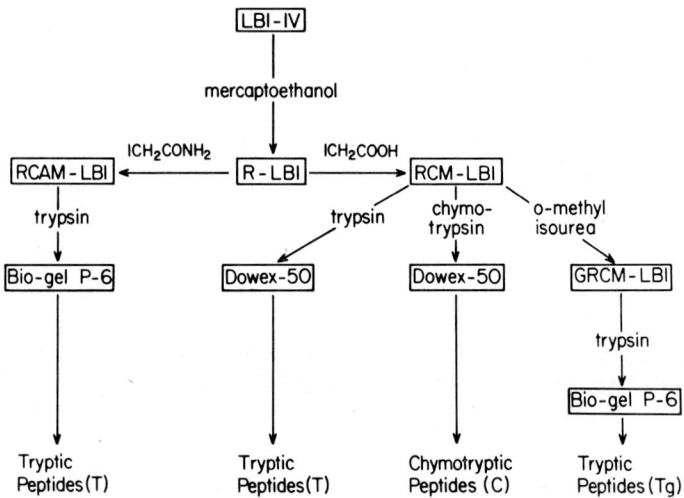

Fig. 2. This figure diagrammatically illustrates how the peptides used for sequence analysis were obtained. Where needed, the peptides were further purified by gel filtration, ion exchange chromatography, paper electrophoresis, paper chromatography or a combination of these. R: reduced; RCM: reduced carboxymethylated;- RCAM: reduced carboxamidomethylated; GRCM: reduced, carboxymethylated and guanidinated.

identical except for a threonine-serine replacement in one position and a leucine-phenylalanine replacement in another position. The sum of the yields of peptides T-3 and T-3a is about equal to the other tryptic peptides and this suggested to us that these two tryptic peptides were variants of the same peptide and the result of microheterogeneity in the purified protein. This was later confirmed by the existence of duplicate peptides for the same region, obtained from the chymotryptic digest and also from the tryptic digest of the guanidinated inhibitor.

The peptides were sequenced using classical methods including the substractive Edman degradation, digestion with carboxypeptidases A and B, digestion with leucine aminopeptidase and aminopeptidase M and, where needed, further cleavage of large peptides with proteolytic enzymes of wide ranging specificities. The results are shown in Figure 3.

With one exeption, the cleavage pattern of trypsin and chymotrypsin was in accordance with their published specificities. The Lys^{28}-Ser^{29} peptide bond was cleaved by chymotrypsin and also by trypsin after conversion of the lysine residue to an homoarginine residue by guanidination. By overlapping peptides we have good evidence for the arrangement of residues 29 through 84. However, because of the peculiar behaviour of the Lys^{28}-Ser^{29} peptide bond we

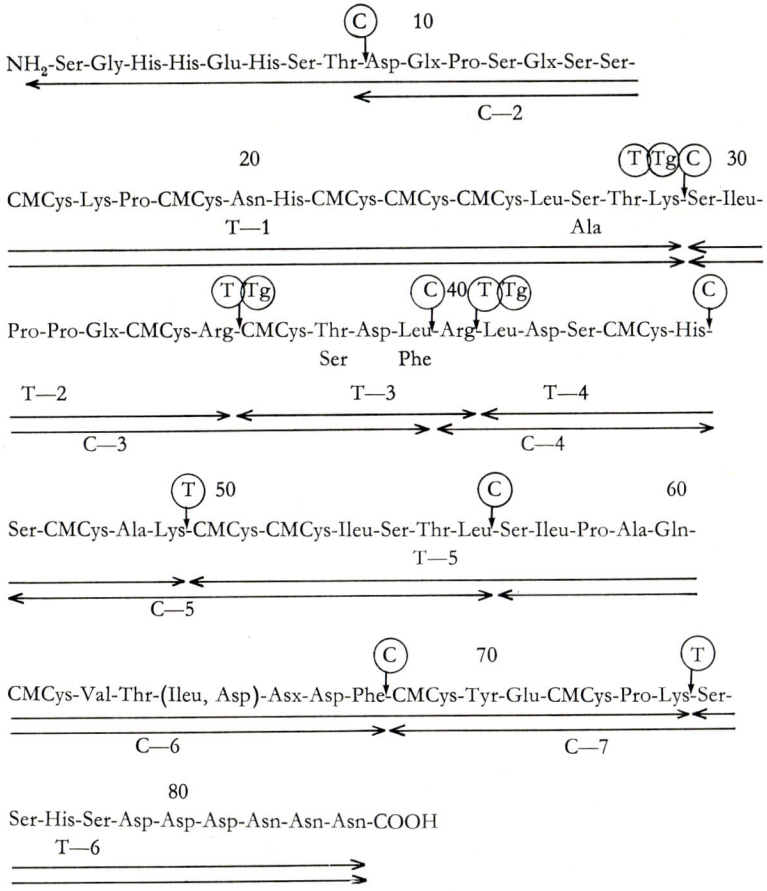

Fig. 3. Amino acid sequence of LBI-IV. The symbols T, C and Tg indicate the sites of cleavage by trypsin, chymotrypsin and by trypsin of the guanidinated derivative, respectively. The tryptic and chymotryptic peptides isolated are designated by T and C and numbered in the order of appearance in the sequence.

have no overlaps linking residues 1—28 to residues 29—84. We have assured ourselves that LBI consists of a single polypeptide chain and on the basis of tryptic specificity it can be argued that the N-terminal is the only possible position for peptide T-1. Circumstantial evidence for this arrangement also comes from the partial proteolysis studies with chymotrypsin carried out on variant III (KRAHN amd STEVENS, 1970).

We have already mentioned evidence for the microheterogeneity of LBI-IV resulting in variation in positions 37 and 39. Our evidence indicates that these two replacements are genetically linked since every peptide isolated from that region contained either Thr^{37} and Leu^{39} or Ser^{37} and Phe^{39}. A further indication of microheterogeneity was found in position 26; sequence studies on T-1 indicated the presence of alanine, while sequence studies on the corresponding chymotryptic peptide as well as the corresponding peptides obtained from a tryptic digest of the guanidinated LBI-IV showed position 26 occupied by a serine. In each case the replacements can be considered conservative from a structural point of view and can be obtained by a single base mutation from an evolutionary point of view.

When the data obtained on variant III are extrapolated to variant IV, it can be seen that the antichymotryptic site of LBI-IV is identified as the Leu-Ser peptide bond between residues 55 and 56.

The Occurrence of Two Homologous Regions in LBI-IV

As illustrated in Figure 4a striking feature of the amino acid sequence of LBI-IV is the occurrence of a repetition in sequence in two separate portions of the polypeptide chain. It can be seen that the sequences from residues 23 through 34 and 50 through 61 are clearly homologous. Without assuming any deletions or additions these two regions are identical in 9 out of 12 positions; there is a conservative replacement

15   22
Ser-**Cys**-Lys-Pro-**Cys**-Asn-His-**Cys**

43   49
Ser-**Cys**-His-Ser-**Cys**-Ala-Lys-D

23   34
Cys-**Cys**-Leu-**Ser**-**Thr**-Lys-**Ser**-**Ile**-**Pro**-Pro-Glx-**Cys**-
     Ala

50   61
Cys-**Cys**-Ile-**Ser**-**Thr**-Leu-**Ser**-**Ile**-**Pro**-Ala-Gln-**Cys**

Fig. 4. The homologus regions of LBI-IV.

of Ile for Leu and the two other replacements involve an alanine-proline and a leucine-lysine substitution respectively. Assuming only one deletion, these homologous regions can be extended further to include residues 15 through 22 on one hand and 43 through 49 on the other hand. Since the cysteine residues of these two regions are homologous it is tempting to speculate that one could be dealing with two homologous disulfide loops.

The presence of repetitive sequences in proteins has been previously observed and is generally interpreted as involving the extension of shorter peptide chains resulting from a process of gene duplication. Some possible mechanisms for achieving this have been reviewed by DIXON (1966). It is generally accepted that the linear arrangement of amino acids in a polypeptide chain predetermines the folding of the protein. It is therefore likely that these two homologous regions of LBI-IV show considerable structural homology.

Of added interest is the fact that one of the homologous regions (residues 50—61) contains the anti-chymotrypsin active site: Leu^{55}-Ser^{56} (KRAHN and STEVENS, 1970). The corresponding position in the other homologous region is occupied by Lys^{28}-Ser^{29}. In view of the fact that trypsin inhibitors contain either an Arg-X or a Lys-X peptide bond in their active site against trypsin (LASKOWSKI, Jr., 1970) and that LBI has been shown to be a trypsin inhibitor of the Lys-X type (HAYNES and FEENEY, 1968; FRITZ et al., 1969; KRAHN and STEVENS, 1970) it is tempting to propose the lysyl-seryl peptide bond located in the other homologous region as the anti-trypsin site of LBI-IV. If the Lys^{28}-Ser^{29} peptide bond is indeed found to be the anti-trypsin active site of LBI-IV it would then be tempting to speculate on the genetic origin of this double headed inhibitor. It may very well be that this "double headed" inhibitor arose by a process of gene duplication followed by divergent evolution, resulting in substituting a lysine residue for a leucine residue thereby changing the site from a chymotrypsin inhibitory site to a trypsin inhibitory site (or vice versa). It is also interesting to point out that when one compares residues 23 through 34 with residues 50 through 61, the leucine to lysine change is the only amino acid substitution requiring a minimum of 2 base changes in the genetic code. If the Lys^{28}-Ser^{29} peptide bond is indeed the anti-tryptic active site this would then indicate that the amino acid sequence in the active site regions against both trypsin and chymotrypsin is very critical since it was maintained for a period long enough to allow the double mutation needed for converting an anti-trypsin active site into an anti-chymotrypsin active site. Alternatively it is also possible that both active site regions evolved from a common ancestor which was not an inhibitor.

Acknowledgements I am most grateful to my students (Dr. CELINE TAN, Mr. JOHN KRAHN and Mr. TUNG-WU LIANG) for their contributions to this work. This work was supported throughout by an operating grant (MA 2907) from the Medical Research Council of Canada and in its later stages also by a grant (G-70-19) from the Life Insurance Medical Research Fund.

Summary

Gel filtration studies on lima bean protease inhibitor have shown that this inhibitor is "double headed" with independent, non overlapping sites for the inhibition of trypsin and chymotrypsin. By partial proteolysis the anti-trypsin site was found to be a Lys-X peptide bond while the anti-chymotrypsin site was identified as a Leu-Ser peptide bond located 29 residues in from the carboxyterminal end of the molecule in the sequence -Thr-Leu-Ser-Ile-Pro.

The amino acid sequence of one of the variants of this inhibitor was determined. During sequence determination evidence for microheterogeneity of the variant was obtained. Examination of the complete sequence revealed the existence of two homologous regions in this protein. One of these regions contains the anti-chymotrypsin site while the other could potentially be the anti-trypsin site. We put forward the hypothesis that this "double headed" inhibitor arose by a process of gene duplication followed by divergent evolution.

References

BIRK, Y., A. GERTLER and S. KHALEF, Biochim. Biophys. Acta **147**, 402 (1967).

DIXON, G. H., in: Essays in Biochemistry, Vol. II, P. N. CAMPBELL and G. F. GREVILLE, Ed., New York, N. Y., Academic Press. p. 128 (1966).

FEENEY, R. E., G. E. MEANS and J. C. BIGLER, J. Biol. Chem. **244**, 1957 (1969).

FRAENKEL-CONRAT, H., R. C. BEAN, E. D. DUCAY and H. S. OLCOTT, Arch. Biochem. Biophys. **37**, 393 (1952).

FRATTALI, V., J. Biol. Chem. **244**, 274 (1969).

FRATTALI, V. and R. F. STEINER, Biochem. Biophys. Res. Commun. **34**, 480 (1969).

FRITZ, H., E. FINK, M. GEBHARDT, K. HOCHSTRASSER and E. WERLE, Hoppe-Seyler's Z. Physiol. Chem. **350**, 933 (1969).

HAYNES, R. and R. E. FEENEY, J. Biol. Chem. **242**, 5378 (1967).

HAYNES, R. and R. E. FEENEY, Biochemistry **7**, 2879 (1968).

JONES, G., S. MOORE and W. H. STEIN, Biochemistry **2**, 66 (1963).

KRAHN, J. and F. C. STEVENS, Biochemistry **9**, 2646 (1970).

LASKOWSKI, M., Jr., in Structure-Function Relationships of Proteolytic Enzymes, P. DESNUELLE, H. NEURATH and M. OTTESEN, Ed., New York, N. Y., Academic Press, p. 89 (1970).

OZAWA, K. and M. LASKOWSKI, Jr., J. Biol. Chem. **241**, 3955 (1966).

STEINER, R. F. and V. FRATTALI, J. Agr. Food Chem. **17**, 513 (1969).

TAN, C. G. L. and F. C. STEVENS, Eur. J. Biochem. **18**, 503 (1971).

TAN, C. G. L. and F. C. STEVENS, Eur. J. Biochem. **18**, 515 (1971).

Chemical Studies on the Site of Interaction between Trypsin and Kunitz Soybean Trypsin Inhibitor (STI)

Irvin E. Liener

Dept. Biochemistry, College of Biological Sciences, University of Minnesota, St. Paul, Minn. 55101, U. S. A.

Two aspects of the interaction of trypsin and STI are now well documented: 1. when the histidine and serine components of the active site of trypsin are chemically modified, combination with STI can no longer be demonstrated [1—3] and 2. an Arg-Ile bond in STI is split by trypsin under certain well defined conditions [4, 5]. In addition there is some spectrophotometric evidence to indicate that some of the tyrosine and tryptophan residues of STI may be at or near the zone of contact [6, 7]. Aside from the observations little is known about the actual size and chemical features of the contact zone wherein these two macromoles interact. In an attempt to gain such information the following approach to this problem was considered:

1. treatment of the trypsin-STI complex with a group specific reagent,
2. dissociation of the modified complex under conditions which permit the resolution of each component,
3. determination of the number of unmodified residues in each component which may have escaped modification while in the form of the complex,
4. treatment of these unmodified residues with same reagent as used in step 1. but now bearing a radioactive or chromophoric label, and
5. isolation and characterization of peptides derived from the labeled regions of trypsin and the inhibitor.

I would like to report the successful accomplishment of the first three of these steps through the use of acetyl imidazole (AcIm), a reagent which selectively acetylates the phenolic groups of tyrosyl residues and, to a lesser extent, the free amino groups of proteins [8]. From such data we were able to assess the number of tyrosyl residues and amino groups which are involved in the site of interaction between trypsin and STI and the contribution made by each component.

Trypsin was purified by chromatography on SE-Sephadex [9] and STI according to the procedure of Frattali and Steiner [10] by chromatography on DEAE-cellulose. The trypsin-STI complex was isolated by gel-filtration on Sephadex G-75 from a mixture of trypsin and STI, the latter in slight molar excess over trypsin (see Fig. 1).

Trypsin, STI, and the complex were acetylated with AcIm according to the procedure of Simpson et al. [8] and Riordan et al. [11] (120-fold molar excess, pH 6.8) in the presence and absence of 8 M urea. The rates at which the tyrosyl residues of trypsin, STI and complex were acetylated are shown in Figure 2. In the absence of urea, all three of the proteins were

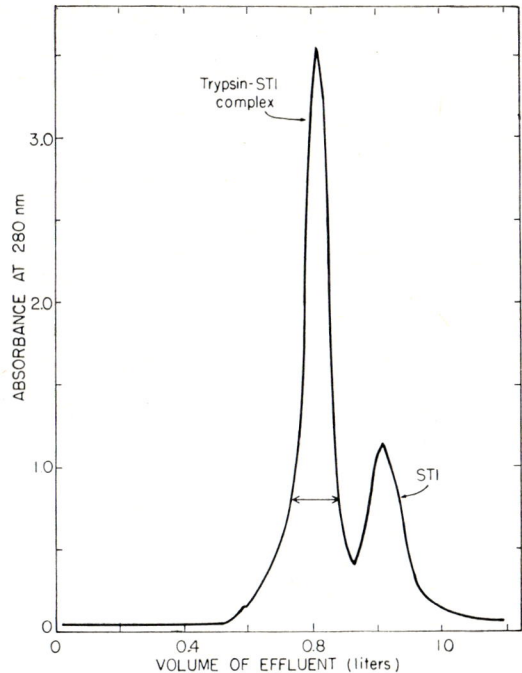

Fig. 1. Isolation of trypsin–STI complex by chromatography on Sephadex G-75.

[12] and STI [13], 10 and 4 respectively, were acetylated within 15 min. A total of 14 tyrosyl residues were acetylated in the complex, the sum of trypsin and STI, probably due to the complete dissociation of the complex in the presence of urea [14]. From these data it may be concluded that trypsin and STI each have 4 tyrosyl residues reactive towards AcIm, but once these two proteins have complexed only 4 out of the total of 8 of these residues are capable of reacting with AcIm.

If one simply measures the free amino groups [15] of trypsin, STI, and the complex using trinitrobenzene sulfonic acid (TNBS) one finds 12 (out of a total of 15) in trypsin, 5 (out of a total of 12) in STI, but only 12 (out of a theoretical 17) in the complex. This means that 5 amino groups were non-reactive towards TNBS in the complex.

The number of amino groups which had become acetylated were measured by the difference in free amino groups at any given time during acetylation and a zero time control. Free amino groups were again determined by reaction with trinitrobenzene sulfonic acid. Figure 3 shows the rate at which the free amino groups of trypsin, STI, and complex are acetylated. Large differences in rate of reaction are to be noted — the amino groups of trypsin react rather sluggishly

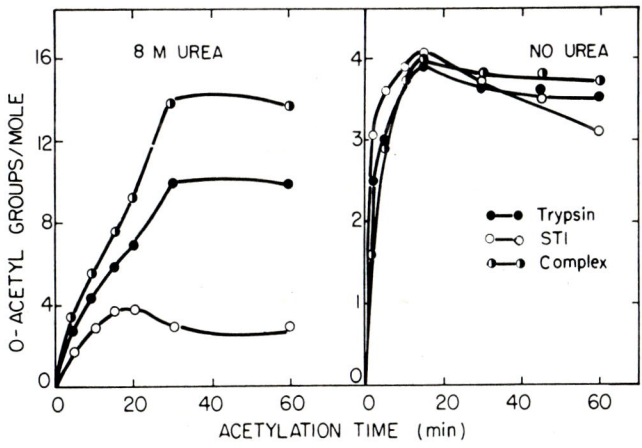

Fig. 2. Rate of reaction of AcIm with tyrosyl residues of trypsin, STI, and trypsin-STI complex in the presence and absence of urea.

acetylated to the same extent, namely 4 tyrosyl residues in about 15 min. followed by a gradual decrease. In the presence of 8M urea, all of the tyrosyl residues known to be present in trypsin

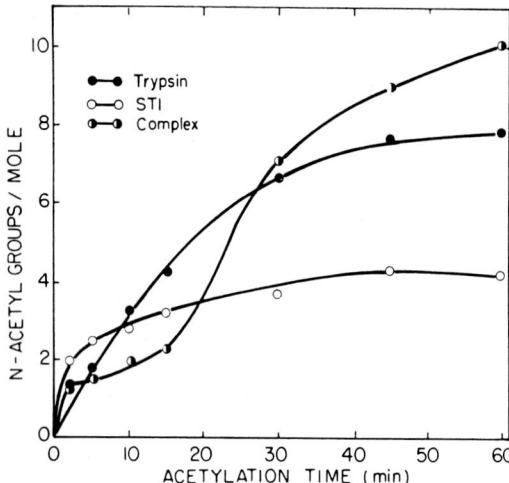

Fig. 3. Rate of reaction of AcIm with amino groups of trypsin, STI, and trypsin–STI complex.

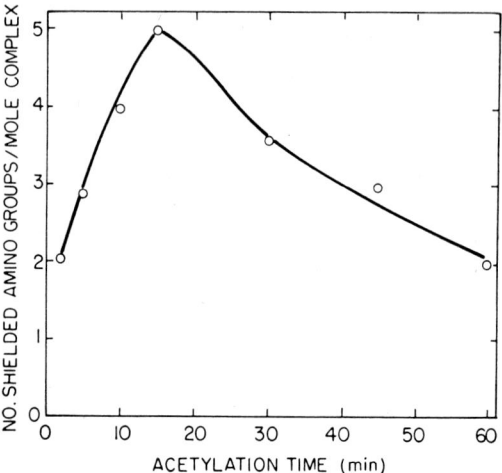

Fig. 4. The number of amino groups in the trypsin–STI complex rendered inaccessible to acetylation as a function of the time of exposure to AcIm.

(8 in about 1 hr), those of STI rather rapidly (4 in 30 to 40 min), and the complex in a fashion which reflects the acetylation of the amino groups of STI in the early part of the reaction and trypsin in the latter stage.

The number of amino groups which have been shielded from reaction with AcIm is obtained at any given time by subtracting the N-acetyl groups of the complex from the sum of the N-acetyl groups of trypsin and STI. A plot of these shielded amino groups vs time of acetylation is shown in Figure 4. This plot reveals that the maximum number of amino groups which were prevented from reacting with AcIm was 5, a value which was attained rather abruptly after 15 min of acetylation. Longer periods of acetylation gave lower values for the number of protected amino groups, indicating that some of the masked amino groups eventually react with AcIm. It may be reasonably concluded that the 5 amino groups which were masked during complexation, as measured by their failure to react with TNBS, are likewise refractory to acetylation provided the complex is not exposed to AcIm for longer than 15 min.

In order to determine how many of the masked tyrosyl and amino groups were derived from trypsin and how many of these were derived from STI, it became necessary to dissociate the acetylated complex and to determine the number of acetylated residues remaining in each component. As shown in Figure 5 dissociation of the non-acetylated complex could be achieved by chromatography on SE-Sephadex, pH 2.6, with a salt gradient from 0 to 0.5M NaCl. In the case of the acetylated complex the STI component could be eluted only after introducing 0.3M acetate buffer, pH 5.6. The complex in this instance had been acetylated for 15 min since the kinetic data had, as already mentioned, indicated that acetylation beyond this period favors deacetylation of O-acetyl groups, and, at the same time, permits the acetylation of masked amino groups in the complex. Both the trypsin and STI derived from this acetylated complex were analyzed for O-acetyl and N-acetyl groups in the manner already described with the results shown in Tables 1 and 2 respectively. Included here is also a summary of data already referred to in connection with acetylation of trypsin, STI, and the complex prior to dissociation.

Trypsin and STI derived from the acetylated complex were found to possess 2.0 and 1.8 O-

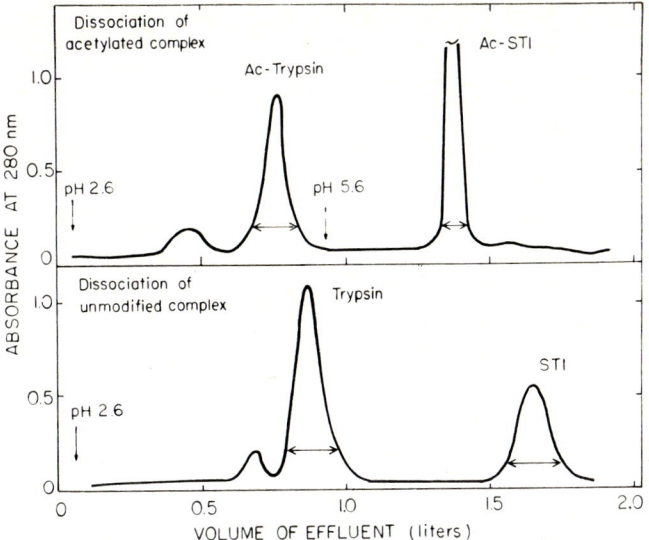

Fig. 5. Dissociation of acetylated (upper curve) and unmodified (bottom curve) trypsin-STI complex by chromatography on sulfoethyl Sephadex C-25.

Table 1. Acetylation of the tyrosyl groups of trypsin, STI, and complex

	Prior to complexation		Trypsin-STI complex	After dissociation of complex	
	Trypsin	STI		Trypsin	STI
Theoretical no. tyrosyl residues	10	4	14	10	4
No. tyrosyl residues acetylated	3.8	4.1	4.0	2.0	1.8[a]
No. tyrosyl residues shielded[b]			3.9 (4)	1.8 (2)	2.3 (2)

[a] Corrected for 50% hydrolysis of acetyl group during chromatographic isolation of STI from acetylated complex.
[b] Closest whole integer shown in parentheses.

Table 2. Acetylation of the amino groups of trypsin, STI, and complex

	Prior to complexation		Trypsin-STI complex	After dissociation of complex	
	Trypsin	STI		Trypsin	STI
Theoretical no. amino groups	15	12	27	15	12
No. TNBS-reactive amino groups in					
unmodified protein	12.0	5.0	12.3	11.2	4.2
acetylated protein	7.9	1.8	10.0		
No. amino groups acetylated	4.1	3.2	2.3	0.8	0.8
No. amino groups shielded[a]			5.0 (5)	3.3 (3)	2.4 (2)

[a] Closest whole integer shown in parentheses.

acetyl groups, respectively, compared to 4 tyrosyl residues in each protein which could be acetylated prior to complex formation. Therefore of the 4 tyrosine residues shielded from reaction with AcIm in the complex, 2 are derived from trypsin and 2 from STI.

Trypsin isolated from the acetylated complex contained only 0.8 N-acetyl groups compared to 4.1 before complexation; therefore, 3.3 amino groups of trypsin are prevented from reacting with AcIm after complexing with STI. STI isolated from acetylated complex likewise contained only 0.8 N-acetyl groups, compared to 3.2 before complexation; therefore, 2.4 amino groups of STI escape acetylation when combined with trypsin. It may be concluded that of the 5 amino groups involved in the combining site, 3 are derived from trypsin and 2 from STI.

The above considerations pertaining to the participation of tyrosyl and amino groups in the contact zone between trypsin and STI are summarized in Figure 6 and 7 respectively.

It is worthwhile noting that acetylation of the tyrosyl and amino groups of trypsin and STI, as performed in these studies, had no significant effect on their activities. Neither did these pro-

TYROSINE RESIDUES

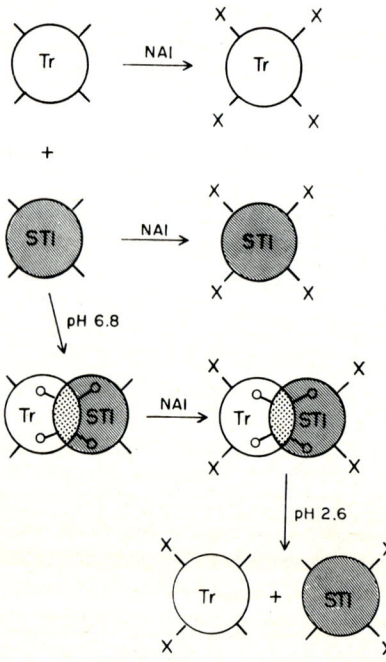

Fig. 6. Scheme showing the involvement of tyrosine residues in the interaction of trypsin (Tr) and STI. –, unmodified tyrosyl residues; X–, O-acetyltyrosyl residues; O–, shielded tyrosyl residues which are unreactive towards AcIm.

AMINO GROUPS

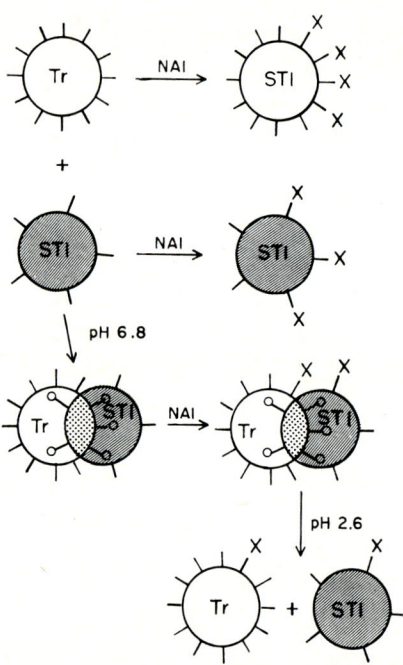

Fig. 7. Scheme showing the involvement of amino groups in the interaction of trypsin (Tr) and STI. –, unmodified amino groups; X–, N-acetylamino groups; O–, shielded amino groups which are unreactive towards AcIm.

teins suffer any appreciable impairment in activity following their dissociation from the acetylated complex. This is strong indication that no gross conformational change had occurred in either protein as a result of complex formation or chemical modification. If a conformational change had indeed occurred during complex formation, it must be completely reversible upon dissociation of the complex. It is conceivable that a conformational change during interaction of the two proteins could lead to a masking of tyrosyl and amino groups not located in the zone of the contact.

Since the ultimate objective of this study is to delineate the regions of the molecules which are involved in the interaction of trypsin and STI, it would not be amiss perhaps to speculate on this point on the basis of the evidence thus far available, at least with respect to trypsin. According to the work of KENNER et al. [16] and HOLEYSOVSKY et al. [17] 4 tyrosine residues of trypsin can be nitrated with tetranitromethane without loss in activity and these can be identified as Tyr 11, 28, 48, and 137. Figure 8 shows the location of 3 of these tyrosyl residues in a 3-dimensional model based on an assumed homology with α-chymotrypsin [18]. It is interesting to note that these tyrosyl residues are closely clustered in a very restricted region of the trypsin molecule, and two of these three tyrosyl residues may indeed be in direct contact with STI. Confirmation of this point should come from labeling experiments which are now in progress.

A more precise delineation of the combining regions of trypsin and STI must await studies

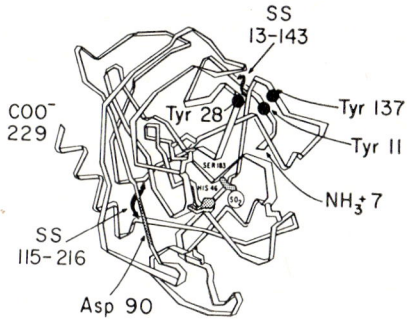

Fig. 8. Hypothetical three-dimensional structure of trypsin deduced from the structure of α-chymotrypsin. See ref. [18].

similar to those reports involving other functional groups, as well as more detailed information on the primary structure of STI.

Acknowledgements: This study was supported by Grant No. AM 13869 from the National Institute of Arthritis and Metabolic Diseases.

References

[1] GREEN, N. M., J. Biol. Chem. **205**, 535 (1953).
[2] ESTERMANN, E. F. and A. D. MCLAREN, Photochem. Photobiolog. **1**, 109 (1962).
[3] FEINSTEIN, G. and R. E. FEENEY, J. Biol. Chem. **241**, 5183 (1966).
[4] OZAWA, K. and M. LASKOWSKI, Jr., J. Biol. Chem. **241**, 3955 (1966).
[5] SEALOCK, R. W. and M. LASKOWSKI, Jr., Biochemistry **8**, 3703 (1969).
[6] EDELHOCH, H. and R. F. STEINER, J. Biol. Chem. **240**, 2877 (1965).
[7] STEINER, R. F., Biochemistry **5**, 1964 (1965).
[8] SIMPSON, R. J., J. F. RIORDAN and B. L. VALLEE, Biochemistry **2**, 616 (1963).
[9] PAPAIOANNOU, S. and I. E. LIENER, J. Chromatog. **32**, 746 (1968).
[10] FRATTALI, V. and R. F. STEINER, Biochemistry **7**, 521 (1968).
[11] RIORDAN, J. F., W. E. G. WACKER and B. F. VALLEE, Biochemistry **4**, 1758 (1965).
[12] DAYHOFF, M. O., Atlas of Protein Sequence and Structure **4**, D-115 (1969).
[13] WU, Y. V. and H. A. SCHERAGA, Biochemistry **1**, 698 (1962).
[14] JACOBBSON, K., Biochim. Biophys. Acta. **16**, 264 (1955).
[15] HAYNES, R., D. T. OSUGA and R. E. FEENEY, Biochemistry **6**, 541 (1967).
[16] KENNER, R. A., K. A. WALSH and H. NEURATH, Biochim. Biophys. Res. Commun. **33**, 353 (1968).
[17] HOLEYSOVSKY, V., B. KEIL and F. ŠORM, FEBS Letters **3**, 107 (1969).
[18] NEURATH, H. and R. A. BRADSHAW, Accounts Chem. Res. **3**, 249 (1970).

The Non-Bond Splitting Mechanism of Action of Inhibitors of Proteolytic Enzymes — the Conservative Interpretation

ROBERT E. FEENEY

Department of Food Science and Technology, University of California, Davis, California 95616

Naturally occurring protein inhibitors of proteolytic enzymes have the capacity to combine with and inactivate proteolytic enzymes. The inhibitor, proteolytic enzyme, and protein substrate, which is also usually present, have very different roles in this relationship. The inhibitor is unique in that it has a capacity to inactivate another protein whose normal function is to hydrolyze and degrade proteins.

There currently appears to be general agreement among the workers in the field that: 1. inhibitors of proteolytic enzymes form highly associated complexes with the enzymes they inhibit, and 2. the inhibitors are resistant to proteolysis, although peptide bond splitting may occur. Small but important interpretations as to the relative importance of various of the events, however, exist. Our laboratory supports a conservative interpretation of the relative importance of these events. This interpretation implies that the formation of an inhibitory complex does not require the primary structure of the inhibitor to be broken and that the principal forces are the ones usually occurring in protein-protein interactions. These would involve the binding and fitting which occur between a biologically active protein and the substance with which it reacts. In the case of the enzyme-inhibitor complex, the selectivity of the binding site on the enzyme for a particular type of side chain on the inhibitor would be important. Catalysis and bond splitting, however, would be secondary and generally not essential.

Earlier Studies

One of the earlier insights into the specificity and mechanisms of actions of inhibitors of proteolytic enzymes was the work of HEINZ FRAENKEL-CONRAT and co-workers (FRAENKEL-CONRAT et al., 1952). They showed that amino groups were important for the inhibitory action of lima bean trypsin inhibitor against trypsin. This observation was thus the forerunner of the many observations that have since been reported on the essentiality of the epsilon amino groups of lysine side chains for the inhibition of trypsin by many trypsin inhibitors. GORINI and AUDRAIN (1952, 1953) first described the enzymatic hydrolysis of an inhibitor by a proteolytic enzyme. The loss of inhibitory action of inhibitors due to enzymatic hydrolysis in the complex was termed "temporary inhibition" (LASKOWSKI and WU, 1953). During this period the excellent researches of the father and son team, MICHAEL J. LASKOWSKI, Sr. and MICHAEL J. LASKOWSKI, Jr., was yielding information on many different inhibitors and their monumental review (LASKOWSKI and LASKOWSKI Jr., 1954) was a milestone in inhibitor chemistry. Their review not only summarized many of the observations of various workers, but also discussed the theories of interaction.

Many other investigators contributed heavily to the knowledge of inhibitors, but one of those which stand out sharply in the light of current considerations of mechanisms is the work of F. F. Nord's laboratory. These workers (Sri Ram et al., 1954) discussed the interaction between chicken ovomucoid and trypsin as follows:

"We would therefore rather consider the inhibitory action of ovomucoid as another instance of competitive substrate inhibition so well known in the case of other enzymes, for example lipases. The rate of action of an enzyme upon different substrates does not have to parallel their Michaelis-Menten constants, and a substrate, which is acted upon only very slowly but forms a stable enzyme-substrate complex will, therefore, act as an inhibitor."

The question of "competitive inhibition" has been argued many times since this early observation and, indeed, is not settled even as of this time.

More Recent Studies

Of the many fine workers that have contributed to the recent understanding of the mechanism of interactions, Michael J. Laskowski Jr.'s laboratory has been one of the most provocative groups. These studies are well known to most of those working in the field and certain of their results are described elsewhere in this volume. Only the bare essentials will therefore be given here. They had observed earlier that there was a sequence of events occurring which indicated that there were several steps in the reaction. They then demonstrated that with chicken ovomucoid and the Kunitz soybean trypsin inhibitor (STI) that there was a cleavage of a peptide bond of the inhibitor by the enzyme. When the new COOH-terminal amino acid was removed by carboxypeptidase B, the inhibitory activity was greatly reduced or lost. They therefore concluded that the trypsin inhibitory reactions consists of a cleavage of one especially sensitive bond in the inhibitor by trypsin (Finkenstadt and Laskowski, Jr., 1915; Ozawa and Laskowski, Jr., 1966). Although Gorini and Audrain had shown proteolysis in the complex and Nord's group had suggested an enzyme substrate relationship as essential for inhibition, the suggestions of Laskowski, Jr. and coworkers placed the enzyme-substrate relationships in sharp focus.

The investigations of Laskowski, Jr.'s laboratory has continued in several intriguing and profitable directions. One of the most striking has been the resynthesis of a peptide bond in STI with a substitution of one trypsin-hydrolyzable residue (lysine) for the original trypsin hydrolyzable residue (arginine) (Sealock and Laskowski, Jr., 1969; Laskowski, Jr., 1970). This was accomplished by first causing proteolysis of STI by trypsin, then removal of the new COOH-terminal arginine by carboxypeptidase B, and finally following this with the reversal of this sequence of reactions starting with ^{14}C labelled lysine. The product apparently contained the ^{14}C labelled lysine in the resynthesized peptide chain of the inhibitor. It had inhibitory activity. Many other investigators have confirmed that there are bonds split in a variety of different inhibitors by the proteolytic enzymes they inhibit and that there are losses of activity when the new COOH-terminal amino acids are removed with carboxypeptidases (Birk et al., 1967; Frattali and Steiner, 1969; Rigbi and Greene, 1968). There thus appears to be no question but that proteolysis does occur and that, at least sometimes, removal of the new COOH-terminal amino acids may result in extensive decreases in inhibitory activities. There are, however, some cases described in which little or no bond splitting is found to occur. One such case is the Kunitz bovine pancreatic trypsin inhibitor (Keil, 1970). In addition, there are other results of different investigators which indicate that proteolysis is not an essential, or even a critical, requirement for inhibition.

Laskowski, Jr.'s laboratory has suggested at this symposium (Laskowski, Jr. et al., 1971) that an acyl-enzyme (acylation of the catalytically active serine by the inhibitor) or a tetrahedral intermediate, involving a covalent bond between inhibitor and enzyme, may be involved in the mechanism of the formation of the complex. Such an intermediate state offers still another attractive possibility for the mechanism. Other workers at this meeting are reporting similar amino acid sequences near the residues which are hydrolyzed in different inhibitors. Perhaps some certain sequences may give some unusual properties which could allow for the formation of some unusual bonds or could cause some unusually high affinity and resistance to proteolysis.

Our laboratory noted that avian ovomucoids were of two types (STEVENS and FEENEY, 1963). The one was represented by chicken ovomucoid; its trypsin inhibitory activity was not abolished by modification of its amino (lysyl) groups. The other was represented by turkey ovomucoid; its trypsin inhibitory activity was abolished by modification of its amino groups. The chymotrypsin-inhibiting activity of turkey ovomucoid was unaffected by modification of its amino groups. This therefore appeared to be a specific reaction involving the inhibition of trypsin. Further studies with several other ovomucoids and the ovoinhibitors, as well as other inhibitors have shown that chemical modification of trypsin inhibitors abolishes the trypsin inhibitory activity when the amino groups are modified in some and when the guanidino (arginine) groups are modified in others (HAYNES et al., 1967; LIU et al., 1968; LIU et al., 1971). Quantitative kinetic studies showed that one particular amino group was the essential residue for inhibitory activity of several of the "lysine type" inhibitors. These studies have thus been in complete agreement with the observations made from the enzymatic hydrolysis approach, namely, there is either an arginine or lysine in an inhibitor which is critical for binding the trypsin molecule. Similar results have been obtained by other investigators as well.

There are, however, at least two lines of investigation which indicate that proteolysis of a particular bond in the inhibitor, or catalytic activity of the enzyme, is not necessary for the formation of an inhibited complex. When several avian ovomucoids were treated with trypsin and then with carboxypeptidase B, they still retained their trypsin-inhibitory activity (FEINSTEIN et al., 1966). Thus, removal of the new COOH-terminal arginine or lysine did not inactivate the inhibitor. In other studies, the newly exposed NH_2-terminal amino acid was modified by substitution of the amino groups and activity was still retained in the case of chicken ovomucoid, but was lost in the case of soybean trypsin inhibitor (HAYNES and FEENEY, 1967). Similar studies have now been done by treatment of penguin ovomucoid and turkey ovomucoid with either α-chymotrypsin or subtilisin. In all four possible combinations, substitution of the newly formed amino groups by dimethylation has not abolished the inhibitory activity against either enzyme (UY and FEENEY, 1971). These results would indicate that inhibitory activity can be retained, even when it is not possible to reform any peptide bonds in the enzyme inhibitor complex as would be the case when the inhibitor has been modified enzymatically and the newly produced NH_2-terminal amino acid is substituted. Identification of the particular amino groups substituted is now underway in our laboratory. This is critical for proof that the newly exposed NH_2-terminal group was blocked by modification. Possible difficulties in modifying these groups are discussed elsewhere in this symposium (FRITZ, 1971). Other principal studies with which we have been associated have utilized catalytically inactive enzymes. In the one case trypsin and chymotrypsin, with the catalytically important histidyl residue substituted, and chymotrypsin with the catalytically inactive serine modified by dehydration, formed complexes with several inhibitors which were competitive with catalytically active enzymes (FEINSTEIN and FEENEY, 1966) (Fig. 1). This work has recently been extended to include complexes of turkey ovomucoid with subtilisin and its derivatives (LIN and FEENEY, 1971). Substitution of the catalytically important serine by tosylation did not abolish the capacity of subtilisin to combine with turkey ovomucoid (Tab. 1). Other workers have also found similar relationships, some with different enzymes and different inhibitors. In fact, the work from our laboratory was stimulated by a preliminary report by

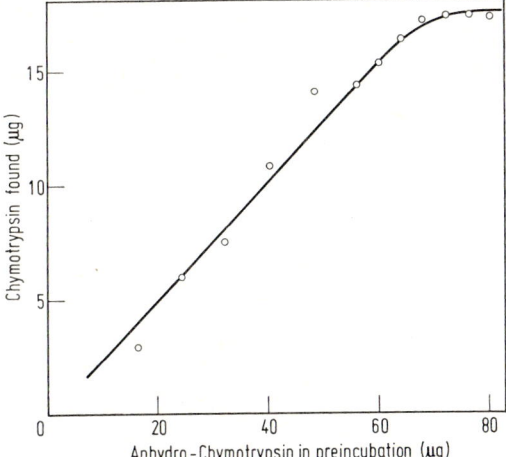

Fig. 1. The interaction of different amounts of anhydro-chymotrypsin with turkey ovomucoid. Competitive enzymatic assays (Method B). Various amounts of anhydro-chymotrypsin were preincubated with 20 μg of turkey ovomucoid at pH 8.2 and 37° for 3 min. Then 20 μg of native α-chymotrypsin were added and the incubation was carried for 3 min under the same conditions. After the substrate-indicator solution was added, the rate of the hydrolysis of substrate was determined spectrophotometrically, (from FEINSTEIN and FEENEY, 1966).

FOSTER and RYAN (1965) describing the competitive complex formation of anhydro-chymotrypsin and α-chymotrypsin with the potato inhibitor. Similar results with another totally different kind of enzyme, a different inhibitor, and a different method of substitution were described by FOSSUM and WHITAKER (1968). Mercury-papain is a highly associated complex which is catalytically inert. These authors showed that the mercury derivative of papain formed a complex with the papain inhibitor isolated from chicken egg white.

Both the results from studies by removing or blocking the newly exposed COOH- or NH$_2$-

Table 1. Catalytically Inactive Enzymes which form Complexes with Proteolytic Enzymes

Inhibitor Name	Enzyme Name	Enzyme Derivative*	Group Modified	Detection Method		Reference
				Electrophoresis	Enzyme Competition	
Potato inhibitor	α-Chymotrypsin	Anhydro-	Serine	?	+	FOSTER and RYAN (1965)
Chicken Ovomucoid	Trypsin	TLCK-	Histidine	+	+	FEINSTEIN and FEENEY (1966)
Turkey Ovomucoid	Trypsin	TLCK-	Histidine	+	+	FEINSTEIN and FEENEY (1966)
Turkey Ovomucoid	α-Chymotrypsin	TPCK-	Histidine	+	+	FEINSTEIN and FEENEY (1966)
Turkey Ovomucoid	α-Chymotrypsin	Anhydro-	Serine	+	+	FEINSTEIN and FEENEY (1966)
Turkey Ovomucoid	Subtilisin	Tosyl-	Serine	+	**	LIN and FEENEY (1971)
Turkey Ovomucoid	Subtilisin	Thio-	Serine	+	**	LIN and FEENEY (1971)
Egg white Papain inhibitor	Papain	Hg-complex	Cysteine	***	**	FOSSUM and WHITAKER (1968)

* Abbreviations: TLCK, 1-chloro-3-tosylamido-7-amino-2-heptanone; TPCK, L-1-tosylamido-2-phenylethyl chloromethyl ketone.
** Tests not performed.
*** Complex demonstrated by separation on Sephadex column.

terminal groups and the results from studies with catalytically inactive enzymes have been found dependent upon the particular enzyme and the particular inhibitors studied. For example, TLCK-trypsin forms a highly associated complex with turkey ovomucoid and chicken ovomucoid but not with STI or lima bean inhibitor. Tosyl-subtilisin complexes with turkey ovomucoid but tosyl-chymotrypsin does not. We prefer to interpret such differences in interactions in the same way that we interpret the fact that both lima bean inhibitor and chicken ovomucoid inhibit bovine trypsin, but only lima bean inhibitor inhibits human trypsin (FEENEY et al., 1969). A catalytically inactive derivative of an enzyme might not form a complex as a result of properties unrelated to its lack of catalytic activity, such as even slight differences in conformation or steric interferences. Whenever a loss of a property is encountered on chemical modification of a protein, these and many other factors must be considered (MEANS and FEENEY, 1971). When the property is retained, interpretations are usually much simpler. However, even here important answers may be unavailable. In the case of the demonstration of the interaction of a catalytically inactive enzyme with an inhibitor, there is aways the possibility that some other side chain groups, either on the enzyme or the inhibitor, could assume the role of the substituted or modified group which is essential for catalytic activity on "normal" substrates.

Transformation of several "lysine inhibitors" into "arginine-type inhibitors" by conversion of the lysines into homoarginines by guanidination did not abolish their trypsin-inhibitory activities (HAYNES and FEENEY, 1968). One of these, guanidinated lima bean inhibitor, was a strong trypsin inhibitor. Although carboxypeptidase B rapidly hydrolyzes COOH-terminal homoarginines (LIN et al., 1969), trypsin is not inhibited by guanidinated casein and apparently does not rapidly hydrolyze homoarginine bonds (FEENEY and LIU, 1971). We therefore reasoned that the inhibitory activity of guanidinated lima bean inhibitor was further evidence for the unessentiality of bond splitting. However, the homoarginine bond in guanidinated lima bean inhibitor is hydrolyzed (STEVENS, 1971). Nevertheless, another interpretation is that the trypsin is combined in the complex in such a manner that a residue with a normal low affinity (homoarginine) has a high affinity for the enzyme.

General Thoughts on Mechanisms

Probably one of the most important future developments may prove to be those showing that there may be several different types of mechanisms. The great diversity of the properties of different inhibitors and different proteolytic enzymes certainly would make this easily possible. The complexities of the interactions are certainly apparent from the detailed studies of LASKOWSKI, Jr.'s laboratory in the kinetics of interaction described elsewhere in this volume. Our results would indicate, however, that for many of the inhibitors and enzymes, catalytic activity of the enzyme and attack on the inhibitor is unnecessary for the inhibitory mechanism. We are therefore still supporting our following recently published interpretation (FEENEY and ALLISON, 1969).

"The proposed general mechanism of action assumes a specific residue for which the enzyme has high affinity. This residue would be a particular lysine or arginine in the case of trypsin inhibitors and a chymotrypsin susceptible residue such as tyrosine, tryptophan, alanine or methionine in chymotrypsin inhibitors. This residue serves as 'the recognition site,' or binding site of the inhibitor to a binding site of the enzyme. In addition, other noncovalent bonds or forces strengthen the association, possibly as a result of a conformational change causing a better fitting. If the peptide bond of a particular residue serving as the binding site on the inhibitor can be cleaved only very slowly, or if the equilibrium is very strongly in favor of the intact bond, then the protein will be an inhibitor for the enzyme."

This mechanism seems to include many of the attributes of those of other investigators.

Most protein chemists today are looking to X-ray diffraction studies of the total structure for answers to many of the unsolved problems of protein chemistry. Such studies are apparently being planned on the enzyme-inhibitor complexes (Huber, 1970) and may be the best way of observing this obviously complicated interaction of proteolytic enzymes and their inhibitors. The careful studies of Laskowski, Jr. et al. (1971) clearly show that the interaction is a kinetically complicated series of events.

References

Birk, Y., A. Gertler and S. Khalef, Further evidence for a dual, independent activity against trypsin and chymotrypsin of inhibitor AA from sobeans, Biochim. Biophys. Acta **147**, 402 (1967).

Feeney, R. E. and R. G. Allison, Evolutionary Biochemistry of Proteins. Homologous and Analogous Proteins from Avian Egg Whites, Blood Sera, Milk, and Other Substances, Wiley-Interscience, New York (1969).

Feeney, R. E. and W. H. Liu, unpublished data (1971).

Feeney, R. E., G. E. Means and J. C. Bigler, Inhibition of human trypsin, plasmin, and thrombin by naturally occurring inhibitors of proteolytic enzymes, J. Biol. Chem. **244**, 1957 (1969).

Feinstein, G. and R. E. Feeney, Interaction of inactive derivatives of chymotrypsin and trypsin with protein inhibitors, J. Biol. Chem. **241**, 5183 (1966).

Feinstein, G., D. T. Osuga and R. E. Feeney, The mechanism of inhibition of trypsin by ovomucoid, Biochem. Biophys. Res. Commun. **24**, 495 (1966).

Finkenstadt, W. R. and M. Laskowski, Jr., Peptide bond cleavage on trypsin-trypsin inhibitor complex formation, J. Biol. Chem. **240**, PC 962 (1965).

Fossum, K. and J. R. Whitaker, Ficin and papain inhibitor from chicken egg white, Arch. Biochem. Biophys. **125**, 367 (1968).

Foster, R. J. and C. A. Ryan, Reactions of potato inhibitor with modified chymotrypsin, Federation Proc. **24**, 473, Abst. 1905 (1965).

Fraenkel-Conrat, H., R. C. Bean, E. D. Ducay and H. S. Olcott, Isolation and characterization of a trypsin inhibitor from lima beans, Arch. Biochem. Biophys. **37**, 393 (1952).

Frattali, V. and R. F. Steiner, Interaction of trypsin and chymotrypsin with a soybean proteinase inhibitor, Biochem. Biophys. Res. Commun. **34**, 480 (1969).

Fritz, H., Specific isolation and modification methods for proteinase inhibitors and proteinases, Proc. of the International Research Conference on Proteinase Inhibitors, this volume, p. 28.

Gorini, L. and L. Audrain, Influence of calcium on the stability of the trypsin-ovomucoid complex, Biochim. Biophys. Acta **8**, 702 (1952).

Gorini, L. and L. Audrain, Ovomucoid trypsin complex. Its proteolytic activity and the roles of certain metal ions on the stabilities of its constituents, Biochim. Biophys. Acta **10**, 570 (1953).

Haynes, R. and R. E. Feeney, Properties of enzymatically cleaved inhibitors of trypsin, Biochim. Biophys. Acta **159**, 209 (1968).

Haynes, R., D. T. Osuga and R. E. Feeney, Modification of amino groups in inhibitors of proteolytic enzymes, Biochemistry **6**, 541 (1967).

Huber, R., D. Kukla, A. Rühlmann, O. Epp, and H. Formanek, The basic trypsin inhibitor of bovine pancreas. I. Structure analysis and conformation of the polypeptide chain, Naturwissenschaften **57**, 389 (1970).

Keil, B., Structural aspects of interaction of trypsin with macromolecular inhibitors, in Structure-Function Relationships of Proteolytic Enzymes, P. Desnuelle, H. Neurath and M. Ottesen, eds., Munksgaard, Copenhagen, pp. 102—112 (1970).

Laskowski, M. and M. Laskowski, Jr., Naturally ocurring trypsin inhibitors, Adv. Protein Chem. **IX**, p. 203, Academic Press, New York (1954).

Laskowski, M. and F. C. Wu, Trypsin inhibition of trypsin, J. Biol. Chem. **204**, 797 (1953).

Laskowski, M. Jr., The chemistry of the reactive site of soybean trypsin inhibitor, in Structure-Function Relationships of Proteolytic Enzymes, P. Desnuelle, H. Neurath, and M. Ottesen, eds., Munksgaard, Copenhagen (1970).

Laskowski, M., Jr., R. W. Duran, W. R. Finkenstadt, S. Herbert, H. F. Hixson, Jr., D. Kowalski, J. A. Luthy, J. A. Mattis, R. E. McKee and C. W. Niekamp, Kinetics and thermodynamics of interaction between soybean trypsin inhibitor (Kunitz) and bovine β trypsin, Proc. of the First International Research Conference on Proteinase Inhibitors, this volume, p. 117.

Lin, Y. and R. E. Feeney, unpublished data (1971).

Lin, Y., G. E. Means and R. E. Feeney, An assay for carboxypeptidases A and B on polypeptides from protein, Anal. Biochem. **32**, 436 (1969).

Liu, W. H., G. E. Means and R. E. Feeney, The inhibitory properties of avian ovoinhibitors against proteolytic enzymes, Biochim, Biophys. Acta. **229**, 176 (1971).

Liu, W. H., G. Feinstein, D. T. Osuga, R. Haynes and R. E. Feeney, Modification of arginines in trypsin inhibitors by 1,2-cyclohexanedione, Biochemistry **7**, 2886 (1968).

Means, G. E. and R. E. Feeney, Chemical Modification of Proteins, Holden-Day, San Francisco (May) (1971).

Ozawa, K. and M. Laskowski, Jr., The reactive site of trypsin inhibitors, J. Biol. Chem. **241**, 3955 (1966).

Rigbi, M. and L. J. Greene, Limited proteolysis of the bovine pancreatic secretory trypsin inhibitor at acid pH, J. Biol. Chem. **243**, 5457 (1968).

Sealock, R. W. and M. Laskowski, Jr., Enzymatic replacement of the arginyl by a lysyl residue in the reactive site of soybean trypsin inhibitor, Biochemistry **8**, 3703 (1969).

Sri Ram, J., L. Terminiello, M. Bier and F. F. Nord, On the mechanism of enzyme action. LVII. Interaction between trypsin and ovomucoid, Arch. Biochem. Biophys. **52**, 451 (1954).

Stevens, F. C., Lima bean trypsin inhibitor: amino acid sequence and active sites against trypsin and chymotrypsin, First International Research Conference on Proteinase Inhibitors, this volume, p. 149.

Stevens, F. C. and R. E. Feeney, Chemical modification of avian ovomucoids, Biochemistry **2**, 1346 (1963).

Uy, R. and R. E. Feeney, unpublished data (1971).

Sequential Studies on Proteinase Inhibitors Isolated by Use of Trypsin Resins

KARL HOCHSTRASSER and EUGEN WERLE

Institut für Klinische Chemie und Klinische Biochemie der Universität München, D-8000 München, Germany

Isolation Methods and Modification Reactions

FRITZ and Coworkers [1] in 1966 for the first time isolated animal protease inhibitors from crude tissue homogenates by selective binding the inhibitors to water insoluble trypsin resins. Using this method we isolated the trypsin inhibitor from maize seeds [2], but we were not satisfied with the product obtained. Firstly, electrophoresis showed the presence of two minor components besides the main inhibitor fraction. However, the minor components were not observed in all preparations. These results correspond to those obtained by FRITZ et al [3] and TSCHESCHE [4], when they investigated the trypsin inhibitor from porcine pancreas isolated in the same manner. Secondly, we found two amino terminal residues by end group analysis of our material, namely serine and leucine in equal amounts. Therefore we tried to separate the inhibitor fractions by ion exchange chromatography, but without success. Finally we were able to separate the components electrophoretically, however no significant differences in the amino acid compositions of the components were found.

Meanwhile LASKOWSKI, Jr. and coworkers [5] published their results of the successful modification of soybean Kunitz trypsin inhibitor by tryptic cleavage at low pH. Modification takes place by splitting a defined Arg-Ala-bond in the inhibitor molecule. After removing the newly formed C-terminal arginine by carboxypeptidase B the inhibitor lost its activity. Therefore the authors assigned this Arg-Ala-sequence as the reactive site of the inhibitor. When we incubated the trypsin inhibitor from maize seeds isolated by the trypsin resin method with carboxypeptidase B, both loss of inhibitor activity and release of arginine occurred. Obviously modification of the inhibitor takes place during contact of the inhibitor with the trypsin resin. Thus the observation of two end groups was explained plausibly. We reduced the inhibitor from maize seeds, carboxymethylated it, and finally isolated two peptide chains, each with only one end group. Later on we isolated a number of inhibitors from plant seeds in the same way, and we were able to identify in all cases a single defined peptide bond as the reactive site [6].

Sequential Studies

Inhibitor from Maize Seeds

Our first structural studies were concerned with the inhibitor from maize seeds isolated by the trypsin resin method. We examined the tryptic peptides from two separated chains of different length obtained from the reduced and carboxymethylated inhibitor [7]. The results were fully reproducible. Adding up the number of amino acid residues of the peptides obtained we only found one third of the molecular weight determined by gel filtration experiments. The tryptic digestion mixture did not contain any insoluble peptides which might have been lost. Therefore we considered the inhibitor investigated to be a polymer of smaller subunits.

In the peptide spectrum of the isolated large chain we found small amounts of peptides having the same amino acid composition as the peptides derived from the short chain. This is explained, if we suppose that the modification reaction is not completed so that the large chain-preparation contains a small amount of the entire peptide chain of the inhibitor.

In Figure 1 the amino acid sequence of the inhibitor from maize seeds is given. The reactive site is an Arg-Leu-bond. It is remarkable that the amino terminal sequence of 19 (!) amino acids does not contain any bond susceptible to tryptic digestion. In the modified inhibitor the sequence in position 20—25 is not linked to the molecule by disulfide bonds, and it might be possible that this sequence is removed partly by the action of trypsin during contact with the trypsin resin. We observed in some preparations low amounts of this particular peptide.

The electrophoretical polymorphism of the inhibitor preparations is now easily explained: The main component is due to the modified inhibitor, the minor components to the native inhibitor or to inhibitors with small peptides lost. The results therefore indicate that a sequence determination is possible with such mixed preparations. For experimental data see l. c. [7]. The modified inhibitor is only very slowly inactivated by carboxypeptidase B. Probably this is explained by steric hindrance due to the polymeric structure. We suppose that for the reaction with trypsin the polymeric inhibitor dissociates and one monomer reacts with one trypsin molecule. This we may conclude from the determination of molecular weights of these complexes which are only a little higher than the value for trypsin. If the inhibitor would react in the polymeric state we should have to expect much higher molecular weights for the enzyme-inhibitor complexes.

Inhibitor from Peanuts

In further investigations we isolated trypsin inhibitors from peanuts [8], which also inhibit chymotrypsin, plasmin, and serum kallikrein. The high value of 8.4 IU/mg inhibitor found for the specific activity does not correspond to the molecular weight of 17.000 estimated by gel filtration. If this inhibitor is also a polymer,

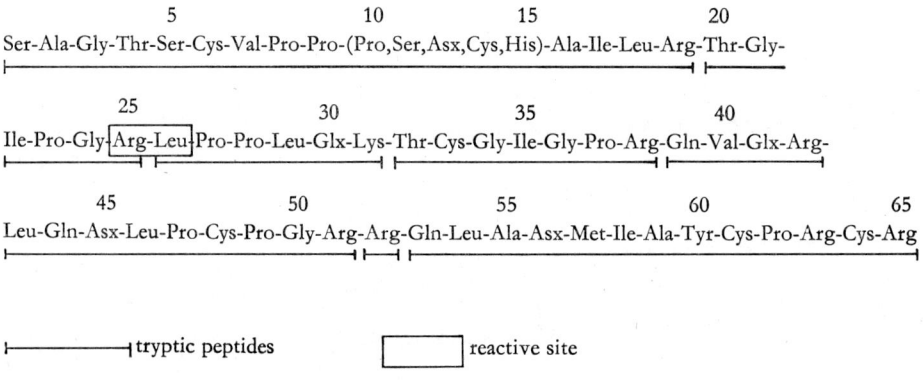

Figure 1. Amino Acid Sequence of the Inhibitor from Maize Seeds.

a relatively short amino acid chain for the primary sequence could be expected.

The isolated inhibitor was not homogeneous electrophoretically. We found a main component, two or three secondary components, and as amino end groups alanine, serine, and a small amount of valine. Nevertheless we tried a sequence analysis with this material.

After reduction and carboxymethylation the inhibitors showed an identical tryptic peptide spectrum. Only two percent of contaminating peptide material was found. Firstly, we assumed that serine is the amino terminal residue and alanine is located in the reactive site. In the tryptic hydrolysates we found peptides with N-terminal serine and alanine respectively. However, serine was excluded as the end group of the inhibitor molecule when we found an overlapping sequence which contained this serine residue. Quantitative determination of the N-terminal alanine by TSCHESCHE using mass spectroscopy revealed that this terminal alanine is only present in 5 percent of the expected amount. In Table 1 the sequence of the tryptic peptides is given.

Further determination of the end groups of the performic-acid oxydized inhibitor revealed cysteine as the main N-terminal residue.

In Figure 2 the sequence of the peanut inhibitor is presented. This inhibitor is not a lysine inhibitor, therefore the reactive sites can only be at the bonds Arg(8)-Cys(9), Arg(15)-Ser(16), and Arg(22)-Ala(23) respectively. Probably the Arg-Ala-bond is the reactive site, if we assume that

Table 1. Amino Acid Sequence of the Tryptic Peptides from Peanut Inhibitor

T I	Thr-Glx-Gly-Arg
T III—1	Ser-Asx-Pro-Pro-Glx-Cys-Arg
T III—2*	Cys-Thr-Asx-Lys-Thr-Glx-Gly-Arg
T V	Cys-Thr-Asx-Lys
T VI	Cys-Pro-Val-Thr-Glx-Cys-Arg
T VII	Ala-Pro-Pro-Tyr-Phe-Glx-Cys-Val-Cys-Asx-Thr-Phe-Asx-His-Cys-Pro-Ala-Ser-Cys-Asx-Ser-Cys-Cys-Thr-Arg

* Peptide T III—2 includes the sequences of peptides T I and T V.

5 percent of the inhibitor is modified and thus contain the additional N-terminal alanine found. The bonds Arg-Ser and Cys-Val may be susceptible to tryptic cleavage, too, without inactivation of the inhibitor, because serine and valine were also found N-terminal in small amounts. The observed electrophoretic polymorphism is explained by these facts. No other explanation is possible as the peptides obtained were extremely pure.

Ovine Pancreatic Inhibitor

We have also investigated the sequence of the ovine pancreatic inhibitor, which was isolated by the same procedure [10, 11]. End group analyses revealed aspartic and glutamic acid in about equal amounts, and in addition, a small amount of leucine. We examined the tryptic

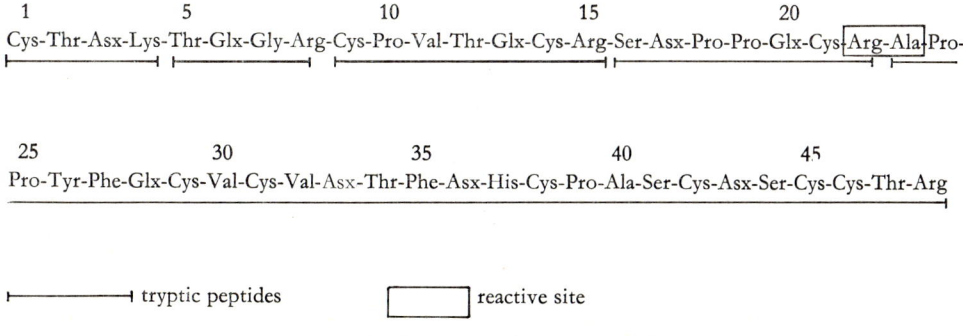

Figure 2. Amino Acid Sequence of the Inhibitor from Peanuts.

peptides. From the peptide with N-terminal aspartic acid only half of the amount of the other peptides was obtained. We assume therefore that the native inhibitor is partly degraded by splitting off an N-terminal pentapeptide during the trypsin resin step as was also demonstrated for porcine inhibitor I [12]. At this state of the investigations we obtained the manuscript from GREENE and GIORDANO, Jr. [13], in which the sequence of the secretory inhibitor from bovine pancreas was presented. So we could prove that the inhibitor from ovine pancreas has nearly the same peptide spectrum as the inhibitor from bovine pancreas. There is only one exception — the serine residue in position 32 of the bovine inhibitor is replaced by alanine in the corresponding tryptic peptide of the ovine inhibitor. In collaboration with TSCHESCHE [14] we found the same amino acid sequences in the tryptic peptides as GREENE has found for bovine pancreatic inhibitor. There is no doubt that the inhomogeneity of the inhibitors isolated by the resin method is caused by modification reactions. Figure 3 shows the sequence of the inhibitor from ovine pancreas. A series of derivatives of the inhibitor can be formed by limited proteolyses, i. e. splitting off an N-terminal peptide or cleavage of the reactive site bond without any loss of the inhibitory activity. Using this isolation method we can thus expect different components with the same basic sequence:

1. Completely intact inhibitor,
2. modified intact inhibitor (reactive site bond split),
3. unmodified inhibitor with a shortened N-terminal chain,
4. modified inhibitor with the shortened N-terminal chain.

Concluding remarks

At this state of the study it is not possible to define an amino acid sequence which is necessary for the inhibitory activity. But all these inhibitors have another remarkable property. In each case the molecules contain long amino acid chains without peptide bonds susceptible to tryptic cleavage. This sequence consists in the inhibitor from maize seeds of 19 amino acids, in the inhibitor from peanuts of 26 amino acids, and in the ovine and bovine pancreatic inhibitor (Kazal-type) of 18 amino acids. These sequences might be located within the inhibitor molecules to protect all the other attackable peptide bonds

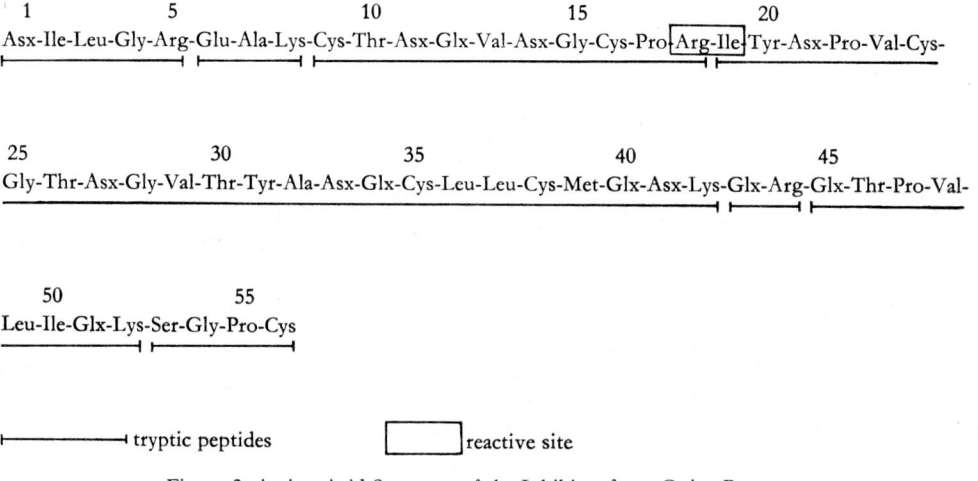

Figure 3. Amino Acid Sequence of the Inhibitor from Ovine Pancreas.

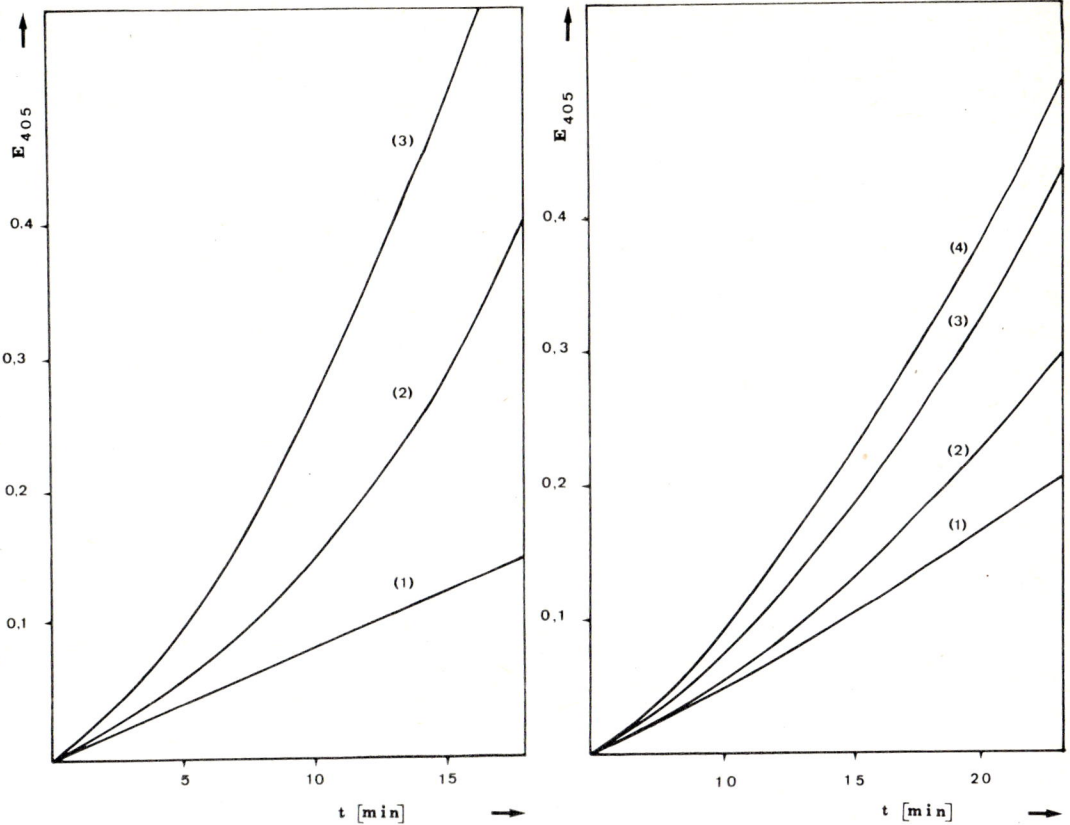

Fig. 4a. Liberation of Trypsin from the Trypsin-Chymotrypsin-Inhibitor Complex. The Inhibitor from Phaseolus vulgaris was used.
Substrate: N^α-Benzoyl-DL-arginine p-nitroanilide, 0.73×10^{-4}M.
Incubation Mixture: 200 μg complex in 2.0 ml 0.2M triethanolamine buffer, pH 7.8, 20 mM $CaCl_2$; mercaptoethanol concentration: (1) 0.05M, (2) 0.1M, (3) 0.2M.
Ordinate: Extinction at 405 nm; Abscissa: Incubation time t.

Fig. 4b. Liberation of Chymotrypsin from the Trypsin-Chymotrypsin-Inhibitor Complex. The Inhibitor from Phaseolus vulgaris was used.
Substrate: N^α-succinyl-L-phenylalanine p-nitroanilide, 4.3×10^{-4}M.
Incubation mixture: 400 μg complex in 2.0 ml 0,2M triethanolamine buffer, pH 7.8, 0.1M $CaCl_2$; mercaptoethanol concentration: (1) 0.5M, (2) 0.1M, (3) 0.15M, (4) 0.2M.
Ordinate: Extinction at 405 nm; Abscissa: Incubation time t.

from the cleavage by trypsin. The only reactive bond should be at the surface of the molecule. This hypothesis is proven in an excellent way for the basic bovine inhibitor by HUBER et al. [15]. The plant inhibitors are different from the secretory inhibitors; they are so-called permanent inhibitors. They were neither digested by the enzymes to which they are linked in the complex nor by excess enzyme molecules present.

Each of the two investigated inhibitors from plant seeds contain five disulfide bridges. This might be one reason for the extreme stability of these inhibitors against proteolysis, the other being the secondary and tertiary structure of the molecules. When a small amount of reducing agent is added to the trypsin-inhibitor complex in a solution free of urea, the secondary structure becomes slightly disturbed so that the inhibitor can be degraded and active trypsin is set free. The same effect is observed with the chymotrypsin-inhibitor complex. Bi valent inhibitors isolated from *phaseolus vulgaris* and *coccinium* form complexes with trypsin and chymotrypsin. For both enzymes different reactive sites may be responsible. In these cases both enzymes are liberated from their complexes in the presence of small amounts of reducing agents [17] (cf. Fig. 4a and 4b).

Our results show that it is possible to determine amino acid sequences using mixtures of a modified and intact inhibitor unless different species of inhibitors are present in the isolated material. Separation of modified products is not necessary in these cases.

References

[1] Fritz, H., H. Schult, M. Hutzel, M. Wiedemann and E. Werle, Z. physiol. Chem. **348**, 308 (1967).
[2] Hochstrasser, K., M. Muss and E. Werle, Z. physiol. Chem. **348**, 1337 (1967).
[3] Fritz, H., I. Hüller, M. Wiedemann and E. Werle, Z. physiol. Chem. **348**, 405 (1967).
[4] Tschesche, H., Z. physiol. Chem. **348**, 1216 (1967).
[5] Ozawa, K., and M. Laskowski, Jr., J. biol. Chem. **241**, 3955 (1966).
[6] Hochstrasser, K. and E. Werle, Z. physiol. Chem. **350**, 249 (1969).
[7] Hochstrasser, K., K. Illchmann and E. Werle, Z. physiol. Chem. **351**, 271 (1970).
[8] Hochstrasser K., K. Illchmann and E. Werle, Z. physiol. Chem. **350**, 929 (1969).
[9] Hochstrasser, K., K. Illchmann and E. Werle, Z. physiol Chem. **351**, 1503 (1970).
[10] Hochstrasser, K., W. Schramm, H. Fritz, S. Schwarz and E. Werle, Z. physiol. Chem. **350**, 893 (1969).
[11] Fritz, H., W. Schramm, B. Greif, K. Hochstrasser, E. Fink and E. Werle, Z. physiol. Chem. **351**, 145 (1970).
[12] Tschesche, H., E. Wachter and G. Kallup, Z. physiol. Chem. **350**, 1662 (1969).
[13] Greene, L. J. and I. S. Giordano, Jr., J. biol. Chem. **244**, 285 (1969).
[14] Tschesche, H., R. Obermeier and K. Hochstrasser, this volume, p. 207.
[15] Huber, R., D. Kukla, A. Rühlmann and W. Steigemann, this volume, p. 56.
[16] Hochstrasser, K. and H. Reich, unpublished results.
[17] Hochstrasser, K., S. Schwarz, K. Illchmann and E. Werle, Z. physiol. chem. **349**, 1449 (1968).

Studies on the Structure and Function of Chymotrypsin Inhibitor *I* in the Solanaceae Family*

C. A. Ryan and L. K. Shumway

The Department of Agricultural Chemistry and the Department of Botany and Program in Genetics, Washington State University, Pullman, Wash.

In the plant kingdom, certain storage organs such as seeds from the Leguminosae and Graminae families, and tubers from the Solanaceae family are excellent sources of proteinase inhibitors [1]. These inhibitors are quite diverse in number and in specificity towards various proteolytic enzymes. There are sometimes several different kinds of inhibitors present in a single tissue, for example, soybeans [1] and the potato tuber [1], and in some instances such as in the lima bean [2], the inhibitors are present in multiple forms.

The physiological significance of plant proteinase inhibitors has been questioned for a long time. Several explanations for their existence have been offered [1, 3, 4], but experimental evidence has been lacking, primarily because of the absence of suitable techniques to study the individual inhibitors within the tissues. Not only has the function of the inhibitors gone unproven, but little is known of the nature of the processes within the plants that control the existence of any proteinase inhibitor at any given time.

Within the last few years we have been seeking the function within the potato plant of chymotrypsin inhibitor *I* protein isolated by Ryan and Balls in 1962 [5]. We began the research with the conviction that we must thoroughly understand the chemistry of the inhibitor *I* molecule if we are ever to fully understand its function at the molecular level within the plant tissues. In the following pages we will first describe what we have learned of the purification and properties of inhibitor *I* and then we will describe our studies concerned with its physiology and function within the Solanaceae family. These studies have revealed that the protein is stored and utilized by the plant in a controlled manner that is intimately involved with the establishment of new growth centers and their development.

I. Purification and Properties of Inhibitor *I*

Our earliest purification of inhibitor *I* from potato tuber juice involved a laborious procedure that yielded hexagonal crystals from 30% magnesium sulfate solutions [6]. The crystals were difficult to obtain from the partially purified preparations but once crystallized they could be easily recrystallized from solutions of sodium acetate within a few minutes. A frustrating feature of this method of purification was that some partially purified preparations would not

---

* This work was supported in part by grants from the U. S. Public Health Service (GM 12505 and GM 2-K3-17059) from the U. S. Department of Agriculture (Grant No. 915-15-29) and from the Washington State Medical and Biological funds.

yield crystals even though immunological and inhibition assays indicated that 90% of their protein content was inhibitor I.

We established a purification procedure that utilized an ammonium sulfate step, a heating step, and finally chromatography on Sephadex G-75 to produce the protein in 90% purity with yields of 50% from crude potato juice [7]. In Figure 1 is shown the elution profile of the Sephadex G-75 chromatography [7]. Ordinarily

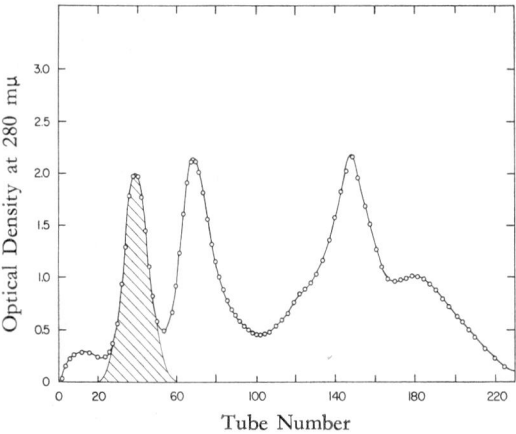

Fig. 1. Chromatography of semi-purified inhibitor I on Sephadex G-75. The column (10 cm × 100 cm) was equilibrated and eluted with 0.05M tris buffer in 0.1M KCl, pH 8.2. Two grams partially purified inhibitor I preparation in 140 ml was applied to the column. Fractions of 25 ml were collected. Flow rate was 600 ml per hour. The shaded second peak contains inhibitor I. (RYAN and MELVILLE, in preparation).

one to two grams of protein was applied to the column per run. The second peak, shaded, is nearly pure inhibitor I. We call this protein $G75\ PI$.

All of our attempts to further purify inhibitor I produced results that suggested inhibitor I was a heterogeneous protein. In general the strongest indicator of heterogenity was the trypsin inhibitory activity. The $G75\ PI$ preparations contained chymotrypsin inhibitory activity that remained fairly constant from preparation to preparation. However, the trypsin inhibitory activity usually varied considerably. On ion exchange resins or preparative gel electrophoresis, the specific trypsin inhibitory activity generally varied in the eluted protein peaks, whereas the chymotrypsin inhibitory activity remained constant [7].

At approximately this time we found that the dissolution of $G75\ PI$ in 6M guanidine hydrochloride drastically reduced its sedimentation coefficient in sedimentation velocity runs on the ultracentrifuge [7]. Therefore, a series of experiments was made to study the effect of increasing guanidine concentrations on the sedimentation velocity of inhibitor I. Figure 2 shows that the

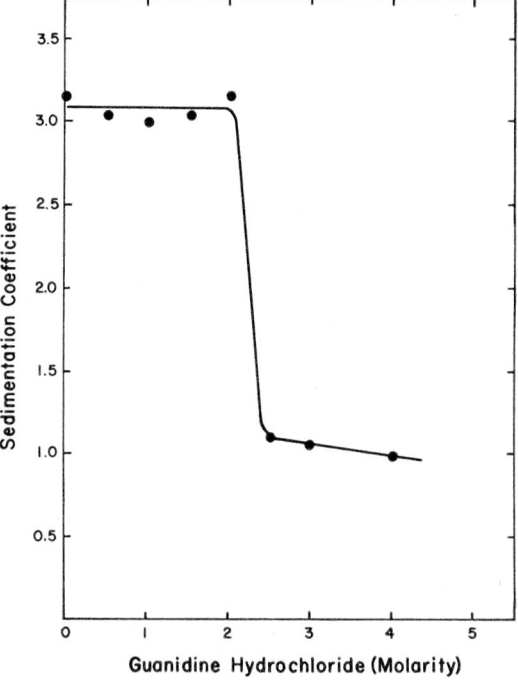

Fig. 2. The effect of guanidine hydrochloride concentration on the sedimentation coefficient of inhibitor I. Sedimentation coefficients were corrected to a solvent density of 1.0 (RYAN and MELVILLE, in preparation).

sedimentation coefficient of inhibitor I, corrected for the density of guanidine at each concentration, dropped from 3.2 to about 1 when guanidine concentrations exceeded 2M. When the dissociated inhibitor was diluted with

Table 1. Summary of Molecular Weight Determinations of Inhibitor I and its Complex with Chymotrypsin

Protein	Method	Molecular Weight or Equivalent Weight
Inhibitor I		
4× crystallized	Ultracentrifuge — Archibald	38800 ± 2400
	Ultracentrifuge — Velocity	37500 ± 2800
	Sephadex G-75, tris buffer, pH 8.2	39000 ± 1000
$G P_2$	Ultracentrifuge — Yphantis	39000 ± 1000
	Sephadex G-75, tris buffer, pH 8.2	39000 ± 1100
Monomers	Sephadex G-75, 4M guanidine, pH 8.5	9800 ± 200
	Sephadex G-75, 8M urea, pH 2.8	10000 ± 500
	Electrophoresis in 0.1% S. D. S.	9300 ± 1100
	Chymotrypsin inhibition (synthetic substrate)	9500 ± 500
	Chymotrypsin inhibition (casein substrate)	10000
Inhibitor I — Chymotrypsin Complex	Ultracentrifuge — Velocity	123000
	Sephadex G-200	140000

water to 1M guanidine or less, it regained its original sedimentation coefficient and had full chymotrypsin inhibitory capacity and immunological properties [7].

We used this knowledge to further purify inhibitor I by dissolving $G75 PI$ in 4M guanidine hydrochloride and passing it through a Sephadex G-75 column equilibrated with 4M guanidine hydrochloride. This separated the small subunits from large, non-dissociating proteins. The small units were recovered, diluted with water to reconstitute them, and then desalted and lyophylized. This reconstituted inhibitor I was called $G PI$ [7]. The preparation was equal to 4 times crystallized inhibitor I in its inhibitory activity toward chymotrypsin, although it could not be crystallized.

We began an intensive effort to establish the molecular weight of the inhibitor in its native and dissociated forms. A summary of this data is shown in Table 1. Non-dissociated inhibitor I, whether 4x crystallized or $G PI$, has a molecular weight of about 39,000 as determined with the ultracentrifuge using three different techniques, or using calibrated Sephadex G-75 columns [8]. The subunits, when dissociated using guanidine hydrochloride, urea or SDS gave a molecular weight of about 10,000 [8]. This agreed quite well with the molecular weight calculated using the combining weight of the inhibitor with chymotrypsin. The cummulative data suggest that inhibitor I is a tetramer of molecular weight 39,000, composed of four monomers of molecular weight of about 10,000.

The combination of inhibitor I with chymotrypsin (M. W. 25,000) to form a large complex of molecular weight approaching 140,000, shown at the bottom of Table 1, indicates that each inhibitor I subunit probably contains a single binding site for chymotrypsin.

A number of attempts were made to isolate the subunits from $G75 PI$ using chromatography in the presence of urea at various pH. With 8M urea and 0.2M formic acid, pH 2.8 and a linear

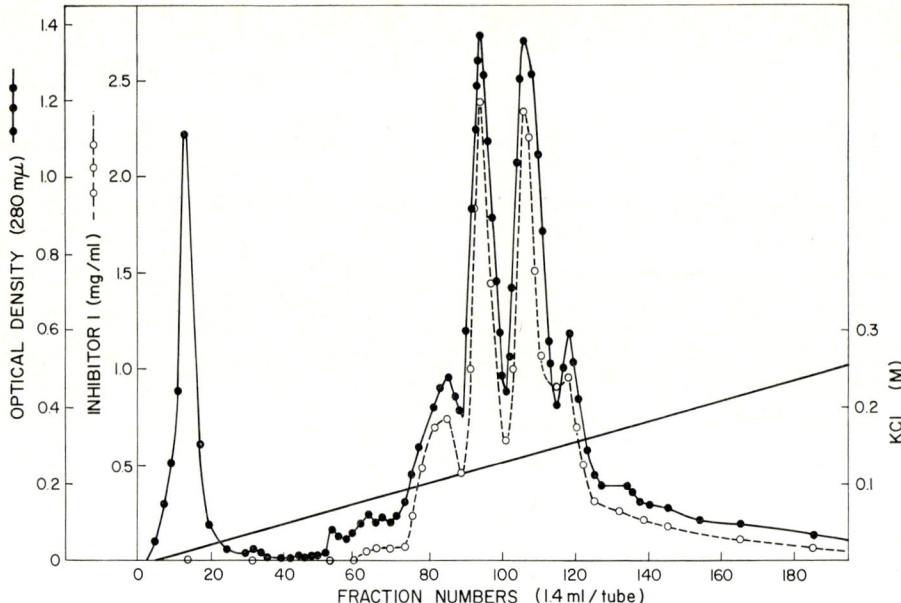

Fig. 3. Chromatography of *G75 PI* from sulfoethyl cellulose column (1 × 10 cm) equilibrated with 8M urea, 0.2M formic acid, pH 2.8, and eluted with a linear KCl gradient, represented as a solid line. The column was charged with 102 mg *G75 PI*. Open circles represent inhibitor *I* as determined by radial diffusion assays; closed circles represent optical density. Fractions of 1.4 ml were collected at 3 minute intervals. (RYAN and MELVILLE, in preparation).

KCl gradient, sulfoethyl cellulose separated dissociated *G75 PI* into four components [7]. This separation is shown in Figure 3. The profile shows two major and two minor components. This is reminiscent of the chromatographic elution profile of the lima bean inhibitor, which is a monomer and can be separated into four fractions [2]. We have called our fractions *A*, *B*, *C*, and *D* in the order they emerged from the column. This procedure is reproducible and has been used successfully with different preparations of *G75 PI*.

Results of the rechromatography of the two major fractions, *B* and *C*, is shown in Figure 4. These data indicate that true separation was achieved. Each peak was eluted essentially as a single peak and at the same salt concentrations as in the original separation.

An electrophoretogram of the four fractions on cellulose acetate paper in 8M urea and 0.2M formic acid, pH 2.7, is shown in Figure 5. Fractions *A* and *D*, the minor peaks, were not repurified before electrophoresis. Peak *D* shows contamination with all of the other components. These four components of inhibitor *I* are compared with the starting material (*G75 PI*) as well as *G PI*, and 7 times crystallized inhibitor *I*. The mobility of crystalline inhibitor is identical with that of peak *C* of the separation. The finding that crystalline inhibitor *I* contains only one kind of subunit component could explain our poor yields of this inhibitor by crystallization. When the fractions from the sulfoethyl cellulose separation were pooled individually and diluted with water or separated from the urea by Sephadex gel filtration, they each reassembled to form tetramers of 39,000 molecular weight. This indicated that the individual subunits did not require the presence of the other subunits for reassociation.

All four tetramers were inhibitors of both chymotrypsin and trypsin as shown in Table 2. The two major fractions *B* and *C* were potent chymotrypsin inhibitors whereas peak *B* was

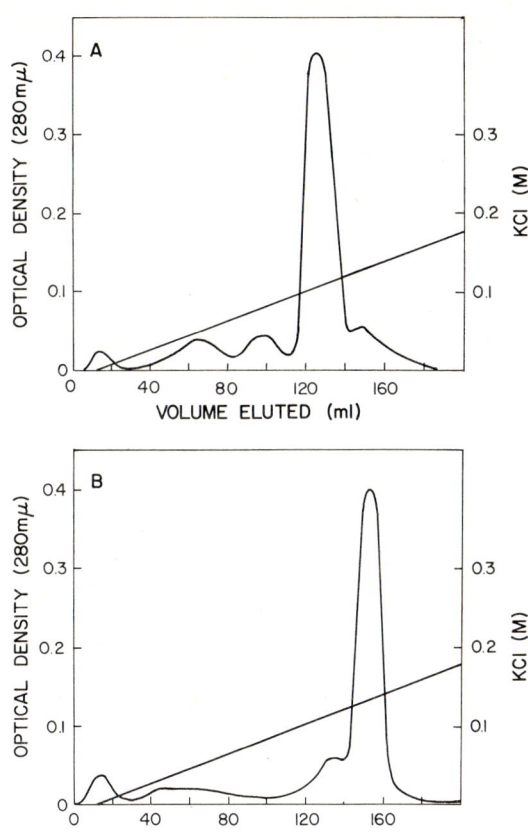

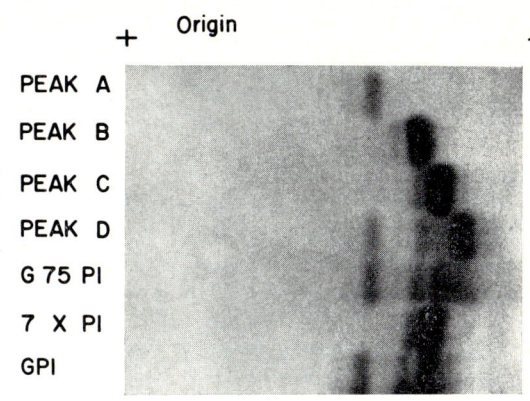

Fig. 5. Cellulose acetate electrophoresis of inhibitor *I* components and purified preparations performed using a buffer of 8M urea, 5% formic acid, pH 3.0. Electrophoresis at 150 volts and 9 mA for 1.25 h. Staining was achieved with 0.25% comassie brilliant blue. Proteins shown from top to bottom: Subunit components *A*, *B*, *C*, and *D* from sulfoethyl cellulose chromatography; *G75 PI*; 7 times crystallized inhibitor *I*; and *G PI*. (RYAN and MELVILLE, in preparation).

Fig. 4. Rechromatography of peak *B* (upper) and peak *C* (lower) from sulfoethyl cellulose chromatography of inhibitor *I*. Conditions were identical to those in Fig. 3 (RYAN and MELVILLE, in preparation).

Table 2. The inhibition of chymotrypsin and trypsin esterase activity by the subunit components of inhibitor *I*

Component	Enzyme Inhibited	
	Chymotrypsin	Trypsin
	mg/mg Inhibitor*	
A	0.72	0.49
B	2.81	1.23
C	2.81	0.22
D	1.67	0.22

* Calculated at 50% inhibition.

The substrate for chymotrypsin was N-acetyl-L-tyrosine ethyl ester and for trypsin tosyl-L-arginine methyl ester using the assay system of HUMMEL [23].

the strongest trypsin inhibitor. Peaks *A* and *D*, the minor components, were weaker inhibitors of both chymotrypsin and trypsin. In peak *C*, the strong chymotrypsin inhibitory activity together with a weak trypsin inhibitory activity again suggests the similarity of peak *C* to the crystalline inhibitor *I*, which has the same properties [9].

Despite the differences among peaks *A*, *B*, *C*, and *D* in inhibitory capacities and electrophoretic mobilities, all four of the components were very similar as judged immunologically [10]. In Figure 6 is shown a double diffusion experiment in which no spurs could be detected between the precipitin lines of the peak components and of *G PI* when challenged with rabbit anti-inhibitor *I* prepared from crystalline inhibitor. No spurs could be detected in these or in similar experiments in which all four fractions were compared with each other. This test is only qualitative and small structural changes can go unseen. However, these results did indicate that the immunologically active structures are very much alike. We have tested this further using quantitative complement fixation assays [11] and

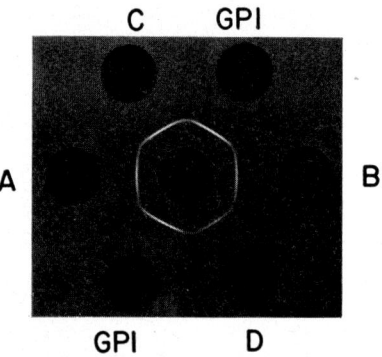

Fig. 6. OUCHTERLONEY double diffusion assays of fractions A, B, C, and D from sulfoethyl cellulose chromatography of $G75$ PI. Sample G PI is a highly pure preparation described in the text. Proteins were dissolved on 0.05M NH$_4$HCO$_3$, pH 7.7 at a concentration of 200 µg/ml. Rabbit antiserum, used in the center well, was diluted tenfold with 0.05M sodium barbital buffer in 0.9% NaCl, pH 8.2. (MELVILLE and RYAN, in preparation).

found that small but significant differences can be detected among the four components. In Table 3 is shown the micro complement fixation index of dissimilarity for each of the four components as compared to G PI. These values

Table 3. Reactivity of the four subunit components of inhibitor I and of G PI with antiserum prepared against crystalline inhibitor I

Protein	index of dissimilarity*
Component A	1.35
Component B	1.18
Component C	0.83
Component D	1.02
$G PI$	1.00

* The relative concentration of antiserum required to produce a complement fixation curve the peak of which was as high as that given by the homologous protein [24].

indicate that all five of the protein preparations shown are very similar, but that distinct and measurable difference are present within those structures of the proteins that determine immunological reactivites. The data here again support our other data indicating peak C is identical to crystalline inhibitor I. The antiserum was prepared against crystalline inhibitor I and it is peak C that reacts the strongest with it. Differences in amino acid composition among the four components have also been found. While precise amino acid analyses are not yet complete we can report that large differences were detected in proline, alanine and isoleucine among the four components, while smaller differences occur among several other amino acids.

The minimum molecular weights of the components, based on preliminary amino acid analyses, do not give values consistent with a molecule of 9,000—10,000. This leads us to suspect that each subunit may consist of several closely related gene products, arising from multiple genes. However, the N-terminal amino acids, determined for all of the fractions by the dansyl method [12] were found to be glutamic acid in each of the four components isolated from sulfoethyl cellulose. N-terminal glutamic acid was also the only N-terminal residue we could detect in $G75$ PI [7].

It was mentioned earlier that all four of the fractions would reassociate to form the tetramer in the absence of the other components. We were interested in seeing if the different components would hybridize with each other. Using immunoelectrophoresis we were able to test this. In Figure 7 is shown a series of immunoelectrophoresis experiments of unhybridized and hybridized inhibitor I component fractions from sulfoethyl cellulose. A_4, B_4, C_4, and D_4 are the immunoelectrophoretic patterns of the four purified components reassociated alone. In the fifth gel from the top, A_4 was mixed with D_4 after each was reassociated separately, and below this is a gel in which A monomer was mixed with D monomer while in 8M urea, 0.2M formic acid, pH 8.2 (to dissociate) and then diluted with water (reassociated) to give a hybrid mixture. It is easily seen that a hybrid was formed. We assumed that the hybrid A_2D_2 is the predominate species because of its migration midway between A_4 and D_4.

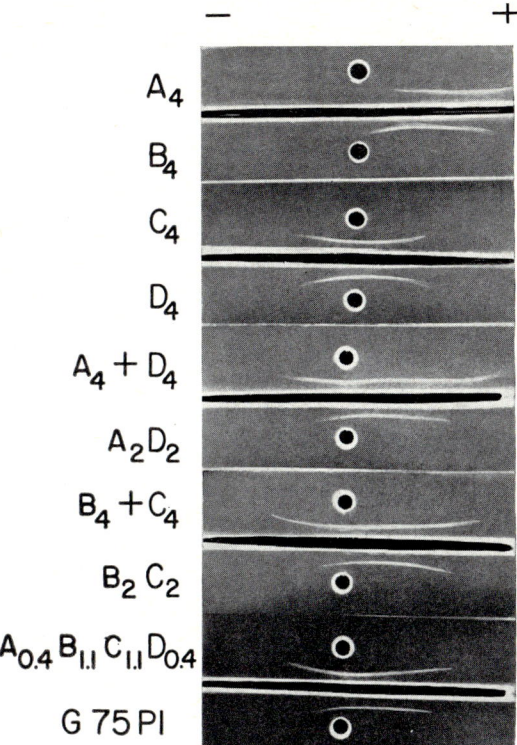

Fig. 7. The demonstration of hybridization of inhibitor I subunit fractions using immunoelectrophoresis in 0.1M sodium barbital, pH 8.6. Samples A_4, B_4, C_4, and D_4 are subunit fractions obtained from sulfoethyl chromatography of $G75 PI$, dissolved in 8M urea and reconstituted by dilution (1—10) with water. In nonhybridizing conditions ($A_4 + B_4$ and $B_4 + C_4$) the individual subunit fractions were mixed in water only. In hybridization experiments (A_2D_2; B_2C_2; and $A_{0.4}B_{1.1}C_{1.1}D_{0.4}$) the individual subunits were dissolved together in 8M urea, 0.2M formic acid, pH 2.8, and after 30 minutes were diluted ten fold with water to promote random hybridization. The concentration of protein in each sample was adjusted so that each diffusion line represented 200 μg inhibitor protein. (MELVILLE and RYAN, in preparation).

The same type of experiment is shown using components B and C. Again hybrid was formed, probably B_2C_2 predominating. At the bottom of the slide is a hybridized mixture of the four components in approximately the same ratios as isolated from sulfoethyl cellulose. In this case the mixture was similar but not identical with the starting material, $G75 PI$. However, $G75 PI$ may not represent the inhibitor as it occurs in the tuber either, since our purification procedure contains at least two steps that might cause dissociation and reassociation of the molecule, that is, acid extraction and heating [7].

II. Physiology and Molecular Biology of Inhibitor I

In order to study inhibitor I in the potato tuber we needed an assay that would specifically quantitate this one protein among the hundreds of proteins present. We turned to immunological techniques and utilized the specifity afforded by the specific rabbit antiserum prepared using injections of 4 times crystallized inhibitor I. Utilizing a concept previously used in quantitating immunologlobulins in animals [13, 14] we developed a quantitative immunological method of radial diffusion in agar gels containing antibodies for use with soluble cellular proteins such as inhibitor I [15]. This method can be used to quantitate inhibitor I in μg quantites in crude juice as well as in solutions of purified or semipurified inhibitor I.

We began to study the location of inhibitor I within the tuber itself. We found that new potatoes had much higher concentrations of inhibitor I in the soluble juice than old potatoes [16], an observation similar to that made several years earlier by WERLE when studying the potato kallikrein inhibitor [17]. Inhibitor I was found concentrated in cortical tissue. In new potatoes its concentration varied among individual tubers and represented from 1% to over 5% of the total soluble protein. The inner pith tissue contained considerably less inhibitor I. In tubers stored at 5°C for several weeks the concentration of inhibitor I became lower at the stem end cortex, but remained high in the apical end cortex. In year old potatoes inhibitor I had virtually disappeared from the entire stem end. When tubers were allowed to sprout and then planted, inhibitor I disappeared from the seed piece within a few days but appeared in the new sprout and young leaves. It could be detected in leaves until about the time the plants set tubers and then virtually disappeared in the vegetative tissues but accumulated in the new tubers.

We looked for the presence of the inhibitor in true potato seeds but were unable to find it. We therefore planted the seeds, confident that inhibitor I would eventually appear in the tubers and perhaps in vegetative tissues. In Figure 8

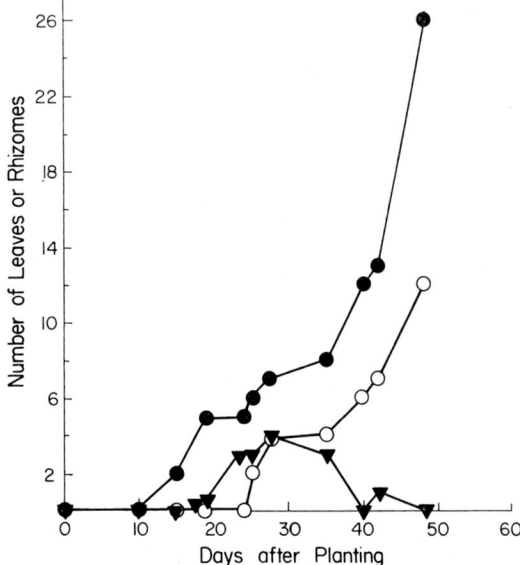

Fig. 9. A plot of the total number of leaflets (●) and rhizomes (○) in each growing young potato plant. Solid triangles (▲) are the number of the leaflets showing the presence of inhibitor I in the immunological double diffusion assays. (RYAN and HUISMAN, 1967). (RYAN and HUISMAN, Nature, 1967).

Fig. 8. Young potato plants grown from true seed. The largest plants are 25 days after planting. Note the large increase in rhizomes between the two largest plants. The difference in the age of these two plants is about 5 days.

is shown some true potato seeds and young potato plants at various stages of growth. The largest plants are about 25 days old. At approximately 20 days rhizomes, or stolens, began to form at or under the soil level. It is on these rhizomes that tubers eventually develop. We tested the leaves immunologically for the presence or absence of inhibitor I every few days. In Figure 9 is shown a graphic representation of the results of these experiments [18]. It can be seen that on about day 20 the leaves began to exhibit the presence of inhibitor I. Within the next day or two they began to produce rhizomes. As more rhizomes appeared and grew larger at a very rapid rate, inhibitor I became difficult to demonstrate in the leaves. It was at this time we discovered a very important phenomenon. If we cut off the rhizomes,

inhibitor I returned to the leaves within a few hours. This led to the observation that when we removed leaflets from the plant, and immersed the petioles in water and placed them in constant light, inhibitor I began immediately to accumulate in the excised leaflets [19].

We began a study of this accumulation process in excised leaflets of potatoes, as well as in leaflets from other species of Solanaceae [19—22]. We found that the same phenomenon occurs in excised tomato leaflets [20]. In fact, tomato leaflets were more suited for study because inhibitor I was very seldom found in normal tomato leaves under our conditions of growth, but they accumulated the protein better than potato leaflets when excised and incubated in light.

We used the excised terminal leaflets, supplied them with water, and incubated them in constant light or in total darkness. To assay for inhibitor I we ground the leaflet using a mortar and pestle and recovered the juice by squeezing with a hand garlic press. An aliquot of the juice (about 20 µl) was used in the radial diffusion method of assay. The content of inhibitor I in excised tomato leaf juice after incubation in constant light or dark is shown in Figure 10. In light, inhibitor I steadily accumulates for several hours. In darkness it does not accumulate at all.

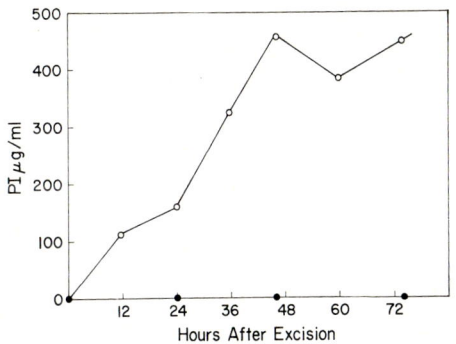

Fig. 10. Synthesis and accumulation of inhibitor I in juice of excised tomato leaflets. Leaflets were excised at 0 time and supplied with water in constant light (○) or constant darkness (●). Inhibitor I concentrations were determined using radial diffusion assays.

This excision-induced accumulation of inhibitor I in light in both tomato and potato leaves is apparently quite specific and is not a reflection of an increase in all proteins of the leaf. In fact, overall total soluble protein generally decreases during the induction [25]. By supplying ^{14}C leucine and ^{14}C lysine through the cut petioles we have estimated that after a 48 hour induction from 13 to 18% of the new soluble proteins in the leaf juice can be accounted for by inhibitor I [19, 22]. The kinetics of this synthesis and accumulation of inhibitor I suggest that either multiple genes are present, or that an unusally stable m-RNA is produced. Known rates of translation from a single bacterial gene would be much too slow to account for so much synthesis. In the excised leaflet system protein synthesis inhibitors such as cycloheximide, puromycin, and actinomycin D each severely inhibit synthesis while chloramphenicol does not [19, 22]. Indole acetic acid, 2, 4, D and DCMU each inhibit accumulation by about 50% [19, 22].

A comparison of the increase in immunologically reactive inhibitor I, with the increase in both trypsin inhibitory and chymotrypsin inhibitory activites, was made for both potato and tomato leaflets and is presented in Table 4. Juice was isolated from mascerated leaflets with a small press and centrifuged at 100,000 ×g for 15 minutes. The inhibitor I content was determined using the radial diffusion assay, wheras total

Table 4. Induction of inhibitors of chymotrypsin and trypsin in excised potato and tomato leaves supplied with water and constant light for 72 hrs

Leaflet	Immunological Assay[1] (µg/ml juice)		Enzymatic Assay[2] Enzyme Inhibited (µg/ml juice)			
	Inhibitor I		Chymotrypsin		Trypsin	
	0 Time	Detached, 72 hr in light	0 Time	Detached, 72 hr in light	0 Time	Detached, 72 hr in light
Potato	0	80	Trace	495	11	138
Tomato	0	127	0	290	33	300

[1] Radial diffusion assay.
[2] Esterase assays by the method of HUMMEL [23] (see Table 2).

chymotrypsin inhibitors and trypsin inhibitors were determined using the Hummel assay [23]. In potato leaflets a greater increase in chymotrypsin inhibitor activity was found than could be accounted for by inhibitor I alone. Inhibitor I inhibits about 2.5 times its weight of chymotrypsin, or in this experiment, about 200 µg of chymotrypsin. The leaflet juice inhibits nearly 500 µg chymotrypsin per ml. Trypsin inhibitor also increases in the juice during this induction process, especially in tomato leaflets. In these leaflets inhibitor I can account for all of the chymotrypsin inhibitory activity, but the increase in trypsin inhibitor activity is larger than we can account for, even if all the inhibitor I were the subunit type that contains strong trypsin inhibitor activity. This evidence suggests that other proteinase inhibitors are accumulating along with inhibitor I in the detached leaflets. We anticipated that the synthesis and accumulation of such a large amount of protein by the leaflets might be associated with ultrastructural changes within the plant cells. To test this possibility we examined fresh leaves and excised leaves induced in the light or kept in the dark. In each of the samples examined (we examined dozens) there was a relationship between protein bodies in the vacuoles of the cells and the presence of inhibitor I. Leaf tissue containing no inhibitor I contained no vacuolar protein bodies. To determine if the relationship was quantitative we designed a time course experiment to quantitate inhibitor I during its induction in light and, at the same time using electron microscopy, to observe the protein bodies in the vacuoles of cells from small sections of these leaflets. For these experiments we were able to eliminate interleaf differences due to age and plant condition. It had previously been observed that opposite "twin" leaflets of a tomato leaf were very similar in the amount of inhibitor protein they would accumulate. We thus detached a leaf and removed all leaflets except one pair. At the end of 4 time periods we removed a longitudinal leaflet half, prepared a small portion for electron microscopy, and tested the juice from the remainder for inhibitor I. We found that, with time, the vacuolar bodies increased in numbers and/or size as the concentration of inhibitor I increased.

As can be seen from Figure 11 the total accumulation of inhibitor I in induced leaflets was quite large in this experiment. Determinations on these 4 leaflet halves were made at 14, 20, 29 and 64 hours. Uninduced tomato laeflets con-

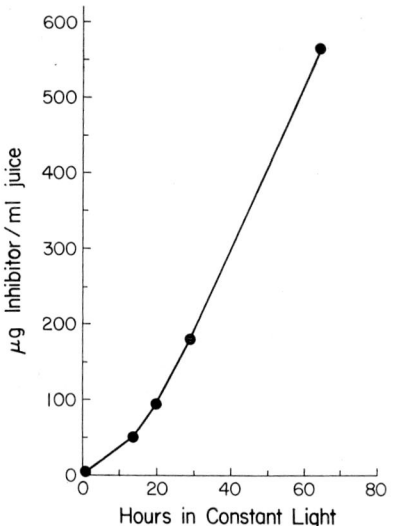

Fig. 11. Changes in inhibitor I concentration with time after detachment. The data shown are from the four halves of the leaflet pairs of one leaf. Electron micrographs of these same tissues are shown in Fig. 13—18. (SHUMWAY, et al. 1970).

tained no inhibitor I, thus the line is extrapolated to 0. Sections of each of these leaflets, at the same times, were fixed and sectioned for electron microscopy.

A tomato leaf in cross-section is shown in Figure 12. The epidermal cells contain very few chloroplasts compared to the mesophyll cells between the two epidermal layers. The columnar palisade mesophyll cells are different in size, shape and orientation from the spongy mesophyll, but both types of cells contain many chloroplasts. The electron micrographs (Figs. 13 to 18) show both types of cells. Both types respond similarly in ultrastructural changes during synthesis of inhibitor I.

Chymotrypsin Inhibitor I in the Solanaceae Family

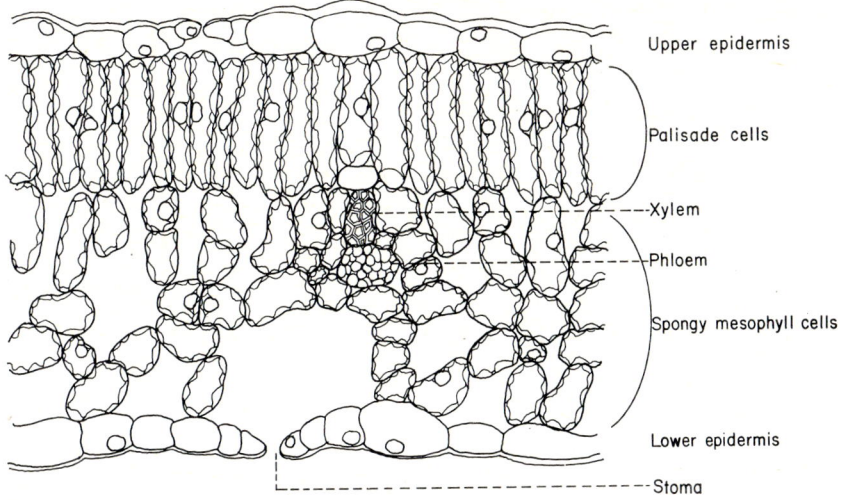

Fig. 12. A drawing of a cross-section of a typical tomato leaf cell.

A mesophyll cell of tomato leaf containing no inhibitor I appears as shown in Figure 13. The large central vacuole is characteristic of mature

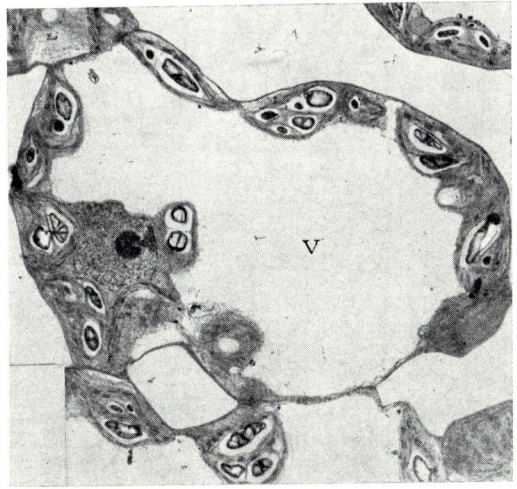

Fig. 13. An electron micrograph showing a non-induced mesophyll cell of a tomato leaf containing no inhibitor I. Fig. 13—17 and 18 are electron micrographs of cross-sections of tomato leaf cells that were fixed in glutaraldehyde followed by osmium tetroxide, dehydrated in ethanol and propylene oxide and embedded in araldite 6005 plastic. Thin sections were stained with $BaMnO_4$. For details of preparation see SHUMWAY, RANCOUR and RYAN, 1970.

plant cells. The vacuole contains no electron dense material. The nucleus and chloroplasts are rich with starch granules in young leaflet cells such as this.

After incubation of the detached leaves in light for 14 hours the leaflet half contained 50 μg of inhibitor I per ml of leaf juice. Electron dense bodies were present in the central vacuole and in small vesicles in the cytoplasm (Fig. 14).

At the end of 20 hours incubation (Fig. 15) there was 95 μg of inhibitor I per ml of leaf juice in the second leaflet half and ultra structural changes were even more obvious. More electron dense material was present in the vacuoles and the chloroplasts contained less starch. The large crystalline bodies in the cytoplasm appeared to be glyoxysomes. However, conditions producing these crystals are not always the same as those that influence formation of inhibitor I and vacuolar protein bodies.

After 29 hours (Fig. 16) the amount of inhibitor I had increased to 160 μg/ml and the vacuolar bodies were filling a considerable portion of the cell vacuoles.

At 64 hours (Fig. 17) inhibitor I concentration was 565 μg/ml and very large bodies were present in the vacuoles. The chloroplasts had lost their starch bodies, the stroma was less electron

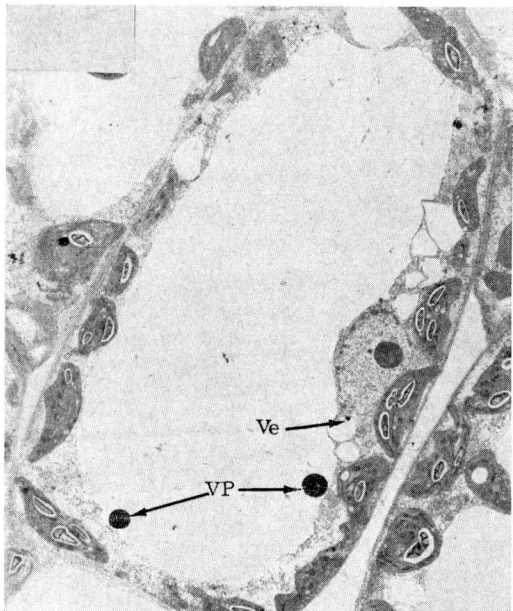

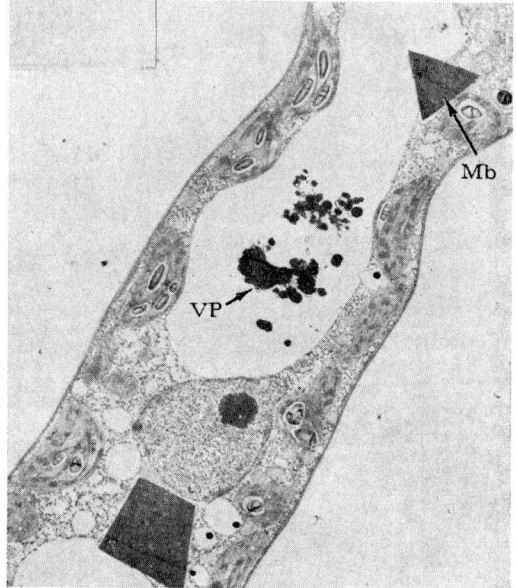

Fig. 15

Fig. 14. Electron micrograph showing a mature tomato-leaf palisade parenchyma cell from a detached leaf after 14 hr inducing conditions (continuous light and 0.01M glutamine). This represents the first of four leaflets-pair halves; see Figs. 15, 16, and 17 for the other three halves. Note the presence of vacuolar protein (VP) and small vesicles (Ve) in the cytoplasm, one of which contains electron-dense material. Protein inhibitor I concentration in this leaflets half was 50 μg/ml. (SHUMWAY, et al. 1970).

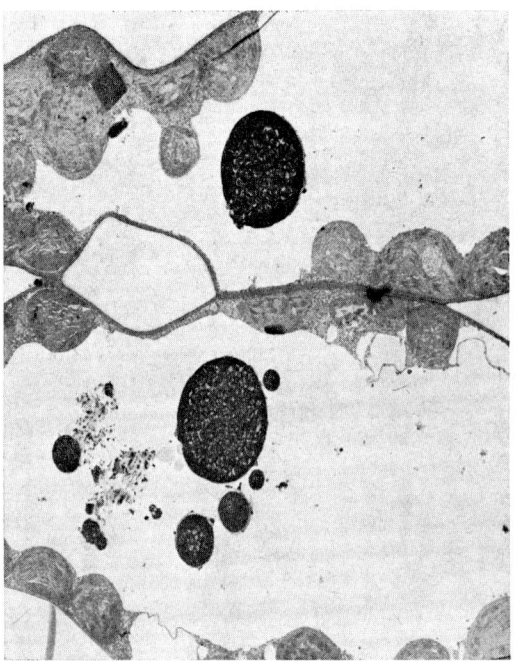

Fig. 15. Electron micrograph showing a palisade parenchyma cell from the second leaflet half (same leaflet as Fig. 6) at the end of 20 hr under inducing conditions. More protein, compared with Fig. 14, is present in the vacuole. Several cytoplasmic vacuoles contain electron-dense material. Two "microbodies" (Mb) are present. Some sectioning compression is evident in the "microbodies" and the vacuolar protein (VP). Protein inhibitor I concentration was 95 μg/ml.

Fig. 16. Electron micrograph showing palisade parenchyma cells from the first half of the second leaflet in the series (twin leaflet to material represented by Fig. 14 and 15) at the end of 29 hr under inducing conditions. Still more protein, compared with Fig. 15, is present in the vacuoles. Protein inhibitor I concentration was 160 μg/ml. (SHUMWAY, et al. 1970).

Fig. 16

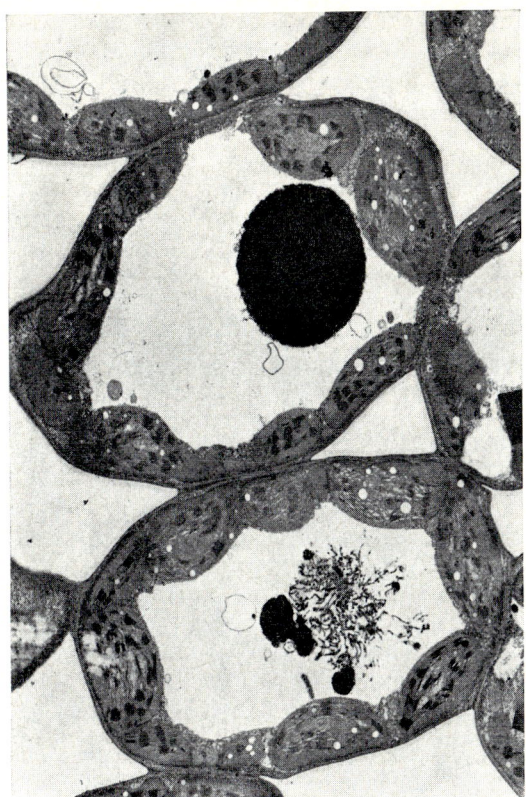

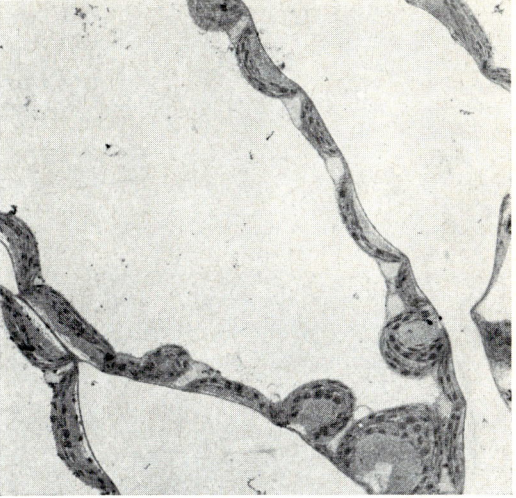

Fig. 18. Electron micrograph showing a mature tomato-leaf palisade parenchyma cell from a detached leaf kept in non-inducing conditions (no light) for 73 hr. No protein inhibitor I was present in the leaf from which this specimen was taken. Some contamination is present, but no electron-dense material comparable to that of Figs. 14—18 is present is the vacuole. (SHUMWAY, et al. 1970).

Fig. 17. Electron micrograph showing spongy parenchyma cells from the other leaflet half (same leaflet as Fig. 16) at the end of 64 hr under inducing conditions. No starch bodies are present in chloroplasts, still more protein, compared with Fig. 16, is present in vacuoles. Protein inhibitor I concentration was 565 μg/ml. (SHUMWAY, et al. 1970).

dense and the chloroplast membranes appeared to be swollen and perhaps degenerating.

Detached leaflets kept in total darkness for 73 hours (Fig. 18) contained no inhibitor I and no bodies were present in the vacuoles.

Using cytochemical tests on paraffin embedded material we have determined that the vacuolar bodies are primarily protein with perhaps a small amount of carbohydrate. We are presently engaged in experiments using ferritin labeled antibodies to determine if the vacuolar protein bodies do in fact contain chymotrypsin inhibitor I.

Vacuolar bodies have been present in every tissue of every member of the Solanaceae in which we have detected inhibitor I immunologically. Because of the above correlations and the transiency of inhibitor I in plants and tubers, we believe that inhibitor I may be a significant part of these vacuolar protein bodies. We suggest that inhibitor I is a special type of reserve protein that is produced at specific times in specific tissues and must serve a specific purpose for the plant. We do not yet know what the role of inhibitor I is but we are now testing the hypothesis that it may be a precursor of part of the cellular membranes. We have not excluded the possibility that the inhibitory activity is involved in function. It is possible that inhibitor I may be involved in the regulation of proteolytic enzymes in the plant or that it might be involved in a protective mechanism of some type within the cell. If inhibitor I is incorporated into membranes it might be involved there in maintaining structural integrity of the cell. It could provide an effective defense against proteolytic digestion by either intracellular or extracellular proteinases. With the information presented here as a beginning, our search for the function continues.

The authors **acknowledge** Dr. J. Curtis Melville for his research on the purification and properties of inhibitor I and to Dr. John Rancour for his contribution to the ultrastructural studies.

References

[1] Vogel, R., I. Trautschold and E. Werle, in Natural Proteinase Inhibitors. Academic Press, New York (1968).
[2] Jones, G., S. Moore and W. Stein, Biochemistry **2**, 66 (1963).
[3] Mansfield, V., A. Ziegelhöffer, Z. Horakova and J. Hladovec, Naturwissenschaften **46**, 172 (1959).
[4] Birk, Y., A. Gertler, Chemistry and Biology of Proteinase Inhibitors from Soybeans. This volume, page 142.
[5] Ryan, C. A. and A. K. Balls, Proc, Natl. Acad. Sci, (U. S.) **48**, 1839 (1962).
[6] Balls, A. K. and C. A. Ryan, J. Biol. Chem. **238**, 2976 (1963).
[7] Ryan, C. A. and J. C. Melville, in preparation.
[8] Melville, J. C. and C. A. Ryan, Arch. Biochem. Biophys. **138**, 700 (1970).
[9] Ryan, C. A., Biochemistry **5**, 1592 (1966).
[10] Melville, J. C. and C. A. Ryan, in preparation.
[11] Wasserman, E. and L. Levine, J. Immunol. **87**, 200 (1961).
[12] Gray, W. R., in Methods in Enzymology Vol. XI, edited by C. H. W. Hirs, Academic Press, New York and London, p. 139 (1967).
[13] Mancini, G., J. P. Vaerman, A. O. Carbonara and J. F. Heremans, in Protides of the Biological Fluids, 11th Colloquium, 1963 (H. Peters, ed.) p. 370 Elsevier, New York—Amsterdam (1964).
[14] Fahey, J. L. and E. M. McKelvey, J. Immunol. **94**, 84 (1965).
[15] Ryan, C. A., Anal. Biochem. **19**, 434 (1967).
[16] Ryan, C. A. and R. W. Van Denbergh, Plant Physiol. **43**, 598 (1968).
[17] Werle, E. and L. Maier, Biochem Z. **323**, 279 (1952).
[18] Ryan, C. A. and O. C. Huisman, Nature **214**, 1047 (1967).
[19] Ryan, C. A., Plant Physiol. **43**, 1859 (1968).
[20] Ryan, C. A., Plant Physiol. **43**, 1880 (1968).
[21] Gurusiddaiah, S., J. C. Melville, T. Kuo and C. A. Ryan, in preparation.
[22] Ryan, C. A. and W. Huisman, Plant Physiol. **45**, 484 (1970).
[23] Hummel, B., Can. J., Biochemistry **19**, 434 (1959).
[24] Sarich, V. M. and A. C. Wilson, Science **158**, 1200 (1967).
[25] Rancour, J. M., Ph. D. Thesis, A Biochemical and Ultrastructural Study of Protein Storage in Leaf Tissue of Tomato and Potato. Washington State University (1970).
[26] Shumway, L. K., J. M. Rancour and C. A. Ryan, Planta **93**, 1 (1970).

Comparative Biochemistry of Avian Egg White Ovomucoids and Ovoinhibitors

Robert E. Feeney

Department of Food Science and Technology, University of California, Davis, California 95616

Introduction

Three inhibitors of proteolytic enzymes with very different properties have been identified in avian egg whites. In chicken egg white these inhibitors have very different properties and exist in very different concentrations (Table 1). These differences are compounded when the homologous inhibitors from different avian species are examined. Indeed, it has sometimes been found necessary to study the physical-chemical properties of extensively purified preparations in order to show that an inhibitor from one egg white is a closely related homolog of the inhibitor from another egg white (Feeney and Allison, 1969). Nevertheless, the variation in inhibitory properties of homologous proteins in different avian egg whites is the very same characteristic that makes egg whites interesting and useful tools for the study of the general protein chemistry of inhibitors of proteolytic enzymes.

The three inhibitors which have been described are ovomucoid, ovoinhibitor, and the ficin or papain inhibitor. Ovomucoid and ovoinhibitor have received extensive study and will be discussed in this report. In contrast, the ficin and papain inhibitor (Fossum and Whitaker, 1968) has not been described since the original report, and that report only concerns the chicken egg white protein. It will therefore not be discussed further in this present report. A more extensive description of the chemical compositions of chicken ovomucoid and ovoinhibitor are presented elsewhere (Lin and Feeney, 1971).

Table 1. Avian Egg White Inhibitors

Protein	Physical-Chemical			Properties Inhibitory			
	M. W. g	I. P.	CHO %	Trypsin	Chymotrypsin-subtilisin[b]	Fungal proteinase	Ficin-papain[c]
Ovomucoid	28000	4.0—4.6	22[a]	+ (—)	+ (—)	—	—
Ovoinhibitor	46000	4.4—5.0	10[a]	+ (—)	+ (—)	+	—
Ficin-papain inhibitor	13000	5—6 (?)	(?)	—	—	—	+

[a] ± 2%, [b] Chymotrypsin and substilisin Compete, [c] Ficin and papain compete.

Ovomucoid

The capacity to inhibit the digestive enzyme bovine trypsin is a biochemical activity in egg white that has been known for many years. Although various workers had presumptive evidence that it might be ovomucoid which had this activity, it remained for LINEWEAVER and MURRAY (1947) to establish that the ovomucoid is the protein responsible for the trypsin inhibitory capacity. Chicken ovomucoid has received a great deal of attention, apparently due to its mucoid nature as well as its inhibitory properties. For over fifteen years after chicken ovomucoid was identified as a main inhibitor in egg white, there were arguments as to whether or not it would inhibit bovine α-chymotrypsin. It was then found by FEENEY et al. (1963) that this ovomucoid was contaminated with one or more other proteins with inhibitory activity. The principal contaminant proved to be the ovoinhibitor, which was responsible for the inhibitory activity of all commercial preparations of ovomucoid (and most other preparations as well) against bovine α-chymotrypsin. But even these more purified preparations of ovomucoid have been found to be very heterogeneous because of the presence of multiple molecular forms (FEENEY et al., 1967). As long ago as 1940 LONGSWORTH, CANNAN and MACINNES (1940) had stated "Ovomucoid, though not electrically separable into more than one component, shows complexity as indicated by reversible boundary spreading". Some further purification has been obtained and there are suggestions that differences in sialic acid content may be responsible for the electrical heterogeneity (FEENEY and ALLISON, 1969; MELAMED 1965). It now appears probable that heterogeneity is also due to other differences such as amide content or carbohydrate content or structure.

Physical Properties of Ovomucoid. RHODES et al. (1960), in one of the first general comparative studies of the physical-chemical properties of avian ovomucoids, reported that ovomucoids from 11 different species had physical properties which were in general similar to one another. These ovomucoids were obtained from the following avian species: turkey, guinea, Peking duck, Khaki Campbell duck, goose, emu, golden pheasant, cassowary, red jungle fowl, chicken, California valley quail, and painted quail. Values of 28,000 $\pm$ 1500 g M. W. and carbohydrate content of 20—25% were found for the ovomucoids of the emu, cassowary, kiwi, ostrich, rhea, tinamou, chicken, and duck (DEUTSCH and MORTON, 1961; OSUGA and FEENEY 1968). The circular dichroism and optical rotatory dispersion of chicken ovomucoid was studied by IKEDA et al. (1968). DONOVAN (1967), in a study of the hydrodynamic properties of chicken ovomucoid, concluded that it is highly hydrated rather than compact or highly asymmetric.

In studies of the absorption of native turkey ovomucoid in the ultraviolet both DONOVAN (1967) and SJOBERG and FEENEY (1968) observed a "blue shift" and a decrease in absorption in acid solution. Reduction of disulfide bridges of both chicken and turkey ovomucoid caused a large (20—25%) decrease in absorption at 280 mμ with a simultaneous decrease in inhibitory activity. Upon air reoxidation, both the absorptivity (Fig. 1) and the inhibitory activity returned to that of the native protein.

Chemical Composition. OSUGA and FEENEY (1968) reported on the amino acid and sugar content of chicken, turkey, cassowary, emu, ostrich, rhea, and tinamou ovomucoids. Our laboratory (OSUGA and FEENEY, 1971) has recently completed analyses of the amino acid and sugar contents of ovomucoids of the adelie penguin, royal penguin, emperor penguin, little blue penguin, and rockhopper penguin. Many differences were noted between the compositions of all the ovomucoids, except among the penguin ovomucoids which were very similar to one another. Table 2 summarizes some of the main differences noted in the amino acid content of different ovomucoids. All the ovomucoids were very high in carbohydrate content, namely nearly 25%, and tryptophan was absent. A high content of cystine was remarkably constant and varied only from a value of 17 to 20 residues per mole.

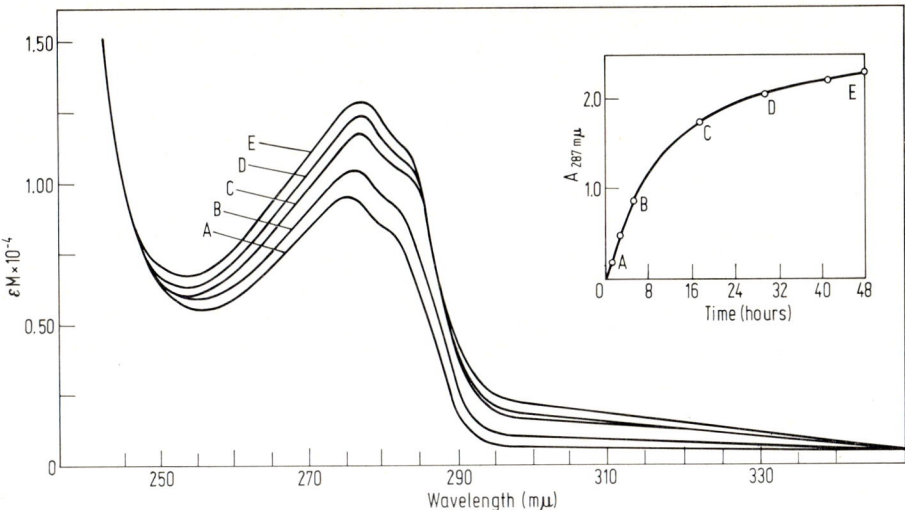

Fig. 1. Changes in ultraviolet absorption spectra of turkey ovomucoid after reduction and after various periods of reoxidation. Protein (0.7 mg/ml as is basis, 8—10% water) was dissolved in 0.006M Tris buffer adjusted to pH 8.3. Incubation was at room temperature for the following times (h): A, zero (starting); B, 4; C, 17; D, 29; E, 48 (from SJOBERG and FEENEY, 1968).

Table 2. Major Differences in Properties of Ovomucoids

Ovomucoid	Inhibitory Activity*		Amino Acids/mole	
	Trypsin	Chymo-trypsin	Methio-nine	Arginine
Chicken	1	0	2	6
Turkey	1	1	2	6
Cassowary	1	0	1	1
Emu	2	1	1	0
Kiwi	1	1	—	—
Ostrich	1	1	1	3
Rhea	1	1	1	0
Tinamou	0	1	0	3
Duck	2	1	8	1

* Against bovine enzymes.

Since no sulfhydryl groups were found in any of the ovomucoids, there appeared to be only 9 or 10 disulfide linkages in all 8 ovomucoids studied. Although other extensive differences were noted, such as a content of serine of only 10 residues in turkey and 18 residues in the ostrich, the differences in the contents of arginine and methionine were the highest on a basis of the percentage of these amino acids present.

These different contents of arginine and methionine have been a valuable tool in our laboratory for investigating chemical reactions for modifying amino acid side chains. In particular, the different contents of arginine made possible the development of a better method for the modification of arginine and for the detection of a side reaction not involving arginine (LIU et al., 1968).

Comparative Inhibitory Activities. Extensive differences among the ovomucoids exist in their inhibitory activities against different proteolytic enzymes (Table 3). The activities against bovine trypsin, bovine α-chymotrypsin, and subtilisin show only one correlation. This correlation is that all of the inhibitors which inhibit chymotrypsin also inhibit subtilisin. In a study of 15 ovomucoids in the author's laboratory the binding sites for trypsin and chymotrypsin have in all cases been found completely independent and not overlapping, while the binding sites for chymotrypsin and subtilisin, when dual inhibitory capacities have been found, are always dependent upon one another and apparently

Table 3. Inhibition of Five Proteinases by Ovomucoids and Ovoinhibitors

	Enzymes[a] Inhibited[b]				
	Human trypsin[c]	Bovine trypsin[d]	Bovine α-chymotrypsin	Subtilisin[d]	Fungal proteinase[e]
Ovomucoids					
Chicken	—	+++	—	—	—
Quail	++	+++	—	—	—
Cassowary	—	++	—	—	—
Ostrich	—	+	—	—	—
Emu	—	++	+	+	—
Rhea	—	++	+	+	—
Pheasant[f]	—	—	++++	++	—
Duck	—	++++	++++	+++	—
Turkey	—	++++	++++	++	—
Tinamou	—	+	++++	++	—
Penguin	—	+	+	++++	—
Ovoinhibitors					
Chicken	—	++++	++++	++++	++++
Quail	?	++++	++++	++++	++++
Turkey	?	++++	++++	++++	++++
Penguin	?	++++	—	++++	++++

[a] Enzymatic activity determined by the casein digestion assay.
[b] The degree of inhibition is indicated as —, for extremely weak or inactive, and +, ++, +++, and ++++, for varying degrees, progressing from weak to strong inhibition.
[c] From FEENEY et al. (1969).
[d] From BIGLER and FEENEY (1971).
[e] Fungal proteinase in alkaline proteinase of *Aspergillus oryzae*.
[f] Golden Pheasant.

overlap. In earlier studies this was clearly shown in the case of duck ovomucoid. Duck ovomucoid binds both trypsin and chymotrypsin. It apparently binds a second trypsin much more weakly than the first, but the binding of trypsin or chymotrypsin does not interfere with the binding of the other enzyme. Studies have been carried further now to the binding of chymotrypsin and subtilisin by penguin ovomucoid. The displacement of chymotrypsin with subtilisin by penguin ovomucoid is readily seen in Figure 2. Penguin ovomucoid has an equilibrium constant (K_{ass}) of approximately 10^6 for chymotrypsin and 10^9 for subtilisin, or at least a thousand-fold stronger affinity for subtilisin than chymotrypsin.

One of the more provocative developments in our laboratory was stimulated by an older preliminary observation from the laboratory of F. F. NORD (BUCK et al., 1962) on the inhibition of human trypsin by chicken ovomucoid. NORD reported that chicken ovomucoid did not inhibit human trypsin. Our data completely substantiated this observation and no relationship was found between the capacity to inhibit bovine and human trypsin using a long list of inhibitors (FEENEY et al., 1969). One ovomucoid, quail ovomucoid, was a comparatively good inhibitor of human trypsin but much weaker than comparatively strong inhibitors such as lima bean trypsin inhibitor and bovine colostrum inhibitor.

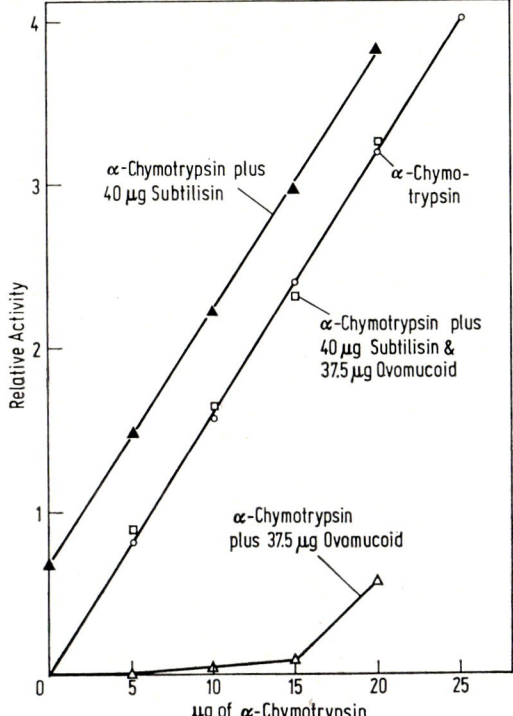

Fig. 2. Displacement of α-chymotrypsin by subtilisin (Nagarse) in complexes with penguin ovomucoid. The plot of the relative activity of the assay mixture is shown with respect to the additions of increasing amounts of α-chymotrypsin, under various conditions which enable the distinction of competition between α-chymotrypsin and subtilisin for inhibition by penguin ovomucoid, using the substrate, BTEE. Enzyme activities are expressed as relative activities and values are actual slopes from charts of recording spectrophotometers. The respective assay conditions were: ○, no additions other than varying amounts of α-chymotrypsin; △, α-chymotrypsin with 37.5 μg of penguin ovomucoid; □, α-chymotrypsin with 37.5 μg of the ovomucoid together with 40 μg of subtilisin; ▲, α-chymotrypsin with 40 μg of subtilisin (from BIGLER and FEENEY, 1971).

Ovoinhibitors

The ovoinhibitors are among the more unique inhibitors of proteolytic enzymes because of their comparatively wide inhibitory spectra. Chicken ovoinhibitor, a good representative of the group, is a strong inhibitor of bovine trypsin, bovine α-chymotrypsin, subtilisin, and the alkaline fungal proteinase of *A. oryzae*. It was first isolated by MATSUSHIMA (1958). Further studies were made on it by RHODES et al. (1960), FEENEY et al. (1963), TOMIMATSU et al. (1966), and DAVIS et al. (1969).

Chemical Composition. Amino acid composition of chicken ovoinhibitor is very different from that of chicken ovomucoid. One exception is the low amount or absence of tryptophan in both proteins and the high amount (nearly identical on a weight basis) of cystine.

Other very recent studies of our laboratory have been on the comparative properties of ovoinhibitors from several avian species (LIU et al., 1971). Quail (*Coturnix*) ovoinhibitor was purified sufficiently to make comparisons of amino acid content and physical properties. Differences between chicken and quail ovoinhibitor were noted but no exceptionally large differences were found.

Physical properties. Sedimentation coefficients of 3.8, 3.8, and 4.4 were obtained for chicken, quail, and turkey ovoinhibitors, respectively, at protein concentrations of 2.5 m*l*/m*l* (LIU et al., 1971). The molecular weight of quail ovoinhibitor as determined by low speed sedimentation equilibrium was 57,000 g as compared to 48,000 g for chicken ovoinhibitor.

Inhibitory properties. Duck, turkey, chicken, and quail ovoinhibitors were all similar in inhibiting trypsin, α-chymotrypsin, subtilisin, and the alkaline proteinase *A. oryzae*. Penguin (*Pygoscelis adeliae*) ovoinhibitor differed by failing to inhibit α-chymotrypsin.

A protein similar to the above described ovoinhibitors was obtained in small quantities from ostrich egg white, but it differed by not inhibiting either bovine trypsin or bovine α-chymotrypsin. It was a strong inhibitor of subtilisin and *A. oryzae* alkaline proteinase.

Essential Amino Acids for Binding Proteolytic Enzymes

The essential residues in avian ovomucoids for trypsin inhibition are usually the lysines. Studies in our laboratory on chemical modifications of arginine and lysine side chains in 12 ovomucoids showed that 11 out of the 12 had lysine as the essential residue for interacting with trypsin (STEVENS and FEENEY, 1963; HAYNES et al., 1967; LIU et al., 1968; FEENEY and ALLISON, 1969). The only ovomucoid in which arginine was the essential residue was chicken ovomucoid (LIU et al., 1968; OZAWA and LASKOWSKI, 1966). The earlier observations that chicken ovomucoid did not have an essential lysine apparently gave the impression to most later workers that the essentiallity of arginine might be the rule rather than the exception in ovomucoid.

Recently studies were undertaken on the essential residues of ovoinhibitors when it was found that chicken ovoinhibitor had an arginine as the essential residue for interacting with trypsin (LIU et al., 1971). This, together with other certain similarities between chicken ovoinhibitor and chicken ovomucoid, suggested that there might be an evolutionary relationship between the two proteins. Chicken, turkey and quail ovoinhibitors were all found to have arginines as the essential residues, which is in contrast to the presence of an arginine in only the chicken ovomucoid but lysines in turkey and quail ovomucoids.

The wide spectra of proteolytic enzymes inhibited by the avian egg white inhibitors and the extensive variations in inhibitory capacities among the homologous inhibitors from different avian species should continue to make the study of these inhibitors a very useful tool. The eggs of many species of birds have not as yet been investigated in this regard and many of these eggs should be available in sufficient quantities to provide adequate material for biochemical studies.

References

BIGLER, J. C. and R. E. FEENEY, Protein inhibitors of bacterial subtilisin, in manuscript (1971).

BUCK, F. F., M. BIER and F. F. NORD, Some properties of human trypsin, Arch. Biochem. Biophys. **98**, 528 (1962).

DAVIS, J. G., J. C. ZAHNLEY and J. W. DONOVAN, Separation and characterization of the ovoinhibitors from chicken egg white, Biochemistry **8**, 2044 (1969).

DEUTSCH, H. F. and J. I. MORTON, Physical-chemical studies of some modified ovomucoids, Arch. Biochem. Biophys. **93**, 654 (1961).

DONOVAN, J. W., A spectrophotometric and spectrofluorometric study of intramolecular interactions of phenolic groups in ovomucoid, Biochemistry **6**, 3918 (1967).

FEENEY, R. E. and R. G. ALLISON, Evolutionary Biochemistry of Proteins. Homologous and Analogous Proteins from Avian Egg Whites, Blood Sera, Milk, and Other Substances, Wiley-Interscience, New York (1969).

FEENEY, R. E., G. E. MEANS and J. C. BIGLER, Inhibition of human trypsin, plasmin, and thrombin by naturally occurring inhibitors of proteolytic enzymes, J. Biol. Chem. **244**, 1957 (1969).

FEENEY, R. E., D. T. OSUGA and H. MAEDA, Heterogeneity of avian ovomucoids, Arch. Biochem. Biophys. **119**, 124 (1967).

FEENEY, R. E., F. C. STEVENS and D. T. OSUGA, The specificities of chicken ovomucoid and ovoinhibitor, J. Biol. Chem. **238**, 1415 (1963).

FOSSUM, K. and J. R. WHITAKER, Ficin and papain inhibitor from chicken egg white, Arch. Biochem. Biophys. **125**, 367 (1968).

HAYNES, R., D. T. OSUGA and R. E. FEENEY, Modification of amino groups in inhibitors of proteolytic enzymes, Biochemistry **6**, 541 (1967).

IKEDA, K., K. HAMAGUCHI, M. YAMAMOTO and T. IKENAKA, Circular dichroism and optical rotatory dispersion of trypsin inhibitors, J. Biochem. (Tokyo) **63**, 521 (1968).

LIN, Y. and R. E. FEENEY, Ovomucoids and ovoinhibitors, in Glycoproteins, 2nd Edition, A. GOTTSCHALK, Ed., Elsevier, New York, in press (1971).

LINEWEAVER, H. and C. W. MURRAY, Identification of the trypsin inhibitor egg white with ovomucoid, J. Biol. Chem. **171**, 565 (1947).

LIU, W. H., G. FEINSTEIN, D. T. OSUGA, R. HAYNES and R. E. FEENEY, Modification of arginines in trypsin inhibitors by 1,2-cyclohexanedione, Biochemistry **7**, 2886 (1968).

LIU, W. H., G. E. MEANS and R. E. FEENEY, The inhibitory properties of avian ovoinhibitors against proteolytic enzymes, Biochim. Biophys. Acta **229**, 176 (1971).

LONGSWORTH, L. G., R. K. CANNAN and D. A. MACINNES, An electrophoretic study of the proteins of egg white, J. Am. Chem. Soc. **62**, 2580 (1940).

MATSUSHIMA, K., An undescribed trypsin inhibitor in egg white, Science **127**, 1178 (1958).

MELAMED, M. D., Ovomucoid, in Glycoproteins, A. Gottschalk, Ed., p. 317, Elsevier, New York (1965).

OSUGA, D. T. and R. E. FEENEY, Biochemistry of the egg white proteins of the ratite group, Arch. Biochem. Biophys. **115**, 536 (1966).

OSUGA, D. T. and R. E. FEENEY, Penguin ovomucoids, in preparation (1971).

OZAWA, K. and M. LASKOWSKI, The reactive site of trypsin inhibitors, J. Biol. Chem. **241**, 3955 (1966).

RHODES, M. B., N. BENNETT and R. E. FEENEY, The trypsin and chymotrypsin inhibitors from avian egg whites, J. Biol. Chem. **235**, 1686 (1960).

SJOBERG, L. and R. E. FEENEY, Reduction and reoxidation of turkey ovomucoid — a protein with dual and independent inhibitory activity against trypsin and α-chymotrypsin, Biochim. Biophys. Acta **168**, 79 (1968).

STEVENS, F. C. and R. E. FEENEY, Chemical modification of avian ovomucoids, Biochemistry **2**, 1346 (1963).

TOMIMATSU, Y., J. J. CLARY and J. J. BARTULOVICH, Physical characterization of ovoinhibitor, a trypsin and chymotrypsin inhibitor from chicken egg white, Arch. Biochem. Biophys. **115**, 536 (1966).

Isolation of a Secretory Trypsin Inhibitor from Human Pancreas*

Lewis J. Greene and Merton H. Pubols**

Biology Department, Brookhaven National Laboratory, Upton, New York 11973

Summary

Human PSTI I, a Kazal-type inhibitor, has been isolated from human pancreas in 45% yield by gel filtration on Sephadex G-75 followed by gradient elution chromatography on DEAE-cellulose and equilibrium chromatography on sulfoethyl Sephadex. Two other components with inhibitor activity were detected but not characterized. The inhibitor contains 56 amino acid residues per molecule: Asx_8, Glx_6, Arg_3, Lys_4, Gly_5, Ala_1, Val_2, Leu_4, Ile_3, Ser_3, Thr_4, Cys_6, Pro_3, Tyr_3, Phe_1.
Human PSTI I inhibits human, bovine and porcine trypsin but does not inhibit bovine α-chymotrypsin.

Introduction

The trypsin inhibitor activity present in human pancreas [1] and pancreatic juice [1—3] is of the Kazal type. Several laboratories have obtained inhibitor in partially purified form [1, 3—5], but none of these preparations were sufficiently homogeneous for detailed enzymatic or structural studies. We have fractionated extracts of human pancreas using the chromatographic systems developed in our laboratory for the isolation of trypsin inhibitors from bovine [6] and porcine [7] pancreatic juice. In this communication we report the isolation and chemical characterization of human PSTI I[1], one of three active chromatographic components present in human pancreas.

Methods

Determination of Trypsin, Trypsin Inhibition and Chymotrypsin — The activity of trypsin was measured in a pH-stat at pH 7.8, 25°, using 0.01M TAMe, 0.1M KCl, 0.005M Tris-HCl, 0.020M $CaCl_2$. Chymotrypsin activity was also measured by the pH-stat method with 0.01M ATEe, 0.1M KCl, 0.005M Tris-HCl, 0.020M $CaCl_2$. The details of the assay and the procedure for measuring trypsin inhibition are given in Greene, Rigbi and Fackre [6]. One unit of inhibition activity is the amount of inhibition that caused a reduction of TAMe hydrolysis by 1 μmole per minute. Inhibitor specific activity is defined as inhibitor units per $A^{1\,cm}_{280\,m\mu}$. Inhibitor activity in column effluents was determined with bovine

---

* Research carried out at Brookhaven National Laboratory under the auspices of the U. S. Atomic Energy Commission.

** Visiting Biochemist, 1968—1969; supported by NIH Special Fellowship GM 40285-01. Permanent address: Department of Animal Sciences, Washington State University, Pullman, Washington 99163.

---

[1] The abbreviations used are: PSTI, pancreatic secretory trypsin inhibitors; TAMe, p-toluenesulfonyl-L-arginine methyl ester; ATEe, N-acetyl-L-tyrosine ethyl ester.

trypsin (Worthington TRL 6257). Human trypsin [8] used in the titration given in Figure 4 was provided by Dr. JAMES TRAVIS, University of Georgia, Athens, Georgia.

Extraction of Tissue — Individual postmortem human pancreata were homogenized in a Waring blender with 5 volumes of chilled 10^{-4}M DFP. The pH of the homogenate was adjusted to pH 4.5 with perchloric acid. After centrifugation to remove cell debris and filtration through glass wool to separate fat, the soluble fraction was brought to 0.7 saturation with solid ammonium sulfate. The precipitates from individual glands were stored at $-22°$. Each preparation was assayed for inhibitor activity before being combined with others for gel filtration on Sephadex G-75.

Isolation of Inhibitor — The following chromatographic systems were employed to purify the inhibitor: 1. Gel filtration on Sephadex G-75 (Fig. 1); 2. Gel filtration on Sephadex G-25

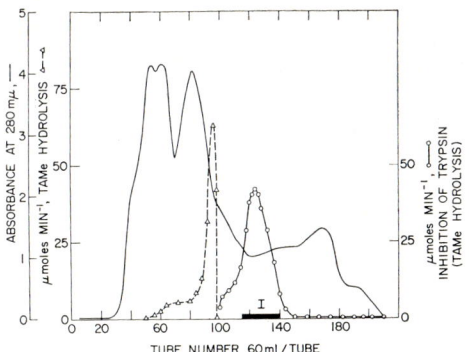

Fig. 1. Gel filtration of ammonium sulfate insoluble fraction (0.7 saturation) of human pancreas on Sephadex G-75. The column (7.6 × 180 cm) was equilibrated and developed with 0.5M potassium chloride, 0.01M Tris-HCl, 1 × 10^{-4}M diisopropyl phosphofluoridate, pH 8.1, 4°, at 150 ml per hour. Sample, 30 $A_{280}^{1\,cm}$ (68 inhibitor units per ml) × 975 ml. ——, absorbance at 280 mµ; O—O, trypsin inhibition (TAMe hydrolysis); △—△, TAMe hydrolysis. The preparation and operation of the column are described in GREENE et al. [6].

(7.6 × 72 cm), equilibrated and developed with 0.05M ammonium bicarbonate buffer, pH 8.1, 4°; 3. Gradient elution chromatography on

DEAE-cellulose (Fig. 2); 4. Gel filtration on Sephadex G-25 (1.8 × 70 cm), equilibrated and developed with 0.05M ammonium bicarbonate buffer, pH 8.1, 4°; 5. Equilibrium chromatography on SE-Sephadex C-25 (Fig. 3); 6. Gel

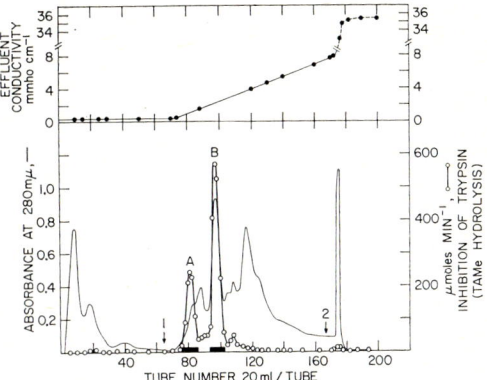

Fig. 2. Gradient elution chromatography of the low molecular weight fraction of human pancreas (region I, Fig. 1) on DEAE-cellulose. The column (1.8 × 45 cm) was equilibrated and developed with 0.028M Tris-HCl, pH 9.0, 4°, at 40 ml per hour. At tube 65 (arrow 1) a linear gradient formed from 2 liters each of starting buffer and 0.028M Tris-HCl, 0.2M KCl, pH 9.0, was used for elution. At tube 168 (arrow 2) a 0.5M KCl was applied to the column. Sample, 8.7 $A_{280}^{1\,cm}$ (900 inhibitor units/ml) × 90 ml. Top, effluent conductivity; bottom, ——, absorbance at 280 mµ; O—O, trypsin inhibition (TAMe hydrolysis). The preparation and operation of the column is given in GREENE et al. [7].

filtration on Sephadex G-50 (0.9 × 200 cm), equilibrated and developed with 0.05M ammonium bicarbonate buffer, pH 8.1, 4°.

The legends to the figures contain the experimental details and literature citations where descriptions of the preparation and operation of the columns may be found.

Amino Acid Analysis — Amino acid analysis of 22- and 72-hour acid hydrolysates was performed by the method of SPACKMAN, STEIN and MOORE [9] as described in ref. [10].

Edman Degradation — A substractive procedure using the method of GRAY [11] was used on unmodified inhibitor as described in GREENE and GIORDANO [10].

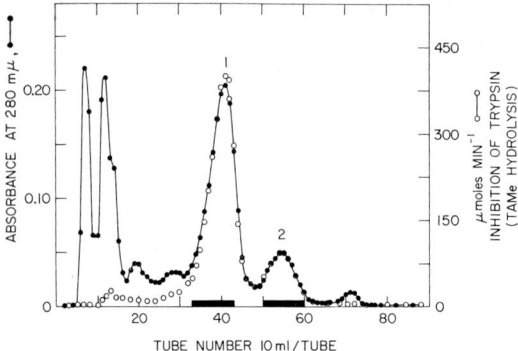

Fig. 3. Equilibrium chromatography of human PSTI (region B, Fig. 2) on sulfoethyl Sephadex C-25. The column (1.8 × 52 cm) was equilibrated and developed with 0.05M ammonium acetate buffer, pH 4.9, 4°, at 15 ml per hour. Sample, 8.4 A_{280}^{1cm} (4.3 × 10³ inhibitor units per ml) × 10 ml. ●—●, absorbance at 280 mμ; ○—○, trypsin inhibition (TAMe hydrolysis). The preparation and operation of the column are given in GREENE et al. [7].

Results

Isolation of Inhibitor — Glands were processed individually through the ammonium sulfate stage of the purification because many tissue samples contained active trypsin. The ammonium sulfate precipitates (1 to 6 mg inhibitor per 100 g tissue) from approximately 15 glands were combined and submitted to gel filtration on Sephadex G-75 in order to separate free inhibitor from trypsinogen and other high molecular weight proteins. A typical elution diagram is presented in Figure 1. Approximately 70% of the inhibitor activity applied to the column was recovered in peak I. This fraction containing low molecular weight substances was approximately 1% inhibitor on a weight basis and was free of trypsinogen. The presence of trypsin-like activity (TAMe hydrolysis) in tubes 60 through 99 demonstrates the importance of the trypsin inhibitor for preventing the activation of the proteolytic zymogens.

Two chromatographic peaks containing inhibitor activity were observed when the low molecular weight fraction was chromatographed on DEAE-cellulose (Fig. 2). Peak A contained 30% and peak B contained 60% of the activity applied to the column. The material corresponding to peak A in Fig. 2 was not studied. When peak B (Fig. 2) was chromatographed on sulfoethyl Sephadex, two peaks with approximately equal inhibitor specific activity were obtained (cf. Fig. 3). The recovery of activity was 75%. The major component, peak 1, was submitted to gel filtration on Sephadex G-50. A slight increase in specific activity was obtained, but more important, a small amount of impurity detectable by amino acid analysis was removed. The inhibitor after gel filtration on Sephadex G-50 is denoted human PSTI I. It is the major component of three chromatographic forms observed thus far in extracts of a pool of glands from twenty individuals.

Amino Acid Composition — The amino acid composition of human PSTI I is given in Table 1. The integral molar ratios of the constitutive amino acids may be taken as strong evidence for the homogeneity of the polypeptide. The inhibitor contains 56 amino acid residues per molecule, the same number of residues as found for the homologous secretory inhibitors from bovine [6], porcine [7], and ovine [12] pancreatic juice or pancreas. There are many similarities in the amino acid compositions of the inhibitors from these three species. However, these data do indicate that there is a *minimum* of ten differences in amino acid sequence between human and bovine PSTI, and twenty differences in sequence between human and porcine PSTI I.

Amino Terminal Sequence — The results of subtractive Edman degradation of human PSTI-I are given in Table 2. The values given in bold face in the table represent the only significant changes in amino acid content after each step of the Edman degradation. On this basis, the amino terminal sequence of the inhibitor is Asx-Ser-Leu-Gly.

Inhibitor Activity — The specific activity of the inhibitor, 1900 μmoles/min/A_{280}, corresponds to an inhibitor to bovine trypsin molar ratio of 1:1 at equivalence. This value for the specific

Table 1. Amino Acid Composition of Pancreatic Secretory Trypsin Inhibitors

Amino Acid	Species			
	Human[a]	Cow[b]	Sheep[c]	Pig-I[d]
	moles per mole peptide			
Aspartic acid	7.99	8	7	7	4
Glutamic acid	6.10	6	7	7	7
Arginine	2.84	3	3	3	2
Lysine	4.09	4	3	3	4
Glycine	4.90	5	5	5	4
Alanine	1.28	1	1	2	1
Valine	2.02	2	4	4	4
Leucine	4.01	4	4	4	2
Isoleucine	2.89	3	3	3	3
Serine	2.73	3	2	1	6
Threonine	3.82	4	4	4	6
Half-cystine	5.15	6	6	6	6
Proline	3.16	3	4	4	5
Tyrosine	2.56	3	2	2	2
Phenylalanine	1.05	1	0	0	0
Methionine	<.05	0	1	1	0
Total		56	56	56	56

[a] Values for 22-hour acid hydrolysate, [b] GREENE et al. [6], [c] FRITZ et al. [12], [d] GREENE et al. [7].

activity based on absorbance at 280 mµ is approximately two-thirds of that found for the bovine PSTI. The difference in specific activity is due to the fact that human inhibitor contains one phenylalanine and one tyrosine residue per molecule more than the bovine inhibitor. Human PSTI I is less stable than the bovine and porcine inhibitors to lyophilization. The inhibitor specific activity decreases by approximately 20% after lyophilization of a solution of the inhibitor in 0.05M ammonium bicarbonate buffer, pH 8.1.

Human PSTI I inhibits human, bovine and porcine trypsin. The titration curve and stoichiometry of the interaction with bovine and porcine trypsin is similar to that given in Figure 4 for human trypsin. A significant difference among the pancreatic secretory inhibitors from these species is that human PSTI I is the only inhibitor that effectively inhibits human trypsin.

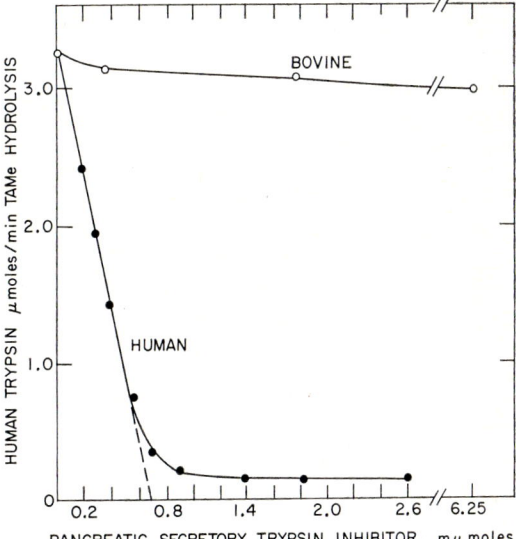

Fig. 4. Titration of human trypsin with human PSTI I (●—●) and bovine PSTI (○—○). The procedure is given in Methods. Inhibitor concentration was determined by amino acid analysis of a 22-hour acid hydrolysate.

Table 2. Subtractive Edman Degradation of Human PSTI I

No. of cycles	Asx	Ser	Leu	Gly	Recovery[a] %
	moles per mole protein				
0	7.85	2.83	3.91	4.83	(100)
1	**7.10**	2.93	3.88	4.91	87
2	7.11	**2.42**	3.84	5.10	94
3	7.36	2.51	**3.33**	5.08	74
4	7.19	2.58	3.26	**4.57**	73

[a] The amount of protein recovered relative to starting material (in parenthesis).

The result given in Figure 4 shows that bovine PSTI has only a small inhibiting effect on human trypsin in the assay system employed (cf. ref. [13]). The inhibition of human trypsin by porcine PSTI I was somewhat greater than that obtained with bovine PSTI. Feeney, Means and Bigler [14] were the first to report that bovine and porcine PSTI do not inhibit human trypsin.

Human PSTI I has been characterized as a Kazal-type inhibitor because it inhibits bovine trypsin but not bovine α-chymotrypsin. The demonstration of temporary inhibition and lack of activity toward porcine kallikrein is necessary to complete the assignment [15]. However, it may be necessary to modify the definition of Kazal-type inhibitors in view of the species differences in the properties of trypsins and inhibitors observed thus far by several investigators [8, 14, 16, 17].

Discussion

Human PSTI I, the major trypsin inhibitor present in human pancreas, has been isolated as a homogeneous polypeptide by column chromatographic techniques. Two additional, chromatographically separable, components with inhibitor activity have been identified but not characterized.

Human PSTI I is a Kazal-type inhibitor, and like the other pancreatic secretory trypsin inhibitors from cow [6], pig [7] and sheep [12], it contains 56 amino acid residues per molecule. The amino-terminal sequence Asx-Ser-Leu-Gly is different from the bovine inhibitor Asn-Ile-Leu-Gly, and the porcine inhibitor I, Thr-Ser-Pro-Gln [7]. The determination of the amino acid sequence of human PSTI I and comparison with the structure of porcine PSTI I should provide useful information about those regions of the molecules required for inhibitor activity because there are a minimum of 20 differences in the amino acid sequences of these homologous inhibitors.

The most striking feature of the inhibitory properties of human PSTI I is that it is the only pancreatic secretory trypsin inhibitor examined thus far which is capable of inhibiting human trypsin. The effect of a variety of polypeptide inhibitors on the activity of human trypsin has been studied by Travis and Roberts [8], Feeney et al. [14] and Travis and Coan [16]. These species differences in the pancreatic trypsin secretory-trypsin inhibitor system are significant both from the point of view of general problems of physiologically important protein-protein interactions as well as for our understanding of human pancreatic enzymes and inhibitors.

References

[1] Fritz, H., F. Woitinas and E. Werle, Z. Physiol. Chem. **345**, 168 (1966).

[2] Fritz, H., M. Hutzel, I. Hüller, M. Wiedemann, H. Stahlheber, P. Lehnert and M. M. Forell, Z. Physiol. Chem. **348**, 1575 (1967).

[3] Keller, P. J. and B. J. Allan, J. Biol. Chem. **242**, 281 (1967).

[4] Fritz, H., I. Hüller, M. Wiedemann and E. Werle, Z. Physiol. Chem. **348**, 405 (1967).

[5] Figarella, C., F. Clemente and O. Guy, FEBS Letters **3**, 351 (1969).

[6] Greene, L. J., M. Rigbi and D. S. Fackre, J. Biol. Chem. **241**, 5610 (1966).

[7] Greene, L. J., J. J. Dicarlo, A. J. Sussman, D. C. Bartelt and D. E. Roark, J. Biol. Chem. **243**, 1804 (1968).

[8] Travis, J. and R. C. Roberts, Biochemistry **8**, 2884 (1969).

[9] Spackman, D. H., W. H. Stein and S. Moore, Anal. Chem. **30**, 1190 (1958).

[10] Greene, L. J. and J. S. Giordano, J. Biol. Chem. **244**, 285 (1969).

[11] Gray, W. R. and J. R. Smith, Anal. Biochem. **33**, 36 (1970).

[12] Fritz, H., W. Schramm, B. Greif, K. Hochstrasser, E. Fink and E. Werle, Z. Physiol. Chem. **351**, 145 (1970).

[13] Rigbi, M. this volume, p. 74.

[14] Feenney, R. E., G. E. Means and J. C. Bigler, J. Biol. Chem. **244**, 1957 (1969).

[15] Vogel, R., I. Trautschold and E. Werle, Natural proteinase inhibitors, Academic Press, New York, p. 95 (1968).

[16] Travis, J. and M. H. Coan, this volume.

[17] Peanasky, R. J. and G. M. Abu-Erreisch, this volume, p. 281.

Disulfide Bridges of the Bovine Pancreatic Secretory Trypsin Inhibitor — Kazal's Inhibitor*

Lewis J. Greene and Odette Guy†

Biology Department, Brookhaven National Laboratory, Upton, New York 11973

Summary

Bovine pancreatic secretory trypsin inhibitor was hydrolyzed with thermolysin at pH 6.5, 48 hours, 37°. Peptides containing disulfide bridges were isolated by chromatographic procedures using Sephadex G-75, Sephadex G-25, and Dowex 50-X2. The disulfide bridges of the inhibitor are I—V, II—IV and III—VI. The structure is based on cystine-containing peptides isolated in 41, 33 and 68% yield, respectively.

Introduction

Bovine pancreatic secretory trypsin inhibitor contains 56 amino acid residues per molecule [1]. The inhibitor is a single polypeptide chain whose amino acid sequence has been determined [2, 3]. The experiments summarized in this communication demonstrate the effectiveness of the use of thermolysin at pH 6.5 for the digestion of the native inhibitor. These results permit the identification of the three disulfide bridges of bovine PSTI.*

---

* The abbreviations used are: PSTI, pancreatic secretory trypsin inhibitor; MES, 2(N-morpholino)ethane sulfonic acid.

Methods

Hydrolysis with Thermolysin — Bovine PSTI, purified by equilibrium chromatography on DEAE-cellulose, was the same preparation used for the determination of the primary structure [2]. The inhibitor (5 μmoles, 31 mg) was incubated with 0.3 mg thermolysin (Calbiochem) in 10 ml of 0.1M MES buffer, pH 6.5, containing 2×10^{-3}M $CaCl_2$ at 37° for 48 hours. The reaction was stopped by adjusting the pH to 3.0 with glacial acetic acid.

Isolation of Peptides — Columns of Sephadex G-25 (0.9 × 400 cm) and Sephadex G-75 (1.8 × 100 cm) were prepared and operated as described previously [2] using 0.2M pyridine-acetate buffer, pH 3.1 [4] at 23°. Gradient elution chromatography on Dowex 50-X2 was carried out by the method of Schroeder [4] as described in Ferreira, Bartelt and Greene [5]. Chromatographic elution patterns were determined by subjecting aliquots of column effluent to alkaline hydrolysis followed by reaction with nin-

---

* Research carried out at Brookhaven National Laboratory under the auspices of the U. S. Atomic Energy Commission.
† Recipient of NATO Postdoctoral Fellowship, 1969—1970.

hydrin [6]. High voltage electrophoresis was performed with an Electrophorator D apparatus (Gilson Medical Electronics) on Whatmann 3 MM paper using pH 6.5 buffer (25 ml pyridine, 1 ml acetic acid, and 225 ml H$_2$O). Peptides I-1-A and I-1-B, prepared by performic acid oxidation of peptide I-1, were isolated by electrophoresis at 52 V/cm for 30 minutes. Peptides II-3-A and II-3-B after oxidation were separated by electrophoresis at 73 V/cm for 40 minutes. Guide strips were developed with ninhydrin (0.5% in acetone) and the peptides were eluted with 50% aqueous pyridine.

Characterization of Peptides — Samples containing 0.05 to 0.1 μmole of peptide were hydrolyzed in an evacuated sealed tube with 1 ml of twice distilled constant boiling HCl for 22 hours at 110°. Amino acid analysis of hydrolysates was performed by the method of SPACKMAN, MOORE and STEIN [7] on an automatic instrument with provisions for multiple sample application [8]. Edman degradation was performed by the method of GRAY [9] using a subtractive procedure [2].

Performic Acid Oxidation — The cystine and methionine content of peptides were determined as cysteic acid and methionine sulfone, respectively, after oxidation and acid hydrolysis by the procedure of MOORE [10]. The method of HIRS [11] was used to oxidize peptides for structural studies.

Calculation of Recovery of Peptides — The yields of peptides are based on the results of amino acid analysis. They were corrected for material used for detection and amino acid analysis but were not corrected for chromatographic losses. The recoveries reported for disulfide bridges given in Tables 1, 2 and 3 are based on the amount of inhibitor treated with thermolysin. The yield may be multiplied by 1.3 to determine yield based on amount of inhibitor extensively digested by thermolysin (fraction 2).

Results

The flow diagram given in Figure 1 summarizes the procedures used for the isolation and

Table 1. Disulfide Bridge I—V

Compositions are expressed in terms of the molar ratios of constituent amino acids obtained by analysis of 22-hour acid hydrolysates after oxidation with performic acid

Amino acid	Peptide		
	II—8[a]	II—8—A	II—8—B
	moles amino acid per mole peptide		
Cysteic acid	1.99	1.07	1.07
Leucine	0.90	0.92	
Lysine	1.00		1.05
Alanine	0.98		0.89
Elution pH (Dowex 50)	4.15	3.15	3.55
NH$_2$-terminus		Leu	Ala
Residue Nos.		37—38	7—9
Yield	41%		

[a] II—8 also contained Leu-Gly-Arg-Glu (cf. Results).

Table 2. Disulfide Bridge II—IV

Compositions are expressed in terms of the molar ratios of constituent amino acids obtained by analysis of 22-hour acid hydrolysates after oxidation with performic acid

Amino acid	Peptide		
	I—4	I—4—A	I—4—B
	moles amino acid per mole peptide		
Cysteic acid	1.99	1.07	1.10
Aspartic acid	1.84	0.95	1.01
Glutamic acid	1.14	0.98	
Serine	0.65	0.69	
Glycine	1.10		1.23
Proline	1.03		1.05
Valine	0.93		0.94
Arginine	0.97		0.93
Elution pH (Dowex 50)	3.95	3.12	3.45
NH$_2$-terminus		Ser	Val
Residue Nos.		32—35	13—18
Yield	33%		

Table 3. Disulfide Bridge III—VI

Compositions are expressed in terms of the molar ratios of constituents of amino acids obtained by analysis of 22-hour acid hydrolysates after oxidation with performic acid

Amino acid	Peptide					
	I—1	I—1—A	I—1—B	II—3	II—3—A	II—3—B
	moles amino acid per mole peptide					
Cysteic acid	1.96	0.88	1.11	1.99	0.93	0.98
Serine	0.76	0.97		0.79	0.70	
Proline	1.08	1.04		0.98	1.02	
Glycine	2.71	1.11	2.01	2.04	1.04	1.00
Valine	0.99		0.86	0.98		1.02
Aspartic acid	1.23		1.13			
Threonine	0.78		0.83			
Elution pH (Dowex 50)	3.55			3.70		
NH$_2$-terminus		Ser	Val		Ser	Val
Residue Nos.		52—56	23—28		52—56	23—25
Yield	23%			45%		

characterization of the disulfide bridges of bovine PSTI.

Hydrolysis with Thermolysin — The products of the digestion of the inhibitor by thermolysin (Fig. 1, line 1) were separated into two fractions by gel filtration on Sephadex G-75 (Fig. 1, line 2). On the basis of amino acid analysis and the results of subtractive Edman degradation, fraction 1 (200—240 ml) was shown to be a mixture of peptides closely related to starting material, having 1 or 2 peptide bonds hydrolyzed per molecule. The sites of cleavage were Ile-Leu (residues 2—3) and/or Glu-Val (residues 12—13). The amino terminal peptides, residues 1—12 (or residues 3—12), were still attached to the remainder of the inhibitor molecule through a disulfide bridge at residue 9 (cf. Fig. 2). Fraction 2 (255—320 ml) was a complex mixture of peptides resulting from the extensive hydrolysis of bovine PSTI by thermolysin. The amino acid content of this mixture of peptides accounted for all the amino acids present in the bovine PSTI plus ~0.6 moles aspartic acid and isoleucine derived from the amino terminal dipeptide Asn-Ile. Under the conditions employed, 75% of the starting

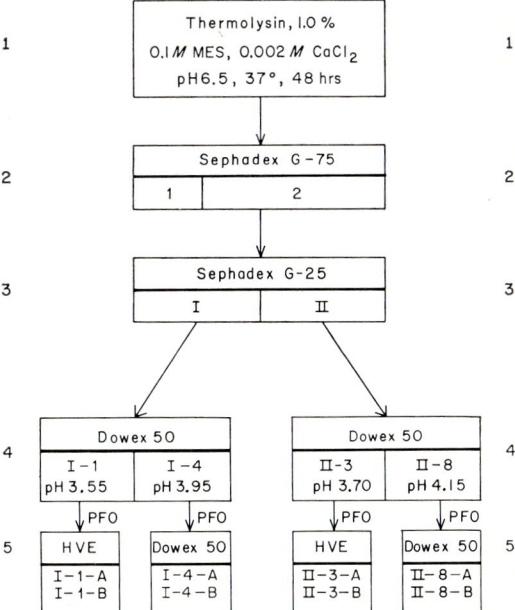

Fig. 1. Flow diagram for the preparation of peptides containing cystine from bovine PSTI. The peptides were oxidized with performic acid (PFO) by the method of HIRS [11].

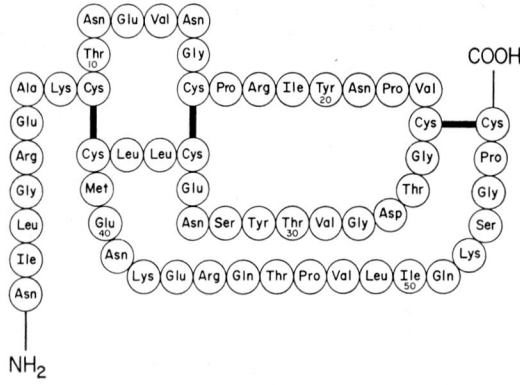

Fig. 2. A two-dimensional schematic diagram of the structure of bovine PSTI showing the arrangement of disulfide bonds and the sequence of the amino acid residues.

material was extensively hydrolyzed to small peptides (fraction 2) after 48 hours of digestion. Peptides used for the characterization of the disulfide bridges of bovine PSTI were isolated from fraction 2.

Isolation of Peptides — Fraction 2 was subjected to gel filtration on Sephadex G-25 (Fig. 1, line 3). More than 88% of the cystine was recovered in the two most rapidly eluted ninhydrin positive peaks*. These mixtures of peptides, peak I (150—180 ml) and peak II (182—200 ml) were further fractionated on Dowex 50-X2 (Fig. 1, line 4). Although the elution diagrams for Dowex 50-X2 chromatograms contained several peaks, three out of the four cystine peptides used for disulfide bridge assignments were homogeneous at this stage of purification. The four cystine-containing peptides were oxidized with performic acid [11] and each pair of peptides containing cysteic acid was isolated by high voltage electrophoresis or by chromatography on Dowex 50-X2 (Fig. 1, line 5).

Assignment of Disulfide Bridges — The structures of peptides given below are based on amino acid composition, identification of amino-terminal

---

* A mixture of low molecular weight peptides devoid of cystine was eluted from the Sephadex G-25 column in effluent corresponding to 210 to 250 ml.

residues and the amino acid sequence of bovine PSTI [2, 3].

Disulfide Bond I—V**: Peptide II-8 was coeluted from the Dowex 50-X2 column at pH 4.15 with the tetrapeptide Leu-Gly-Arg-Glu (Fig. 1, line 4). Peptides II-8-A and II-8-B were separated from the contaminating tetrapeptide on Dowex 50-X2 after oxidation with performic acid. The yield of peptide II-8 was 41%. The amino acid compositions of peptides II-8, II-8-A, and II-8-B are given in Table 1. Peptide II-8-A corresponds to residues 37 through 38 and II-8-B to residues 7 through 9.

$$
\text{Peptide II-8} \quad \begin{array}{c} \text{I} \\ \overrightarrow{\text{Ala-Lys-Cys}} \\ | \\ \overrightarrow{\text{Leu-Cys}} \\ \text{V} \end{array}
$$

Disulfide Bond II—IV: Peptide I-4 was isolated as a homogeneous peptide after chromatography on Dowex 50-X2 in 33% yield. After performic acid oxidation two cysteic acid-containing peptides, I-4-A and I-4-B, were isolated by chromatography on Dowex 50-X2. The analytical data are given in Table 2. Peptide I-4-A corresponds to residues 32 through 35 and I-4-B corresponds to residues 13 through 18.

$$
\text{Peptide I-4} \quad \begin{array}{c} \text{II} \\ \overrightarrow{\text{Val-Asn-Gly-Cys-Pro-Arg}} \\ | \\ \overrightarrow{\text{Ser-Asn-Glu-Cys}} \\ \text{IV} \end{array}
$$

Disulfide Bond III—IV: Two cystine-containing peptides, I-1 and II-3, were isolated in 23 and 45% yield, respectively. These peptides were homogeneous after chromatography on Dowex 50-X2. The oxidized forms of the peptides were separated by high voltage electrophoresis. The data presented in Table 3 give amino acid compositions and amino terminal residue identi-

---

** The half-cysteine residues have been assigned Roman numerals starting from the amino terminal portion of the molecule.

fication for these peptides. The smaller peptide, II-3, was produced by an additional peptide bond cleavage by thermolysin between Gly^{25} and Thr^{26}.

Peptide II-3
$$\begin{array}{c} \text{III} \\ \rightarrow \\ \text{Val-Cys-Gly} \\ \rightarrow \quad \quad | \\ \text{Ser-Gly-Pro-Cys} \\ \text{VI} \end{array}$$

Peptide I-1
$$\begin{array}{c} \text{III} \\ \rightarrow \\ \text{Val-Cys-Gly-Thr-Asp-Gly} \\ \rightarrow \quad \quad | \\ \text{Ser-Gly-Pro-Cys} \\ \text{VI} \end{array}$$

The structure of bovine PSTI showing the arrangement of the disulfide bonds deduced from these experiments is given in Fig. 2.

Discussion

Thermolysin proved to be an effective proteolytic enzyme for the hydrolysis of the native polypeptide in this study of the disulfide bridges of bovine PSTI. Cystine-containing peptides of manageable size were released in high yields even though the digestion was carried out at pH 6.5 [12], two pH units below the optimum for thermolysin. The digestion of bovine PSTI by thermolysin proceeded very slowly and selectively for the hydrolysis of the first three bonds in each molecule, and then very rapidly and essentially completely after the third peptide bond was cleaved. Thermolysin hydrolyzed peptide bonds at the amino-terminus of Ile, Leu, Val, Met, Ser, Thr, and Ala. These sites of hydrolysis are consistent with the specificity of the enzyme reported by other investigators [13, 14]. The enzyme proved to be particularly useful for hydrolysis of Cys^{35}-Leu-Leu-Cys^{38}. Attempts to hydrolyze peptide bonds in this region of the molecule with pepsin followed by trypsin were unsuccessful. The use of chromatographic procedures with Sephadex G-75, Sephadex G-25 and Dowex 50-X2 facilitated the isolation and purification of cystine-containing peptides in high yield from 5 μmoles of bovine PSTI.

The structure of bovine PSTI given in Figure 2 shows that the carboxyl terminal cysteine residue (residue 56) is linked in a disulfide bond to the cysteine residue at position 24 and indicates that the molecule is probably quite compact. Disulfide bridges I—V and II—IV are sites of attachment of residues 1 through 18 to the remainder of the inhibitor after cleavage at the reactive site Arg-Ile (residues 18—19) [15]. This structure supports the generalization of Ozawa and Laskowski [16] that the reactive site peptide is attached to the inhibitor through a disulfide bridge.

The pairings of disulfide bridges of bovine PSTI are different from those in the basic pancreatic inhibitor, the second trypsin inhibitor present in bovine pancreas [17—19]. This is consistent with differences in primary structure already noted [3] and further supports our conclusion that the bovine acinar pancreas synthesizes two structurally unrelated poplypeptides capable of inhibiting trypsin.

References

[1] Greene, L. J., M. Rigbi and D. S. Fackre, J. Biol. Chem. **241**, 5610 (1966).
[2] Greene, L. J. and J. S. Giordano, J. Biol. Chem. **244**, 285 (1969).
[3] Greene, L. J. and D. C. Bartelt, J. Biol. Chem. **244**, 2646 (1969).
[4] Schroeder, W. A., in C. H. W. Hirs (Editor), Methods in enzymology, Vol. XI, Academic Press New York, p. 351 (1967).
[5] Ferreira, S. H., D. C. Bartelt and L. J. Greene, Biochemistry **9**, 2583 (1970).
[6] Hirs, C. H. W., in C. H. W. Hirs (editor), Methods in enzymology, Vol. XI, Academic Press, New York, p. 325 (1967).

[7] SPACKMAN, D. H, W. H. STEIN and S. MOORE, Anal. Chem. **30**, 1190 (1958).
[8] ALONZO, N. and C. H. W. HIRS, Anal. Biochem. **23**, 272 (1968).
[9] GRAY, W. R. and J. F. SMITH, Anal. Biochem. **33**, 36 (1970).
[10] MOORE, S., J. Biol. Chem. **238**, 235 (1963).
[11] HIRS, C. H. W., in C. H. W. HIRS (editor), Methods in enzymology, Vol. XI, Academic Press, New York, p. 197 (1967).
[12] SPACKMAN, D. H., W. H. STEIN and S. MOORE, J. Biol. Chem. **235**, 648 (1960).
[13] MATSUBARA, H., A. SINGER, R. SASAKI and T. H. JUKES, Biochem. Biophys. Res. Commun. **21**, 242 (1965).
[14] AMBLER, R. P. and R. J. MEADWAY, Biochem. J. **108**, 893 (1968).
[15] RIGBI, M. and L. J. GREENE, J. Biol. Chem. **243**, 5457 (1968).
[16] OZAWA K. and M. LASKOWSKI, Jr., J. Biol. Chem. **241**, 3955 (1966).
[17] KASSELL, B. and M. LASKOWSKI, Sr., Biochem. Biophys. Res. Commun. **20**, 463 (1965).
[18] DLOUHÁ, V., D. POSPÍŠILOVÁ, B. MELOUN and F. ŠORM, Collect. Czech. Chem. Comm. **31**, 346 (1966).
[19] CHAUVET, J. and R. ACHER, Bull. Soc. Chim. Biol. **49**, 985 (1967).

Studies on Structure and Activity of Pancreatic Secretory Trypsin Inhibitors from Pig, Sheep, Dog and Cat

Harald Tschesche, Elmar Wachter*, Sigrid Kupfer, Rainer Obermeier**, Günter Reidel, Gernot Haenisch***, and Michael Schneider

*Organisch-Chemisches Laboratorium der Technischen Universität München,
Lehrstuhl für Organische Chemie und Biochemie, Germany*

Since the early studies of Kazal [1] further investigations on pancreatic proteinase inhibitors have shown that the pancreatic gland of all mammalian species secretes a trypsin inhibitor characterized by a specific inhibition of trypsin. Therefore, the synonymous term specific trypsin inhibitor has been in use as well for this class of proteins. Previous publications from our laboratory were concerned with the characterization [2, 3] of the porcine pancreatic secretory trypsin inhibitors I and II, and they were focused on the determination of the amino acid sequences [3, 4, 5] of both proteins and on studies concerned with activity of the proteins upon modification by trypsin during complex formation and temporary inhibition [6—8]. With this paper we wish to review our new results in this field.

Isolation and Composition of Pancreatic Secretory Trypsin Inhibitors

We have isolated several of these pancreatic secretory trypsin inhibitors from different species of mammals (pig [3], sheep [9, 10], dog [11] and cat [12]). For the isolation crude pancreas extracts were used and subjected to a series of chromatographic procedures [3], starting with gel filtration on Sephadex G-50, chromatography on CM-cellulose and DEAE-Sephadex. Finally the inhibitors were purified by ion exchange chromatography on SE-Sephadex. The main material isolated was identical to the one isolated by affinity chromatography on trypsin resins with respect to amino acid composition, chromatographic behavior and inhibition characteristics [3]. For the latter procedure we adopted the method developed by Fritz and his coworkers [13, 14]. But since the pancreatic secretory trypsin inhibitors are characterized by temporary inhibition, they are subjected to gradual degradation during affinity chromatography, which implies losses in native material. During one step of purification by affinity chromatography between 5 to 15% of inhibitor are modified [7]. Besides this modification by proteolytic attack we have observed partial losses of sensitive amide groups during acid elution of the inhibitor from the column. Though losses are not pronounced they have to be taken into consideration. The isolated inhibitors have to be carefully rechromatographed after isolation by

---

* Present address: Medizinische Universitätsklinik, Luitpoldkrankenhaus, 8 Würzburg, Germany
** Present address: Protein Research Laboratory, University of Pittsburgh, Pittsburgh, Pennsylvania 15213, USA.
*** Present address: Max-Planck-Institut für Eiweiß- und Lederforschung, 8 München, Germany.

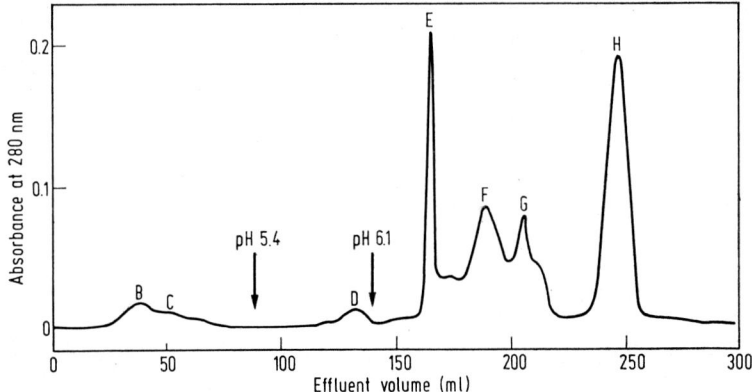

Fig. 1. Rechromatography of ovine pancreatic secretory trypsin inhibitor on SE-Sephadex C-25 (1.5×90 cm) equilibrated with 0.05M ammonium formate pH 4.0. The arrows indicate buffer changes to 0.05M ammonium acetate pH 5.4 and 6.1. Peaks B to H exhibit trypsin inhibiting activity. The amino acid composition of peak H is given in Table 1.

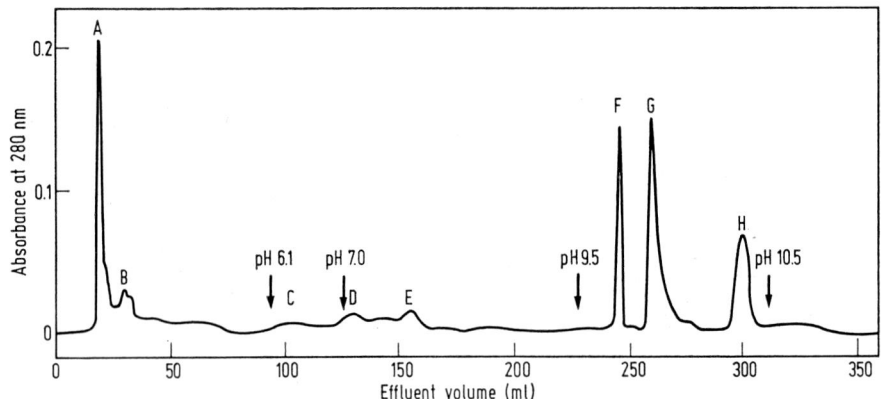

Fig. 2. Rechromatography of canine pancreatic secretory trypsin inhibitor on SE-Sephadex C-25 (0.9×100 cm) equilibrated with 0.05M ammonium formate pH 4.0. The arrows indicate buffer changes to 0.05M ammonium acetate pH 6.1, 7.0, 9.5 and 10.5. Peaks B to H exhibit trypsin inhibiting activity. Peak A represents inactive material. The amino acid composition of peak H is given in Table 1.

affinity chromatography to receive homogenous fractions. This is demonstrated for the rechromatography of the inhibitors from sheep (Fig. 1), dog (Fig. 2) and cat (Fig. 3) by ion exchange chromatography on SE-Sephadex C-25 with 0.05M ammonium formate and acetate buffers of increasing pH (see Fig. and legends). All minor fractions, peaks B to G, showed inhibitory activity when tested against trypsin and N$^\alpha$-benzoyl-arginin-p-nitroanilide. Peaks A were inactive. The native inhibitors were eluted from the columns as the last major peaks H.

Therefore, the most convenient isolation procedure combines affinity chromatography and rechromatography on an ion exchange resin. Using this combined process we isolated the pancreatic secretory trypsin inhibitors from pig, sheep, dog and cat as homogenous proteins. As might be seen from the table of amino acid compositions, all inhibitors contain 56 amino acid residues as expected (Tab. 1). An exception is the porcine inhibitor II, which only contains 52 residues [3, 15]. Both porcine inhibitors I and II have been isolated as well by GREENE

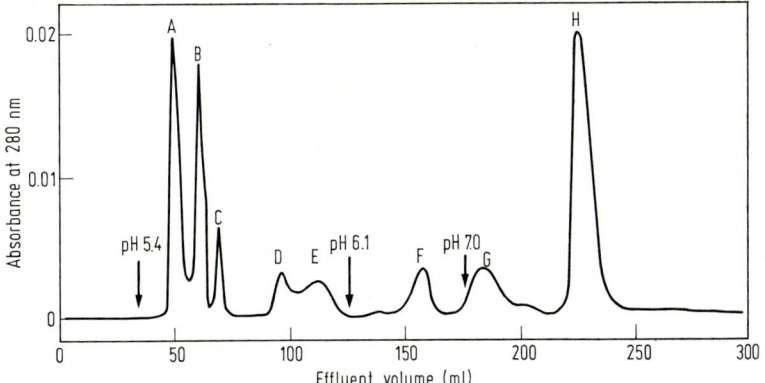

Fig. 3. Rechromatography of the pancreatic secretory trypsin inhibitors from cat on SE-Sephadex C-25 (1.5×90 cm) equilibrated with 0.05M ammonium formate pH 4.0. The arrows indicate buffer changes to 0.05M ammonium acetate pH 5.4, 6.1 and 7.0. Peaks B to H exhibit trypsin inhibiting activity. Peak A represents inactive material. The amino acid composition of peak H is given in Table 1.

Table 1. Amino Acid Compositions of Pancreatic Secretory Trypsin Inhibitors

Amino acid	PIG II		PIG I		DOG		CAT		SHEEP		COW[b]	
	16 hrs	Integer	16 hrs	Integer	16 hrs	Integer	20 hrs	Integer	20 hrs	Integer	22 hrs	Integer
Aspartic acid	4.12	4	3.89	4	7.05	7	6.00	6	6.96	7	7.08	7
Threonine	4.83	5	5.84	6	2.88	3	2.98	3	3.92	4	3.88	4
Serine	4.81	5	5.91	6	3.73	4	3.60	4	1.07	1	1.94	2
Glutamic acid	5.91	6	7.11	7	5.90	6	6.01	6	7.00	7	7.00	7
Proline	4.07	4	4.82	5	2.92	3	4.05	4	3.90	4	4.07	4
Glycine	4.00	4	4.04	4	4.32	4	4.01	4	4.71	5	5.03	5
Alanine	1.00	1	1.00	1	1.93	2	2.24	2	2.01	2	1.05	1
Valine	4.07	4	3.83	4	2.62	3	3.00	3	4.00	4	3.93	4
Half-cystine	5.58	6	5.85	6	5.41	6	4.99	6	5.60	6	5.79	6
Methionine	—		—		—		—		0.64	1	1.02	1
Isoleucine	3.00	3	3.00	3	3.84[a]	4	2.70	3	2.47	3	2.88	3
Leucine	2.02	2	1.95	2	5.00[a]	5	4.37	5	3.69	4	3.92	4
Tyrosine	1.75	2	1.97	2	1.54[a]	2	1.28	2	1.89	2	1.91	2
Lysine	3.96	4	3.94	4	5.02	5	6.07	6	2.99	3	3.06	3
Arginine	1.88	2	1.99	2	1.97	2	1.99	2	3.05	3	2.97	3
Total		52		56		56		56		56		56

[a] Values of 48-hour hydrolysate. [b] GREENE, RIGBI and FACKRE 1966.

et. al. [15] but in a different ratio. The bovine inhibitor (56 residues) was isolated by two groups [16—18]. Big differences were found when the compositions of certain species were compared. There is a minimum of nine differences in the amino acid composition when the inhibitors from dog and pig are compared. Thirteen residues differ between pig and sheep and twelf between pig and cow. Closely related are dog and cat with only two substitutions, while sheep and cow only differ in one residue; since Ala is exchanged for Ser.

When these few data are compared it is striking that the tyrosine content of all inhibitors remains constant while the amount of all other residues is subject to change. Therefore, we considered this minimal amount of two tyrosine residues to be essential for the biological activity. Later on we will come back to this point.

Sequence Data

A number of sequence data has been collected. The bovine sequence has been elucidated by GREENE, GIORDANO and BARTELT [19, 20]. We determined the amino acid sequences of the porcine inhibitors I and II [4, 5] and of the inhibitor from ovine pancreas [21] (Fig. 4). The sequence of the ovine inhibitor could be elucidated very quickly since we were able to degrade the tryptic peptides isolated by HOCHSTRASSER et. al. [9] by our technique using mass spectral identification of the cleaved p-bromophenyl-thiohydantoines [4]. The mass spectral identification of the cleaved amino acids was facilitated by the use of the p-bromo-substituted phenyl-isothiocyanate for the Edman degradation. The characteristic fragment and molecular ions are emphasized by a significant double peak of a mass difference of $\Delta\, m/e = 2$, which is caused by the natural isotope abundance of the p-bromo-substituent ($Br^{79} : Br^{81} = 50.53 : 49.47$) [4]. This technique permitted sequencing of all tryptic peptides with up to thirteen consecutive cycles of Edman degradation. All amide groups could be located unequivocally within the polypeptide chain, since the molecular ions of the amides differ by one mass unit from the ones of the dicarboxylic acid residues.

Porcine inhibitors I and II could be proved to be homologous proteins with identical residues in positions 5 through 56 (inhibitor I) and 1 through 52 (inhibitor II) respectively by complete amino acid sequence analysis of both proteins [4—5]. All tryptic peptides derived from reduced and carboxymethylated inhibitors I and II, none containing more than thirteen residues, could be separated and purified by taking advantage of the size fractionating and ion exchange properties of Bio Gel P-2 in basic medium. The major peptides had to be purified by ion exchange chromatography on SE-Sephadex. They could be ordered into a unique sequence by tryptic overlap peptides obtained in small

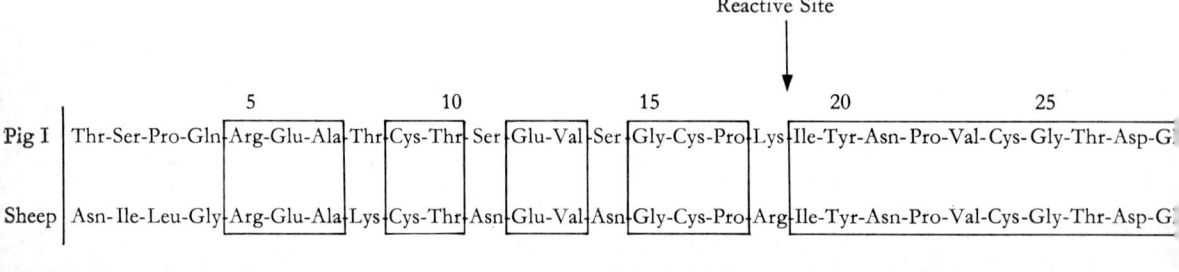

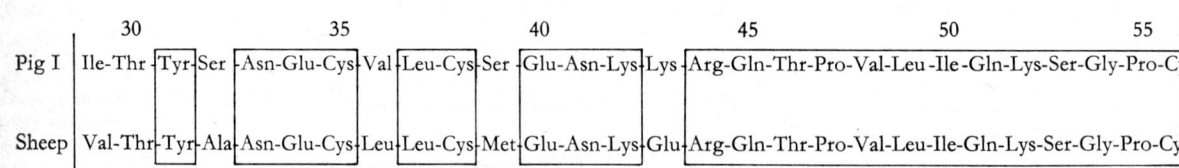

Fig. 4. Linear amino acid sequences of the pancreatic secretory trypsin inhibitors from pig (inhibitor I) and sheep [4,21]. Identical residues in both sequences are framed.

amounts during rechromatography of the major tryptic peptides on SE-Sephadex [5]. In addition tryptic overlap peptides were obtained from the S-carboxymethylated and N^ε-thiocarbamylated inhibitors by arginine-directed tryptic hydrolysis [5]. No difference in amide groups could be found between porcine inhibitors I and II. Thus the question still remains, whether inhibitor II is an artefact produced from inhibitor I by a partial proteolysis reaction effected by enzymes of the type of dipeptidyl aminopeptidases [22, 23]. With aminopeptidase M, however, transformation of inhibitor I into inhibitor II could not be obtained [24]. Therefore, it is quite possible that inhibitor II is a genetic variant derived from allelomorphism or gene dublication. Both porcine inhibitors I and II have been isolated in different ratios of 1.2:1 from two individual animals by our group [3] and in a ratio of 4:1 by GREENE et. al. [15] under nearly identical conditions of collection of the pancreatic juice.

The sequence of the ovine inhibitor was found to be identical with the one determined for the bovine inhibitor [19, 20], except that Ser^{32} was found substituted by an Ala residue in the ovine inhibitor. All tryptic peptides [9] from S-carboxymethylated ovine inhibitor were subjected to Edman degradation without prior preparation of smaller fragments with the exception of peptide X (residues 19 through 42). Peptide X was degraded by chymotryptic cleavage of the bond Tyr^{31} — Ala^{32} (TLCK-treated α-chymotrypsin, 45 mU/mg, 10% by weight, 24 hours, pH 8.0) into two peptides, which could be separated by ion exchange chromatography on SE-Sephadex C-25 with a gradient of 0.05M ammonium formate and acetate buffers of increasing pH (Fig. 5). The peptide X-1 (residues 21 through 31) was eluted prior to peptide X-2 (residues 32 through 42). Edman degradation of peptide X-2 resulted in a p-bromophenyl-thiohydantoine with the characteristic molecular ion peak of m/e 284/286 after the first cycle, thus confirming Ala^{32} as the amino terminal residue of this particular peptide. In the second cycle Asn^{33} (m/e 327/329) was obtained

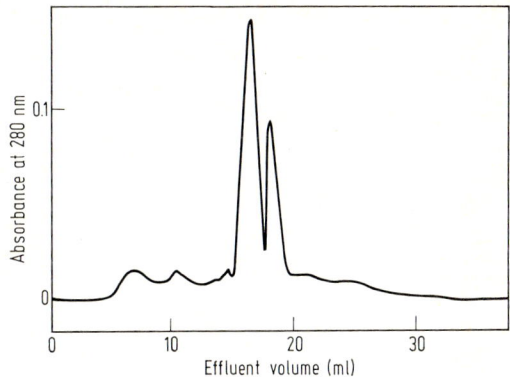

Fig. 5. Elution profile of the peptides X-1 (residues 21 through 31) and X-2 (residues 32 through 42) from chymotryptic cleavage of peptide X (residues 19 through 42) from ovine pancreatic secretory trypsin inhibitor after chromatography on SE-Sephadex C-25 (0.9×60cm) equilibrated with 0.05M ammonium formate pH 4.0. Gradient elution was performed by mixing 0.05M ammonium acetate pH 7.0 to 10 ml of the equilibrating buffer. The dipeptide Ile-Tyr (residues 19—20) was separated previously by gel filtration.

as expected. Detailed investigation of all other positions of the purified tryptic [9] and chymotryptic peptides of the ovine inhibitor were found identical with those of the bovine inhibitor. The tryptic peptides were ordered by homology with the bovine and porcine inhibitors according to their sequences. No additional overlap peptides have been prepared for positioning in this case, as has been done for bovine [15] and porcine inhibitors [3—5, 15].

The ovine inhibitor differs in thirteen positions from the sequence of the porcine inhibitor I (Fig. 4). Identical residues in both sequences are framed. Two sections may be distinguished which are areas of lower mutability, i. e. the inner section containing residues 15 through 28, where only the reactive site residue Arg^{18} is substituted by a Lys residue in the porcine inhibitor, and another long one at the carboxy terminal end containing residues 44 through 56. It may be concluded that these sections are of special importance for the native structure and activity. This would be evident for the reactive site region (residues 15 through 28) but less obvious for the carboxy terminal portion. But since we could establish that this carboxy terminal section is bound to Cys^{24} near the reactive site via the carboxy-terminal Cys^{56}, this area might be important in maintaining the native and biological active structure. Furthermore residues 42 to 44 and 52 may render the main chain sensitive to tryptic cleavage thus perhaps providing a mechanism for inactivation in the course of temporary inhibition. Investigations concerned with the mechanism of this phenomenon are under way in our laboratory (this book, page 299.)

The amino terminal section of the inhibitors indicate a high mutability and are not important for the inhibitory function. This is demonstrated by the porcine inhibitor II which lacks the four amino terminal residues, but otherwise is identical in all positions with inhibitor I and exhibits full inhibitory activity [3, 5]. An inhibitor fragment lacking the five amino terminal residues is still active [8]. The amino acid substitutions observed in this section of the sequence are of a non-conservative type, while all others with the exception of the Ser^{32}/Ala^{32} mutation are conservative (Fig. 4).

Reactive Sites and Function of Tyrosine-20

In addition to the sequence data the reactive sites of these inhibitors have been determined as presented in Table 2. For these investigations several different methods have been in use as tabulated. Some early determinations of the involved basic residues were performed by hydrazinolysis of reactive site modified inhibitors. The results obtained were identification of the newly formed carboxy terminal basic residues [6]. The reactive site peptide bonds were positioned within the sequence by detailed structural investigations on inhibitors previously modified by an acid tryptic, partial proteolysis reaction. Either conventional methods of isolation and investigation of the separated polypeptide fragments [25] were used or the simultaneous sequential investigation of virgin/modified inhibitor mixtures by the method of Edman degradation and quantitative mass spectral assay as has already been described (this book, page 135). Another approach for the classification of inhibitors as to the lysine or arginine group used chemical reactions on the involved amino acid side chains of the reactive site residues [26, 27].

In all pancreatic secretory trypsin inhibitors investigated so far (bovine, porcine and ovine) the sequence at the reactive site reads Gly^{15}-Cys-Pro-Lys (Arg)-Ile-Tyr^{20}-Asn-Pro-Val-Cys-Gly^{25}-Thr-Asp-Gly- (residues 15 through 28, Fig. 4). One of the two Tyr's found in all pancreatic secretory trypsin inhibitors (Tab. 1) is located right next to the reactive site: Tyr^{20}. Therefore, this residue might be involved in the formation of the associated protein-protein complex representing a subside for this reaction.

Table 2. Reactive Sites of Pancreatic Secretory Trypsin Inhibitors (Kazal)

Animal	Residues 18—19	Method
Cow	Arg—Ile	Hydrazinolysis[a]; Limited acid proteolysis[b]
Sheep	Arg—Ile	Limited acid proteolysis[c]
Pig	Lys—Ile	Hydrazinolysis[a]; Limited acid proteolysis[d]
Dog	Lys—Ile	Hydrazinolysis[a]; Limited acid proteolysis[d]
Cat	Lys—	Maleylation[e]

[a] Tschesche 1967. [b] Rigbi and Greene 1968. [c] Hochstrasser, Schramm, Fritz, Schwarz and Werle 1969. [d] Tschesche and Obermeier 1969. [e] Fritz, Fink, Gebhardt, Hochstrasser and Werle 1969.

We investigated the accessibility to chemical reaction of both Tyr residues in the porcine inhibitor I. It could be demonstrated by spectrophotometric titration with base at 295 nm that only one Tyr titrated normal with increasing pH at an apparent pK of 10.8 (Tyr pK 10.05) while the second residue was buried (Fig. 6) [2]. This second residue could only be forced to ionization

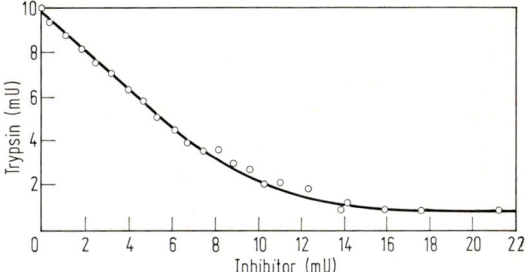

Fig. 7. Reaction of mononitrated porcine pancreatic secretory trypsin inhibitor I with trypsin. The incubation solution was 0.01M $CaCl_2$, 0.2M triethanolamin-HCl, pH 7.8, at 25° for 1 min, then adjusted to 7×10^{-4} molar in N^α-benzoyl-D,L-arginin-p-nitroanilide. The extinction was recorded at 405 nm ($\triangle E = 0.00332$/min corresponds to 1 mIU).

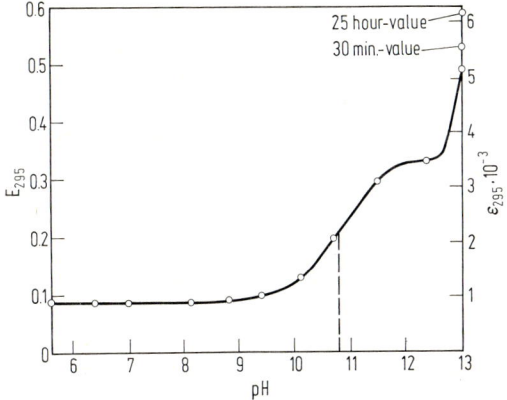

Fig. 6. Spectrophotometric titration curve of porcine pancreatic secretory trypsin inhibitor I (0.95×10^{-4} molar solution) followed at 295 nm. Values for 30 min and 25 hours were obtained in 0.1N NaOH.

when the intact structure had been destroyed by oxidation and disulfide interchange at high pH, thus destroying the inhibitory activity [2]. Therefore, one Tyr is located at the periphery of the inhibitor molecule, obviously Tyr^{20} near the reactive site, while the other, Tyr^{31}, is buried in the interior and perhaps is a functional part in maintaining the native structure.

This result could be supported by nitration experiments with tetranitromethane [28]. Only one Tyr could be nitrated in the free inhibitor (0.05M Tris/HCl, pH 8.0, 20°, 30 min, 60 molar excess TNM, protein conc. 3 mg/ml) [29]. This is demonstrated by the amino acid composition of the isolated inhibitor (Tab. 4). The nitrated residue is necessarily peripheral. The mononitrated inhibitor exhibits full inhibitory activity as can be seen from the reaction with trypsin (Fig. 7). The complex association, however, seems to have decreased with the mononitrated derivative and complete inactivation of trypsin by excess of mononitrated inhibitor could not be accomplished. A residual activity of about 10% persists when the molar ratio of inhibitor to trypsin is 2.0. The residual activity observed with native inhibitor I amounts only to 3—5% [3].

Yet mononitration of the inhibitor cannot be accomplished when the inhibitor is bound to trypsin within the trypsin-trypsininhibitor complex [28]. Tyr^{20} at the reactive site is then buried and no longer accessible to nitration by tetranitromethane. This could be demonstrated when porcine inhibitor I was reacted with 85% molar amount of trypsin and the complex then was nitrated under identical experimental conditions as the free inhibitor (0.05M Tris/HCl, pH 8.0, 20°, 30 min, 60 molar excess of TNM, protein conc. 3 mg/ml). A 15% molar excess of inhibitor was used in order to prevent rapid inactivation of the temporary inhibitor. After 30 min of reaction with TNM this excess of inhibitor had been decreased to 7%. The nitrated complex was purified by rapid gel filtration on Bio Gel P-2 in 0.01N pyridin and dissociated by denaturation of trypsin on heating to 65° for 1 min with 3% perchloric acid. The denatured trypsin was centrifuged and analyzed by amino acid analysis. In trypsin about three tyrosyls out of ten had been nitrated during the 30 min reaction time according to the results obtained with free trypsin [30]. The inhibitor, obtained in solution from the denatured complex, was purified and separated from traces of trypsin by gel filtration on Sephadex G-50 in 0.01N HCl. It was tested for homogeneity by rechromatography on SE-Sephadex C-25 with ammonium acetate pH 5.4. The purified inhibitor was recovered in an overall yield of 25% and contained no nitrotyrosin (0.06 mole/mole inhibitor) as is indicated by the amino acid composition (Tab. 3).

Table 3. Nitration of Porcine PSTI Compositions

Amino acid	A Starting Material	B Free Inhibitor	C Bound Inhibitor
Aspartic acid	4.15	4.10	4.16
Threonine	5.99	5.91	6.00
Serine	5.97	5.90	5.94
Glutamic acid	7.15	7.00	6.98
Proline	4.87	5.00	5.05
Glycine	3.99	4.04	4.17
Alanine	1.04	1.03	1.08
Valine	3.98	4.00	3.99
Half-cystine	5.50	5.75	4.68
Isoleucine	3.00	2.99	3.00
Leucine	2.00	2.04	2.08
Tyrosine	1.74	0.88	1.35
NO_2-Tyrosine	—	0.61	0.06
Lysine	4.13	3.96	4.13
Arginine	2.05	2.08	1.96

The nitrated Tyr^{20} chromophore may be investigated by spectrophotometric titration in the visible spectrum without disturbance by other Tyr residues [29]. The still active mononitrated inhibitor shows normal titration behavior with the nitro-Tyr having a pK of 6.9 (Fig. 8). Titration in the trypsin-complex however, resulted in a titration curve shifted to higher pH values with an apparent pK of the nitro-Tyr^{20} at 7.5 (Fig. 8). The actual pK shift in the complex may still be somewhat higher, since the temporary inhibitor is subject of continuous degradation [6]. This process regenerates free

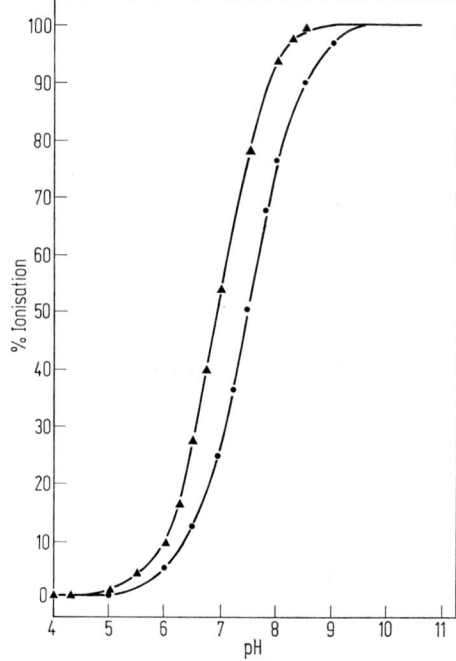

Fig. 8. Spectrophotometric titration curve of ▲ free mononitrated porcine pancreatic secretory trypsin inhibitor I and ● of an 85% equimolar mixture of this inhibitor and trypsin. The apparent pK of the free inhibitor is 6.9 and of the inhibitor bound in the complex 7.5

nitro-Tyr^{20} in non-bonded inactive inhibitor and indicates a higher degree of ionization than would correspond to the inhibitor bound in the complex. Though initially this effect cannot be regarded as essential in the presence of a small excess of inhibitor, it tends to decrease the pK shift measured in the complex and should be taken into consideration. But there is, however, no doubt that the pK of nitro-Tyr^{20} is shifted by at least 0.6 pK units to higher values upon complex formation and that Tyr^{20} is buried at the reactive site between inhibitor and trypsin.

Activity and Disulfide Pairing

The native structure of the inhibitors is maintained by three disulfide links. Their trypsin inhibiting activity is destroyed upon denaturation in urea [7] and/or by reduction of the disulfide bonds [31]. None of the three disulfide bridges is exposed and can be subjected to selective cleavage by sodium borohydride [32], a reaction practicable with the basic pancreatic trypsin inhibitor [33] (Kunitz inhibitor) which is identical with the kallikrein inactivator [34]. The definite three-dimensional structure allied with the biological activity is destroyed upon reduction by dithiothreitol as is indicated by the ultraviolet cotton effect [31]. The curves of the optical rotatory dispersion (0.1M phosphate buffer, pH 6.0, 20°) of native inhibitor I, of the reduced protein in 8M guanidin and of the air-reoxidized material are shown in Figure 9. The native and

helix content of the proteins may be estimated from the depth of the trough [35]. For native porcine inhibitor I a mean residue rotation of $[R']_I = -4280$ was calculated which roughly indicated 26% α-helix content. The corresponding values for the fully reduced inhibitor after air oxidation amounted to $[R']_{\text{red.-reox.}} = -4160$, i.e. 25% α-helix content. The absolute values calculated for α-helix content should be accepted with reservation, because they might be altered considerably by β-structure [36] and because of the general uncertainty whether it is correct to assume that a protein is a simple mixture of α-helical and random coiled regions. But it is obvious that the optical rotatory dispersions of native inhibitor I and of reduced inhibitor I after reoxidation are in excellent agreement within experimental error.

Since the native three-dimensional structure is preferentially regenerated upon air-reoxidation after full reduction, the specific inhibitory activity is restored completely (100%) within a minimum period of 80 min for the reduced inhibitors I and II [31]. The course of the reaction plotted

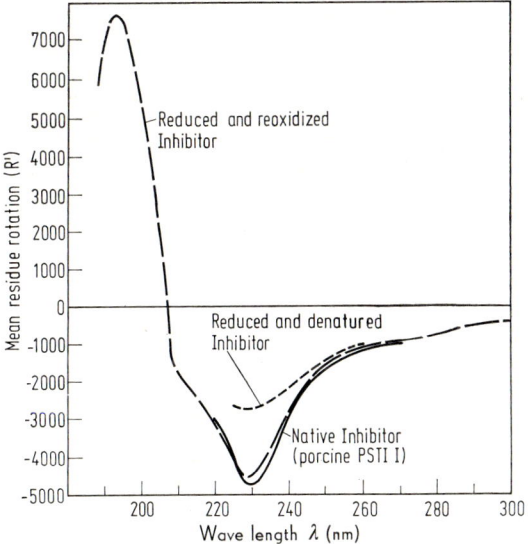

Fig. 9. Optical rotatory dispersion of porcine pancreatic secretory trypsin inhibitor I. The curves were measured in 0.1M phosphate buffer pH 6.0, at 20°. —— native inhibitor I, 3.15 mg/ml; - - - - reduced inhibitor I, 0.865 mg/ml, solution 8M in guanidin; — — — reduced and reoxidized inhibitor I, 3.04 mg/ml.

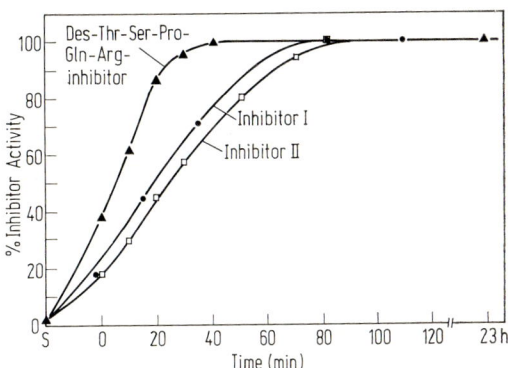

Fig. 10. Regeneration of trypsin inhibiting activity of reduced pancreatic secretory trypsin inhibitors I, II and des Thr-Ser-Pro-Gln-Arg-inhibitor I upon air-reoxidation. The inhibitors were reduced with dithiothreitol in 8M urea, subjected to gel filtration, dissolved in 0.02M phosphate buffer pH 8.0 and ventilated by bubbling air through the solution. At indicated times samples of 0.2 ml were withdrawn, diluted to 10 ml and tested for inhibitory activity. The lowest activity (<2%) was found after elution of the reduced inhibitors from the Bio Gel column (time S).

the reoxidized inhibitor exhibit nearly identical optical rotatory dispersions typical for proteins. A deep trough was found at 230 nm, which was almost absent in the denatured and reduced inactive inhibitor. The reduced inhibitor therefore exhibits a less ordered structure. The α-

against time is shown in Figure 10. The inhibitor fragment lacking the amino-terminal pentapeptide (des-Thr-Ser-Pro-Gln-Arg-inhibitor I) generates its total activity within a minimum period of only 40 min upon air-reoxidation under the same experimental conditions [31]. Therefore, no additional template nor the complete amino acid sequence were necessary for this process. Cleavage of the first four amino acid residues (inhibitor II) [3, 4, 5, 15] from the amino terminal end had no significant influence on time and course of the process of regeneration upon reoxidation, though the reaction was slowed down slightly. It seemed likely that formation of ion pairs from the positively charged side chain of the arginine (residue 5) and negatively charged residues of the dicarboxylic acid residues would render the process of correct folding and coiling more difficult. Therefore, elimination of the first five amino terminal residues including this arginine indeed facilitated the regeneration of the activity during reoxidation. This is demonstrated by the time required for the regeneration of total activity of des-Thr-Ser-Pro-Gln-Arg-inhibitor I (Fig. 10).

Disulfide Bridges

The three disulfide bridges were located within the single polypeptide chain of porcine pancreatic secretory trypsin inhibitor [37]. Trypsin inhibiting activity of the native inhibitor I is rapidly destroyed by the action of pepsin. Therefore, the inhibitor was first digested with pepsin (50 mg inhibitor I, 1% solution titrated to pH 2.0 with 0.1N HCl, 3 mole % pepsin, 37°, 24 hours) and after adjusting the pH to 6.7 by titration with 0.1N NaOH was then subjected to tryptic hydrolysis (2 mole % TPCK-treated trypsin, 37°, 24 hours). The peptide mixture obtained was resolved by gel filtration on a series of three columns of Bio Gels P-6 (0.9 ×15 cm), P-4 (0.9 ×30 cm) and P-2 (0.9 ×90 cm) connected in series by polyethylene tubing and equilibrated with 0.1N acetic acid. The chromatographic elution pattern was continuously monitored at 280 nm by a flow cell device (Fig. 11).

The first fraction eluted contained the major peptides and was further fractionated on SE-Sephadex C-25 (0.9 ×90 cm, equilibrated with 0.05M ammonium formate pH 4.0 and eluted

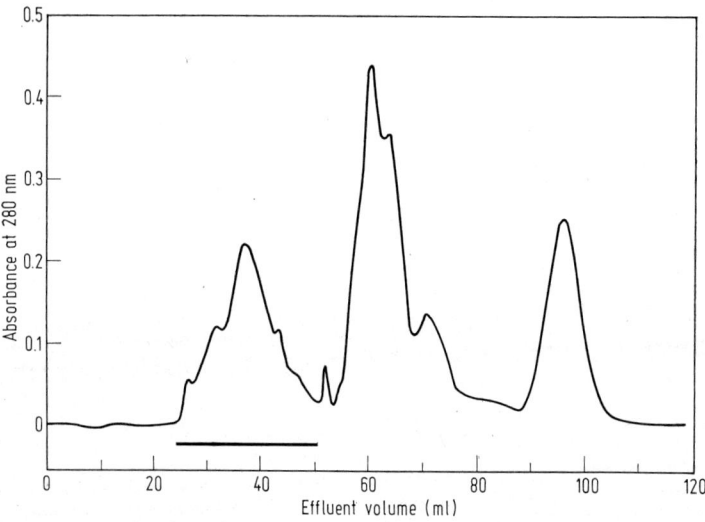

Fig. 11. Elution profile of the peptides from peptic and tryptic hydrolysis of porcine pancreatic secretory trypsin inhibitor I after gel filtration on three columns of Bio Gels P-6 (0.9 ×15 cm), P-4 (0.9 ×30 cm) and P-2 (0.9 ×90 cm) connected in series by polyethylene tubing. Equilibrating solvent was 0.1N acetic acid. Peak I indicated by the solid bar was subjected to further separation (Fig. 12).

with 0.05M ammonium acetate pH 5.4) (Fig. 12) according to the flow diagram (Fig. 13). Two homogenous disulfide-containing fragments were obtained when peaks I and II were rechromatographed supplying peptides I-1 and I-2 in 32% and 17% yield, respectively.

Peptide I-1 still contained two disulfide bridges and was degraded further by digestion with thermolysin. The amino acid composition is given in Table 5.

Peptide I-1

Val-Ser-Gly-Cys^{16}-Pro-Lys
|
Asn-Glu-Cys^{35}—Val Ala-Thr-Cys^9-Thr-Ser-Glu
       Leu |
          Cys^{38}-Ser-Glu-Asn-Lys

The disulfide bridge Cys^{24}—Cys^{56} was found in peptide I-2 containing residues 19 through 32 and 53 through 56. The amino acid composition is presented in Tab. 4. The peptide was oxidized with performic acid [38] and the pair of cysteic acid containing peptides separated by gel filtration on Bio Gel P-2 (0.9×60 cm, 0.1N acetic acid) and analyzed for amino acid composition. In addition a smaller peptide I-3 containing the same disulfide bridge (residues 19 through 28 and 53 through 56) was eluted from the SE-Sephadex column in minor amounts right after peptide I-2 during rechromatography using 0.05M ammonium acetate pH 4.7. This peptide derived from limited peptic hydrolysis of the bond Gly-Ile (residues 28—29). The corresponding tetrapeptide (residues 29—32) was isolated in 9% yield from peak IV as a homogenous peptide.

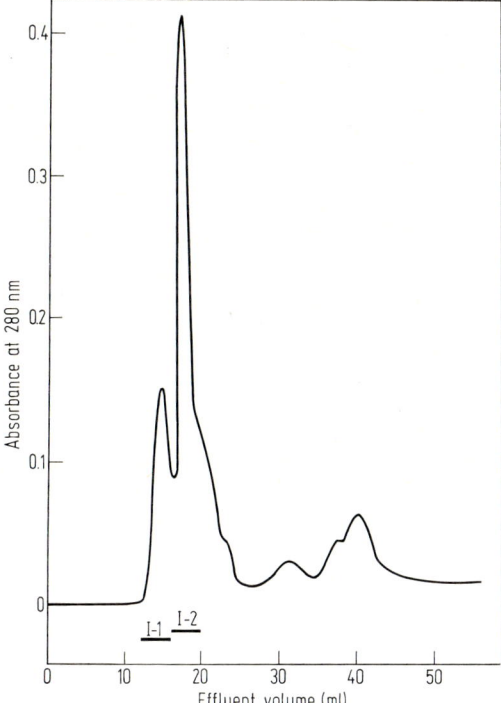

Fig. 12. Chromatographic elution pattern of peak I (Fig. 11) from SE-Sephadex C-25 (0.9×90 cm) equilibrated with 0.05M ammonium formate pH 4.0 and developed with 0.05M ammonium acetate pH 5.4. Peaks containing peptides I-1 and I-2 are indicated by solid bars.

Peptide I-2   Ile-Tyr-Asn-Pro-Val-Cys^{24}-Gly-Thr-Asp-Gly-Ile-Thr-Tyr-Ser
                                    |
               Ser-Gly-Pro-Cys^{56}

Peptide I-3   Ile-Tyr-Asn-Pro-Val-Cys^{24}-Gly-Thr-Asp-Gly
                                    |
               Ser-Gly-Pro-Cys^{56}

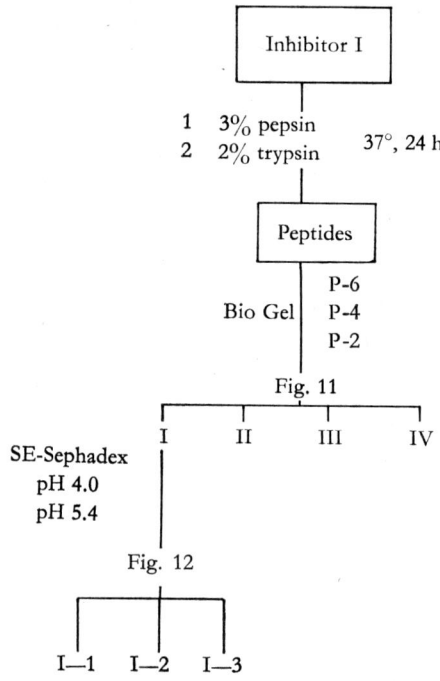
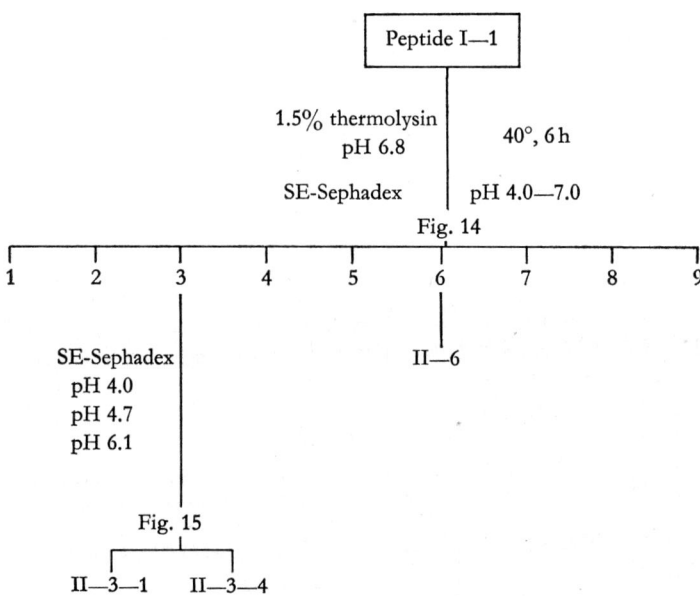

Fig. 13. Flow diagram for the purification of the disulfide-containing peptides described in the text.

Table 4. Amino acid compositions of disulfide-containing peptides

Amino acid	Peptide I—1	Peptide I—2	Peptide I—3	Peptide II—3—1	Peptide II—3—4	Peptide II—6
Asp	1.69	2.08	1.72	1.12	0.82	1.00
Thr	2.18	1.96	1.09	1.65		
Ser	3.02	1.71	1.24	1.92	1.18	1.06
Glu	2.91			2.51	1.07	1.01
Pro	0.98	2.12	2.00		0.86	0.94
Gly	1.43	3.17	2.03		0.94	1.00
Ala	1.22			0.85		
Val	2.05	1.30	0.93	1.00	0.80	2.00
Half-Cys	3.81	1.56	1.91	1.87	1.59	1.58
Leu	1.09			0.68		
Ile		1.89	0.83			
Tyr		(0.81)	0.65			
Lys	2.01			0.99	1.11	1.00
Yield	32%	17%	4%	7%	4%	6%

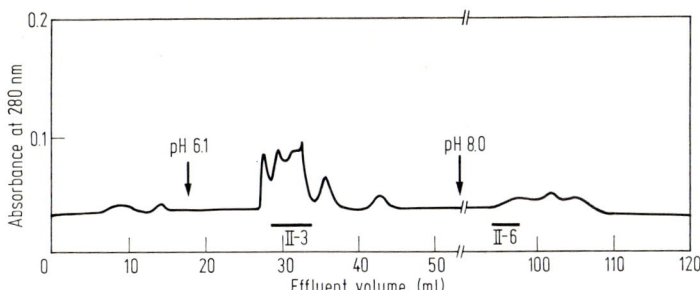

Fig. 14. Chromatographic elution pattern from SE-Sephadex C-25 (0.9 × 45 cm) of the peptides obtained from digestion of peptide I-1 by thermolysin. The column was equilibrated with 0.05M ammonium formate pH 4.0, developed with 18 ml of this buffer, then with a gradient of 0.05M ammonium acetate pH 6.1 mixed to 15 ml of equilibrating buffer up to 55 ml effluent volume, and finally eluted with the last buffer adjusted to pH 8.0 (effluent pH of peptide II-6 was 7.0).

Peptide I-1 contained two disulfide bridges linked by the sequence -Cys^{35}-Val-Leu-Cys^{38}-. Unfortunately, the bond Val-Leu was not split by pepsin as was expected from the cleavage of this particular bond (Val-Leu, residues 113—114) in the β chain of human hemoglobin [39]. This peptide was further degraded by enzymatic cleavages using thermolysin (1.5% thermolysin by weight, 1 ml 0.05M ammonium acetate pH 6.8, 40°, 6 hours). The complex mixture of peptides was subjected to gradient elution chromatography on SE-Sephadex C-25 (Fig. 14) (0.9 × 50 cm, developed with 8 ml 0.05M ammonium formate pH 4.0, 25 ml 0.05M ammonium acetate pH 6.1 mixed to 15 ml pH 4.0 buffer and finally eluted with 0.05M ammonium acetate buffer adjusted to pH 8.0). Although the hydrolysis was incomplete and starting peptide was recovered in 10% yield, three peptides each containing a single disulfide bridge were obtained from fractions II-3 and II-6 on SE-Sephadex C-25 (Fig. 14 and 15) according to the flow diagram (Fig.13).

The disulfide bridge Cys^9—Cys^{38} was found in peptide II-3-1 (Fig. 15). The peptide corresponds to residues 7 through 12 and 36 through 42. The yield was 7% based on the amount of peptide

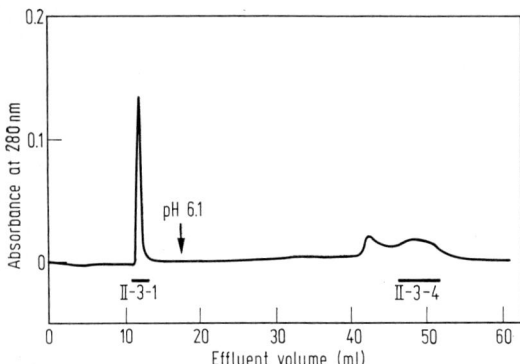

Fig. 15. Chromatographic elution pattern of fraction II-3 on SE-Sephadex C-25 (0.9×45 cm) equilibrated with 0.05M ammonium formate pH 4.0 and developed with 0.05M ammonium acetate pH 4.7. The arrow indicates a change of the buffer pH to 6.1. The peak containing the peptides II-3-1 and II-3-4 are marked by solid bars.

I-1 digested by thermolysin (not corrected for chromatographic losses during purification). The amino acid composition is given in Table 4.

Peptide II-3-1   Ala-Thr-Cys^9-Thr-Ser-Glu
                                    |
                  Val-Leu-Cys^{38}-Ser-Glu-Asn-Lys

The disulfide bridge Cys^{16}—Cys^{35} was obtained in the peptides II-3-4 (Fig. 15) and II-6 (Fig. 14). Peptide II-3-4 corresponds to residues 13 through 18 and 33 through 35. This peptide was homogenous, but the yield based on the disulfide-containing peptide was only about 4% (uncorrected for chromatographic losses). Peptide II-6 contained the residues 13 through 18 and 33 through 36 and was obtained in about 6% yield, it was however contaminated with the starting peptide I-1 and was not homogenous (yield and amino acid composition corrected). By one cycle of Edman degradation it was confirmed that the amino terminal residues were Val and Asx. The amino acid compositions of the peptides are given in Table 4.

Peptide II-3-4   Val-Ser-Gly-Cys^{16}-Pro-Lys
                                    |
                  Asn-Glu-Cys^{35}

Peptide II-6   Val-Ser-Gly-Cys^{16}-Pro-Lys
                                    |
                  Asn-Glu-Cys^{35}-Val

Using pepsin, trypsin and thermolysin for a series of enzymatic cleavages, one peptide containing two disulfide bridges, and five peptides, each containing a single disulfide bond were isolated from porcine pancreatic secretory trypsin inhibitor I. The structures of the peptides were based on the amino acid sequence of the porcine inhibitor [4] and on their amino acid composition. Three of the purified peptides were further characterized by identification of the amino-terminal residues or by performic acid oxidation and isolation of the separated pair of cysteic acid-containing peptides. The isolation and purification of peptide I-1, containing the disulfide bridges Cys^9—Cys^{38} and Cys^{16}—Cys^{35}, proved to be useful prior to further degradation by thermolysin, since the thermolytic cleavage of the particular bonds within the sequence Cys^{35}-Val-Leu-Cys^{38} was slow and, within the limited period of digestion, was incomplete. The complex mixture of peptides could, however, be resolved. Although the yields of the homogenous peptides after purification were low, unequivocal assignment of all three disulfide bridges could be obtained.

The complete covalent structure of the porcine pancreatic secretory trypsin inhibitor I is presented in Figure 16. The positioning of the disulfide bridges is in full agreement with the structure of the bovine pancreatic secretory trypsin inhibitor (Kazal's inhibitor) [19, 20] pre-

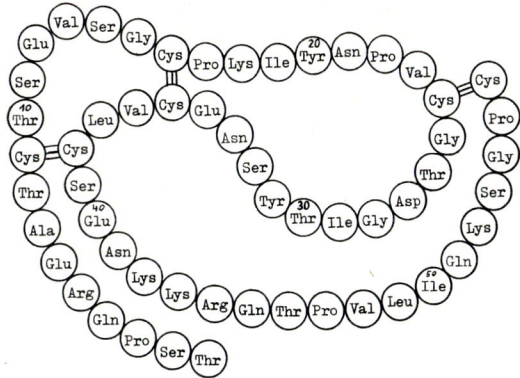

Fig. 16. The complete covalent structure of the porcine pancreatic secretory trypsin inhibitor I documenting the positioning of the disulfide bridges within the sequence of the amino acid residues.

sented first by GREENE and GUY at this meeting [40].
This result indicates that the disulfide bridges of the pancreatic secretory trypsin inhibitors (Kazal typ) are different from those of the basic bovine pancreatic trypsin inhibitor (KUNITZ inhibitor, kallikrein inactivator) [41—43]. It supports the prediction of OZAWA and LASKOWSKI, Jr. [44], that the reactive site peptide bond (Lys^{18}—Ile^{19}) [4, 45] is embedded in a disulfide loop (Cys^{16}—Cys^{35} and Cys^{24}—Cys^{56}) thus retaining the biologically active structure of the modified (Lys^{18}—Ile^{19} bond open) inhibitor [6, 45].

Acknowledgements. We wish to thank Miss C. FRANK for her assistance in carrying out all the amino acid analyses, Miss I. BRUCKMEIER for the preparation of the ovine inhibitor and Mr. M. VANDREY for the isolation of the secretory trypsin inhibitor from canine pancreas.
We are indebted to Mr. K. AICHER, Mr. N. HUMS and Miss I. SCHMID for the recording of the mass spectra.
We are grateful to the Farbenfabriken Bayer AG, Wuppertal-Elberfeld, Germany, for their assistance and the supply of crude extracts from porcine pancreas.
We express our thanks to the Deutsche Forschungsgemeinschaft, Bad Godesberg, Germany, for their support with grants and materials.

References

[1] KAZAL, L. A., D. S. SPICER and R. A. BRAHINSKY, J. Amer. chem. Soc. **70**, 3034 (1948).
[2] TSCHESCHE, H., Hoppe-Seyler's Z. Physiol. Chem. **348**, 1653 (1967).
[3] TSCHESCHE, H., E. WACHTER, S. KUPFER and K. NIEDERMEIER, Hoppe-Seyler's Z. Physiol. Chem. **350**, 1247 (1969).
[4] TSCHESCHE, H. and E. WACHTER, Europ. J. Biochem. **16**, 187 (1970).
[5] TSCHESCHE, H. and E. WACHTER, Hoppe-Seyler's Z. Physiol. Chem. **351**, 1449 (1970).
[6] TSCHESCHE, H., Hoppe-Seyler's Z. Physiol. Chem. **348**, 1216 (1967).
[7] TSCHESCHE, H. and H. KLEIN, Hoppe-Seyler's Z. Physiol. Chem. **349**, 1645 (1968).
[8] TSCHESCHE, H., E. WACHTER and G. KALLUP, Hoppe-Seyler's Z. Physiol. Chem. **350**, 1662 (1969).
[9] HOCHSTRASSER, K., W. SCHRAMM, H. FRITZ, S. SCHWARZ and E. WERLE, Hoppe-Seyler's Z. Physiol. Chem. **350**, 893 (1969).
[10] TSCHESCHE, H., S. KUPFER and I. BRUCKMEIER, unpublished results from this laboratory.
[11] TSCHESCHE, H., S. KUPFER and M. VANDREY, unpublished results from this laboratory.
[12] TSCHESCHE, H. and S. KUPFER, Hoppe-Seyler's Z. Physiol. Chem. **352**, 764 (1971).
[13] FRITZ, H., H. SCHULT, M. NEUDECKER and E. WERLE, Angew. Chem. **78**, 775 (1966); Angew. Chem. Internat. Edit. **5**, 735 (1966).
[14] FRITZ, H., H. SCHULT, M. HUTZEL, M. WIEDEMANN and E. WERLE, Hoppe-Seyler's Z. Physiol. Chem. **348**, 308 (1967).
[15] GREENE, L. J., J. J. DICARLO, A. J. SUSSMAN, D. C. BARTELT and D. E. ROARK, J. Biol. Chem. **243**, 1804 (1968).
[16] FRITZ, H., F. WOITINAS and E. WERLE, Hoppe-Seyler's Z. Physiol. Chem. **345**, 168 (1966).
[17] FRITZ, H., I. HÜLLER, M. WIEDEMANN and E. WERLE, Hoppe-Seyler's Z. Physiol. Chem. **348**, 405 (1967).

[18] GREENE, L. J., M. RIGBI and D. S. FACKRE, J. Biol. Chem. **241**, 5610 (1966).
[19] GREENE, L. J. and J. S. GIORDANO, Jr., J. Biol. Chem. **244**, 285 (1969).
[20] GREENE, L. J. and D. C. BARTELT, J. Biol. Chem. **244**, 2646 (1969).
[21] TSCHESCHE, H., R. OBERMEIER and K. HOCHSTRASSER, unpublished results from this laboratory.
[22] MCDONALD, J. K., B. B. ZEITMAN, T. J. REILLY and S. ELLIS, J. Biol. Chem. **244**, 2693 (1969).
[23] MCDONALD, J. K., P. X. CALLAHAN, B. B. ZEITMAN and S. ELLIS, J. Biol. Chem. **244**, 6199 (1969).
[24] TSCHESCHE, H. and S. KUPFER, unpublished results from this laboratory.
[25] RIGBI, M. and L. J. GREENE, J. Biol. Chem. **243**, 5457 (1968).
[26] HAYNES, R. and R. E. FEENEY, Biochemistry **7**, 2879 (1968).
[27] FRITZ, H., E. FINK, M. GEBHARDT, K. HOCHSTRASSER and E. WERLE, Hoppe-Seyler's Z. Physiol. Chem. **350**, 933 (1969).
[28] TSCHESCHE, H. and G. REIDEL, unpublished results from this laboratory.
[29] RIORDAN, J. F., M. SOKOLOWSKY and B. L. VALLEE, Abstracts 152nd National Meeting of the American Chemical Society, Sept. 1966. New York, N. Y., and J. Amer. chem. Soc. **88**, 4104 (1966).
[30] VINCENT, I. P., M. LADUNSKY and M. DELAAGE, Europ. J. Biochem. **12**, 250 (1970).
[31] TSCHESCHE, H. and HAENISCH, FEBS letters **11**, 209 (1970).
[32] TSCHESCHE, H. and S. KUPFER, unpublished results from this laboratory.
[33] KRESS, L. F. and M. LASKOWSKI, Sr., J. Biol. Chem. **242**, 4925 (1967).
[34] ANDERER, F. A., Z. Naturforsch. **20b**, 462 (1965).
[35] FORMANEK, H. and J. ENGEL, Biochim. Biophys. Acta **160**, 151 (1968).
[36] JIZUKA, E. and J. T. YANG, Proc. Natl. Acad. Sci., U. S. **55**, 1175 (1966).
[37] TSCHESCHE, H. and M. SCHNEIDER, unpublished results from this laboratory.
[38] HIRS, C. H. W., J. Biol. Chem. **219**, 611 (1956).
[39] KÖNIGSBERG, W., J. GOLDSTEIN and R. J. HILL, J. Biol. Chem. **238**, 2028 (1963).
[40] GREENE, L. J. and O. GUY, First International Conference on Proteinase Inhibitors, Munich (1970).
[41] KASSELL, B. and M. LASKOWSKI, Sr., Biochem. Biophys. Res. Commun. **20**, 463 (1965).
[42] KASSELL, B. and M. LASKOWSKI, Sr., Acta Biochim. Polonica **13**, 287 (1966).
[43] ANDERER, F. A. and S. HÖRNLE, J. Biol. Chem. **241**, 1568 (1966).
[44] OZAWA, K. and M. LASKOWSKI, Jr., J. Biol. Chem. **241**, 3955 (1966).
[45] TSCHESCHE, H. and R. OBERMEIER, First International Research Conference on Proteinase Inhibitors, Munich 1970.
[46] TSCHESCHE, H. and C. FRANK, J. Chromatogr. **40**, 296 (1969).

Porcine Pancreatic Secretory Trypsin Inhibitor I, Amino Acid Sequence

L. J. GREENE and D. C. BARTELT

Biology Department, Brookhaven National Laboratory, Upton, N. Y. 11973

We have determined the amino acid sequence of porcine pancreatic secretory trypsin inhibitor I [GREENE, L. J., J. J. DICARLO, A. J. SUSSMAN, D. C. BARTELT and D. E. ROARK, J. Biol. Chem. **243**, 1804 (1968)] by a combination of selective trypsin and chymotrypsin hydrolysis reactions and conventional methods for the sequence determination of the small peptides. As shown in Figure 1, the limited number of intermediate peptides (usually two or three) prepared at each stage of the degradation procedure were ordered on the basis of the identification of the amino- and carboxyl-terminal residues. In this way the "overlap" information required to order the peptides was obtained at the same time the peptides were prepared. Arginine directed trypsin cleavage at pH 8 was achieved by acylating the ε- and ω-amino groups of lysine and S-2-aminoethyl cysteine with maleic anhydride (line 2). After demaleation, selective hydrolysis of lysyl bonds in peptides containing S-2-aminoethyl cysteinyl bonds was accomplished by trypsin hydrolysis at pH 11 (line 3). Limited hydrolysis with α-chymotrypsin cleaved the peptide bonds between Tyr-Ser (residues 31—32) and Leu-Cys (Ae) (residues 37—38) in high yield while the Tyr-Asn bond (residues 20—21) was not hydrolyzed (lines 7 and 8). The amino acid sequences of the small peptides were determined by Edman degradation using a substractive procedure. The amide forms of the dicarboxylic acids were identified by a subtractive procedure after total enzymatic hydrolysis with aminopeptidase M. The yields of the peptides, 38 to 93%, compared favorably to those obtained by the conventional trypsin hydrolysis of bovine pancreatic secretory trypsin inhibitor [L. J. GREENE and J. S. GIORDANO, Jr., J. Biol. Chem. **244**, 285 (1969)].

The amino acid sequence of porcine pancreatic secretory trypsin inhibitor I exhibits 12 differences in 56 amino acid residues per molecule from that of the bovine pancreatic secretory trypsin inhibitor [L. J. GREENE and D. C. BARTELT, J. Biol. Chem. **244**, 2646 (1969)].

TSCHESCHE, WACHTER, KUPFER and NIEDERMEIER [Z. Physiol. Chem. **350**, 1662 (1969)] announced a structure for porcine pancreatic secretory inhibitor I and recently presented a revised sequence with data documenting the structures of the small peptides [H. TSCHESCHE and E. WACHTER, European J. Biochem. **16**, 187 (1970)]. The structure proposed by our laboratory is in complete agreement with that reported by TSCHESCHE's group in 1970. A detailed report of our study is published elsewhere [D. C. BARTELT and L. J. GREENE, J. Biol. Chem. **246**, 2218 (1971)].

Research supported by U.S. Atomic Energy Commission.

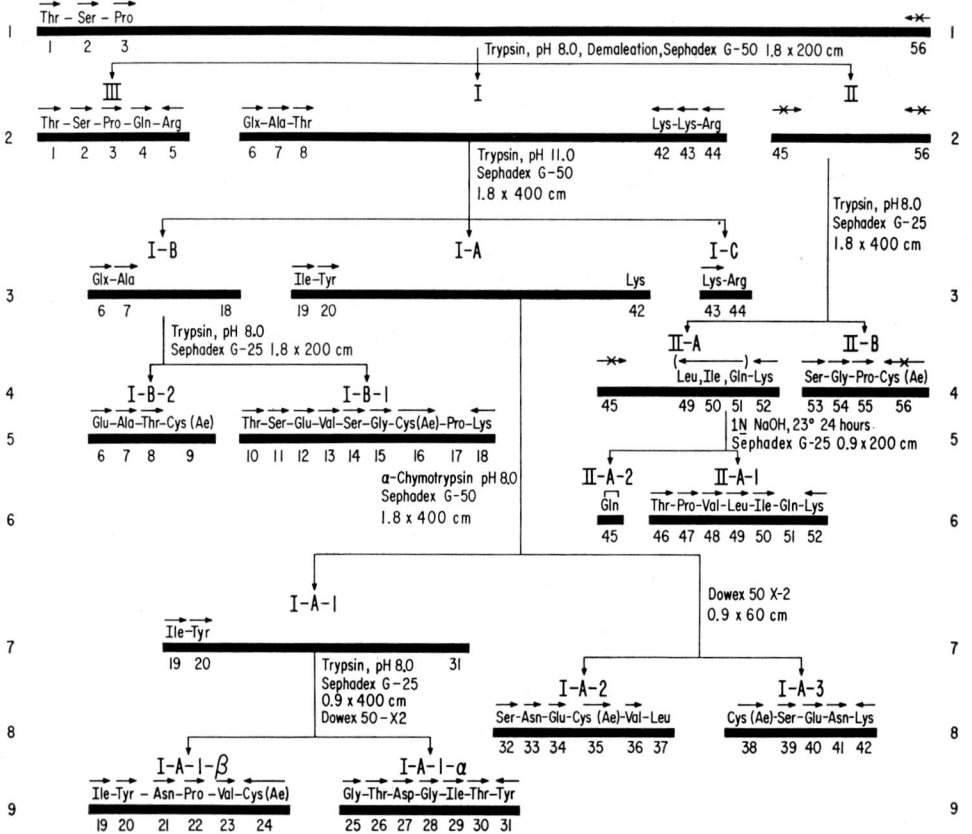

Fig. 1. Flow diagram for the selective enzymatic cleavage and amino acid sequence determination of maleyl-S-2-aminoethyl cysteinyl-porcine pancreatic secretory trypsin inhibitor I. Arrows above the amino acid residues indicate the results of Edman degradation (→) and digestion with carboxypeptidase A and B (←). Arrows with crosses (-x→, ←x-) show unsuccessful attempts to degrade the peptide by the procedure indicated by the direction of the arrow, [Bartelt, D. C. and L. J. Greene, J. Biol. Chem. 246, 2218 (1971)].

Protein Proteinase Inhibitors in Male Sex Glands

Edwin Fink, Gernot Klein, Fritz Hammer, Gottfried Müller-Bardorff
and Hans Fritz

*Institut für Klinische Chemie und Klinische Biochemie der Universität München,
D-8 München 15, Germany*

Summary

Trypsin-specific and trypsin-plasmin inhibitors were isolated from seminal vesicles of guinea pigs. Two different procedures were used: 1. Inhibitor material obtained from perchloric acid extracts was purified by affinity chromatography (using water insoluble trypsin resin) and gradient elution chromatography on Sulfoethyl Sephadex. Mainly two very similar trypsin-specific inhibitors and five somewhat different trypsin-plasmin inhibitors were obtained. (The amino acid compositions are given in Table 3). 2. Also by avoiding the trypsin resin step several inhibitor fractions were obtained which were differing considerably in their amino acid compositions.

Inhibitors containing a lysine residue in the reactive site are reversibly inactivated by acylation with maleic anhydride; arginine inhibitors are inactivated by reaction with a butandion-2,3 reagent.

In the reactive site of the trypsin-specific inhibitor the sequence Arg-Ile is present. The modified inhibitor (Arg-Ile bond is broken) is inactivated by incubation with carboxypeptidase B or reaction with excessive maleic anhydride. The native inhibitor (Arg-Ile bond intact) is converted into the modified form both during contact with the trypsin resin and by incubation with 2.3 mole percent trypsin.

From acidic extracts of boar seminal plasma a trypsin-plasmin inhibitor was isolated by affinity chromatography. The preliminary amino acid and amino sugar composition is given in Table 8. The calculated molecular weight (11,607) is in good agreement with the value found by gel filtration experiments (12,000). In the reactive site of the inhibitor an arginine residue is present.

Introduction

Inhibitors for trypsin in male sex glands and in seminal plasma were first discovered in 1965 by Haendle [1], when he was looking for the origin of inhibitory activity in urine [2]. In Table 1 trypsin-inhibiting activities found in sexual glands and seminal plasma of man and of some animals are shown [3]. From these values it is obvious why we have first chosen guinea pig for our studies: Seminal vesicles of this animal contain by far the highest inhibitory activity.

One inhibition unit, IU, is the amount of inhibition that causes the reduction of enzymatic BAPNA hydrolysis by 1 μmole per minute.

The abbreviations used are: BAEE, N^α-benzoyl-L-arginine ethyl ester; BAPNA, N^α-benzoyl-D,L-arginine p-nitroanilide hydrochloride; TRA, triethanolamine hydrochloride.

This article contains parts of E. Fink, Dissertation, Naturwissenschaftliche Fakultät der Universität München, 1970.

Table 1. Trypsin Inhibition Activities (mIU) in Male Sex Glands and their Secretions

Species	mIU per g tissue or ml plasma			
	Testes	Epididymis	Glandula vesicul.	Seminal plasma
Man	70—100	50—80	50—100	150—330
Cattle	40—70	50—80	900—1500	2400—3100
Pig	90—120	70—110	500—1000	800—1200
Sheep	—	—	250—500	—
Rat	100—200	90—130	1400—1600	—
Mouse	90—130	100—200	2200—2700	—
Guinea pig	100—220	300—400	3500—5000	—
Hamster	60—90	80—120	300—600	—

The values were adopted from H. Haendle [3]. For definition of mIU see Fritz et al. [19]. One mIU inhibits the activity of about 1 g trypsin Novo.

Inhibitors from Guinea Pig Seminal Vesicles

Inhibitors Isolated from Acidic Extracts Using Trypsin Resin. The seminal vesicles were freshly collected and immediately frozen. Greater portions were thawed and extracted with ice water. Proteins of higher molecular weights in these extracts were precipitated by addition of perchloric acid to a final concentration of 3% (w/w). After neutralisation of the acidic supernatants with potassium carbonate, the inhibitors were adsorbed from the extracts onto water insoluble trypsin resin [4]. After dissociation of the insoluble complex, the degree of purity of the resulting inhibitor preparations was about 60%, calculated from the specific activity of desalted and lyophilized fractions [4, 5].

The inhibitor material was subsequently chromatographed on Sulfoethyl Sephadex, using an ammoniumacetate concentration gradient with constant pH 5.4 (Fig. 1). In the molarity range 0.05 to 0.15, there are mainly two peaks due to a trypsin-specific inhibitor, and from 0.15 to 0.4 there are about eight peaks due to a trypsin-plasmin inhibitor. We [5, 6] have studied the last three fractions (d, e and f) thoroughly. Fraction d is the component with the lowest molecular weight while fractions e and f both contain two components. The components e_1, e_2, f_1 and f_2 are derived from component d by addition of 1 to 5 amino acids at the N-terminal residue valine of component d (see Tab. 2). Component d might be produced from the other fractions during the isolation step with the trypsin resin.

Both components, a and b, of the trypsin-specific inhibitor have the same amino acid composition (Tab. 3); but as shown later in fraction a the Arg-X-bond in the reactive site is split.

Comparison with Inhibitors Isolated from Aqueous Extracts Avoiding the Trypsin Resin Step. In order to avoid loss of inhibitory activity by acidification of the aqueous extracts with perchloric acid as well as modification reactions during

Table 2. N-Terminal Amino Acid Sequences of Five Components of the Trypsin-Plasmin Inhibitor Isolated by Use of Trypsin Resin

N-Terminal sequence	Inhibitor fraction[a]
Val	d
Lys-Val	f_1
Ser-Lys-Val	f_2
Ala-Pro-Ser-Lys-Val	e_1
Phe-Ala-Pro-Ser-Lys-Val	e_2

[a] See Fig. 1.
The sequences indicated were obtained by investigation of the tryptic peptides of the inhibitor components. Experimental details are given in ref. [6].

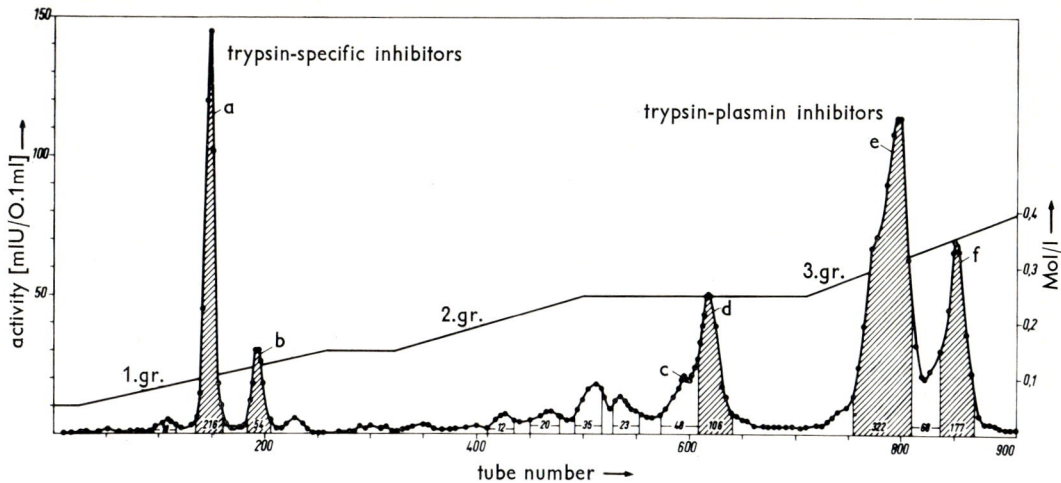

Fig. 1. Fractionation on Sulfoethyl Sephadex C-25 of an Inhibitor Preparation Isolated by Affinity Chromatography (Use of Trypsin Resin [4, 5]).
Left ordinate: Trypsin inhibition in mIU; right ordinate: Molarity of the elution buffer. The numbers in the peak areas indicate the total inhibitory activities found in these fractions.
10.9 IU (trypsin inhibition) of the inhibitor preparation were applied to the SE-Sephadex C-25 column (1.8 × 120 cm) equilibrated with 0.05M ammonium acetate buffer, pH 5.4. Elution was done with buffers and/or linear concentration gradients of ammonium acetate, pH 5.4, as shown in the figure. Elution rate: 10 cm/h; volume per fraction: 10 ml.

the trypsin resin step, part of the seminal vesicles were processed under more moderate conditions: The aqueous extracts obtained after centrifugation of the homogenates were concentrated in vacuo and fractionated by gel filtration on Sephadex G-50 equilibrated and developed with collidine acetate buffer, pH 7.8. The inhibitor-containing fractions were lyophilized. The material thus obtained was separated into

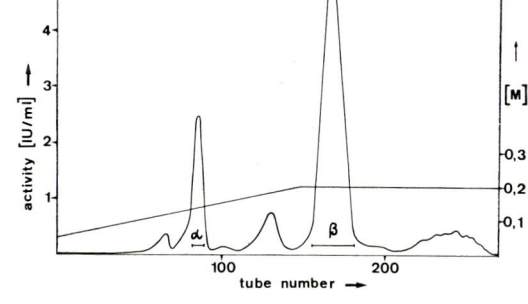

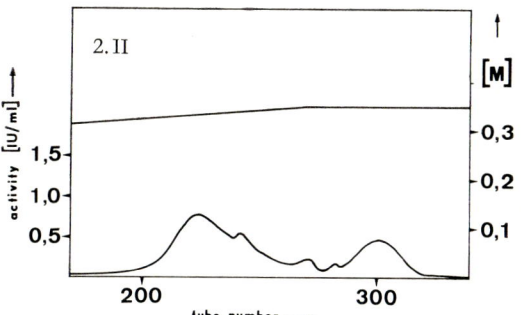

Fig. 2.I and 2.II. Fractionation on Sulfoethyl Sephadex C-25 of an Inhibitor Preparation Isolated by Conventional Chromatographic Methods [6].
For description of the figure and performance of the experiment see Fig. 1.
Fig. 2. I. 1880 IU (trypsin inhibition) of the trypsin-specific inhibitor preparation (see text) were applied to the column. In fraction α were eluted 297 IU, in fraction β 919 IU, and in other fractions 450 IU.
Fig. 2. II. 500 IU of the trypsin-plasmin inhibitor preparation were applied to the column. 482 IU were eluted in all fractions together.

15*

Table 3. Amino Acid Composition of Trypsin-Specific Inhibitors from Guinea Pig Seminal Vesicles
Samples were hydrolyzed in 6N HCl for different times at 110°C and analyzed in the "Beckman Unichrom" analyzer in the two-column system

Isolation	With trypsin resin						Without trypsin resin					
Fraction	a (Fig. 1)			b (Fig. 1)			α (Fig. 2. I)			β (Fig. 2. I)		
Amino acid	Residues per molecule											
	20 hrs	120 hrs	Integer	20 hrs	120 hrs	Integer	20 hrs	120 hrs	Integer	20 hrs	120 hrs	Integer
Cysteic acid[a]	5.90		(6)	5.75		(6)	3.44			3.96		
Methionine sulfone[a]	0.00		0	0.00		0	0.35			0.33		
Aspartic acid	5.96	6.01	6	5.98	5.95	6	6.03	6.11	6	5.01	5.01	5
Threonine	1.04	1.01	1	1.17	1.11	1	1.82	1.70	2	1.06	1.01	1
Serine	2.15	1.91	2	2.26	2.01	2	6.63	5.17	7	2.89	2.41	3
Glutamic acid	10.26	10.26	10	10.00	10.05	10	16.93	16.81	17	9.32	9.22	9
Proline	5.16	5.14	5	5.29	5.19	5	5.58	5.19	5—6	4.16	4.12	4
Glycine	6.10	6.26	6	5.97	6.01	6	8.01	7.98	8	5.24	5.20	5
Alanine	1.03	1.11	1	1.14	1.12	1	1.92	1.93	2	1.10	1.18	1
1/2 Cystine	5.16	4.22	6	5.27	3.38	6	3.13	1.46	6[b]	3.64	3.32	6[b]
Valine	1.82	2.96	3	1.76	2.56	3	5.47	6.97	7	2.31	3.18	3
Methionine	—	—	0	—	—	0	—	—	0	—	—	0
Isoleucine	2.57	3.87	4	2.31	4.02	4	2.36	3.57	4	1.95	3.01	3
Leucine	5.11	5.14	5	4.87	5.08	5	4.67	4.58	5	4.01	3.90	4
Tyrosine	1.68	1.32	2	1.80	1.47	2	1.09	0.85	1	1.22	1.13	1
Phenylalanine	0.00	0.00	0	0.00	0.00	0	0.96	0.92	1	0.26	0.28	0
Lysine	0.91	1.03	1	1.14	1.17	1	4.88	5.06	5	2.11	2.20	2
Histidine	1.61	1.25	(1)2[c]	1.75	1.70	2[c]	5.10	5.12	5	2.57	2.58	2—3
Arginine	5.96	5.94	6	6.00	6.00	6	9.19	9.01	9	5.88	6.16	6
Tryptophan[d]	0.2		0	0.2		0	0.2		0	0.2		0
Total			(59)60			60			86—91			55—56
Molecular weights[e]			6772			6772			9931—10254			6275—6412

[a] After performic acid oxidation.
[b] Related to the number of aspartic acid residues in this fraction and in fractions a and b only a value of 6 half-cystine residues is possible. The low yield by performic acid oxidation cannot be easily explained.
[c] By analyzing this fraction shortly after isolation a value near 2 was found, by analyzing the same sample 1 year after storage (4°C) only about 1.25 histidine residues per molecule could be detected. Part of the histidine therefore may be destroyed during storage.
[d] Spectrophotometric determination [20].
[e] Amide content has not been considered.

two fractions — the trypsin-specific inhibitors and the trypsin-plasmin inhibitors — by chromatography on Sulfoethyl Sephadex C-25 under the conditions described in the legend of Figure 1, but with more specific gradients. Afterwards both fractions were desalted and rechromatographed individually under the conditions given in Figures 2. I and 2. II. Surprisingly Figure 2. I now shows the presence of at least four components of the trypsin-specific

Table 4. Amino Acid Compositions of Trypsin-Plasmin Inhibitors from Guinea Pig Seminal Vesicles
The inhibitors were isolated by the trypsin resin procedure and then fractionated as shown in Fig. 1. Samples were hydrolyzed in 6N HCl for different periods at 110°C and analyzed in the "Beckman Unichrom" analyzer in the two-column system

Fraction (Fig. 1)	d			e*			f*		
Amino acid	Residues per molecule								
	20 hrs	120 hrs	Integer	20 hrs	120 hrs	Integer	20 hrs	120 hrs	Integer
Cysteic acid[a]	5.92		(6)	5.79		(6)	5.92		(6)
Methionine sulfone[a]	0.93		(1)	0.92		(1)	0.96		(1)
Aspartic acid	6.00	6.05	6	6.01	6.00	6	6.00	5.99	6
Threonine	3.91	3.74	4	3.75	3.67	4	3.84	3.58	4
Serine	5.00	4.26	5	5.73	4.79	6	5.14	3.91	5
Glutamic acid	4.11	4.28	4	4.11	4.14	4	4.07	4.19	4
Proline	2.11	2.10	2	3.01	3.14	3	2.27	2.36	2
Glycine	5.14	5.27	5	4.96	5.01	5	4.89	4.92	5
Alanine	0.16	0.16	0	1.41	1.45	1	0.01	0.02	0
1/2 Cystine	5.17	4.96	6	5.20	5.29	6	4.85	3.03	6
Valine	2.63	2.99	3	2.23	2.98	3	2.45	2.90	3
Methionine			1			1			1
Isoleucine	0.96	1.36	1	0.98	1.09	1	0.85	1.03	1
Leucine	3.13	3.15	3	3.07	3.03	3	3.03	2.96	3
Tyrosine	3.66	3.42	4	4.05	3.67	4	3.49	3.11	4
Phenylalanine	2.67	2.92	3	3.78	3.59	4	2.59	2.90	3
Lysine	4.13	4.04	4	4.65	5.32	5	5.38	4.99	5
Histidine	2.89	2.94	3	2.78	3.08	3	2.96	2.90	3
Arginine	3.93	3.98	4	3.83	4.19	4	4.06	3.98	4
Tryptophan[b]	0.20		0	0.20		0	0.20		0
Total			58			63			59
Molecular weights[c]			6687			7217			6815

* Mixture of 2 inhibitors (see Tab. 2).
[a] After performic acid oxidation.
[b] Spectrophotometric determination [20].
[c] Amide content has not been considered.

inhibitor. The amino acid compositions of the two main components α and β are very different (Tab. 3); yet the compositions of component β and the trypsin inhibitors obtained using the resin method (Tab. 3, components a and b) are very similar. Either component α is very sensitive to acidification and therefore was lost during the isolation procedure previously mentioned, or peptides were split off during the resin step to yield fraction a or b. From these results it can be deduced that there are at least two different trypsin-specific inhibitors in seminal vesicles of guinea pigs.

Figure 2. II shows the elution diagram of the trypsin-plasmin inhibitors gained in this procedure. The amount of inhibitors obtained is much lower than in the former procedure. Further studies of these fractions are in progress.

By comparing the two described isolation procedures striking differences are noticed: The activity ratio of trypsin-specific inhibitors to trypsin-plasmin inhibitors changes drastically from 10:25 (following perchloric acid precipitation and the trypsin-resin step) to 10:3 (when avoiding these steps). Loss of the big amount of trypsin-specific inhibitors during the first procedure is, as we know now, mainly due to their low solubility in acidic solutions. The last mentioned ratio therefore reflects the amounts of inhibitors existing in the tissue of the vesicles. We cannot exclude that the several components of the trypsin-specific inhibitor (as well as some components of the trypsin-plasmin inhibitors shown in Figure 1 and Table 2) shown in Figure 2.I are produced from one component by proteolysis during the first isolation steps. But it is more likely that some of them are synthesized by the action of mutated gens. Several similar examples are mentioned in the course of this conference [7—10].

reacts with one molecule of the enzyme to form the complex. The inhibition curves of the different components of each inhibitor are identical.

Reactive Sites of Proteinase Inhibitors

Influence of Lysine- and Arginine-modifying Reagents. By acylation of an inhibitor protein with maleic anhydride it can be elucidated whether a lysine or an arginine residue is located in the reactive center [11]. Lysine-inhibitors were inactivated by this reaction (Tab. 5), but not the arginine-inhibitors (Tab. 6). If the polymaleyl derivative of an inhibitor was still active, we incubated this derivative with the butandion reagent of Grossberg and Pressman [12]; this reagent reacts rapidly with the guanidino group of arginine (Fig. 4).

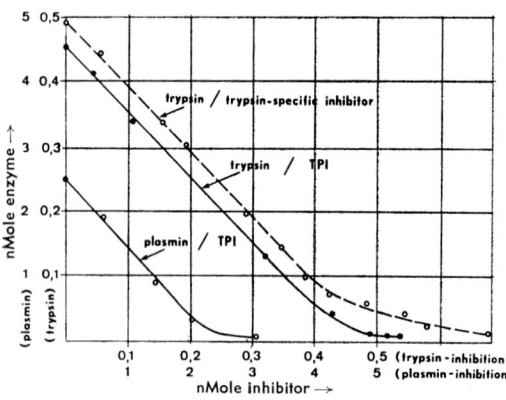

Fig. 3. Inhibition of Trypsin and Plasmin by Inhibitors from Guinea Pig Seminal Vesicles: To constant amounts of the enzymes increasing amounts of inhibitors (abscissa) were added and the remaining activities (ordinate) tested. As substrates served N^α-benzoyl-DL-arginine p-nitroanilide hydrochloride for trypsin and plasmin and also p-nitrophenyl p'-guanidinobenzoate for trypsin. For experimental details see ref. [5]. The following inhibitor preparations were used (see Fig. 1): Trypsin-specific inhibitor, fraction a; trypsin-plasmin inhibitor (TPI), fraction d. With the other inhibitor fractions identical curves were obtained.

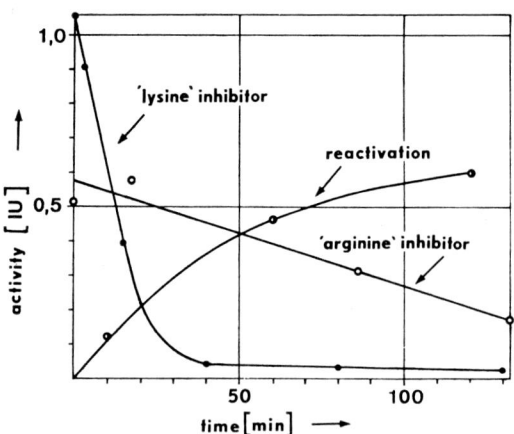

Fig. 4. Inactivation Rate of Lysine- and Arginine-Inhibitors by a Butandion-(2,3) Reagent and Reactivation Rate of a Maleylated Lysine-Inhibitor: The inhibitory activities of the samples (ordinate) were tested after different incubation periods (abscissa). A trypsin inhibitor with lysine (from Phaseolus vulgaris) and another with arginine (from sheep pancreas) in the reactive site were incubated with the butandion-(2,3) reagent used by Grossberg and Pressman [12]. — The maleylated basic pancreatic trypsin inhibitor with a lysine residue in the reactive site was incubated in perchloric acid (3%, w/w; temperature: 60°C). For experimental details see ref. [11].

Figure 3 illustrates the relationship between inhibitor concentration and inhibition of trypsin and plasmin. One molecule of each inhibitor

In all cases investigated [11] inhibitors which were fully active after maleylation lost their activity when they were treated with the butandion reagent. With inhibitors inactivating besides

Table 5. Lysine-Inhibitors[a]

Trypsin inhibitor from	Reversible inactivation by maleylation of the inhibitory activity for
Pancreas: pig, dog, cat	trypsin
Seminal vesicles of guinea pigs:	
Modified[b] trypsin-specific inhibitor TI′	trypsin
Trypsin-plasmin inhibitors	trypsin plasmin
Leeches[c] A	trypsin plasmin
B	trypsin plasmin
Lung[d]: cattle, sheep	trypsin plasmin chymotrypsin kallikrein

[a] Inhibitors bearing a lysine residue in the reactive site; see also reference [10].
[b] See Table 7.
[c] See reference [9].
[d] Identical with the basic pancreatic Kunitz inhibitor.

Table 6. Arginine-Inhibitors[a]

Trypsin inhibitor from	Maleylated inhibitor + butandion-(2,3) reagent: Loss of activity against
Pancreas: sheep, cattle	trypsin
Seminal vesicles of guinea pigs:	
native[b] trypsin-specific inhibitor TI	trypsin
Boar seminal plasma	trypsin plasmin
Submandibular glands of dogs	trypsin (not chymotrypsin)
Human serum: inter-α-trypsin inhibitor	trypsin
Sea anemone (actiniaria)	trypsin plasmin chymotrypsin kallikrein

[a] Inhibitors bearing an arginine residue in the reactive site; see also reference [10].
[b] See Table 7.

trypsin other proteinases as well, the following results were found: The loss of activity against trypsin after one of these modification reactions is accompanied also by the loss of activity against the other proteinases (Tab. 5 and 6). Only the inhibitor from submandibular glands of dog did not lose its activity against chymotrypsin when the arginine residues of the polymaleyl derivative were modified. This result is discussed more extensively by FRITZ et al. [10].

The trypsin-plasmin inhibitors of seminal vesicles of guinea pigs are inactivated by reaction with maleic anhydride whether they are isolated by use of trypsin resin or not; i. e. they bear a lysine residue in their reactive center.

Reactive Site of the Trypsin-Specific Inhibitor from Seminal Vesicles.

Of the trypsin-specific inhibitors from seminal vesicles isolated with the trypsin resin method the main fraction (component a in Fig. 1) is inactivated by maleylation; noticeably, for complete inactivation a large excess of reagent is necessary (Tab. 7). When the same fraction is incubated with carboxypeptidase B the loss of inhibitory activity is paralleled by the liberation of arginine (0.9 mole per mole inhibitor); i. e. in component a the Arg-X-bond of the reactive center is split. Inactivation of this inhibitor fraction by maleylation is therefore due to acylation of the α-amino group of the X-

Table 7. Reactive Site of the Trypsin-Specific Inhibitor
MA = maleic anhydride
TI = native inhibitor (Arg-X-bond intact)[a]
TI' = modified inhibitor (Arg-X-bond broken)[b]

Inhibitor	Treatment with	Remaining activity %
TI	MA	104
TI'	MA (big excess)	7
TI	Cpdase B	96
TI'	Cpdase B	6
TI	trypsin-resin + Cpdase B	10

[a] Fraction β from Figure 2. I.
[b] Fraction a from Figure 1.

residue in the reactive site. These findings are in full agreement with the results of LASKOWSKI, Jr. [13] obtained in similar experiments with the modified form of the soybean trypsin inhibitor. In the light of these results our former conclusions drawn from similar experiments with plant proteinase inhibitors [11] may not be correct. We failed to determine the exact amount of modified forms in these inhibitors.

In the same experiments, but using the trypsin-specific inhibitors isolated by avoiding the resin step (Fig. 2. I), different results were obtained: Inactivation did not occur by reaction with excess maleic anhydride or carboxypeptidase B. The same was true for component b in Figure 1. Samples of none-modified inhibitor (fraction β of Fig. 2. I) were applied to a column filled with a trypsin-polyacrylamide resin [14]. After dissociation of the complex in acidic solution 90% of the eluted inhibitor were modified for they lost their activity during maleylation or incubation with carboxypeptidase B. TSCHESCHE by means of mass spectrometry found that the newly formed α-amino group in the reactive site of this inhibitor belongs to isoleucine [16]. The reactive site bond of the none-modified inhibitor is also split — with different velocities depending on the acidity of the incubation mixture — in the presence of only 2.4 mole % trypsin [5, 6].

The reaction mechanism of the modification reactions will not be discussed; it is most elegantly elucidated in the lecture of LASKOWSKI, Jr. et al. [13] preceding this paper. The fact that the trypsin-specific inhibitors are obtained in almost completely modified form using the resin method and the trypsin-plasmin inhibitors are unmodified may be explained by different velocity constants of the individual modification steps (see ref. [13]). The reasons for this may be found in differences of the tertiary and primary structure of the reactive sites of these inhibitors.

Inhibitor from Seminal Plasma of Boar

Isolation Procedure. From acidic extracts of boar seminal plasma we isolated a trypsin-plasmin inhibitor using a trypsin-polyacrylamide resin. These studies are still in progress and we wish to point out that we can present preliminary results only.

Seminal fluid, collected at weekly periods, was pooled in aqueous perchloric acid (3%, w/w). After centrifugation the supernatants were neutralized, concentrated and fractionated on Sephadex G-25 columns, equilibrated and developed with aqueous acetic acid (5%, w/w). The eluted clear inhibitor fraction contained 90% of the original activity. The next steps were: Adsorption of the inhibitor to Sulfoethyl Sephadex and desorption with sodium chloride solutions (5%, w/w), pH 7, in 85—90% yield, dialyzation against deionized water (6 hrs), concentration and fractionation on Sephadex G-50 columns, equilibrated and developed with aqueous acetic acid. Lyophilisation of the eluted inhibitor fraction yielded a white powder with a specific activity which varied with different preparations from 50 to 150 mIU/mg.

This material was further purified by the trypsin-resin method [17]: The inhibitor (300 IU in 10 ml) was adsorbed to a trypsin-polyacrylamide resin column [15] from 0.4M NaCl, 0.1M TRA-buffer solution, pH 7.8. After washing the resin with the same salt-buffer solution the inhibitor-trypsin complex was then dissociated and the inhibitor eluted with 0.4M KCl/HCl buffer, pH 2, in 85% yield. The resulting inhibitor solution was neutralized, concentrated by ultrafiltration and fractionated on Sephadex G-50, equilibrated and developed with aqueous acetic acid. After lyophilization of the eluted inhibitor fraction a white powder with a specific activity of 2.0 IU/mg (trypsin inhibition) was obtained. This material was used for the following investigations. (Some attempts to further purification by ion exchange chromatography were unsuccessful.)

Inhibition Properties and Composition. It can be seen from Figure 5 that the inhibitor reacts stoichiometrically with both trypsin and plasmin. Dependent on the degree of inhibition 5—8 minutes are necessary to reach inhibition equilibrium. The inhibitor has no effect on the activity of chymotrypsin (substrate: N-3-[carboxy-propionyl]-L-phenylalanine p-nitroanilide), thrombin (substrate: BAEE) and kallikrein (substrate: BAEE) from pig pancreas.

From the specific activity a molecular weight of about 13,000 is calculated. In gel filtration experiments in slightly acidic (pH 2.2) buffer solution the molecular weight was found to be approximately 13,500, in neutral solution near 12,000. The molecular weight calculated from

Table 8. Amino Acid and Amino Sugar Composition[a] of the Trypsin-Plasmin Inhibitor from Boar Seminal Plasma
Samples were hydrolyzed in 6N HCl at 110°C and analyzed in the "Beckman Unichrom" analyzer in the two-column system

Amino acid, amino sugar	Residues per molecule			
	20 hrs	70 hrs	122 hrs	Integer
Cysteic acid	7.2[b]			
Methionine sulfone	1.4[b]			
Aspartic acid	10.75	10.75	10.84	11
Threonine	5.95	5.71	5.60	6
Serine	6.82	6.00	5.34	7
Glutamic acid	8.26	8.11	8.09	8
Proline	4.56	4.08	4.28	4
Glycine	7.82	7.92	7.89	8
Alanine	4.32	4.32	4.32	4
1/2 Cystine	5.85	8.16	2.04	8
Valine	1.54	1.87	1.88	2
Methionine		1.10		2[c]
Isoleucine	2.83	3.97		4
Leucine	5.47	4.18	3.87	4
Tyrosine	4.12	4.37	3.05	4
Phenylalanine	7.25	7.34	7.33	7
Lysine	8.00	8.04	7.99	8
Histidine	3.98	3.97	3.90	4
Arginine	8.25	8.13	8.10	8
Tryptophan	(1.57)[d]			2
Glucosamine	2.3	1.1	0.6	3
Galactosamine	1.5	0.7	0.3	2
Total[a]				104
Mol. weight[a]				11607

[a] Investigations for other residues (e.g. neutral sugars, sialic acid etc.) were not performed.
[b] After performic acid oxidation.
[c] Calculated from the oxidized form of the inhibitor.
[d] Determined by a spectrophotometric method [20].

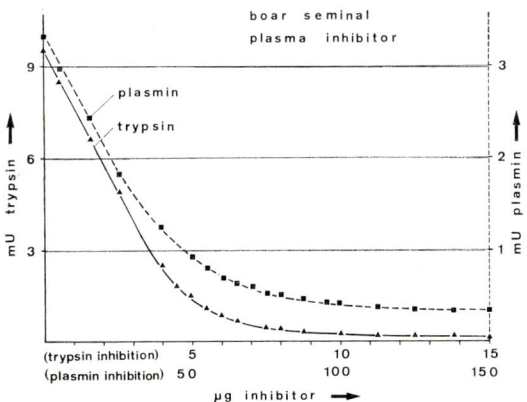

Fig. 5. Inhibition of Trypsin and Plasmin by a Seminal Plasma Inhibitor from Boar. To constant amounts of the enzymes increasing amounts of inhibitor (abscissa) were added and the remaining activities (ordinate) tested. As substrate served N^α-benzoyl-DL-arginine p-nitroanilide hydrochloride. Enzyme and inhibitor were incubated (for trypsin inhibition: 8 min, for plasmin inhibition: 10 min) in 2.0 ml 0.2M buffer solution, pH 7.8, afterwards 1.0 ml substrate solution was added, see ref. [5].

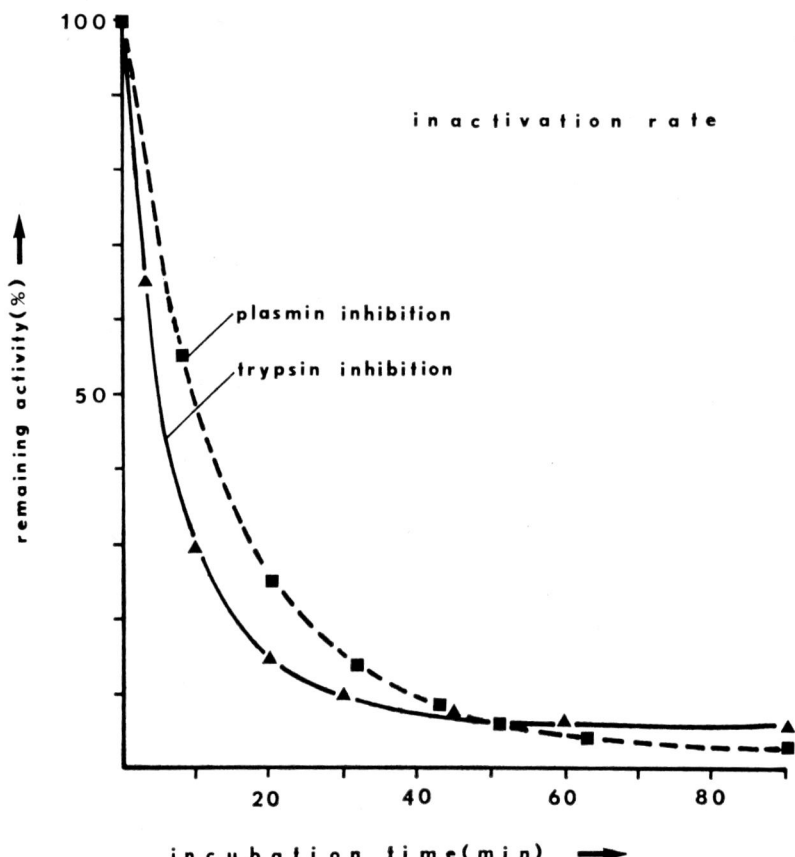

Fig. 6. Inactivation of Maleylated Inhibitor by a Butandion-(2,3) Reagent. The inhibitor from boar seminal plasma was maleylated as described in ref. [11]. To the maleylated inhibitor the butandion-(2,3) reagent was added; the decrease in inhibitory activity (ordinate) against trypsin and plasmin was tested after different incubation periods (abscissa). For methods see Fig. 4 and ref. [11].

the amino acid and amino sugar composition given in Table 8 is in good agreement with these values. Nevertheless, the calculation of other compositions may also be possible. We have not yet investigated if the inhibitor contains also some other residues (e. g. neutral sugar, amino sugar and sialic acid residues) not mentioned in Table 8.

The inhibitor is not inactivated by maleylation. After reaction with the butandion reagent, the polymaleyl inhibitor lost its activity against both trypsin and plasmin in a manner characteristic of arginine-inhibitors (Fig. 6). The inhibitor must be in the unmodified form (Arg-X-bond intact in the reactive site), because maleylation or incubation with carboxypeptidase B has no effect on its inhibition ability. No temporary inhibition could be demonstrated.

From the physiological point of view it is interesting that we only found the trypsin-plasmin inhibitor in seminal plasma of boar. The inhibitor with similar properties from seminal vesicles of guinea pigs seems to be more active in inhibition of fertilization than the trypsin-specific inhibitor as shown by Dr. ZANEVELD [18]. Further investigations are necessary to clear up whether only the trypsin-plasmin inhibitor exists in seminal fluid and the trypsin-specific inhibitor is a non-excretable constituent of the vesicle cells. But it is also possible that our group and Dr. ZANEVELD [18] precipitated the trypsin-specific inhibitor by acidification or that this inhibitor is not existing in the species under investigation.

The inhibitor isolated by Dr. ZANEVELD et al. [18] from boar seminal plasma differs significantly in amino acid composition and molecular weight from the one presented in this paper. We assume that both groups isolated different inhibitors. But as we have not yet examined further criteria for homogeneity of the inhibitor isolated in our laboratory, we cannot exclude that further chromatographic procedures yield an inhibitor with somewhat different analytical data.

Experiments which show the significance of proteinase inhibitors from seminal plasma in fertilization are elegantly demonstrated and discussed in detail by Dr. ZANEVELD et al. [18].

Acknowledgements This research was supported by Sonderforschungsbereich-51, München and Behringwerke AG, Marburg.

We wish to thank Fräulein FRIEDERIKE BÜRGER for collecting the seminal vesicles.

We are also grateful to Prof. Dr. Dr. E. WERLE for generously supporting these investigations.

References

[1] HAENDLE, H., H. FRITZ, I. TRAUTSCHOLD and E. WERLE, Z. physiol. Chem. **343**, 185 (1965).

[2] ASTRUP, T., O. K. ALBRECHTSEN, Scand. J. clin. Lab. Invest. **9**, 233 (1957). For further references see also R. VOGEL, I. TRAUTSCHOLD and E. WERLE: Natural Proteinase Inhibitors. Academic Press, New York—London, p. 73—76 (1968).

[3] HAENDLE, H., Dissertation, I. Medizin. Fakultät der Universität München (1969).

[4] FRITZ, H., M. GEBHARDT, E. FINK, W. SCHRAMM and E. WERLE, Z. physiol. Chem. **350**, 129 (1969); see also H. FRITZ, B. BREY, M. MÜLLER and M. GEBHARDT, this volume, p. 28.

[5] FRITZ, H., E. FINK, R. MEISTER and G. KLEIN, Z. physiol. Chem. **351**, 1344 (1970).

[6] FINK, E., Dissertation, Naturwissenschaftliche Fakultät der Universität München 1970. Clinical Enzymologie **2**, 74 1970, Karger, Basel.

[7] PEANASKY, R. J. and G. M. ABU-ERREISH, this volume, p. 281.

[8] RYAN, C. A. and L. K. SHUMWAY, this volume, p. 175.

[9] FRITZ, H., M. GEBHARDT, R. MEISTER and E. FINK, this volume, p. 271.

We are indebted to Dr. O. HAEGER, Besamungshauptstelle München, for gifts of boar semen.

[10] FRITZ, H., E. JAUMANN, R. MEISTER, P. PASQUAY, K. HOCHSTRASSER and E. FINK, this volume, p. 257.

[11] FRITZ, H., E. FINK, M. GEBHARDT, K. HOCHSTRASSER and E. WERLE, Z. physiol. Chem. **350**, 933 (1969).

[12] GROSSBERG, A. L. and D. PRESSMAN, Biochemistry **7**, 272 (1968).

[13] LASKOWSKI, M., Jr., R. DURAN, W. R. FINKENSTADT, S. HERBERT, H. F. HIXSON, Jr., D. KOWALSKI, J. A. LUTHY, J. A. MATTIS, R. E. MCKEE, C. W. NIEKAMP, this volume, p. 117.

[14] INMAN, J. K. and H. M. DINTZIS, Biochemistry **8**, 4074 (1969).

[15] FRITZ, H., M. GEBHARD, R. MEISTER and H. SCHULT, Z. physiol. Chem. **351**, 1119 (1970).

[16] TSCHESCHE, H. and R. OBERMEIER, this volume, p. 135.

[17] An example is given by FRITZ, H., M. GEBHARDT, R. MEISTER and H. SCHULT, Z. physiol. Chem. **351**, 1119 (1970), Table 2.

[18] ZANEVELD, L. J. D., K. L. POLAKOSKI, R. T. ROBERTSON and W. L. WILLIAMS, this volume, p. 236.

[19] FRITZ, H., G. HARTWICH and E. WERLE, Z. physiol. Chem. **345**, 150 (1966).

[20] GOODWIN, T. W., and R. A. MORTON, Biochem. J. **40**, 628 (1946).

Trypsin Inhibitors and Fertilization

L. J. D. Zaneveld, K. L. Polakoski, R. T. Robertson and W. L. Williams

Department of Biochemistry, University of Georgia, Athens, Georgia 30601

Spermatozoa are produced in the testis and stored in the epididymis. During ejaculation they come in contact with seminal plasma and are deposited into the vagina from where they migrate through the cervix into the uterus. Here, mammalian sperm must reside for a period of time (at least 6 hours in case of the rabbit) to become fertile. This process is called capacitation [1, 2] and is under hormonal control [3]. Spermatozoa that do not undergo this process are unable to penetrate the outer investments of the ovum including the zona pellucida. Penetration of this layer is accomplished through the lytic action of acrosin, a protease that resembles trypsin and plasmin (Tab. 1) [4, 5, 6]. Acrosin is located in the acrosome, a sack-like cap over

Table 1. Properties of Acrosin and other Mammalian Trypsin-like Enzymes [6]

Properties	Acrosin	Pancreatic Trypsin	Plasmin	Thrombin	Kallikrein		
					Plasma	Organs	Urine
Molecular Weight	55,000	23,800	82,900	33,000	97,000	33,000	36,300
Autoproteolysis	+	+	+	+	—	—	—
Stability at low pH	+	+	+	—	—	—	—
Optimum pH	8	8	8	8	8.5	9	8.5
Calcium Activation	+	+	+	+	—	—	—
Relactive Activity on Substrates:							
BAEE	++	+	++	++	++	++	++
TAME	+	++	++	++	+	+	+
ATEE	—	—	—	—	±	±	±
Inhibitors:							
TLCK	+	+	+	+			—
TPCK	—	—	—	—			
DFP	+	+	+	+	+	+	+
Soybean Trypsin Inhibitor	+	+	+	—	+	—	—
(Kunitz) Pancreatic Trypsin Inhibitor	+	+	+	—	+	+	+
(Chicken) Ovomucoid Trypsin Inhibitor	±	+	—	—	+	—	—
Seminal Vesicle Trypsin Inhibitors:							
TI	+	+	—	—	—	—	—
TPI	+	+	+	—	—	—	—

the anterior portion of the sperm head [7]. No visible morphological changes occur in rabbit sperm during capacitation, but during sperm penetration of the ovum layers the outer acrosomal membrane and plasma membrane vesiculate and disappear [8] exposing acrosin so that lysis of the zona pellucida can occur. In our laboratory we have obtained this enzyme in an almost pure form by selective removal of the acrosomes with detergents, alcohol precipitation of the acrosomal extracts and further purification by DEAE cellulose and Sephadex G-75 or G-100 column chromatography [9]. Some of the properties of acrosin as compared to other mammalian trypsin-like enzymes are shown in Table 1. Acrosin has a molecular weight of 55,000 and appears to consist of equal subunits. It is stable at pH 2—3 and can be maintained at that pH for weeks without loss of activity. At neutral pH the enzyme is highly autoproteolytic. Acrosin has an optimum pH of 8.0, requires calcium for maximum activity on synthetic substrates and hydrolyzes BAEE 2 to 3 times faster than TAME. It does not hydrolyze ATEE. In the pure form the enzyme is unstable to slow freezing but stable if lyophilized. It digests oxidized RNAse less completely than pancreatic trypsin. This is either caused by the high degree of autoproteolysis or by the inherent inability of the enzyme, similar to plasmin and thrombin, to hydrolyze all the arginine and lysine bonds. RNAse digested with acrosin can be further hydrolyzed using pancreatic trypsin. Acrosin is inhibited by the synthetic inhibitors tosyl lysine chloromethyl ketone (TLCK) and diisopropyl fluorophosphate (DFP), but not by tosyl phenylalanine chloromethyl ketone (TPCK), a specific chymotrypsin inhibitor. Soybean trypsin inhibitor, Kunitz pancreatic trypsin inhibitor and the seminal vesicle trypsin inhibitors TI and TPI isolated by Fritz et al. [10] inhibit acrosin. In contrast to pancreatic trypsin, acrosin is poorly inhibited by the chicken ovomucoid trypsin inhibitor [6].

Acrosomal extracts of epididymal and capacitated sperm have a high acrosin activity (Tab. 2). This activity is virtually absent in ejaculated sperm until after DEAE cellulose column chromatography of the extracts, most likely due to the removal of an inhibitor [11]. The presence of an acrosin-inhibitor complex in these sperm was further investigated by incubating acrosomal extracts either at pH 5.9 or at pH 7.6. The extracts were then added to BAEE at pH 8.0. As control acrosomal extracts of epididymal sperm were also incubated at pH 7.6. Figure 1 shows that the epididymal and ejaculated sperm acrosomal extracts incubated at alkaline pH hydrolyze BAEE at a constant rate, acrosin activity being much greater in epididymal than in ejaculated sperm extracts. The ejaculated sperm extracts incubated at pH 5.9 cause initially a fast rate of change at OD_{253} but this decreases rapidly until the final rate is the same as that of extracts incubated at pH 7.6. Since trypsin-trypsin inhibitor complexes dissociate at low pH and combine again at neutral pH it appears that a similar complex is present in ejaculated sperm. Apparently such a complex is absent in epididymal or capacitated sperm.

Incubation of epididymal sperm with seminal plasma causes a significant decrease in acrosin activity (Tab. 3) indicating that the acrosin-inhibitor complex is formed during ejaculation by penetration of an inhibitor from the seminal plasma through the outer acrosomal membrane of the sperm head [12]. The presence of such an inhibitor in the male accessory glands and fluids

Table 2. Acrosin Activity of various Acrosomal Extracts [11]

Sperm Extract	m Units/mg Protein
Capacitated	80—130
Epididymal	84—430
Ejaculated (crude)	0—10
Ejaculated (after DEAE column chromatography)	40—150

Acrosin activity was measured using BAEE as substrate in borate buffer at pH 8.0. One m Unit is defined as that amount of acrosin causing a rate of change at OD_{253} of 0.001/min.

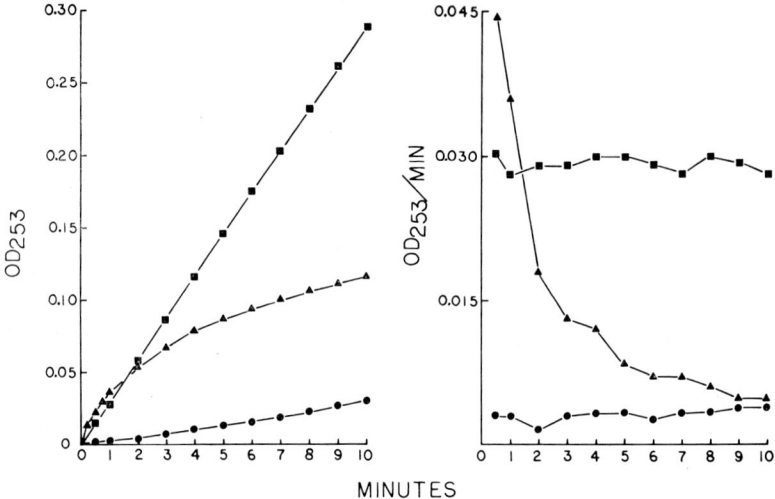

Fig. 1. Acrosin-Inhibitor Complex in Ejaculated Sperm

Rabbit sperm acrosomal extracts were incubated either in HCl at pH 5.9 or in 0.05M Borate buffer, pH 7.6 and added to BAEE in 0.05M Tris-HCl buffer, pH 8.0 containing 0.05M $CaCl_2$.

■—■—■ Acrosomal extracts of epididymal sperm incubated at pH 7.6.
●—●—● Acrosomal extracts of ejaculated sperm incubated at pH 7.6.
▲—▲—▲ Acrosomal extracts of ejaculated sperm incubated at pH 5.9.

Table 3. Acrosin Activity of Epididymal Sperm after Treatment with Seminal Plasma [12]

Epididymal sperm incubated with the following before preparation of acrosomal extracts	m Units/mg protein*
Krebs-Ringer Phosphate Buffer	151
Seminal Plasma	8

See Table 2 for assay conditions.

* Average amounts from 8 experiments.

Table 4. Amino Acid Analyses of Boar Seminal Plasma Trypsin Inhibitor (SPTI) [17] and the Trypsin Inhibitor (TI) and Trypsin-Plasmin Inhibitor (TPI) isolated from Guinea Pig Seminal Vesicles [10]

Amino acid	SPTI	TI	TPI
Alanine	3	1	0
Arginine	4	6	4
Aspartic acid	6	6	6
Half-cystine	6	6	6
Glutamic acid	5	10	4
Glycine	6	6	5
Histidine	2	1	3
Isoleucine	2	4	1
Leucine	3	5	3
Lysine	6	1	4
Methionine	1	0	1
Phenylalanine	4	0	3
Proline	6	5	2
Serine	1	2	5
Threonine	2	1	4
Tryptophan	0	0	0
Tyrosine	2	2	4
Valine	1	3	3
Total	58	59	58
Molecular Weight	6.781	6.635	6.687

was first reported by Haendle et al. in 1965 [13] and WALDSCHMIDT et al. in 1966 [14]. It was further shown that the inhibitors in these tissues are hormone dependent [13, 15]. Two inhibitors were isolated from the seminal vesicles, both having a molecular weight of approximately 6700 but differing in enzyme specificity (Tab. 1) and amino acid composition (Tab. 4) [10]. One inhibits trypsin only (TI) whereas the other inhibits both plasmin and trypsin (TPI).

In our laboratory, we have partially purified the acrosin inhibitor from rabbit seminal plasma (rabbit SPTI) by the use of an insoluble trypsin-copolymer [16] and employed this in the antifertility assays described later (Tab. 5). Boar SPTI was isolated by treatment of seminal plasma with 2,5% TCA, precipitation with 50—90% $(NH_4)_2SO_4$ and SE-Sephadex and Sephadex G-50 column chromatography and shows only one band by polyacrylamide disc electrophoresis [17]. Both trypsin inhibitors are heat and acid stable and inhibit acrosin and pancreatic trypsin but do not effect chymotrypsin. Amino acid analyses of the boar SPTI show that it has a molecular weight of 6781 and is high in lysine, aspartic acid, proline, cysteine and glycine but does not possess any tryptophan (Tab. 4). Estemations by Sephadex chromatography showed a molecular weight of 13,000 [18]. This discrepancy is either caused by the presence of carbohydrates or by dimer formation during chromatography, the latter being most plausible.

It seems likely that the seminal plasma trypsin inhibitor is produced in the seminal vesicles. Boar SPTI elutes from SE-Sephadex at a salt concentration of 0.25M, similar to TPI [10], inhibits both trypsin and plasmin [18] and has an amino acid composition that resembles TPI more than that of TI (Tab. 4). Preliminary results further indicate that TPI, but not TI is an effective antifertility agent when added to capacitated sperm (Tab. 6) and it appears therefore that SPTI and TPI are related. This relationship may be similar to that of the Kunitz and Kazal inhibitors of the bovine pancreas in that only the Kazal inhibitor is found in the pancreatic secretions. Trypsin inhibitors are also present in the seminal plasma of the human, ram, horse, bull, chicken and turkey [13, 17, 19].

The acrosin-inhibitor complex is dissociated during capacitation (Tab. 2) so that penetration of sperm through the zona pellucida can occur. A similar enzyme-inhibitor relationship in which the inhibitor is removed during capacitation

Table 5. Effect of Mammalian Trypsin Inhibitors on Fertilization *in vivo* [16]

Additions per 10^5 Sperm	No. of Rabbits	Cleaved Ova/ Total No. of Ova	Percent Fertilization
Pancreatic Trypsin Inhibitor (200 µg)	10	8/49	16.3
None	10	28/37	75.7
Pancreatic Trypsin Inhibitor (200 µg) (centr.)	7	2/22	9.1
None	7	15/22	68.2
Seminal Plasma Trypsin Inhibitor (250 µg)	5	1/14	7.1
None	5	17/18	94.4

After incubation for 20 min at 37°C, the mixture of trypsin inhibitor and capacitated sperm (0.05 ml, 5×10^4 sperm) was inseminated into one of the oviducts of a rabbit 12.5 hr after administration of HCG, except in the second group of experiments in which the trypsin inhibitor was first removed from the sperm by centrifugation at 300 g for 5 min, discarding the supernatant solution and resuspending the sperm in KRP. Untreated spermatozoa were inseminated into the contralateral oviduct.

Table 6. Effect of Seminal Vesicle Trypsin Inhibitors [10] on the Fertilizing Ability of Capacitated Rabbit Spermatozoa

Additions per 10^5 Sperm	Units of Inhibitor Activity (ImU)	No. of Rabbits	No. of Eggs	Percent Fertilization
100 μg TPI*	170	4	14	14.3
0	0		10	100
250 μg TI*	750	5	15	73.5
0	0		10	90

* TPI — Trypsin-Plasmin Inhibitor; TI — Trypsin Inhibitor.

Capacitated sperm were treated with inhibitor for 20 min at 37°C and 0.05 ml of the mixture (5 × 10^4 sperm) was inseminated into the oviducts of rabbits 12.5 hr after administration of HCG. The contralateral oviducts were inseminated with control sperm that were treated the same except that no inhibitor was added.

exists between the acrosomal enzyme involved in the penetration of sperm through the corona radiata (CPE) and the decapacitation factor (DF) [20]. DF is also present in seminal plasma and although it occurs naturally as a large polypeptide it is active in a low molecular weight form [21]. DF does not inhibit pancreatic trypsin or acrosin and appears to be specific for CPE. Part of capacitation therefore appears to be the activation of acrosomal enzymes necessary for ovum penetration. The mechanism of DF and SPTI removal from the acrosomal enzymes is not known. An attractive hypothesis can be made, however, if we consider that if sperm acrosomes are stained with tetracycline HCl, this dye is removed within 2 hours after addition of the sperm to the uterus [22] indicating that a change in the plasma membrane and the outer acrosomal membrane has occurred. Further, trypsin-trypsin inhibitor complexes are reversible and have a half-life of approximately 5 hours [23] depending on the complex. When these times are added together they approximate the period required for capacitation in the rabbit uterus (6 hours). We can therefore speculate that during capacitation the plasma and outer acrosomal membranes are modified rendering them semi-permeable, after which the acrosin-inhibitor complex dissociates and the inhibitor is removed by diffusion because of its low molecular weight. Similarly, removal of DF from the acrosome may also occur by diffusion or by proteolytic destruction [24]. The significance of these enzyme-inhibitor complexes is not understood although it is conceivable that since capacitated sperm become penetrating bodies, not only penetrating ova but also uterine cells, tissues and phagocytes [25], these inhibitors function to prevent sperm from penetrating cells, and tissues in the male with possible pathological results.

A high molecular weight [26] and a low molecular weight [27] trypsin inhibitor are present in cervical mucus. These inhibitors may function in the control of sperm migration through the cervix since it appears that an enzyme in the seminal plasma that resembles chymotrypsin is involved in the migration process [28].

Inhibition of Fertilization by Natural and Synthetic Protease Inhibitors

Since acrosin has to be in the active form before sperm can penetrate the zona pellucida and since low molecular weight trypsin inhibitors pass through the outer acrosomal membrane (Tab. 3), addition of trypsin inhibitors to capacitated, i. e. fully fertile sperm, should prevent fertilization under conditions where recapacitation is not possible. Large amounts of soybean trypsin inhibitor indeed prevent fertility *in vitro* if added to the sperm-egg solutions [29]. These large amounts are probably necessary because of the high molecular weight of this inhibitor. Of greater significance is the finding that fertilization *in vivo* can be inhibited by addition of pancreatic trypsin inhibitor or partially purified rabbit SPTI to capacitated sperm (Tab. 5) [16]. These inhibitors effect the sperm and not the ova since washing the sperm after treatment with pancreatic trypsin inhibitor does not result in a higher rate of fertility. Preliminary results show an interesting difference between the two inhibitors isolated from the seminal vesicles in that TPI in much lower quantities than TI causes a significant decrease in fertilization whereas TI has virtually no antifertility effect (Tab. 6). Both these inhibitors inhibit acrosin activity using BAEE as substrate, however, and this difference cannot be explained.

Mammalian trypsin inhibitors form reversible complexes with their respective enzymes and since the acrosin-inhibitor complex dissociates during capacitation it did not appear likely that natural trypsin inhibitors would be of use as contraceptive agents. It has indeed been shown by Schumacher [30] that incubation of ejaculated sperm with Trasylol or soybean trypsin inhibitor does not prevent fertilization, nor does the systemic treatment of rabbits with these inhibitors. We therefore tested the antifertility effect of synthetic protease inhibitors that bind irreversibly to the histidine or serine of the active site [31, 32]. On incubation with capacitated sperm, TLCK causes complete inhibition of fertilization whereas TPCK has essentially no effect (Tab. 7) correlating with the effect of these inhibitors on acrosin (Tab. 1). These results further show that synthetic trypsin inhibitors but not chymotrypsin inhibitors are effective antifertility agents. These inhibitors do not interfere with sperm metabolism since sperm incubated in TLCK and control sperm maintain motility for the same length of time *in vitro* [31].

To be of practical use, it was essential to show that these synthetic inhibitors are not removed during capacitation. Ejaculated sperm were therefore incubated with various synthetic trypsin inhibitors and allowed to capacitate normally. It was found that TLCK and NPGB (nitrophenyl p-guanidinobenzoate) but not EPGB (ethyl-p-guanidinobenzoate) cause a decrease in fertility under these conditions (Tab. 8). These results again correlate with the effect of these inhibitors on trypsin or plasmin since NPGB reacts approximately 1000 times faster with these

Table 7. Effect of TLCK and TPCK on the Fertilizing Ability of Capacitated Rabbit Spermatozoa [31]

Inhibitor	Treated Sperm			Control	
	Amount (μg) Per 10^5 Sperm	No. of Ova	Percent Fertilization	No. of Ova	Percent Fertilization
TLCK	5	13	15.6	10	100
TLCK	15	16	0	16	87.5
TPCK	15	7	86	6	100

The inhibitor was added to capacitated sperm, incubated for 30 min at 37°C and removed by centrifugation. Fertilization experiments were performed using 5×10^4 sperm in 0.05 ml as described in Table 6. Control sperm were treated the same except that no inhibitor was added.

Table 8. Effect of Synthetic Trypsin Inhibitors on the Fertilizing Ability of Ejaculated Rabbit Spermatozoa [31]

Inhibitor	Treated Sperm			Control	
	Amount (μg) Per 10^5 Sperm	No. of Ova	Percent Fertilization	No. of Ova	Percent Fertilization
TLCK	3	20	9	14	100
TLCK	10	25	0	19	100
EPGB	10	13	100	17	100
NPGB	10	9	21	10	100

Each incubation mixture consisted of 100 times the indicated amount of inhibitor and 10^7 ejaculated sperm in 10 ml. The treated sperm were centrifuged, excess inhibitor removed and 2.5×10^6 sperm in 0.25 ml were inseminated into non-ligated uteri. Control sperm were treated the same as the test sperm except in the absence of inhibitor and were inseminated into ligated uteri so that migration of the sperm into the contralateral uteri was not possible.

enzymes than EPGB [33]. The duration of incubation was apparently not long enough for EPGB to penetrate the outer acrosomal membrane and react with acrosin.

Since TLCK decreased fertilization more successfully than NPGB (Tab. 8) it was used to test the effect of these inhibitors if deposited into the vagina. At a level of 3 mg/ml, TLCK completely prevents fertilization when mixed with K—Y sterile lubricant (Tab. 9). Delfen, the most effective vaginal cream available on the market, only causes a 47.5% decrease in fertilization when added to the vagina of rabbits (Tab. 10) confirming previous reports [34]. Addition of TLCK

Table 9. Antifertility Effect of TLCK in the Vagina [31]

TLCK mg/ml	Rabbits	No. of Ova	Percent Fertilization
0	5	37	100
1	5	34	38.3
2	4	32	6.2
3	5	28	0

The TLCK was dissolved in 2.5 ml KRP, mixed with 2.5 ml K-Y sterile lubricant and deposited into the vagina of mature rabbits 5 min before breeding.

to Delfen increases the antifertility effect of this cream significantly. These results show that synthetic trypsin inhibitors can be effective vaginal contraceptives either alone or mixed with vaginal contraceptives available on the market today.

Table 10. Contraceptive Activity of Delfen Vaginal Cream Mixed with TLCK

TLCK mg/ml	Rabbits	No. of Ova	Percent Fertilization	Range
0	7	40	52.5	0—100
3	5	48	14.6	0—33
5	4	28	10.7	0—25

Twice the indicated amount of TLCK was dissolved in 0.2 ml H$_2$O, mixed with 2 ml Delfen Cream and deposited into the vagina of mature rabbits 5 min before breeding.

Acknowledgements

This research was supported by Training Grant No. 5-T01-HD 00140 from the National Institute of Child Health and Human Development; Career

Development Award No. 2-K3-GM 4831 from the National Institute of Medical Sciences; Contracts NIH-70-2147 and NIH-69-2103 with the National Institutes of Health, Department of Health, Education and Welfare; and a Ford Foundation Grant.

References

[1] Austin, C. R., Observations on the penetration of the sperm into the mammalian egg. Austr. J. Sci. Res. B4, 581 (1951).

[2] Chang, M. C., Fertilizing capacity of spermatozoa deposited into the Fallopian tubes. Nature 168, 697 (1951).

[3] Chang, M. C., Hormonal regulation of sperm capacitation. Adv. Biosciences 4, 13 (1969).

[4] Hartree, E. F. and P. N. Srivastava, Chemical composition of the acrosomes of ram spematozoa. J. Reprod. Fert. 9, 47 (1965).

[5] Stambaugh, R. and Buckley, J. Identification and subcellular localization of the enzymes effecting penetration of the zona pellucida by rabbit spermatozoa. J. Reprod. Fert. 19, 423 (1969).

[6] Zaneveld, L. J. D., K. L. Polakoski and W. L. Williams, Properties of acrosin, a proteolytic enzyme from rabbit sperm acrosomes. In manuscript.

[7] Gaddum, P. and R. J. Blandau, Proteolytic reaction of mammalian spermatozoa on gelatin membranes. Science 170, 749 (1970).

[8] Bedford, J. M., Morphological aspects of sperm capacitation in mammals. Adv. Biosciences 4, 35 (1969).

[9] Polakoski, K. L., L. J. D. Zaneveld and W. L. Wiiliams, Purification of acrosin, a proteolytic enzyme from rabbit sperm acrosomes. In manuscript.

[10] Fritz, H., E. Fink, R. Meister and G. Klein, Isolierung von Trypsin-Inhibitoren und Trypsin-Plasmin-Inhibitoren aus den Samenblasen von Meerschweinchen. Z. Physiol. Chem. 351, 1344 (1970).

[11] Zaneveld, L. J. D., P. N. Srivastava and W. L. Williams, (1969) Relationship of a trypsin-like enzyme in rabbit spermatozoa to capacitation. J. Reprod. Fert. 20, 337 (1969).

[12] Zaneveld, L. J. D., P. N. Srivastava and W. L. Williams, Inhibition by seminal plasma of acrosomal enzymes in intact sperm. Proc. Soc. Exp. Biol. Med. 133, 1172 (1970).

[13] Haendle, H., H. Fritz, I. Trautschold and E. Werle, Über einen hormonabhängigen Inhibitor für proteolytische Enzyme in männlichen Geschlechtsdrüsen und im Sperma. Hoppe Seyler's Zeitschr. Physiol. Chem. 343, 185 (1965).

[14] Waldschmidt, M., B. Hoffmann and H. Karg, Untersuchungen über die tryptische Enzymaktivität in Geschlechtssekreten von Bullen. Zuchthygiene 1, 15 (1966).

[15] Ingrisch, H., H. Haendle and E. Werle, Über die Konzentration des Trypsin-Inhibitors in Sperma von Gesunden und andrologisch Kranken und über ihre Beziehung zu anderen Parametern des Spermas. Andrologie 2, 103 (1970).

[16] Zaneveld, L. J. D., R. T. Robertson, M. Kessler and W. L. Williams, Inhibition of fertilization in vivo by pancreatic and seminal plasma trypsin inhibitors. J. Reprod. Fert. 25, 387.

[17] Polakoski, K. L., L. J. D. Zaneveld and W. L. Williams, Unpublished results.

[18] Fink, E., G. Klein, F. Hammer, G. Müller-Bardorff and H. Fritz, Protein Proteinase Inhibitors in Male Sex Glands, this volume, p. 225.

[19] Feeney, R. E. and R. G. Allison, Evolutionary Biochemistry of Proteins (Wiley), 213 (1969).

[20] Zaneveld, L. J. D. and W. L. Williams, A sperm enzyme that disperses the corona radiata and its inhibition by decapacitation factor. Biol. Reprod. 2, 363 (1970).

[21] Williams, W. L., R. T. Robertson and W. R. Dukelow, Decapacitation factor and capacitation. Adv. Biosciences 4, 61 (1969).

[22] Vaidya, R. A., J. M. Bedford, R. H. Glass and J. M. Morris, Evaluation of the removal of tetracycline fluorescence from spermatozoa as a test for capacitation in the rabbit. J. Reprod. Fert. 19, 483 (1969).

[23] Laskowski, M., Jr., Personal communications (1970).

[24] Robertson, R. T., V. K. Bhalla and W. L. Williams, Unpublished results.

[25] Austin, C. R., Entry of spermatozoa into the Fallopian tube mucosa. Nature 183, 908 (1959).

[26] Schumacher, G. F. B. and M. Y. Pearl, Alpha$_1$-antitrypsin in cervical mucus. Fert. Steril. 19, 91 (1968).

[27] HAENDLE, H., H. INGRISCH and E. WERLE, Über einen neuen Trypsin-Chymotrypsin-Inhibitor im Cervixsekret der Frau. Hoppe-Seyler's Zeitschr. Physiol. Chem. **351**, 545 (1970).

[28] SYNER, F. N. and K. S. MOGHISSI, Properties of proteolytic enzymes isolated from human seminal plasma and spermatozoa. Abstract presented at the Second Annual Meeting of the Soc. Study Reprod., Davis, Cal. 16 (1969).

[29] STAMBAUGH, R., B.G. BRACKETT and L.MASTROIANNI, Inhibition of in vitro fertilization of rabbit ova by trypsin inhibitors. Biol. Reprod. **1**, 223 (1969).

[30] SCHUMACHER, G. F. B., Personal communications (1970); see this volume, p. 247.

[31] ZANEVELD, L. J. D., R. T. ROBERTSON and W. L. WILLIAMS, Synthetic enzyme inhibitors as antifertility agents. FEBS Letters. In press.

[32] ROBERTSON, R. T., L. J. D. ZANEVELD and W. L. WILLIAMS, Antifertility effect of enzyme active site-directed reagents. Submitted to Biol. Reprod.

[33] SHAW, E., Selective chemical modification of proteins. Physiol. Reviews **50**, 244 (1970).

[34] CHANG, M. C., Effect of commercial contraceptive jellies placed in the vagina before mating. Fert. Steril. **11**, 109 (1960).

Discussion Remarks: **Inhibition by the Trypsin-Plasmin-Inhibitor from Sperm Plasma of the Dispersion of the Corona Radiata and Zona Pellucida by a Trypsin-like Enzyme from Spermatozoa**

H. INGRISCH, H. HAENDLE und E. WERLE, *Munich*

In 1965 a trypsin-plasmin-inhibitor was detected by H. HAENDLE and coworkers in sperm plasma [1]. This inhibitor later on was isolated and studied in detail in our laboratory [2]. Trypsin inhibitors are found in the vesicular glands and semen of many mammals, including man [2a]. The inhibitory activities found in the seminal vesicles of various species were as high as 5000 mIU per gram of fresh tissue. On the concentration of these inhibitors in sperm of healthy man and of those with andrological deseases and its relationship to other parameters of sperm see H. INGRISCH, H. HAENDLE and E. WERLE, 1970 [3].

Seminal vesicles of guinea pigs contain at least 2 different polypeptide inhibitors, a trypsin-specific inhibitor and an inhibitor for trypsin and plasmin. The bacterial proteinase from Streptomyces griseus is also inhibited, whereas chymotrypsin and the kallikreins are not [1].

In addition to these inhibitors, the seminal vesicles of the bull contain the "basic bovine inhibitor" too. This is paralleled by the findings in bovine pancreas, which contains the basic pancreatic trypsin inhibitor and the secretory trypsin-specific inhibitor. The basic bovine inhibitor can be readily separated from the other inhibitors on Sephadex G-50, because only the bovine inhibitor forms a complex with chymotrypsin.

The occurence of the seminal plasma inhibitors is hormone controlled. The synthesis of the inhibitors in mice vesicular glands can be stimulated by the administration of testosterone. On the other hand, castration leads to a significant decrease in the inhibitor level, while subsequent administration of testosterone leads to a rapid rise to above the normal value [1].

We suggested that the physiological function of these inhibitors could be the inhibition of a trypsin-like enzyme located in the acrosomes of spermatozoa, described by STAMBAUGH and BUCKLEY, 1968 [4] and by ZANEVELD and WILLIAMS, 1969 [5].

To study this question we isolated ova of rabbits and incubated them with an aqueous extract of washed spermatozoa of rabbits a) in absence and b) in presence of the purified sperm plasma trypsin inhibitor from guinea pigs*. In case a) the corona radiata and zona pellucida were destroyed, in case b) preserved (see Fig. 1 and 2 in ref. [6]).

Our observations are in agreement with those of ZANEVELD and WILLIAMS [7] who found that acrosomal extracts, obtained from ejaculated rabbit and human sperm, contain enzymes which remove the zona radiata and the zona pellucida. In contrast to the sperm plasma inhibitor the soybean trypsin inhibitor prevented according to ZANEVELD and WILLIAMS [7] the removal of the zona but not of the corona by this preparation.

---

* For experimental details see ref. [6].

References

[1] HAENDLE, H., H. FRITZ, I. TRAUTSCHOLD and E. WERLE, Hoppe Seylers Z. Physiol. Chem. **343**, 185 (1965).
[2] FRITZ, H., E. FINK, R. MEISTER and G. KLEIN, Hoppe Seylers Z. Physiol. Chem. **351**, 1344 (1970). FINK, E., in J. FREI and M. JEMELIN, Clinical Enzymology **2**, 74 (1970), Karger, Basel/München/Paris/New York 1970; FINK, E., G. KLEIN, F. HAMMER, G. MÜLLER-BARDOFF and H. FRITZ, this volume, p. 225; [2a] HAENDLE, H., Dissertation, I. Medizinische Fakultät der Universität München (1969).
[3] INGRISCH, H., H. HAENDLE and E. WERLE, Andrologie **2**, 103 (1970).
[4] STAMBAUGH, R. and J. BUCLEY, Science **161**, 585 (1968).
[5] ZANEVELD, L. J. D., P. N. SRIVASTAVA and W. L. WILLIAMS, J. Reprod. Fert. **20**, 337 (1969).
[6] HAENDLE, H., H. INGRISCH and E. WERLE, Klin. Wschr. **48**, 824 (1970).
[7] ZANEVELD, L. J. D. and W. L. WILLIAMS, Biol. Reprod. **2**, 363 (1970); ZANEVELD, L. J. D., K. L. POLAKOSKI, R. ROBERTSON and W. L. WILLIAMS: this volume, p. 236.

Discussion Remarks: α_1-Antitrypsin in Seminal Fluid

GEBHARD F. B. SCHUMACHER,
Chicago, Illinois

In order to complete the picture of protease inhibiting components of seminal fluid I would like to mention that human seminal fluid does contain also aplha$_1$-antitrypsin. We had a chance to investigate a number of specimens of different quality using the radial immuno-

Table 1. Alpha$_1$-antitrypsin values in seminal fluid specimens of different quality. The values were placed in the next lower category when only one morphological parameter indicated a lower quality. (From G. F. B. SCHUMACHER, J. Reprod. Med. **5**, 3 (1970) [2].)

Good Count: > 60 mill/ml Motil: > 60% Morph: >80% Oval Forms		Fair Count: 20—60 mill/ml Motil: 40—60% Morph: 60—80% Oval Forms		Poor Count: < 20 mill/ml Motil: < 40% Morph: <60% Oval Forms		Pathologic Specimens Azoospermia and/or White Blood Cells and/or Bacteria Present	
Spec. #	α_1-Antitryp. μg/ml	Spec. #	α_1-Antitryp. μg/ml	Spec. #	α_1-Antitryp. μg/ml	Spec. #	α_1-Antitryp. μg/ml
159	79	160	52	164	27	154	53
160 B	105	176	105	174	52	299	160
265	79	159	79	175	52	421	75
366	100	251	78	213	79	399	180
369 B	70	379 A	50	179	64	412	94
380	90	432	80	214	110	433	72
402	70	439	55	401 A	100	431	78
414	80	442	78	423	80	430	75
426	72	448	140	444	120	440	59
428	94	449	90	457	90	443	84
437	120	—	—	—	—	450	140
446	78	—	—	—	—	451	90
Mean	86	Mean	81	Mean	77	Mean	97
SD	16	SD	27	SD	29	SD	41
SE	6	SE	9	SE	9	SE	12

diffusion technique [1]. Values between 50 $\mu g/ml$ and 180 µg/ml were obtained. These are indeed low concentrations if compared with alpha$_1$-antitrypsin levels of 2—5 mg/ml in human serum. The following table shows a total of 44 alpha$_1$-antitrypsin values in seminal fluid specimens of different quality (indicated in the table).

The different groups show slightly decreasing values from the "good" to the "poor" category, a finding which is similar to the pattern of the low molecular weight protease inhibitors in human semen, which has been elaborated by WERLE, HAENDLE and INGRISCH [3].

The alpha$_1$-antitrypsin values in human seminal fluid are slightly higher if compared with the concentration in human cervical mucus at time of ovulation but considerably lower than in cervical mucus specimen during the luteal phase or after administration of gestagen when cervical mucus is hostile against sperms.

The alpha$_1$-antitrypsin concentration in seminal fluid represents only approximately 20% of the total trypsin inhibitor capacity of seminal fluid, calculated on the basis of the inhibitor values observed by WERLE, HAENDLE and INGRISCH [3]. It has not yet been clarified whether alpha$_1$-antitrypsin reacts with acrosomal proteases or seminal fluid proteases.

References

[1] MANCINI, G., A. O. CARBONARA and J. F. HEREMANS, Immunochemical quantitation of antigens by single radial immunodiffusion. Immunochemistry **2**, 235 (1965).

[2] SCHUMACHER, G. F. B., Alpha$_1$-antitrypsin in genital secretions.
a) in E. T. TYLER, Progress in Conception Control, J. B. Lippincott Co., Philadelphia and Toronto, pp. 13 (1969).
b) J. Reprod. Med. **5**, 3 (1970).

[3] WERLE, E., H. HAENDLE, und INGRISCH, Über die Konzentration des Trypsin-Inhibitors im Sperma von Gesunden und andrologischen Kranken und über ihre Beziehung zu anderen Parametern des Spermas. Andrologie **2**, 103 (1970).

Fertility Experiments in Mice and Rabbits with the Trypsin-Kallikrein Inhibitor from Bovine Lung[1]

GEBHARD F. B. SCHUMACHER, JOSEPH R. SWARTWOUT and FREDERIK P. ZUSPAN
with technical assistance of BRUNO DRAGOJE and WILLIAM JEMISON

*Department of Obstetrics and Gynecology, The University of Chicago, Pritzker School of Medicine
The Chicago Lying-in Hospital, Chicago, Illinois*

Introduction

Recent observations of STAMBAUGH et al. [9, 10, 11] and ZANEVELD et al. [15—19] show that proteases in the sperm acrosome are essential for the fertilization processes. These enzymes as well as seminal fluid proteases may also be involved in the penetration of sperms into the cervical mucus [6, 7, 8]. It was, therefore, of interest to study, whether or not a polyvalent proteinase inhibitor such as the trypsin-kallikrein inhibitor described by WERLE et al [12]* would interfere with reproductive processes. This substance inhibits also chymotrypsin, plasmin and cathepsin D [2, 4, 12, 14]. Its toxicity and antigenicity is extremely low and the pharmacological properties are known [1, 4, 12, 14]. The substance is being used clinically in acute pancreatitis and other conditions such as activation of the fibrinolytic system, where the inhibition of proteolytic activity is of importance [4, 12, 13, 14]. Since the trypsin-kallikrein inhibitor excretion is relatively fast [12, 14], the animal experiments were designed in a way that supposedly high blood levels were maintained over the period of mating or before and after the artificial insemination and the subsequent days. The amount per kg per day injected was calculated on the basis of 5—20 times the amount, which had been successfully used in clinical applications [4]. One mg inhibitor substance equals 5,000 kallikrein inhibitor units.

1. Experiments with Mice

a) *Treatment Schedule.* Ten to twelve weeks old mice were adapted to a reversed nocturnal cycle for a week and kept at a temperature of 24° C in the animal quarters.
During the experiment two females were kept together with one male during the day in the dark animal room. The males were separated from the females in the evening, since injections could not be continued during the night (light). Figure 1 shows the treatment schedule for mice which were injected with different doses of the trypsin-kallikrein inhibitor three times a day intraperitoneally and once in the evening subcutaneously (depot). Two groups treated with different doses of the inhibitor dissolved in saline, one group treated with the same volume of saline and one group without treatment were

---

* We would like to express our gratitude and appreciation to Prof. Dr. E. AUHAGEN and Dr. E. TRUSCHEIT, Research Laboratories, Farbenfabriken Bayer AG in Wuppertal-Elberfeld, W.-Germany, for making sufficient amounts of the purified inhibitor from bovine lung available for these experiments. "TRASYLOL" is the registered trademark of this inhibitor.

[1] Supported by grant 690—0108 from the Ford Foundation, New York, N. Y.

Treatment schedule

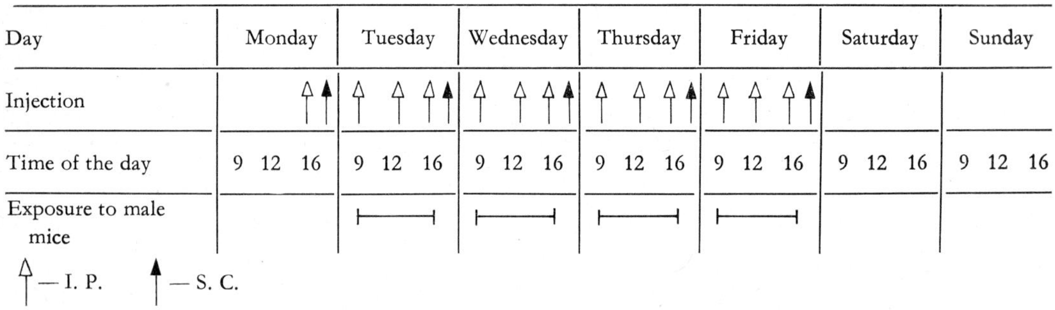

Day	Monday	Tuesday	Wednesday	Thursday	Friday	Saturday	Sunday
Injection	⇧⇧	⇧ ⇧⇧⇧	⇧ ⇧⇧⇧	⇧ ⇧⇧⇧	⇧ ⇧⇧⇧		
Time of the day	9 12 16	9 12 16	9 12 16	9 12 16	9 12 16	9 12 16	9 12 16
Exposure to male mice		⊢――⊣	⊢――⊣	⊢――⊣	⊢――⊣		

⇧ — I. P.   ↑ — S. C.

Fig. 1. Treatment schedule for mice with trypsin-kallikrein inhibitor
open arrows = intraperitoneal injections
solid arrows = subcutaneous injections
⊢――⊣ = indicates time of exposure to male during the day.

Fertility in Mice
under Treatment with Trypsin-Kallikrein Inhibitor

	Females Treated with T—K Inhibitor							
	A		B		C		D	
	# Pregnant	# Mice	# Pregnant	# Mice	# Pregnant	# Mice	# Pregnant	# Mice
Dosage	12 mg/kg/day		6 mg/kg/day		saline control		untreated control	
Exp. #3	1	10	2	10	3	11	1	10
Dosage	20 mg/kg/day		10 mg/kg/day		saline control		untreated control	
Exp. #5	6	10	7	10	7	10	5	10
	Males Treated with T—K Inhibitor							
Dosage	—		10 mg/kg/day		saline control		untreated control	
Exp. #4	—		5	10	5	10	5	12

Fig. 2. Results of fertility studies in mice treated with trypsin-kallikrein inhibitor.

observed in the experiments. The females were sacrificed 16 days later and the uteri removed for evaluation and recording of the experiment. In one experiment male mice were treated according to the same schedule (Fig. 1) and bred with untreated females under the same conditions.

b) Results. The results of three different experiments on four different dose levels are demonstrated in Figure 2. The table is self-explanatory. The mice treated with trypsin-kallikrein inhibitor do not show any significant differences from the saline control or from the untreated control.

Treatment schedule

Date: 1/12/70   Exp. No. 14

Total mg per kg/week	Total mg per kg/day		T—K Inhibitor			T—K Inhibitor in 0.9% NaCl mg/kg	Saline Control	
			Time of the day	Apli-cation	Volume ml/kg		Apli-cation	Volume ml/kg
43	3	Monday	16	S. C.	1.5	3	S. C.	1.5
		Tuesday	9	I. V. / I. P.	2.0	4	I. V. / I. P.	2.0
			10	Insemination + chorionic Gonad. 75 U			Insemination + chor. Gonad. 75 U	
			12	S. C.	1.0	2	S. C.	1.0
			14	I. P.	0.5	1	I. P.	0.5
	10		16	S. C.	1.5	3	S. C.	1.5
			9	I. V. / I. P.	2.0	4	I. V. / I. P.	2.0
		Wednesday	12	S. C.	1.0	2	S. C.	1.0
			14	I. P.	0.5	1	I. P.	0.5
	10		16	S. C.	1.5	3	S. C.	1.5
			9	I. V. / I. P.	2.0	4	I. V. / I. P.	2.0
		Thursday	12	S. C.	1.0	2	S. C.	1.0
			14	I. P.	0.5	1	I. P.	0.5
	10		16	S. C.	1.5	3	S. C.	1.5
			9	I. V. / I. P.	2.0	4	I. V. / I. P.	2.0
		Friday	12	S. C.	1.0	2	S. C.	1.0
			14	I. P.	0.5	1	I. P.	0.5
	10		16	S. C.	1.5	3	S. C.	1.5

Fig. 3. Treatment schedule for rabbits with trypsin-kallikrein inhibitor.

2. Experiments with Rabbits

a) Artificial Insemination. After several unsatisfactory results in using natural breeding methods, artificial insemination technique was successfully applied. This technique is also more convenient for this type of experiment, since weekly schedules between Monday and Friday can be followed more easily by the laboratory personnel. Mature New Zealand white rabbits of proven fertility have been used. They were kept at a room temperature of 21° C in the animal quarters. Semen was collected from bucks with the help of an artificial vagina and sperm count and motility was determined. The semen was 10 to 15 times diluted with saline solution for insemination. One to two ml of this suspension was deposited deep in the vagina before the vaginal portion of the uterus with the help of an angled pipette. 20—40 million spermatozoa were inseminated. The rabbits received 75 units of chorionic gonadotrophin intravenously one to two hours prior to the insemination for induction of ovulation.

b) *Treatment Schedule*

1. Systemic Treatment. A detailed description of the treatment schedule is demonstrated in Figure 3. The inhibitor was dissolved in saline solution. The experiment was started on Monday afternoon injecting 3 mg/kg of the trypsin-kallikrein inhibitor subcutaneously. 4 mg/kg were injected the next morning half intravenously and half intraperitoneally one hour prior to insemination. Subsequent injections were given subcutaneously and intraperitoneally. The last injection of 4 mg/kg was given as a subcutaneous depot. The same pattern of injections was applied during the 3 subsequent days. The control animals were simultaneously injected with the same volume of saline solution.

2. Treatment of the Ejaculate. Ejaculated rabbit semen was divided into two portions. One portion was incubated at room temperature for 50—60 minutes with the 10-fold volume of inhibitor in saline solution, the other portion was treated in the same way with saline solution as a control. Both dilutions were tested for sperm mobility and used for insemination.

c) *Results*

1. Systemic Treatment. The results of three experiments involving a total number of 16 does are shown in Figure 4. Experiment 9 was conducted under the condition of natural mating. Artificial insemination was performed in the other experiments. The pregnancy rates in the treated group and the control groups do not show a significant difference. The summary indicates 6 pregnancies in 8 does of the saline treated control group and 7 pregnancies within the group of 8 rabbits treated with the trypsin-kallikrein inhibitor. The litter size appears to be almost the same in both groups. No abnormalities and malformations could be detected.

2. Treatment of the Ejaculate. The results of 4 experiments involving a total number of 24 does are shown in Figure 5. The pregnancy rates in the groups inseminated with sperms treated with trypsin-kallikrein inhibitor and soybean trypsin inhibitor on two different dose levels are not significantly different from those in the group inseminated with sperms treated with saline solution as a control. The summarized results indicate 9 pregnancies in the group of 12 does inseminated with inhibitor treated

Fertility in Rabbits

	Group I T—K Inhibitor			Group II Saline control		
	# Pregnant	# Does	Litter size	# Pregnant	# Does	Litter size
Mating exp. 9	2	2	8, 9	2	2	10, 10
Insemination exp. 14	2	3	5, 8, 0	3	3	6, 8, 8
Insemination exp. 15	3	3	7, 1, 7	1	3	0, 2, 0
Total	7	8	45	6	8	44

Fig. 4. Results of fertility studies in rabbits, treated with trypsin-kallikrein inhibitor by combined intravenous, intraperitoneal and subcutaneous injections.

Fertility in Rabbits

	Group I Inhibitor			Group II Saline control		
	# Pregnant	# Does	Litter size	# Pregnant	# Does	Litter size
Exp. # 11 Trypsin-Kallikrein Inhibitor 4 mg/ml	3	3	4, 1, 5	3	3	4, 8, 9
Exp. # 16 Soybean Trypsin Inhibitor 9 mg/ml	2	3	4, 3	2	3	4, 7
Exp. # 17 Trypsin-Kallikrein Inhibitor 10 mg/ml	2	3	4, 5	0	3	...
Exp. # 18 Soybean Trypsin Inhibitor 22.5 mg/ml	2	3	5, 3	2	3	8, 6

Fig. 5. Results of fertility studies in rabbits after treatment of ejaculate with trypsin inhibitors prior to insemination

sperms and 7 pregnancies in the group of 12 does inseminated with sperms treated with saline as a control.

Discussion

Under the conditions of the described experiments with mice and rabbits, no interference of the trypsin-kallikrein inhibitor after systemic application in the females was observed. Equally, neither the systemic treatment of male mice nor the treatment of rabbits' semen with trypsin-kallikrein inhibitor or soybean trypsin inhibitor prior to artificial insemination in rabbits interfered with fertility.

Several reasons for this negative result may be discussed.

There is no direct evidence yet* that this trypsin-kallikrein inhibitor preparation reacts with acrosomal or seminal fluid proteases. However, the result of recent experiments of ZANEVELD [15, 16] indicate that a low molecular weight bovine pancreas trypsin inhibitor as well as soybean trypsin inhibitor (STAMBAUGH [9, 10, 11]) does inhibit acrosin (trypsin-like enzyme) from rabbit sperm acrosomes. Low molecular weight trypsin inhibitors have also been found in seminal fluid of several species, including rabbits [3, 12, 14, 18]. Acrosomal extract of ejaculated rabbit sperms contain in contrast to epididymal or capacitated sperms an acrosin-inhibitor complex, which seems to be dissociated during the process of capacitation in the uterus according to ZANEVELD [15, 16].

These findings offer an explanation for the negative outcome of the previously conducted experiments, where the incubation of rabbit semen with high concentrated trypsin-kallikrein inhibitor and soybean inhibitor prior to vaginal insemination did not result in fertilization inhibition. The inhibitors are probably being removed during the capacitation process. Only the application of pancreatic trypsin inhibitor or seminal fluid inhibitor after capacitation and the insemination** of the inhibitor treated capacitated sperms results apparently in fertilization inhibition. The fact that the trypsin-kallikrein inhibitor does not interfere with fertility after combined intraperitoneal, intravenous and subcutaneous injection of high doses may have its reason in an insufficient level in tubal secretions, the environment where fertilization occurs. Interference with fertilization should be expected, therefore, if the level of inhibitor in tubal secretions could be increased.

It is of interest to note in this connection that a synthetic inhibitor of trypsin and acrosin,

---

* see addendum
** into the oviduct

tosyl-lysine-chloromethyl-ketone (TLCK), is apparently not removed from the sperm acrosome during capacitation. Moreover, it seems to be effective on ejaculated sperms, which have been already exposed to the seminal fluid inhibitors according to ZANEVELD's experiments [15]. This mechanism is not well understood. The question is whether the TLCK has a higher affinity to the acrosomal enzymes or whether it inhibits the capacitation process in utero. However, the biochemistry of proteinase inhibitors and the biology of reproduction show fascinating new aspects after the recent developments which will stimulate further research in both areas.

Addendum. Recent experiments in our laboratory revealed a definite inhibitory activity of the trypsin-kallikrein inhibitor from bovine lung towards acrosomal protease prepared from rabbit epididymal spermatozoa. BAEE and gelatin has been used as substrate. The inhibition is of the immediate type.

References

[1] ANDERER, F. A. and S. HOERNLE, Chemical studies on kallikrein inactivator from bovine lung and parotid gland. Ann. N. Y. Acad. Sci. **146**, Art. 2, 381 (1968).

[2] BARNHART, M. I., C. QUINTANA, H. L. LENON and G. B. BLUHM, Proteases in inflammation. Ann. N. Y. Acad. Sci **146**, Art. 2, 527 (1968).

[3] FINK, E., G. KLEIN, F. HAMMER, G. MÜLLER-BARDORFF and H. FRITZ, Protein Proteinase Inhibitors in Male Sex Glands, this volume, p. 225.

[4] GROSS, R. und G. KRONEBERG, (eds.) Neue Aspekte der Trasylol Therapie. F. K. Schattauer Verlag, Stuttgart (1966).

[5] LASKOWSKI, M., Sr., The mechanism of formation of a trypsin — trypsin inhibitor complex. Ann. N. Y. Acad. Sci. **146**, Art. 2, 374 (1968).

[6] MOGHISSI, K. S., Sperm migration in the human genital tract. J. Reprod. Med. **3**, 156 (1969).

[7] MOGHISSI, K. S. and F. N. SYNER, The effect of seminal protease on sperm migration through cervical mucus. Int. J. Fertil. **15**, 43 (1970).

[8] SCHUMACHER, G. F. B., Alpha$_1$-antitrypsin in genital secretions.
  a) In E. TYLER (ed.), Progress in Conception Control, J. B. Lippincott Co., Philadelphia & Toronto, p. 13 (1969).
  b) J. Reprod. Med. **5**, 3 (1970).

[9] STAMBAUGH, R., B. G. BRACKETT and L. MASTROIANNI, Inhibition of in vitro fertilization of rabbit ova by trypsin inhibitors. Biol. Reprod. **1**, 223 (1969).

[10] STAMBAUGH, R. and J. BUCKLEY, Zona pellucida dissolution enzymes of the rabbit sperm head. Science **161**, 585 (1968).

[11] STAMBAUGH, R. and J. BUCKLEY, Identification and subcellular localization of the enzymes effecting penetration of the zona pellucida by rabbit spermatozoa. J. Reprod. Fert. **19**, 423 (1969).

[12] VOGEL, R., I. TRAUTSCHOLD and E. WERLE, Natural proteinase inhibitors. Academic Press, New York and London (1968).

[13] WERLE, E., Proteinase inhibitors in clinical studies, this volume, p. 23.

[14] WERLE, E., I. TRAUTSCHOLD, H. HAENDLE and H. FRITZ, Physiologic, pharmacologic and clinical aspects of proteinase inhibitors. Ann. N. Y. Acad. Sci. **146**, Art. 2, 464 (1968).

[15] ZANEVELD, L. J. D., K. L. POLAKOSKI, R. T. ROBERTSON and W. L. WILLIAMS, Trypsin inhibitors and fertilization, this volume, p. 236.

[16] ZANEVELD, L. J. D., R. T. ROBERTSON, M. KESSLER and W. L. WILLIAMS, Inhibition of fertilization in vivo by pancreatic and seminal plasma trypsin inhibitors.
  a) Fed. Proc. **29**, 644, Abstr. 2243 (1970),
  b) J. Reprod. Fert. **25**, 387 (1971).

[17] ZANEVELD, L. J. D., P. N. SRIVASTAVA and W. L. WILLIAMS, Relationship of a trypsin-like enzyme in rabbit spermatozoa to capacitation. J. Reprod. Fert. **20**, 337 (1969).

[18] ZANEVELD, L. J. D., P. N. SRIVASTAVA and W. L. WILLIAMS, Inhibition by seminal plasma of acrosomal enzymes in intact sperm. Proc. Soc. Exp. Biol. Med. **133**, 1172 (1970).

[19] ZANEVELD, L. J. D. and W. L. WILLIAMS, A sperm enzyme that disperses the corona radiata and its inhibition by decapacitation factor. Biol. Reprod. **2**, 363 (1970).

Alpha$_1$-Antitrypsin in Uterine Secretions*

GEBHARD F. B. SCHUMACHER

Department of Obstetrics and Gynecology, The University of Chicago, Pritzker School of Medicine, The Chicago Lying-in Hospital, Chicago, Illinois

The mucoid material in the cervical canal of the human uterus represents a mechanical and biochemical barrier against intruding organisms. Even spermatozoa negotiate the cervical mucus successfully only under certain conditions which are determined by sex hormones. Preovulatory mucus under the influence of estrogens is penetrable for sperms, whereas the cervical mucus of the postmenstrual and luteal phase (after ovulation) represents a barrier for sperms. This can be imitated by the sequential application of estrogenic and progestagenic steroid hormones (sequential hormonal contraceptives). It is not clear, whether the structure of the mucus gel, or the biochemical composition of the fluid phase are determining factors for this phenomenon. It is conceivable that both factors are involved.

Cervical mucus can be depolymerized by proteolytic enzymes [3, 4]. Sperm acrosomes as well as seminal fluid contain this type of enzymes [3, 4, 13]. The presence of inhibitors of proteolytic enzymes in cervical mucus may be an essential factor for the penetrability of the mucus for sperms.

A low molecular weight proteinase inhibitor has been recently described by HAENDLE, INGRISCH and WERLE [1]. However, no information is available at present on cyclical changes of this inhibitor. Alpha$_1$-antitrypsin has been found in human cervical mucus dependent on the hormonal situation [5, 6, 8, 9, 12]. Radial immunodiffusion technique has been applied for quantitative determination [2, 7]. Figure 1 shows the course of an ovulatory cycle, where the alpha$_1$-antitrypsin levels are at a minimum prior to ovulation, the time, where sperm can penetrate the mucus easily. Twelve ovulatory cycles have been investigated thus far. Figure 2 shows the synopsis of clinical observations and the course of alpha$_1$-antitrypsin in cervical secretions. The values of alpha$_1$-antitrypsin are extremely low one to two days prior to ovulation. They increase immediately after ovulation significantly. The sequential application of hormonal contraceptives (estrogen from day 5 to day 19, and estrogen plus progestagens from day 20 to day 24) produces essentially the same pattern [5, 6, 8, 9].

Very little is known about the mechanism of secretion in the human uterus. It seems, however, interesting to note that most of the alpha$_1$-antitrypsin is being secreted in the lower part of the cervical canal. Figure 3 shows the distribution of diffusible alpha$_1$-antitrypsin in the tissue of the inner surface of the uterus. Small tissue cylinders from different areas of the uterus had been subjected to radial immunodiffusion in a similar way which had been previously applied for investigations on lysozyme [10, 11].

---

* supported by USPH Grant $\neq$ HD 2682 and HD 03696.

There seems no doubt, that the secretory tissue of the lower part of the cervical canal contains significantly more alpha$_1$-antitrypsin if compared with the endometrium of the uterine cavity. It may be, therefore, secreted mainly in that area. It is not known at present, whether or not the

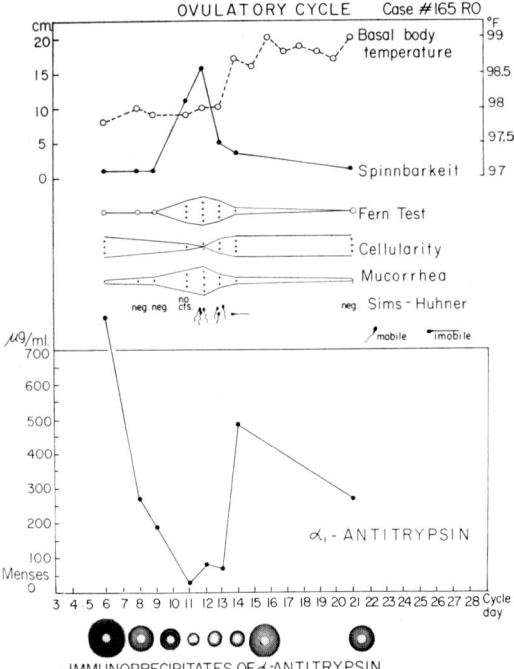

Fig. 1. Alpha$_1$-antitrypsin values in cervical mucus during a menstrual cycle. The 13th day as the last day of low basal body temperature is probably the day of ovulation. The mucus properties such as spinnbarkeit, positive ferning (crystallization phenomenon) and a low cell content indicate high estrogenic activity between the 11th and 13th day. The results of the Sims-Huhner test show motile sperms in the cervical mucus only during the immediate preovulatory period. The minimum of alpha$_1$-antitrypsin values coincides with this period. (From G. F. B. SCHUMACHER and M. J. PEARL, Fertil. Steril. **19**, 91 (1968) [9]).

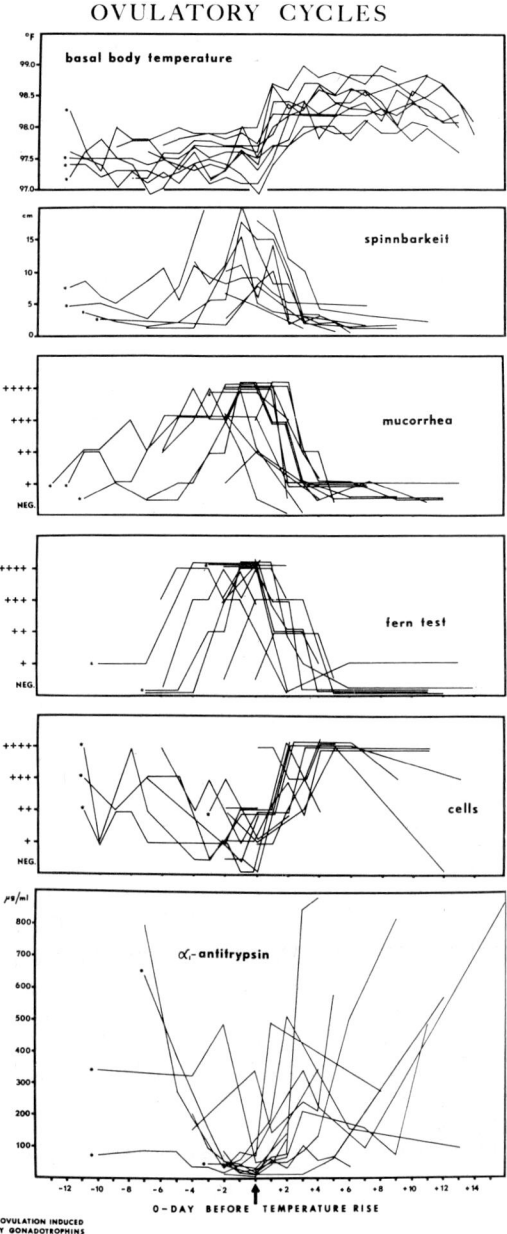

Fig. 2. Alpha$_1$-antitrypsin in cervical mucus observed during 12 menstrual cycles. Changes in basal body temperature, spinnbarkeit, mucorrhea, fern test and cell content show the typical pattern for ovulatory cycles. The last day of low basal body temperature has been determined as day zero (ovulation day) in order to relate the other parameters as closely as possible to ovulation. The minimum of alpha$_1$-antitrypsin values correlates with high spinnbarkeit, mucorrhea, positive ferning and low cell content prior to ovulation. These changes have to be considered as estrogen effects followed by the counteracting activity of progesterone.

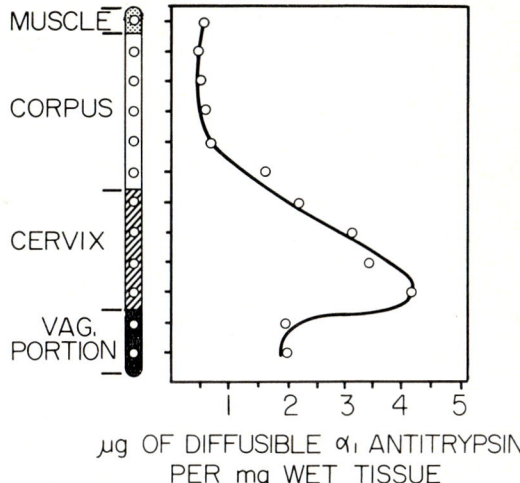

Distribution of α_1-Antitrypsin on the inner Surface of the Uterus (Posterior Wall)

μg OF DIFFUSIBLE α_1 ANTITRYPSIN PER mg WET TISSUE

Patient J. P., 32 Years, Adhes. after 3 Ces. Scts. Hysterectomy, 26TH Day of Enovid Cycle

Fig. 3. Diffusible alpha$_1$-antitrypsin per mg wet weight of small tissue cylinders from the inner surface of the uterus and the portio vaginalis (method: radial immunodiffusion, see SCHUMACHER [7, 10, 11]). The endometrium shows little diffusible alpha$_1$-antitrypsin, whereas the glandular tissue in the cervical canal releases approximately 10 times the amount of alpha$_1$-antitrypsin. The values in the lower part of the cervical canal are in the range of 3—4 mg per gram which is comparable to serum levels. The concentration in the mucus itself is considerably lower. This inhibitor as well as other proteins may be accumulated in this area of mucus producing glands.

presence of alpha$_1$-antitrypsin in cervical secretions is of biological significance for reproductive processes. Further studies on sperm-mucus interaction and its biochemical mechanisms may elucidate the role of proteinase inhibitor in genital secretions.

Addendum. Recent experiments in our laboratory revealed a definite inhibitory activity of human alpha$_1$-antitrypsin towards acrosomal protease prepared from rabbit epididymal spermatozoa. BAEE and gelatin has been used as substrate. The inhibition is of the progressive type, similar to the plasmin inhibition described by HEIMBURGER and to the kallikrein inhibition described by FRITZ.

References

[1] HAENDLE, H., H. INGRISCH und E. WERLE, Über einen neuen Trypsin-Chymotrypsin-Inhibitor im Cervixsekret der Frau. Hoppe-Seyler's Z. Physiol. Chem. **351**, 545 (1970).

[2] MANCINI, G., A. O. CARBONARA and J. F. HEREMANS, Immunochemical quantitation of antigens by single radial immunodiffusion. Immunochemistry **2**, 235 (1965).

[3] MOGHISSI, K. S., Sperm migration in the human female genital tract. J. Reprod. Med. **3**, 156 (1969).

[4] MOGHISSI, K. S. and F. N. SYNER, The effect of seminal protease on sperm migration through cervical mucus. Int. J. Fertil. **15**, 43 (1970).

[5] SCHUMACHER, G. F. B., Alpha$_1$-antitrypsin in genital secretions.

a) J. Reprod. Med. **5**, 3 (1970).

b) In E. TYLER (ed.), Progress in Conception Control, J. B. Lippincott Co., Philadelphia and Toronto, pp. 13—21 (1969).

[6] SCHUMACHER, G. F. B., Biochemical and biophysical properties of cervical mucus in different hormonal states. In G. RASPÉ (ed.), Advances in Biosciences 4, Schering Symposium "Mechanism Involved in Conception" Berlin, 1969, Pergamon Press Ltd. and Vieweg, Oxford, New York and Braunschweig, pp. 95—119 (1969).

[7] SCHUMACHER, G. F. B., Protein analysis of secretions of the female genital tract. J. Reprod. Med. (Lying-in) **1**, 61 (1968).

[8] SCHUMACHER, G. F. B., Biochemistry of cervical mucus. Fertil. Steril. **21**, 697 (1970).

[9] SCHUMACHER, G. F. B. and M. J. PEARL, Alpha$_1$-antitrypsin in cervical mucus. Fertil. Steril **19**, 91 (1968).

[10] SCHUMACHER, G. F. B. and M. J. PEARL, Cyclic changes of muramidase (lysozyme) in cervical mucus. J. Reprod. Med. **3**, 171 (1969).

[11] SCHUMACHER, G. F. B. and M. J. PEARL, Muramidase (lysozyme) in cervical secretion, in Protides of Biological Fluids. Pergamon Press Ltd., Oxford, **16**, pp. 525—534 (1969).

[12] SCHUMACHER, G. F. B., E. K. STRAUSS and G. L. WIED, Serumproteins in cervical mucus. Amer. J. Obstet. Gynec. **91**, 1035 (1965).

[13] ZANEVELD, L. J. D., K. L. POLAKOSKI, R. T. ROBERTSON and W. L. WILLIAMS, Trypsin inhibitors and fertilization, this volume, p. 236.

Discussion Remarks: A Trypsin-Chymotrypsin Inhibitor in Human Female Cervix Secretion

H. HAENDLE, H. INGRISCH and E. WERLE,

Munich

Besides α_1-antitrypsin [1, 2], a proteinase inhibitor of higher molecular weight, the cervix secretion of women contains also a trypsin-chymotrypsin inhibitor of lower molecular weight [3]. The average inhibitory activity of the secretions of 11 women was 137 ± 34 mIU (trypsin inhibition, substrate: BAPA) per g secret. By acidification with perchloric acid high molecular material is precipitated whereas the latter inhibitor remains in solution. The inhibitor is heat stable (100° C, pH 7) and dialysable. The inhibition of trypsin is instantaneous and permanent. The molecular weight of the inhibitor determined by gel filtration is very low (about 2200). The inhibitor is inactive against plasmin, thrombin and pancreatic kallikrein. Experimental details are given in reference [3]. The inhibitor is not yet studied in more detail. Its physiological function is unknown.

References

[1] SCHUMACHER, G. F. B. and M. J. PEARL, Fertility and Sterility **19**, 91 (1968).

[2] SCHUMACHER, G. F. B., this volume, p. 253.

[3] HAENDLE, H., H. INGRISCH and E. WERLE, Z. Physiol. Chem. **351**, 545 (1970).

Proteinase Inhibitors from Dog Submandibular Glands — Isolation, Amino Acid Composition, Inhibition Spectrum

Hans Fritz, Eugen Jaumann, Renate Meister, Peter Pasquay, Karl Hochstrasser and Edwin Fink

Institut für Klinische Chemie und Klinische Biochemie der Universität München, D—8 München 15, Germany

Summary

Proteinase inhibitors were isolated from aqueous and acidic extracts of dog submandibular glands employing gradient elution and equilibrium chromatography on CM-cellulose. Mainly four inhibitors were obtained which differ only slightly in amino acid composition (cf. Table 2 and 3). Molecular weights of 12750 up to 12878 were calculated for the different components.

The following enzymes are strongly inhibited: bovine trypsin and α-chymotrypsin, subtilisin Novo, porcine pancreatic elastase, Aspergillus Oryzae protease, plasmin (pig), and part of the proteolytic activity of pronase. No inhibition of collagenase was found. The dog submandibular inhibitor (DSI) is double-headed; the two different reactive sites do not overlap: The DSI-chymotrypsin and the DSI-subtilisin complex — both complexes contain equimolar amounts of enzyme and inhibitor — are able to bind an additional trypsin molecule so that ternary complexes are formed.

The potential physiological function as well as the significance of DSI for medical problems are discussed.

Introduction

The physiological role of pancreatic trypsin inhibitors, described in some foregoing papers [1—4], seems to be obvious: Inhibition of premature activation of trypsinogen with all its severe consequences in the gland. The high concentrations of proteinase inhibitors found in other organs of various animals [5] may protect them against the action of trypsin-like and chymotrypsin-like proteinases.

The highest concentration of a proteinase inhibitor found until now in animal tissues exists in submandibular glands of dogs [5]. Depending on the state of the gland 1 g of fresh tissue contains maximal 5 mg inhibitor. This inhibitor was discovered by Trautschold, Werle, Haendle and Sebening [6] and described in more detail by Trautschold [7] and Haendle [8]. Some of the results were presented in earlier lectures given by Trautschold [9, 10]. Similar to the pancreatic trypsin inhibitors [1—4] the dog submandibular proteinase inhibitor is also a secretory protein [8].

Abbreviations: BAPNA, N^{α}-benzoyl-DL-arginine p-nitroanilide; BPTI, basic pancreatic trypsin inhibitor (Kunitz-type) = trypsin-kallikrein inhibitor from bovine organs; CM, carboxymethyl; CPPN, N-3-(carboxypropionyl)-L-phenylalanine p-nitroanilide; M. W., molecular weight; TRA, triethanolamine; TRIS, trishydroxymethyl aminomethane.

Methods

Isolation Procedure

Extraction and Purification Steps: Dog submandibular glands from Peele Freeze, Biologicals (USA), containing 10—14 IU (trypsin inhibition) per g tissue, were thawed and homogenized in deionized ice water ($2\,l$ for 100 g glands). After centrifugation the supernatant was adjusted to pH 6.0—6.5 and stirred with 100 g CM-cellulose (H^+-form) for two hours in an ice bath. The CM-cellulose adsorbate was washed 3 times with 500 ml 0.01M sodium acetate, pH 5.0. In order to elute the inhibitor the adsorbate was suspended in 5% (w/w) NaCl, 0.01M TRA-HCl, pH 8.0, for 10 minutes. By repeated elution 90—95% of the inhibitory activity found in the homogenate was recovered in the supernatant eluates of the cellulose.

The main salt portions were separated by dialysation: 4 hours, deionized water, 0—4°C. Concentration (evaporation in vacuo) was followed by fractionation on Sephadex G-50 columns equilibrated and developed with aqueous (5%, V/V) acetic acid. Lyophilisation of the inhibitor containing eluates yielded a white powder with a specific activity of 1.4 up to 1.8 IU (trypsin inhibition) per mg. Loss of about 18% of the inhibitor was observed during these steps.

Another part of the inhibitor material was isolated from acidified extracts (perchloric acid, 3%, w/w) of the homogenates [6, 7, 8]. In this case the supernatant of the precipitated proteins was neutralized with 5M K_2CO_3 solution. Precipitated $KClO_4$ was separated by filtration, and the inhibitor solution was diluted with water 1:5 before adding CM-cellulose.

Chromatographic Separations: Depending on the foregoing isolation procedure somewhat different methods were employed.

1) The following chromatographic systems were used for further purification of the inhibitor material isolated from *perchloric acid extracts* of the homogenates:

1—1) Gradient elution chromatography on CM-cellulose (Fig. 1);

1—2) Gel filtration on Biogel P-2 (5.0×30 cm), equilibrated and developed with 0.01M acetic acid (followed by lyophilisation of the inhibitor-containing eluate);

1—3) Rechromatography on CM-cellulose of each of the two fractions I* and II* shown in Figure 1 under the same conditions as given in the legend of Figure 1, except that the slope of the NaCl-gradient was only 2/3 of the one described;

1—4) Gel Filtration on Biogel P-2 (for conditions see step 1—2) followed by lyophilisation of the eluted inhibitor containing fractions.

The amino acid composition of the two inhibitor fractions I* and II* thus obtained is given in Table 2.

2) For further purification of the inhibitor material isolated from *aqueous extracts* the following systems were employed:

2—1) Gradient elution chromatography on CM-cellulose as shown in Figure 1 yielding fractions 2-I and 2-II (a somewhat modified gradient was used: 0,5M NaCl to 0.05M NaCl, each in the elution buffer.);

2—2) Ultrafiltration of the inhibitor fractions in Amicon cells (membrane: UM-2) by repeated dilution with deionized water;

2—3) Equilibrium chromatography on CM-cellulose of fraction 2-I (Fig. 2) and fraction 2-II (Fig. 3) in separate runs;

2—4) Ultrafiltration of the inhibitor fraction A_2 shown in Figure 2 followed by rechromatography under identical conditions as given in Figure 2, except the length of the column used (twice as high, complete separation of fractions A_1 and A_2, cf. legend of Fig. 2, section 2-2).

2—5) Gel filtration of the rechromatographed inhibitor fraction A_2 and of the inhibitor fraction C (shown in Fig. 3) on Sephadex G-50 equilibrated and developed with aqueous acetic acid (5%, V/V) followed by lyophilisation of the inhibitor-containing fractions.

The amino acid composition of the two inhibitor fractions A_2 and C thus obtained is given in Table 3.

Determination of Enzyme Activity and Enzyme Inhibition

Trypsin: The activity of trypsin and trypsin inhibition was measured with N^{α}-benzoyl-DL-arginine p-nitroanilide (BAPNA) as substrate. Details are given in ref. [11, 12]. One mU corresponds to about 1 μg trypsin; bovine trypsin (Novo Industri A/S) was used throughout. One unit of inhibition activity (IU) causes the reduction of BAPNA hydrolysis by 1 μmole per minute, one mIU the 10^{-3} fold amount. The molarity of the trypsin solutions used in the titration experiments (Fig. 5) was determined according to CHASE and SHAW [13] and by inhibition tests with an inhibitor (BPTI) solution of known molarity [14].

Chymotrypsin: N-3-(carboxypropionyl)-L-phenylalanine p-nitroanilide (CPPN) was applied as substrate. Details are given in ref. [12, 15]. Definitions correspond to the ones mentioned above. Bovine α-chymotrypsin (Novo Industri A/S, 1100 NF per mg) was used throughout. One mU corresponds to about 20 μg α-chymotrypsin.

The molarity of the chymotrypsin solutions used in the titration experiments (Fig. 7) was determined by inhibition tests with an inhibitor (BPTI) solution of known molarity [14].

Plasmin: Plasmin activity and plasmin inhibition was measured with BAPNA as substrate. Assay conditions were the same as for trypsin [11, 12] except for the presence of 0.05M L-lysine in the buffer solution. Increase in extinction was observed for 10 minutes [12]. Plasmin from pig (batch 25-S-68, 2.68 Novo units per mg) was a gift from Novo Industri A/S. For the stock solution 10 mg plasmin were solved in 4.0 ml 0.0025N HCl. The molarity of the plasmin solution used in the titration experiments (Fig. 6) was determined by inhibition tests with an inhibitor (BPTI) solution of known molarity [14].

Subtilisin, Aspergillus Oryzae Protease, Pronase: Proteinase activity and enzyme inhibition was measured with azo-casein (Pentex-PP 6262, Fluka AG) as substrate. Constant amounts of the enzymes were incubated with increasing amounts of inhibitor in 1.0 ml 0.1M sodium potassium phosphate buffer, pH 7.6, for 5 minutes at 30°C. Afterwards 2.0 ml azo-casein solution (2%, w/w) in the same buffer was added and the mixture incubated for 10 minutes at 30°C. The enzymatic reaction was stopped by addition of 3.0 ml aqueous trichloracetic acid (5%, w/w). After 30 minutes at room temperature the extinction of the supernatant was read against a blank at 366 nm. The assay procedure is described in more detail in ref. [12], p. 1029.

Subtilisin (Crystalline Bacterial Proteinase, 22.0 Anson trypsin units per g, Batch 50-2) was a gift from Novo Industri A/S. Pronase E (lyophil., 70000 PUK/g from Streptomyces griseus) was purchased from Merck AG and alkaline Aspergillus Oryzae Protease (highly purified, 3500 PU (pH 8)/mg protein) from Röhm & Haas GmbH, Darmstadt.

Elastase: The activity of elastase from pig pancreas (cryst., suspension, 15 E/mg, from Merck AG) was measured according to the method published by SACHAR et al. [16] with elastin-orcein (Merck AG) as substrate. Elastase inhibition was determined in the following manner: A mixture of 0.15 ml of the elastase suspension (containing about 0.75 mg elastase in 0.2M TRIS-HCl, pH 8.8) and the inhibitor solution was filled up to 1.50 ml with 0.2M TRIS-HCl, pH 8.8. This incubation mixture was briefly (5 minutes) shaken and admixed with 20 mg elastin-orcein. The test sample was vigorously shaken for 30 minutes at room temperature. The enzymatic reaction was stopped by addition of 2.0 ml 0.5M phosphate buffer, pH 6.0. After centrifugation the extinction of the supernatant was read against a blank at 578 nm.

Results and Discussion

Isolation of Inhibitors

The developement of simpler isolation methods as described for dog submandibular inhibitor [6—10] was necessary in order to obtain enough material for sequential studies [18]. The method presented includes only a few steps, repeated chromatography on CM-cellulose and gel filtration or ultrafiltration, each with high yield in inhibitory activity.

From *perchloric acid* extracts of the glands — in which all proteases are inactivated — two inhibitor fractions were obtained by gradient elution chromatography (Fig. 1) in about equal amounts (Tab. 1, 1ᶜ). Both fractions differ only slightly in amino acid composition: Fraction I* contains 1 more glutamic acid residue and 1 lysine residue less than fraction II* (Tab. 2).

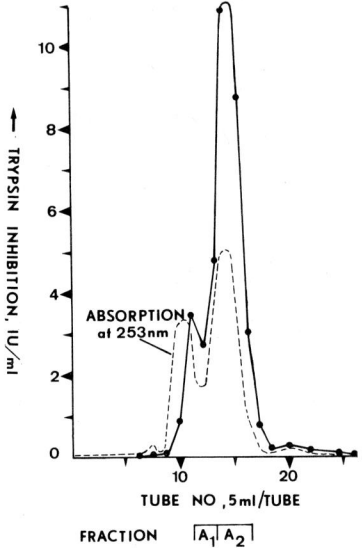

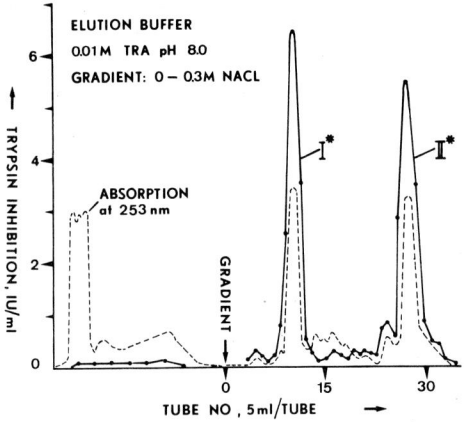

Fig. 1. Gradient Elution Chromatography on CM-Cellulose. Inhibitor material isolated from perchloric acid extracts (see Methods) was employed. 100—140 mg, dissolved in 1.0 m*l* of the starting buffer, were applied to the column (1.6 × 30 cm) which was equilibrated and developed with 0.01M TRA-HCl, pH 8.0, at 10.5 m*l* per hour. As soon as the protein content and the inhibitory activity in the eluate decreased, a linear gradient formed from 0.5 liters each of starting buffer and 0.01M TRA-HCl, 0.3M NaCl, pH 8.0, was used for elution. Inhibitor fraction I* appeared in the eluate at a NaCl concentration of about 0.04M, inhibitor fraction II* at 0.08M. Yields are given in Table 1.

Using the same chromatographic procedure from *aqueous* extracts of the glands an inhibitor fraction I (termed "2-I") was obtained in an amount that was about twice as high as that of fraction II (termed "2-II", cf. Tab. 1, 1ᵈ), however, both fractions were not homogeneous. Further purification by equilibrium chromatography was necessary (Fig. 2 and 3):

Fig. 2. Equilibrium Chromatography of Fraction 2-I, obtained by Gradient Elution Chromatography. Conditions were the same as given in Fig. 4; however columns with deviating dimensions were employed.

2-1) 228 IU (trypsin inhibition) were applied to the column (2.0 × 30 cm). The elution curve is shown in the Figure. Elution rate: 12 m*l* per hour.

2-2) Complete separation of fractions A_1 and A_2 is achieved by using a longer column (2.0 × 65 cm) and a smaller elution rate (6 m*l* per hour, cf. Table 1).

For distribution of inhibitory activity in the eluted fractions and yields see Table 1.

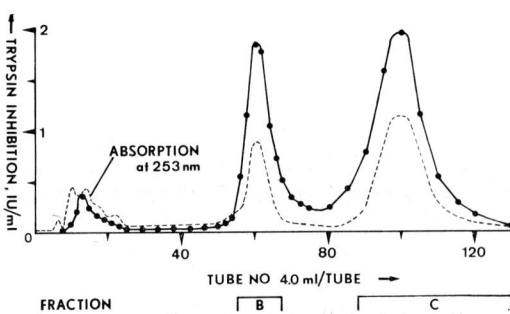

Fig. 3. Equilibrium Chromatography of Fraction 2-II, obtained by Gradient Elution Chromatography. The column and the conditions described in Fig. 4 were used. 255 IU (trypsin inhibition) were applied. For distribution of inhibitory activity in the eluted fractions see Table 1.

Table 1. Distribution of Inhibitory Activity among the Fractions obtained by Gradient Elution and Equilibrium Chromatography

Important intermediate fractions are put in parenthesis

Fractions from Fig.	Number of runs	Percent[a] of inhibitory activity (trypsin inhibition) found in the given fractions					Total yield[b] [%][a]
		Gradient elution chromatography					
		I		II			
1[c]	3	32—38*	(10—12)	26—28*	(12—15)		97—100
1[d]	3	45—51	(9—25)	22—29	(4—8)		94—100
		Equilibrium chromatography					
		A_1	A_2	B	C		
4	1	14	50	7	28		99
2[e]	3	17—27	53—71				87—97
2[f]	3	1—2	87—99*				95—100
3[g]	1		8	27	54*		95

[a] Related to the inhibitory activity applied to the column.
[b] Including all fractions containing inhibitory activity.
[c] Inhibitors isolated from perchloric acid extracts of the glands (see Methods).
[d] Inhibitors isolated from aqueous extracts of the glands (see Methods).
[e] Fraction I (from 1[d]) served as starting material; the short column described in Figure 2 was used.
[f] Fraction I (from 1[d]) and fraction A_2 (from 2[e]; rechromatography!) served as starting material; the long column described in Figure 2 was used.
[g] Fraction II (from 1[d]) served as starting material.
* These inhibitor fractions were used, after gel filtration (Sephadex G-50, 5% (w/w) acetic acid) and lyophilisation, for further investigations.

1) After complete separation of an inactive contamination A_1 from fraction I (see legend of Fig. 2 and Tab. 1, 2[e] and 2[f]) an inhibitor A_2 was obtained which has the same amino acid composition (Tab. 3) as inhibitor fraction I* which was isolated from acidic extracts.

2) Inhibitor C, obtained from fraction II (cf. Fig. 3 and Tab. 1, 3[g]) was found to lack a single residue each of glycine and proline, when compared to inhibitor fraction II[1]. Probably glycine and proline are split off by exopeptidases* in the aqueous extracts of the glands, whereas the glutamic acid residues is also absent in inhibitor fraction II* isolated from acidic extracts.

---
[1] Glycine was determined as the only N-terminal residue by H. Tschesche and E. Fink.

Therefore the assumption is logical that submandibular glands of dogs contain *two* very similar inhibitors, one of them synthesized by a mutated gene, which differ only in one Glu and one Lys residue. This is astonishing in so far as the glands were collected from different breeds. The occurrence of isoinhibitors is also described by other authors [19—24].

In order to find out the proportions of the inhibitor fractions in the material obtained from aqueous extracts a sample was subjected directly to equilibrium chromatography (Fig. 4). The portion of the main fractions A_2 and C, which are identical with the corresponding inhibitor fractions shown in Figures 2 and 3, amounts to 92% of the total inhibitory activity found in this sample (cf. Tab. 1; the inhibitory activity in fraction A_1 belongs to A_2, see 2[e] and 2[f] in Tab. 1

Table 2. Amino Acid Composition (Residues per Molecule) of DSI-Fractions Isolated from Acidic Extracts Using Gradient Elution Chromatography

Fraction	I* (Fig. 1, Table 1)			II* (Fig. 1, Table 1)		
	20 hrs	70 hrs	Integer	20 hrs	70 hrs	Integer
Cysteic acid	12.06[a]		(12)	12.19[a]		(12)
Methionine sulfone	3.16[a]		(3)	2.84[a]		(3)
Aspartic acid	13.33	13.18	13	13.17	13.09	13
Threonine	6.71	6.59	7	6.70	6.45	7
Serine	7.81	6.99	8	7.64	6.79	8
Glutamic acid	8.88	9.33	9	8.02	8.16	8
Proline	5.87	6.03	6	5.84	6.27	6
Glycine	8.91	9.24	9	8.79	9.09	9
Alanine	6.07	6.19	6	5.86	6.14	6
Half-cystine	10.56	10.10	12[b]	11.60	9.87	12[b]
Valine	2.66	4.00	4	2.74	3.74[c]	4
Methionine	2.28	2.90	3[b]	2.22	2.47	3[b]
Isoleucine	4.33	5.27	5	4.45	5.24[d]	5
Leucine	5.96	6.23	6	5.82	6.18	6
Tyrosine	4.95	4.54	5	4.93	4.25	5
Phenylalanine	3.96	4.09	4	3.86	4.00	4
Lysine	9.90	9.87	10	10.91	10.89	11
Histidine	3.08	3.08	3	2.80	3.04	3
Arginine	4.84	5.17	5	4.69	4.84	5
Tryptophan[e]	0.3		0	0.4		0
Total			115			115
Mol. weight[f]			12750			12750

[a] After performic acid oxidation.
[b] Calculated from the values of the oxidized inhibitor (cf. a).
[c] 120 hrs: 3.89 residues.
[d] 120 hrs: 5.06 residues.
[e] Spectrophotometric determination [17].
[f] Degree of amidination is not considered.

and Fig. 2). Fraction B is not yet further investigated. These results show that enzymatic degradation in the aqueous extracts of the glands is limited and causes no serious disadvantages. On the other hand some of the inhibitor is adsorbed by the precipitated protein and must be eluted by repeated extractions if the extracts are acidified with perchloric acid.

Amino Acid Composition

The amino acid compositions of the isolated inhibitor fractions are shown in Tables 2 and 3. The small differences in the content of glutamic acid, lysine, glycine and proline are already discussed in the preceding paragraph. Furthermore, the following should be mentioned: When based on the molecular weight the number of disulfide bridges corresponds to that of many other inhibitors obtained from animal organs (e. g. pancreas glands, bovine organs, seminal vesicles, etc.). Values obtained by the one-column method were used to coordinate the numbers of basic and neutral or acidic amino acid residues. Release of arginine and isoleucine is finished only after a hydrolysis time of 70

Table 3. Amino Acid Composition (Residues per Molecule) of DSI-Fractions Isolated from Aqueous Extracts Using Gradient Elution and Equilibrium Chromatography

Fraction	A₂ (Fig. 2, Table 1)			C (Fig. 3, Table 1)		
	20 hrs	70 hrs	Integer	20 hrs	70 hrs	Integer
Cysteic acid	11.68[a]		(12)	11.75[a]		(12)
Methionine sulfone	2.90[a]		(3)	2.98[a]		(3)
Aspartic acid	13.13	13.17	13	13.22	13.18	13
Threonine	6.95	6.74	7	6.96	6.80	7
Serine	7.90	7.06	8	7.78	7.53	8
Glutamic acid	9.05	8.92	9	8.12	8.34	8
Proline	6.19	5.92	6	4.86	5.05	5
Glycine	8.99	8.97	9	8.25	8.43	8
Alanine	5.99	5.97	6	5.98	6.15	6
Half-cystine	11.41	9.89	12[b]	10.59	10.99	12[b]
Valine	2.75	3.82	4	2.90	3.84	4
Methionine			3[b]			3[b]
Isoleucine	4.64	5.12	5	4.90	5.08	5
Leucine	6.15	6.17	6	6.03	6.15	6
Tyrosine	4.82	4.46	5	4.68	4.53	5
Phenylalanine	3.85	3.93	4	3.75	4.02	4
Lysine	9.97	10.08	10	11.10	11.30	11
Histidine	3.06	3.09	3	2.98	3.09	3
Arginine	4.88	4.98	5	4.57	5.02	5
Tryptophan[c]	0.35		0			
Total			115			113
Mol. weight[d]			12750			12595

[a] After performic acid oxidation.
[b] Calculated from the values of the oxidized inhibitor.
[c] Spectrophotometric determination [17].
[d] Degree of amidination is not considered.

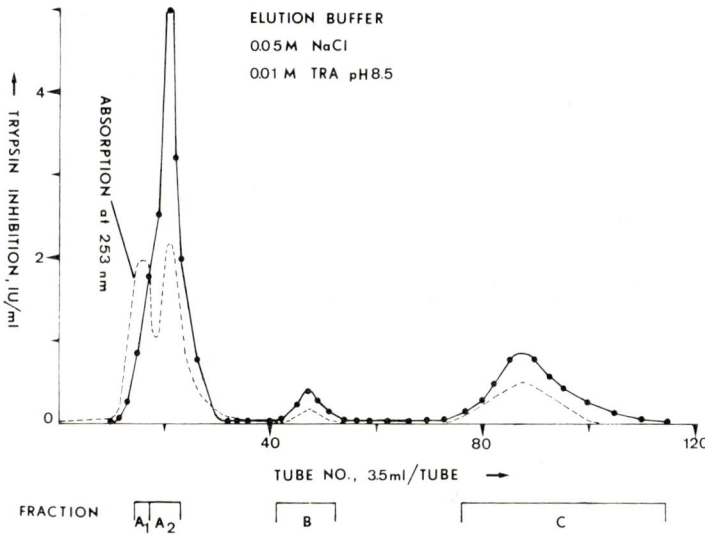

Fig. 4. Equilibrium Chromatography on CM-Cellulose. Inhibitor material isolated from aqueous extracts was employed. 100—150 mg (dissolved in 5.0 ml of the elution buffer) were applied to the column (2.0 × 30 cm), which was equilibrated and developed with 0.05M NaCl, 0.01M TRA-HCl, pH 8.5, at 12 ml per hour. Distribution of inhibitory activity in the eluted fractions and yields are presented in Table 1.

hours. Due to methodical difficulties the values of tryptophan and proline must be verified in the course of sequential studies. The presence of carbohydrate residues in the DSI-molecules was not observed.

Molecular Weight

For the molecular weight of DSI-fraction A_2 the following values were obtained by different methods:

1) Calculated from amino acid composition (Tab. 3) 12750
2) From gel filtration experiments (Tab. 4) 12000
3) From ultracentrifuge studies [25] 11900
4) Calculated from the specific activity (2.5 IU per mg, trypsin inhibition) 13200

The values are in good agreement with each other and with those reported in the literature [9, 10, 26, 27]. DSI forms a ternary complex with trypsin and chymotrypsin (Tab. 4). Dissociation of DSI into subunits during complex formation may therefore be excluded; it is also improbable regarding the amino acid composition.

Some observations indicate that less pure preparations of DSI may bind two trypsin molecules per molecule inhibitor. Perhaps in this state the reactive site for chymotrypsin can bind a trypsin molecule, too. But we have no evidence that the pure DSI-molecule binds two trypsin molecules as might be deduced from the specific activity of earlier obtained preparations [7—10].

Inhibition Spectrum

DSI inhibits strongly the following proteinases: Bovine *trypsin* (Fig. 5) and α-*chymotrypsin* (Fig. 7), *subtilisin* Novo (Fig. 8), porcine pancreatic *elastase* (Fig. 9) and alkaline *A. oryzae protease* (Fig. 10). In these cases *one* enzyme molecule reacts with *one* inhibitor molecule to form the complex under the conditions employed. This conclusion is based on the values given in Table 5 and — especially for elastase — on the linear shape of the titration curves. On the other hand from the titration curves it is possible to

Table 4. Estimation of M. W. by Gel Filtration of DSI and DSI-Enzyme Complexes

The chilled (10°C) Sephadex G-75 column used was equilibrated and developed with 0.05M TRA-HCl, 0.15M NaCl, pH 7.0. Of each protein or protein complex (containing equimolar amounts of enzyme(s) and inhibitor) about 3 mg were applied. Absorption at 253 nm as well as enzyme inhibition (after dissociation of the complexes with aqueous perchloric acid) were measured in the eluates

Substance	V/V_0	M. W.
Dextran blue	(V_0)	
Albumin (human)	1.27	68000
Chymotrypsinogen	1.57	25000
Cytochrom c	1.86	12000
BPTI	2.30	6500
DSI	1.86[a]	12000
DSI + trypsin	1.52	29000[c]
DSI + chymotrypsin	1.42	39000
DSI + trypsin and chymotrypsin	1.30[b]	60000

[a] Trypsin and chymotrypsin inhibition paralleled exactly in the eluted fractions.
[b] Degradation products of lower m. w. were also observed.
[c] Non additive molecular weights were also found with other examples [26].

calculate the amount of active enzyme molecules present in the enzyme preparations, e. g. in subtilisin Novo, A. oryzae protease, and elastase (cf. Tab. 5). It is remarkable that the activity of subtilisin and elastase was not decreased by high amounts (0.3—0.5 mg) of BPTI under the same conditions.

Pronase also contains proteinases which are strongly inhibited by DSI: One third of the azocasein-splitting activity of pronase was inhibited by titration with DSI in a manner characteristic for 1:1 complex formation (Fig. 11). That is why it is possible to calculate that 1 mg of the pronase preparation employed contains 14 n mole of DSI-reactive proteinases.

Porcine *plasmin* is also inhibited by DSI, but much more weaker than the above mentioned enzymes (Fig. 6). The complex is already highly

Table 5. Enzyme/Inhibitor Ratio in the Complex Calculated from the Titration Curves

Enzyme	Fig.	Substrate used	Amount of enzyme titrated		Theoretical amount of inhibitor necessary for complete inhibition
			n mole	µg	n mole
Trypsin	5	BAPNA	0.28		0.26
Chymotrypsin	7	CPPN	0.78		0.78
Subtilisin	8	Azo-casein		25[a]	0.80
Elastase	9	Elastin-orcein		750[b]	0.80
A. oryzae protease	10	Azo-casein		25[c]	0.58
Pronase	11	Azo-casein		50[d]	0.70

[a] Assuming a m. w. of 27500 [43] the subtilisin preparation contains 22 µg of active enzyme.
[b] Assuming a m. w. of 28500 [43] the elastase used contains 23 µg of active enzyme.
[c] Assuming a m. w. of 19600 [43] the protease used contains 11 µg of active enzyme.
[d] Assuming a m. w. of 19000 [44] the pronase preparation contains 13 µg of an enzyme inhibited by DSI.

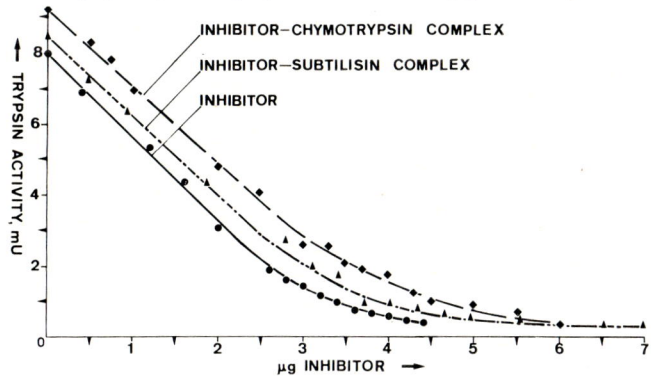

Fig. 5. Titration of Bovine Trypsin with Dog Submandibular Inhibitor (DSI), DSI-Chymotrypsin Complex, and DSI-Subtilisin Complex. DSI-fraction A_2 (cf. Fig. 2 and Table 1) was used throughout.
1) *Titration with DSI:* Constant amounts of trypsin, 0.28 n mole (i. e. about 10 µg by weight), titrated with BPTI, were incubated with increasing amounts of DSI in 2.0 ml 0.2M TRA-HCl (without $CaCl_2$), pH 7.8, for 8 minutes at 25°C. The enzymatic reaction was started by addition of 1.0 ml substrate solution (1 mg BAPNA in deionized water, cf. [11, 12]). Measured at 90% inhibition, no time dependence of the degree of inhibition was observed using preincubation periods from 1 up to 15 minutes.
2) *Titration with DSI-Chymotrypsin Complex:* Equimolar amounts of DSI (120 µg) and α-chymotrypsin (about 240 µg), calculated from the titration curve in Fig. 7 from measurements at 50% inhibition, were preincubated in 5.0 ml 0.2M TRA-HCl, pH 7.8, for 15 minutes and longer at 0°C. Constant amounts (0.32 n mole) of trypsin were incubated with increasing amounts of the DSI-chymotrypsin complex in 2.0 ml TRA-HCl for 8 minutes at 25°C; the enzymatic reaction was started as described above. No time dependence of the degree of inhibition of trypsin (by the DSI-chymotrypsin complex) was observed using preincubation periods from 1 up to 15 minutes. On the *abscissa* only the amount of DSI bound in the complex is given.
3) *Titration with DSI-Subtilisin Complex:* Equimolar amounts of DSI (103 µg) and subtilisin (about 250 µg), calculated from the titration curve in Fig. 8 from measurements at 50% inhibition, were preincubated in 2.24 ml phosphate buffer, pH 7.6, for 10 minutes and longer at 0°C. Constant amounts (0.30 n mole) of trypsin were incubated with increasing amounts of DSI-subtilisin complex in 2.0 ml 0.2M TRA-HCl, pH 7.8, for 8 minutes at 25°C. The enzymatic reaction was started by addition of the BAPNA-substrate solution (cf. Methods and trypsin titration). The amount of subtilisin employed caused no BAPNA-hydrolysis. On the *abscissa* only the amount of DSI bound in the complex is given.

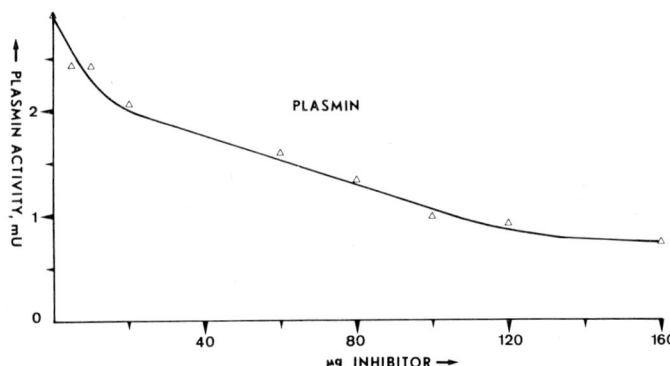

Fig. 6. Plasmin Inhibition. Constant amounts (250 μg, 1.8 n mole) of plasmin were incubated with increasing amounts of DSI (fraction A_2, cf. Fig. 5) in* 0.2M TRA-HCl, 0.05M L-lysine, pH 7.8, for 8 minutes at 25°C. The enzymatic reaction was started by addition of 1.0 ml BAPNA-substrate solution (cf. Methods and Fig. 5, trypsin titration). Measured at 71% inhibition, no time dependence of the degree of inhibition was observed using preincubation periods from 1 up to 15 minutes. * (2,0 ml).

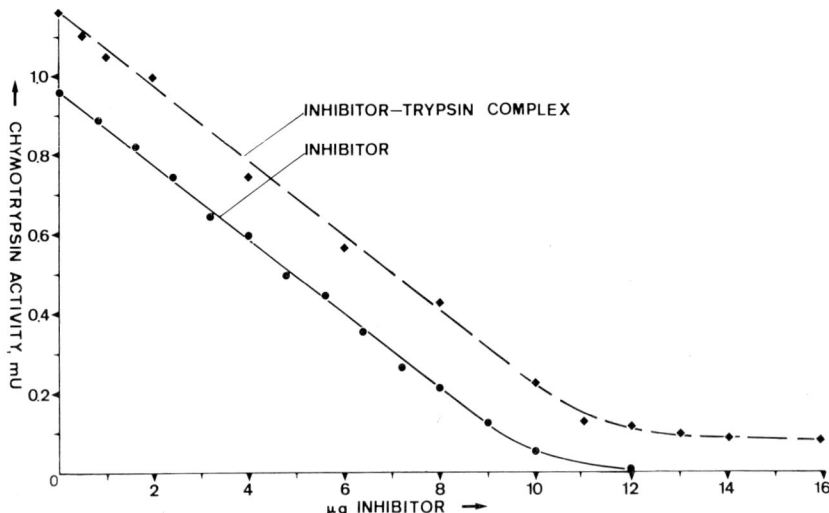

Fig. 7. Titration of Bovine α-Chymotrypsin with DSI and DSI-Trypsin Complex. DSI-fraction A_2 (cf. Fig. 2 and Table 1) was used throughout.

1) *Titration with DSI:* Constant amounts of α-chymotrypsin, 0.78 n mole (i. e. about 20 μg by weight), titrated with BPTI, were incubated with increasing amounts of DSI in 2.0 ml/ 0.2M TRA-HCl, 0.02M $CaCl_2$, pH 7.8, for 8 minutes at 25°C. The enzymatic reaction was started by addition of 1.0 ml substrate solution (5 mg CPPN in 1.0 ml/0.2M TRA-HCl, pH 7.8), cf. [12, 15]. Measured at 90% inhibition, the degree of inhibition is constant at preincubation periods from 1 up to 15 minutes.

2) *Titration with DSI-Trypsin Complex:* Equimolar amounts of DSI (80 μg) and trypsin (about 200 μg), calculated from the titration curve in Fig. 5 from measurements at 50% inhibition, were preincubated in 6.0 ml/0.2M TRA-HCl, 0.02M $CaCl_2$, pH 7.8, for 10 minutes and longer at 0°C. Constant amounts (0.96 n mole) of α-chymotrypsin were incubated with increasing amounts of the DSI-trypsin complex for 8 minutes at 25°C; the enzymatic reaction was started by addition of the substrate solution (see above). Measured at 90% inhibition, no time dependence of the degree of inhibition of α-chymotrypsin (by the DSI-trypsin complex) was observed using preincubation periods from 1 up to 15 minutes.

On the *abscissa* only the amount of DSI bound in the complex is given.

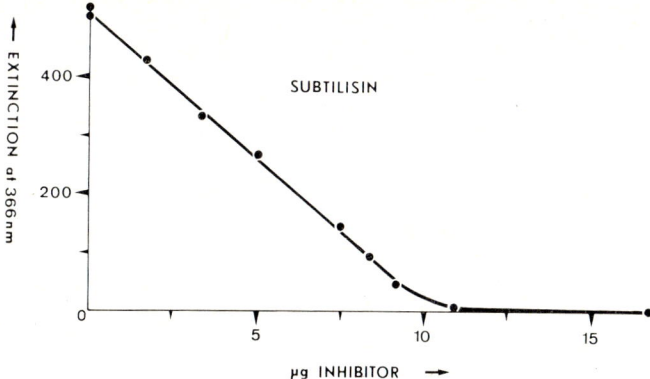

Fig. 8. Inhibition of Subtilisin Novo. To constant amounts (25 μg) of subtilisin increasing amounts of DSI (fraction A_2, cf. Fig. 5) were added. The procedure is given in the Methods. Employed substrate: Azo-casein.

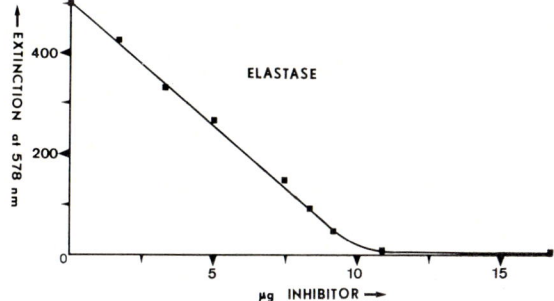

Fig. 9. Inhibition of Pancreatic Elastase. To constant amounts (0.75 mg) of elastase increasing amounts of DSI (fraction A_2, cf. Fig. 5) were added. The procedure is given in the Methods. Employed substrate: Elastin-orcein.

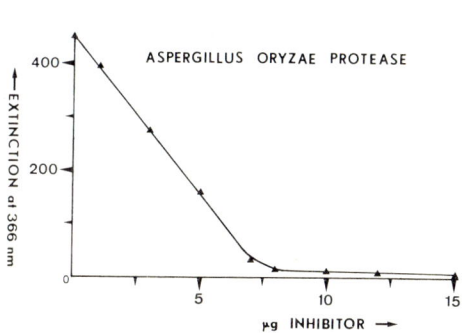

Fig. 10. Inhibition of Aspergillus Oryzae Protease. To constant amounts (25 μg) of the protease increasing amounts of DSI (fraction A_2, cf. Fig. 5) were added. The procedure is given in the Methods. Employed substrate: Azo-casein.

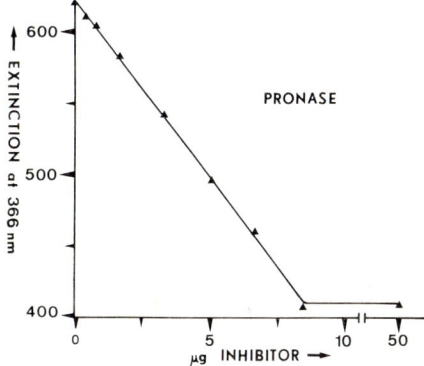

Fig. 11. Inhibiton of Proteolytic Activity of Pronase. To constant amounts (50 μg) of Pronase increasing amounts of DSI (fraction A_2, cf. Fig. 5) were added. The procedure is given in the Methods. Employed substrate: Azo-casein. Note, that only part of the proteolytic activity (ordinate) is inhibited

dissociated in the presence of BAPNA which is a substrate with only a small affinity to plasmin. The steeper slope of the titration curve at the beginning may be due to a small degree of contamination with trypsin which was used for the activation of plasminogen.

No inhibition of *collagenase* could be demonstrated. The activity of 0.1 mg of collagenase (from Worthington: 159 U/mg; substrate: p-phenylazobenzyloxycarbonyl-L-pro-L-leu-gly-L-pro-D-argOH from Fluka; method according to [28]) was neither diminished in the presence of 0.1 mg DSI nor by 0.5 mg BPTI. HAENDLE [8] reported that porcine *pancreatic kallikrein* is also not inhibited by DSI.

The results show clearly that DSI is a strong inhibitor of functional serine proteinases belonging to different families of this group [29]: In addition to proteinases of the trypsin-chymotrypsin family of mammals other serine proteinases from bacteria or mold fungus bearing no structural resemblance to the afore mentioned ones are also inhibited. It is most interesting that the inhibition spectra of ovoinhibitors [19] and of inhibitors from potatoes [30] are very similar to that of DSI. (See also the following paragraph.)

Reactive Sites

Different reactive sites on the DSI-molecule are responsible for the inhibition of trypsin and chymotrypsin. No decrease in activity against both enzymes is observed after exhaustive maleylation; however, if the maleylated DSI-derivative is reacted with the arginine-modifying butandion-2,3 reagent [31] it retains only its inhibitory activity for chymotrypsin. Therefore an arginine residue is located in the reactive site for trypsin inhibition [32].

The formation of ternary complexes is unambiguously demonstrated by the titration curves given in Figures 5 and 7 and the results of the gel filtration experiments (Tab. 4). The binding of trypsin therefore does not interfere with the binding of chymotrypsin or subtilisin, consequently the DSI-molecule is double-headed with not overlapping reactive sites [33]. Very similar properties were reported for the ovoinhibitors from egg white [19, 34], the inhibitor AA from soybeans [35], and the lima bean protease inhibitor LBI [21].

The subtilisin-DSI complex inhibits trypsin but not chymotrypsin. Therefore and from the specifity requirements of the proteinases [36] it may be deduced that the chymotrypsin-reactive site and the subtilisin-reactive site on the DSI-molecule are identical; this same site may also be responsible for the inhibition of elastase [36] and perhaps A. oryzae protease. Unfortunately DSI is not so easily modified in slightly acidic solutions as the Kunitz trypsin inhibitor [37], the Bowman-Birk inhibitor AA [35] from soybeans, and the inhibitor from lima beans [21]; investigations with altered conditions are in progress.

The results mentioned show that the inhibition of chymotrypsin by BPTI is a special case [38, 39]. The wide inhibition spectrum of DSI as well as the similar inhibition spectra of the ovoinhibitors are mainly caused by the chymotryptic reactive sites of these inhibitor molecules.

Physiological Function

The assumption that DSI protects mucosa cells in mouth and esophagus against the action of proteinases ingested with the food [5, 7—9] is supported by our findings: The inhibition of subtilisin, elastase and mold proteases by DSI. But also a special function of DSI in connection with proteinases found in submaxillary glands [40] etc. of some animals is possible.

The inhibition of elastase by DSI may be an important fact for the application of this inhibitor for medical therapy in future: The destruction of connective tissue cells caused by elastases during acute pancreatitis [41] or α_1-antitrypsin deficiency [42] is possibly prevented by DSI.

Acknowledgements: This work was supported by Sonderforschungsbereich-51, Munich.

We thank Novo Industri A/S for gifts of trypsin and subtilisin.

We are grateful to Prof. Dr. Dr. E. WERLE for generously supporting these investigations.

References

[1] GREENE, L. J. and M.H. PUBOLS, this volume, p. 196.
[2] GREENE, L. J. et al., this volume, p. 223 and 201.
[3] TSCHESCHE, H., H. KLEIN and G. REIDEL, this volume, p. 299.
[4] TSCHESCHE, H., E. WACHTER, S. KUPFER, R. OBERMEIER, G. REIDEL, G. HAENISCH and M. SCHNEIDER, this volume, p. 207.
[5] VOGEL, R., I. TRAUTSCHOLD and E. WERLE, Natural Proteinase Inhibitors. Academic Press, New York—London (1968).
[6] TRAUTSCHOLD, I., E. WERLE, H. HAENDLE and H. SEBENING, Z. physiol. Chem. **332**, 328 (1963).
[7] TRAUTSCHOLD, I., Habilitationsschrift, I. Medizin. Fakultät der Universität München (1965).
[8] HAENDLE, H., Habilitationsschrift, I. Medizin. Fakultät der Universität München (1969).
[9] TRAUTSCHOLD, I., H. FRITZ and E. WERLE in E. G. ERDÖS, N. BACK and F. SICUTERI, Hypotensive Peptides. Springer Verlag, New York (1966).
[10] FRITZ, H., I. TRAUTSCHOLD, H. HAENDLE and E. WERLE, Ann. N. Y. Acad. Sciences **146**, 400 (1968).
[11] FRITZ, H., G. HARTWICH and E. WERLE, Z. physiol. Chem. **345**, 150 (1966).
[12] FRITZ, H., I. TRAUTSCHOLD and E. WERLE in H. U. BERGMEYER, Methoden der enzymatischen Analyse. Verlag Chemie, Weinheim/Bergstr., p. 1021 (1970).
[13] CHASE, Jr., T. and E. SHAW, Biochem. biophysic. Res. Commun. **29**, 508 (1967).
[14] FRITZ, H., R. MEISTER and M. GEBHARDT, unpublished results.
[15] FRITZ, H., F. WOITINAS and E. WERLE, Z. physiol. Chem. **345**, 168 (1966).
[16] SACHAR, L. A., K. K. WINTER, M. SICHER and S. FRANKEL, Proc. Soc. Exp. Biol. Med. **90**, 323 (1955).
[17] BAILEY, L., Techniques in Protein Chemistry. Elsevier Publishing Company, Amsterdam—London—New York, p. 342f. (1967).
[18] HOCHSTRASSER, K., and coworkers, in preparation.
[19] FEENEY, R. E., this volume, p. 189.
[20] PEANASKY, R. J. and G. M. ABU-ERREISH, this volume, p. 281.
[21] STEVENS, F. C., this volume, p. 149.
[22] FINK, E., G. KLEIN, F. HAMMER, G. MÜLLER-BARDORFF and H. FRITZ, this volume, p. 225.
[23] FRITZ, H., M. GEBHARDT, R. MEISTER and E. FINK, this volume, p. 271.
[24] ČECHOVÁ, D., V. JONAKOVA-ŠVESTKOVÁ and F. ŠORM, Coll. Czechoslov. Chem. Commun. **35**, 3085 (1970); LASKOWSKI, Sr., M. et al., this volume, p. 66.
[25] BEIER, G., Max-Planck-Institut für Eiweiß- und Lederforschung, Munich.
[26] FRITZ, H., I. TRAUTSCHOLD and E. WERLE, Z. physiol. Chem. **342**, 253 (1965).
[27] GEOKAS, M. C., P. SILVERMAN and H. RINDERKNECHT, Gastroenterology **58**, 949 (1970).
[28] WÜNSCH, E. and H.-G. HEIDRICH, Z. physiol. Chem. **333**, 149 (1963).
[29] WALSH, K. A. and P. E. WILCOX, Methods in Enzymology **19**, 31 (1970).
[30] RYAN, C.A. and L.K. SHUMWAY, this volume, p. 175; RYAN, C. A., Biochemistry **5**, 1592 (1966).
[31] GROSSBERG, A. L. and D. PRESSMAN, Biochemistry **7**, 272 (1968).
[32] FRITZ, H., E. FINK, M. GEBHARDT, K. HOCHSTRASSER and E. WERLE, Z. physiol. Chem. **350**, 933 (1969).
[33] FEENEY, R. E. and R. G. ALLISON: Evolutionary Biochemistry of Proteins. Wiley Interscience, New York, p. 199f. (1969).
[34] DAVIS, J. G., J. C. ZAHNLEY and J. W. DONOVAN, Biochemistry **8**, 2044 (1969).
[35] BIRK, Y. and A. GERTLER, this volume, p. 142.
[36] SHOTTON, D., this volume, p. 47.
[37] LASKOWSKI, Jr., M., R. DURAN, W. R. FINKENSTADT, S. HERBERT, H. F. HIXSON, Jr., D. KOWALSKI, J. A. LUTHY, J. A. MATTIS, R. E. McKEE and C. W. NIEKAMP, this volume, p. 117.
[38] RIGBI, M., this volume, p. 74.
[39] FRITZ, H., H. SCHULT, R. MEISTER and E. WERLE, Z. physiol. Chem. **350**, 1531 (1969).
[40] LEVY, M., L. FISHMAN and I. SCHENKEIN, Methods in Enzymology **19**, 672 (1970).
[41] NAGEL, W. and F. WILLIG, Klin. Wschr. **42**, 400 (1964); MANDL, N., Collagenases and Elastases. Adv. Enzymology **23**, 163 (1961).
[42] Lectures presented at the *International Symposium on Proteolysis and Pulmonary Emphysema* under the auspices of the City of Hope Medical Center on January 4—6, 1971 in Pasadena, USA. LIEBERMANN, J., Digestion of antitrypsin deficient lung by leukoproteases; JANOFF, A., Elastase-like proteases of human granulocytes and elveolar macrophages; KELLER, S. and I. MANDL, Qualitative differences between normal and emphysematous human lung elastin; ADAMSON, J., R. BRUCE and J. PIERCE,

Collagen and elastin in antitrypsin emphysema; SENIOR, R., Paul HUEBNER and J. PIERCE, Elastase inhibitory capacity and elastin agar gel; TURINO, G. and R. LOURENCO, Role of connective tissue in lung mechanics; KIMBEL, P., V. MARCO, B. MASS, D. MERANZE and G. WEINBAUM, Emphysema in dogs induced by leucocyte contents; MANDL, J., S. KELLER, Y. HOSANNAH and C. BLACKWOOD, Induction and prevention of experimental Emphysema.

[43] Values reported in Methods in Enzymology **19** (1970): Proteolytic Enzymes.

[44] WÄHLBY, S., Biochim. Biophys. Acta **151**, 394 (1968).

Trypsin-Plasmin Inhibitors from Leeches
Isolation, Amino Acid Composition, Inhibitory Characteristics

Hans Fritz, Maria Gebhardt, Renate Meister and Edwin Fink

Institut für Klinische Chemie und Klinische Biochemie der Universität München, D-8 München 15, Germany

Summary

Two groups of trypsin-plasmin inhibitors were isolated from commercially available "Hirudin" samples, extracts of the leech Hirudo medicinalis. The trypsin-plasmin inhibitors, named Bdellins*, were separated from the thrombin-specific inhibitor, named Hirudin, by chromatography (stepwise elution) on DEAE-cellulose. The Bdellins were further purified by affinity chromatography using trypsin resins and subsequently separated into two inhibitor fractions, named Bdellin A and B, by stepwise elution from DEAE-cellulose. Each fraction was separated into several components by equilibrium chromatography on DEAE-cellulose.

The amino acid compositions of the Bdellin-A inhibitors (Table 3) differ significantly from those of the Bdellin-B inhibitors (Table 4), whereas the amino acid values obtained for the different A (Fig. 1) and B (Fig. 2) components are partly very similar. The inhibitors form 1:1 complexes with bovine trypsin and porcine plasmin; the titration curves are shown in Figures 4 and 5. Some physiological and medical problems concerning the Bdellins are discussed.

Introduction

Markwardt et al. [1] and de la Llosa et al. [2] isolated a thrombin-specific inhibitor from salivary glands of the leech Hirudo medicinalis. The polypeptide nature of this inhibitor, named

---
* This name was suggested by R. Marx, Munich.

Hirudin, has been well established by these authors; its molecular weight is near 10,000. Commercially available samples of Hirudin contain, in addition to the thrombin-specific inhibitor, appreciable amounts of trypsin-plasmin inhibitors [3]. The purification, the amino acid compositions, and some of the inhibitory characteristics of these inhibitors are presented in this paper.

Methods and Results

Determination of Enzyme Activity and Enzyme Inhibition

The procedures described in the foregoing paper [4] were used for determination of the activity of bovine trypsin, bovine α-chymotrypsin, porcine plasmin, and subtilisin Novo as well as enzyme inhibition. The enzyme samples employed were also identical with the ones described [4]. The following substrates were

---
Abbreviations: Ac, acetate; DEAE, diethylaminoethyl; EMA, ethylene maleic acid copolymer; TRA, triethanolamine. IU (inhibition unit): 1 IU causes the reduction of substrate hydrolysis by 1μ mole per minute; cf. ref. [4].

used: N^α-Benzoyl-DL-arginine p-nitroanilide for trypsin and plasmin, N-3-(carboxypropionyl)-L-phenylalanine p-nitroanilide for chymotrypsin. Inhibition equilibrium was achieved at every degree of inhibition within the preincubation (enzyme plus inhibitor) period used (5 minutes, cf. Fig. 4 and 5).

Isolation Procedures and Properties

Separation of the Trypsin-Plasmin Inhibitors from Hirudin: DEAE-cellulose (acetate-form) columns, 4.2 × 44 cm, equilibrated and developed with 0.2M sodium acetate, pH 6.0, at 60 m*l* per hour were employed. 4.5 g "Hirudin" from Medimpex, Budapest, 270 ATE (anti thrombin units) and 0.9—1.0 IU (trypsin inhibition, cf. ref. [4]) per mg, dissolved in 6 m*l* of the elution buffer, were applied to each column. In three typical runs 94—95% of the trypsin-inhibiting activity applied were eluted in the inhibitor-containing fractions in a total buffer volume of 1.0—1.2 *l*.

Subsequently the thrombin inhibitor, which is completely retained on the cellulose under the conditions employed, was eluted using a linear gradient formed from 2.0 liters each of the starting buffer and 0.6M NaCl, 0.6M sodium acetate, pH 6.0. 74% of the inhibitory activity applied were recovered.

The trypsin-plasmin inhibitor fractions obtained from three column runs were combined and concentrated in vacuo until precipitation of salts started. The main portion of the salts was separated by dialysis for 4—5 hours against deionized water; afterwards concentration was continued, followed by dialysis as soon as salt crystals were formed. Thus a final concentration of 25 IU per m*l* was attained (total yield: 12,500 IU in 500 m*l*). Aliquots were diluted with 0.2M TRA-HCl, pH 7.8, to get the concentration desired for the trypsin-resin step.

Affinity Chromatography Using Water-Insoluble Trypsin Resins: The polyamphoteric trypsin EMA-resin applied was synthesized starting from 10 g trypsin according to the method given in ref. [5]. The isolation procedure described below was repeated four times using this same resin.

The trypsin resin was suspended in 70 m*l* of the trypsin-plasmin inhibitor solution containing 1000 IU (cf. the foregoing paragraph). The chilled (0—4°C) mixture was vigorously stirred for 60—120 minutes, followed by centrifugation. Calculated from the inhibitory activities in the supernatants, amounts of inhibitors corresponding to 910 up to 960 IU were bound to the trypsin resin. The insoluble complex was washed free from contaminating material by repeated suspension in chilled 0.2M TRA-HCl, pH 7.8 (cf. ref. [5]).

For dissociation the complex was suspended in 80—95 m*l* chilled 0.2M KCl-HCl, pH 2.0, for 60—90 minutes, followed by centrifugation. The supernatants contained 61—68% of the inhibitors bound to the resin. This procedure repeated yielded in each case another 8% of the bound inhibitor so that the overall recoveries amounted to 69—76%.

The combined acidic inhibitor solutions were adjusted to pH 6.0, concentrated by evaporation in vacuo, and desalted by gel filtration. Using Sephadex G-50 columns, equilibrated and developed with 0.02M ammonium acetate, pH 6.0, 82% of the inhibitory activity applied was found in the eluted inhibitor fractions; using Biogel P-2 columns, equilibrated and developed with 0.01M acetic acid, 83—88% of the inhibitory activities applied were recovered. The inhibitor-containing fractions were lyophilized. For the material thus obtained the following specific activities were found: 3.5—3.8 IU per mg using Biogel and 2.7—2.9 IU per mg using Sephadex. In the latter case the ammonium acetate was not sublimed off quantitatively during lyophilisation.

Recently, we repeated this isolation step applying 10 g of the trypsin-cellulose resin ("bovine trypsin polymer bound to CM-cellulose, 7—10 U/mg") from Merck AG, Darmstadt. The resin was suspended for 20 minutes in about 400 m*l* chilled 0.5M NaCl, 0.05M TRA-HCl, pH 7.8, containing 790—1070 IU of the trypsin-plasmin inhibitors. For separation of contaminations the insoluble complex was suspended five times in the mentioned 0.55M salt-buffer solution and afterwards one time in 0.05M NaCl, 0.005M TRA-HCl, pH 7.8. Dissociation of the complex was achieved by suspending it three times in 300 ml chilled 1.0M KCl-HCl, pH 2.5, each time for 10 minutes. 82—71% of the inhibitor amount bound to the resin (878—526 IU) were thus recovered. The combined neutralized inhibitor solutions were desalted by repeated ultrafiltration using Amicon cells with UM-05 membranes or by gel filtration on Sephadex G-25. Specific activity of the inhibitor material obtained after lyophilisation: 4.2 IU per mg.

240 IU of the Bdellins were bound from 1 g of the trypsin resin (sample "2851 C") from Röhm & Haas, Darmstadt, under the same conditions; 66% of it were recovered after dissociation.

Separation of Bdellin Fraction A from Fraction B: DEAE-Cellulose (OH$^-$-form) columns, 1.6 × 60 cm, equilibrated and developed with 0.1M sodium acetate, pH 6.0, at 65 ml per hour were employed. In 5 runs performed 350—539 mg of the Bdellin-A, B mixture (specific activity: 2.7—3.8 IU per mg), dissolved in about 3 ml of the equilibration buffer were applied. The inhibitor fractions eluted with the equilibration buffer in a total volume of 500—600 ml amounted in all runs to 55% of the inhibitory activity applied. The inhibitors contained in these fractions were named *Bdellin A*.

In each run 45% of the inhibitory activity applied were retained on the DEAE-cellulose column. These inhibitors, named *Bdellin B*, were subsequently eluted with 0.1M sodium acetate, 0.4M NaCl, pH 6.0, in a total volume of 170—240 ml. Cf. Table 2.

The inhibitor-containing fractions were concentrated by evaporation in vacuo or by ultrafiltration and desalted using Sephadex G-25 columns, equilibrated and developed with aqueous acetic acid (5%, V/V). The inhibitor fractions eluted (yields: 80—90% of the inhibitory activities applied) were lyophilized. The following specific activities were found: 3.3—3.5 IU/mg for Bdellin fraction A and 4.2—4.8 IU/mg for Bdellin fraction B.

Equilibrium Chromatography of Bdellin Fraction A: The DEAE-cellulose (OH$^-$-form) columns (1.2 × 24 cm) employed were equilibrated and developed with 0.05M ammonium acetate, pH 6.3, at 6.5 ml per hour. 30 mg of Bdellin fraction A, dissolved in 1—2 ml of the elution buffer, were applied to each column. The distribution of the inhibitory activity among the fractions eluted is shown in Table 1, the elution diagram in Figure 1. Yields are given in Table 2. Equilibrium chromatography was repeated with the desalted (by repeated ultrafiltration) fractions 2, 3, and 4 shown in Figure 1 in separate runs using the same conditions. As expected from the form of the absorption line, fractions 2 and 3 turned out to be identical.

Table 1. Distribution of the Inhibitory Activity among the Bdellin Fractions obtained by Equilibrium Chromatography

	Bdellin A				
Fraction (cf. Fig. 1)	1, (1a)	2, 3		4	5
% of inhibitory activity[a]	11—16	**33—40**		9—17	12—15
	Bdellin B				
Fraction (cf. Fig. 2)	1	2	3	4, 5, 6	
% of inhibitory activity[a]	5—7	14—20	**29—36**	9—15	

[a] Related to the total inhibitory activity applied to the columns. Four columns running under identical conditions were evaluated.

Table 2. Isolation of the Trypsin-Plasmin Inhibitors Bdellins from Commercially Available "Hirudin" Samples

Purification step	Specific activity[a] IU/mg	Yield[b] %
Commercial samples	0.9—1.0	
Separation from Hirudin (thrombin inhibitor) on DEAE-cellulose (acetate-form): With 0.2M NaAc, pH 6.0, only the Bdellins were eluted.		94—95
Affinity chromatography using polyamphoteric trypsin EMA-resins	3.5—3.8[c]	69—76
Separation of Bdellin fractions A and B on DEAE-cellulose (OH⁻-form): Elution of Bdellin A with 0.1M NaAc, pH 6.0, Bdellin B with 0.1M NaAc, 0.4M NaCl, pH 6.0	3.3—3.5 4.2—4.8	55 ⎫ 45 ⎭ 100
Fractionation of Bdellin A and B on DEAE-cellulose (OH⁻-form): Equilibrium chromatography of Bdellin A with 0.05M NH$_4$ Ac, pH 6.3, Bdellin B with 0.1M NaAc, pH 5.5	3.8 (fractions 2, 3, 4) 5.1 (fraction 3)	75—85[d] 67—73[d]

[a] Trypsin inhibition; the values were determined of the desalted and lyophilized material.
[b] Related to the inhibitory activity applied to the column or resin; loss during desalting by gel filtration and ultrafiltration is not considered, see text.
[c] Material containing some ammonium acetate (see text): 2.7—2.9.
[d] All fractions together, cf. Table 1.

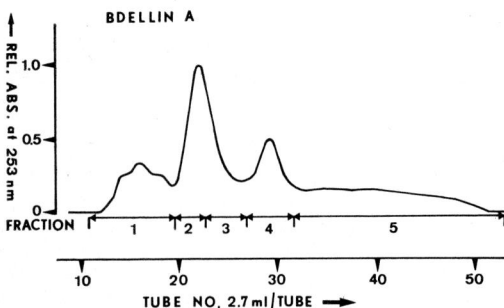

Fig. 1. Equilibrium Chromatography of Bdellin Fraction A.

For experimental details see Methods and Results. The inhibition curve paralleled that of the relative absorption, measured at 253 nm (ordinate).

Fraction 1a (cf. Table 3 and 1) corresponds to the last third of fraction 1, subsequent to fraction 2.

The inhibitor fractions thus obtained were desalted by repeated ultrafiltration employing Amicon cells with UM-05 membranes. About 5% of the inhibitor amount applied was lost during this step. The inhibitors obtained after lyophilisation with a specific activity of 3.8 IU per mg were used for all further investigations.

Equilibrium Chromatography of Bdellin Fraction B: 50 mg of Bdellin fraction B, dissolved in 1—2 m*l* of the equilibration buffer, were applied to each of the DEAE-cellulose (OH⁻-form) columns (1.2 × 55 cm). The columns were equilibrated and developed with 0.1M sodium acetate, pH 5.5, at 16.5 m*l* per hour. After elution of fraction B-3 (cf. Fig. 2) the residual inhibitor retained was eluted from the column with a buffer solution of higher molarity, 0.4M NaCl, 0.1M sodium acetate, pH 5.5. See Table 1 for distribution of the inhibitory activity among the fractions eluted and Table 2 for yields. After desalting equilibrium chromatography of Bdellin fraction B-3 was repeated under identical conditions.

The inhibitor fractions eluted were concentrated by ultrafiltration and desalted by gel filtration as described above. The inhibitor B-3 obtained after lyophilisation with a specific activity of 5.1 IU per mg was used for further investigations.

In order to obtain enough material for sequential studies the isolation procedure was repeated on a larger scale. Fig. 3 shows the distribution of the inhibitory activity among the Bdellin fractions obtained employing inhibitors with 4.2 IU per mg purified with the trypsin-cellulose resin from Merck AG. The last fraction shown in Figure 2 can be separated into three fractions (B-4,-5,-6) using a linear gradient formed from 350 ml each of the equilibration buffer (0.1M sodium acetate, pH 5.5) and 0.6M NaCl, 0.1M sodium acetate, pH 5.5 (cf. Fig. 3). By total yields of 82—84%, based on the inhibitory activity applied, 25—29% of the inhibitory activity were found in fraction B-1, 2, 35% in fraction B-3, 11—15% in fraction B-4, 5, 6 and 8—10% in the intermediate fractions. Inhibitor B-3 is used for the sequential studies under investigation.

Amino Acid Composition: The amino acid compositions of the different inhibitor components separated by equilibrium chromatography and of the starting mixtures, Bdellin fractions A and B, are presented in Tables 3 and 4. The inhibitors obtained after rechromatography were applied for analysis of components A-2, 3, A-4, and B-3. At first, inhibitor A-2, 3 was divided into two fractions, A-2 and A-3, however both fractions turned out to be identical in their behaviour during chromatography and in the amino acid composition.

Only the nearest integer values are given for the other fractions not further purified by rechromatography, provided that the deviations from the values of the corresponding main fraction were not larger than ± 0.1; otherwise the values found are shown.

The presence of carbohydrate residues and free sulfhydryl groups was not observed in the pure Bdellin fractions. Only about 91% of the cystine content of the Bdellins were recovered as cysteic acid after performic acid oxidation; some observations indicate that this is probably due to a small contamination of non-protein character which influences the oxidation reaction.

The inhibitor molecule of Bdellin B-3 contains no residue of the following amino acids: proline, methionine, isoleucine, phenylalanine, and tryptophan. For complete release of valine and lysine from the Bdellin-A fractions a longer hydrolysis time (70 hours) was necessary.

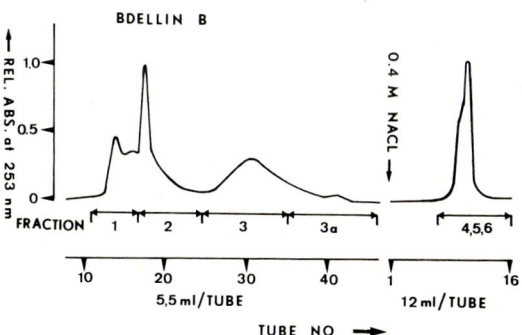

Fig. 2. Equilibrium Chromatography of Bdellin Fraction B.
For experimental details see Methods and Results. The absorption of the eluate was measured at 253 nm (ordinate).

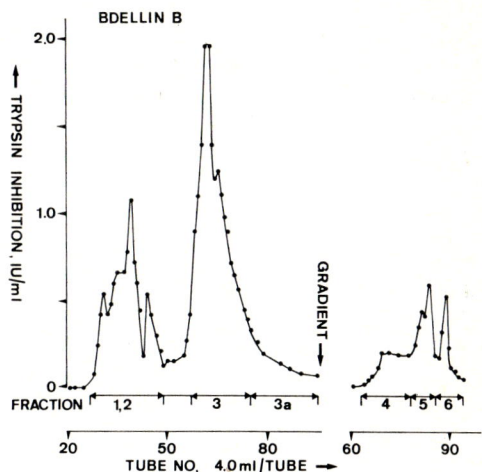

Fig. 3. Equilibrium and Gradient Elution Chromatography of Bdellin Fraction B.
Experimental details are given in Methods and Results. The gradient was started with tube number 1.

Table 3. Amino Acid Compositions of Bdellin-A Fractions

Bdellin-A: Mixture of inhibitors used as starting material for separation by equilibrium chromatography (cf. Fig. 1).
Bdellin A-1, A-1a (cf. legend of Fig. 1): Only the values deviating from the corresponding ones of inhibitor A-2, 3 are given.
Bdellin A-2, 3 and A-4: The inhibitors obtained after rechromatography were used for analysis

Bdellin fraction	A	A-1	A-1a	A-2, 3			A-4		
Hours hydrolyzed	20	20, 70	20, 70	20	70	Integer	21	70	Integer
Cysteic acid				9.10[a]		(10)	9.16[a]		(10)
Methionine sulfone				0.81[a]		(1)	0.95[a]		(1)
Aspartic acid	7.19	6	6	7.85	7.98	8	7.17	7.24	7
Threonine	3.01	2.6		3.01	2.88	3	2.91	2.82	3
Serine	2.97			3.06	2.64	3	2.90	2.77	3
Glutamic acid	6.45	6	6	5.18	5.27	5	5.04	5.06	5
Proline	2.80	2.3	2	2.95	2.73	3	2.76	2.74	3
Glycine	4.73	5	5	4.19	4.00	4	4.11	3.96	4
Alanine	3.96	3		3.98	3.96	4	3.83	3.88	4
Half-cystine	9.41			9.14	8.56	10	7.87	3.68	10
Valine	4.19[b]	3.4	4	4.61	4.95	5	4.35	4.89	5
Methionine	0.42			0.62	0.68	1	0.66	0.48	1
Isoleucine	1.01[b]			0.87	1.07	1	0.94	1.03	1
Leucine	1.58	2	2	1.19	1.12	1	1.14	1.12	1
Tyrosine	1.21			0.94	0.86	1	0.88	0.77	1
Phenylalanine	1.95		1	1.88	1.77	2	1.67	1.80	2
Lysine	4.82[b]	3	3	4.58	4.96	5	4.35	4.78	5
Histidine	2.80	2		2.95	2.94	3	2.93	2.83	3
Arginine	1.05	1.6	1	0.17	0.13	0	0.17	0.19	0
Tryptophan[c]	0.12			0.12		0			0
Total			56			59			58
Molecular weight[d]			5965			6339			6224

[a] After performic acid oxidation,
[b] Value from 70 hours hydrolysate,
[c] Spectrophotometric determination [7],
[d] Degree of amidation is not considered.

Molecular Weight: The molecular weights calculated from the amino acid compositions (cf. Tab. 3 and 4) are in good agreement with the values obtained by gel filtration experiments. Using Sephadex G-75 columns, equilibrated and developed with 0.15M NaCl, 0.05M citric acid, pH 2.5, at 17 ml per hour, the molecular weight found for Bdellin A (mixture) amounted to 7000, for Bdellin B (mixture) to 5600. From the specific activities similar values are calculated.

Inhibitory Characteristics: Both Bdellin groups A and B are strong inhibitors for bovine trypsin and porcine plasmin. The titration curves with Bdellin A-4 and Bdellin B-3 are shown in Figure 4 and Figure 5 respectively. In each case *one* enzyme molecule reacts with *one* inhibitor molecule to form the complex. No differences were observed applying the other Bdellin fractions in the titration experiments.

In both Bdellin groups lysine residues are located in the reactive site of the inhibitors: More than

Table 4. Amino Acid Compositions of Bdellin-B Fractions

Bdellin-B: Mixture of inhibitors used as starting material for separation by equilibrium chromatography (cf. Fig. 2).
Bdellin B-1, B-2, B-4, 5, 6: Only the values deviating from the corresponding ones of inhibitor B-3 are given.
Bdellin B-3: The inhibitor obtained after rechromatography was used for analysis.

Bdellin fraction	B	B-1	B-2	B-3			B-4, 5, 6
Hours hydrolyzed	20	20	20	20	70	Integer	20
Cysteic acid				5.60[a]		(6)	
Methionine sulfone				<0.05[a]		(0)	
Aspartic acid	5.21			5.13	5.14	5	7
Threonine	3.39	3	3.7	3.84	3.80	4	
Serine	1.93		2.7	1.97	1.96	2	4
Glutamic acid	5.81	5		5.93	6.02	6	8
Proline	0.60	1.4	1	<0.05	<0.05	0	1
Glycine	3.71			4.10	4.10	4	
Alanine	3.42	3.5		3.92	3.92	4	
Half-cystine	4.32			5.27	3.72	6	
Valine	4.08			4.09	4.17	4	
Methionine	0.19			<0.05	<0.05	0	
Isoleucine	0.19		0.3	0.16	0.09	0	0.3
Leucine	1.69	1.6		1.98	2.10	2	
Tyrosine	0.95			0.99	0.98	1	
Phenylalanine	0.35	0.4		0.09	0.10	0	0.5
Lysine	1.75	2	1.5	1.12	1.10	1	2
Histidine	6.15			4.94	5.02	5*	
Arginine	1.01		0.6	1.11	0.94	1	
Tryptophan[b]	0.12			0.25		0	
Total						45	
Molecular weight[c]						4830	

[a] After performic acid oxidation,
[b] Spectrophotometric determination [7],
[c] Degree of amidation is not considered.

* In Bdellin B-3, isolated recently (Fig. 3), 5.80 His were found. His may be destroyed to some extent during storage (cf. this volume, p. 228, Table 3, footnote c).

96% of the inhibitory activity against trypsin and plasmin are lost during maleylation [8]. The inhibitory activity is restored by deacylation in acidic solution [8]. The inhibitors are modified (reactive site bond split) to a smaller extent: The Bdellin fractions A, A-4, B, and B-3 lose 15—20% of their inhibitory activity during incubation with carboxypeptidase B under the following conditions: 0.3 mg inhibitor and 0.05 mg carboxypeptidase B in 0.43 ml TRIS-HCl, pH 8.0, 25°C, 29 hours (and 53 hours). The modification reaction occurs probably during contact with the trypsin resin [6]. In gel filtration experiments the formation of ternary complexes was not observed.

The following proteinases are not inhibited by the Bdellins: Porcine pancreatic kallikrein, bovine α-chymotrypsin, and subtilisin Novo. However, the commercial Hirudin samples contain about 25 mg (assuming a molecular weight near 6000) of an inhibitor for bovine α-chymotrypsin. Small amounts of this inhibitor are still determinable in both Bdellin mixtures, 2% by weight in Bdellin B and 5% in Bdellin A.

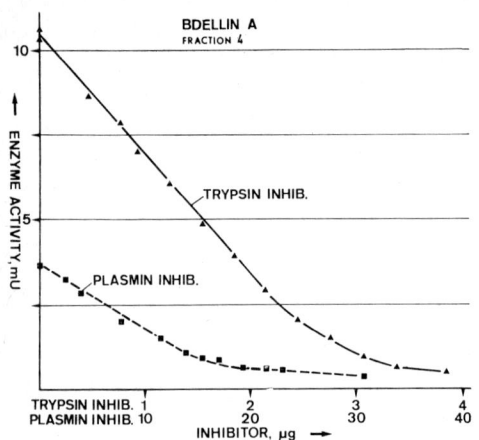

Fig. 4. Titration of Bovine Trypsin and Porcine Plasmin with Bdellin-A.

Bdellin fraction A-4 (cf. Fig. 1 and Table 3) with a specific activity of 3.8 IU (trypsin inhibition) per mg was used. Details of the procedure are presented in the foregoing paper [4] in Methods and in the legends of Figures 5 and 6.

1) *Titration of Trypsin*: Constant amounts of trypsin, 0.39 n mole titrated [4], were incubated with increasing amounts of the inhibitor in 2.0 m/ 0.2M TRA-HCl, pH 7.8, for 5 minutes at 25° C. The enzymatic reaction was started by addition of 1.0 m/ of the substrate solution. Measured at 94% inhibition, the degree of inhibition was constant using preincubation periods from 5 up to 15 minutes.

2) *Titration of Plasmin*: Constant amounts of plasmin, 2.5 n mole titrated [4], were incubated with increasing amounts of the inhibitor in 2.0 m/ 0.2M TRA-HCl, 0.05M L-lysine, pH 7.8, for 5 minutes at 25° C. The enzymatic reaction was started by addition of 1.0 m/ of the substrate solution.

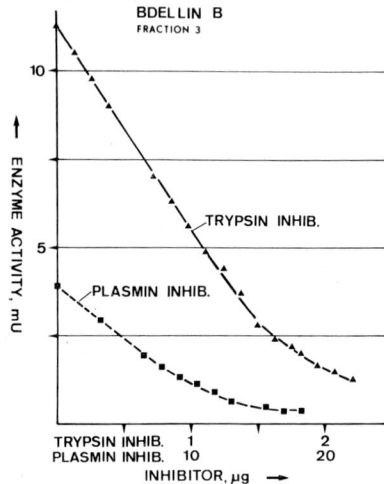

Fig. 5. Titration of Bovine Trypsin and Porcine Plasmin with Bdellin-B.

Bdellin fraction B-3 (cf. Fig. 2 and Table 4) with a specific activity of 5.1 IU (trypsin inhibition) per mg was used. Details of the procedure are presented in the foregoing paper [4] in Methods and in the legends of Fig. 5 and 6; cf. legende of Fig. 4.

1) *Titration of Trypsin*: Constant amounts of trypsin, 0.41 n mole titrated [4], were incubated with increasing amounts of the inhibitor in 2.0 m/ 0.2M TRA-HCl, pH 7.8, for 5 minutes at 25° C. The enzymatic reaction was started by addition of 1.0 m/ of the substrate solution.

2) *Titration of Plasmin*: Constant amounts of plasmin, 2.6 n mole titrated [4], were incubated with increasing amounts of the inhibitor in 2.0 m/ 0.2M TRA-HCl, 0.05M L-lysine, pH 7.8, for 5 minutes at 25° C. The enzymatic reaction was started by addition of 1.0 m/ of the substrate solution.

Discussion

Purification

A series of purification steps was necessary to obtain the trypsin-plasmin inhibitors Bdellins in highly purified form from commercially available "Hirudin" samples.

One gram of the starting material from Medimpex (270 ATE per mg) contains about 230 mg of the Bdellins and only 27 mg of the thrombin inhibitor Hirudin. This proportion to the disadvantage of Hirudin is probably the reason why the Bdellins were not easily separated during purification of Hirudin: We calculated from the results of our measurements (trypsin inhibition, electrophoresis) that one gram of the sterile "Hirudin" samples from Serva (3400 ATE per mg) contains, besides 300 mg Hirudin, still about 75 mg of the Bdellin-A, B mixture. The purest Hirudin samples isolated by MARKWARDT et al. [1] are free of Bdellins [3].

Separation of Bdellins and Hirudin is easily and completely achieved with satisfying yields by stepwise elution of the inhibitors from the acetate form of DEAE-cellulose. The recovery of Hirudin is much lower if the OH⁻-form of DEAE-cellulose is employed. The Hirudin thus obtained was separated by gradient elution chromatography into four fractions, two main and two smaller fractions; the conditions used are described elsewhere.

The purification of the Bdellins is decisively simplified by the trypsin resin step. Many attemps to obtain pure Bdellins using only ion exchange chromatography were not successful. The recoveries obtained with the two different resins employed were similar, however, the specific activity of the Bdellins isolated with the trypsin-cellulose resin was higher (4.2 IU per mg instead of 3.5—3.8 IU per mg using a polyamphoteric EMA-resin). Probably this is due to reduction of unspecific adsorption by the cellulose carrier and to the use of 1.0M salt-buffer solutions for the binding and washing steps. If the complex was dissociated at pH 2 (EMA-resin), only two extractions were necessary, whereas at pH 2.5 (cellulose resin) some more extractions were made. In this case 63, 12, 4, and 1% of the bound inhibitors were recovered.

Loss of inhibitory activity was very low (only about 5%) during desalting of the inhibitor solutions by repeated (4—5 times) ultrafiltration in Amicon cells fitted with UM-05 membranes. The only disadvantage is that small soluble portions of the resin and of DEAE-cellulose are not separated during this procedure so that subsequent gel filtration may be necessary.

Multiple Inhibitor Forms

Both Bdellin groups A and B contain a relative high amount of disulfide bridges, a finding which is common to most of the proteinase inhibitors known till now [9]. The high content of histidine residues in the Bdellins is also remarkable. On the other hand, the amino acid compositions of the A-components (Table 3) are so different from those of the B-components that obviously no genetic relationship exists between the Bdellin components A and B. However, the differences within both inhibitor groups may be caused by enzymatic degradation during storage of perished animals and the first isolation steps; but also the synthesis of somewhat different components by mutated gens is possible [10]. It was expected that the number of bonds hydrolyzed during contact with the trypsin resin should be very small [11].

The isolation of at least four different components of the thrombin inhibitor may be explained in the same way.

The sequence of Bdellin B-3 is now under investigation. To our knowledge this is the first proteinase inhibitor which contains no proline residue [12]. It is noteworthy that lack of isoleucine, methionine, phenylalanine, and tryptophan is observed in this inhibitor. The only lysine residue present in Bdellin B-3 is located in the reactive site [8]. Therefore, the influence of chemical modification of this residue on trypsin and plasmin inhibition can be uniquely assigned.

Bdellin B-3 is the smallest naturally occuring inhibitor for trypsin and plasmin found until now. The inhomogeneity of Bdellin fraction B-3 shown in Figures 2 and 3 (shoulder in the right part of the peak!) is caused by modified inhibitor forms in which peptide bonds within the molecule are split without loss of the inhibitory activity.

Physiological and Medical Problems

The Bdellins occur in all zones of the body of the leech, but the highest concentration is found in the region of the outer sexual organs [13]. We assume therefore that the function of these inhibitors has to be seen in connection with the trypsin-plasmin inhibitors found in seminal vesicles and seminal plasma of many animals and of man [14, 15]. The Bdellins are strong inhibitors of the plasmin-like activity occurring in human sperm plasma [13]. Certainly, the Bdellins will also inhibit the trypsin-like proteinase isolated from rabbit sperms by ZANEVELD et al. [15] which has an important function in fertilization.

The possible use of Bdellin-A in medical therapy (concerning plasmin inhibition) is under investigation [13]. An advantage may be the secretion of the Bdellins in the urine in contrast to the basic trypsin-kallikrein inhibitor which is stored in the kidneys shortly after intraveneous injection [16]. The reason why Hirudin is not widely used in medical therapy is perhaps caused by the fact that the purest samples available for medical use contain considerable amounts of the Bdellins so that both effects, thrombin inhibition and plasmin inhibition, supperimpose and thus no consistent results are obtained.

Acknowledgements: This work was supported by Sonderforschungsbereich-51, München. We are grateful to Farbenfabriken Bayer, Elberfeld, for gifts of commercial Hirudin samples. We are deeply indebted to Prof. Dr. Dr. E. WERLE for generously supporting our investigations.

References

[1] MARKWARDT, F., in Methods in Enzymology **19** ("Proteolytic Enzymes"), 924 (1970); further references are given there.

[2] DE LA LLOSA, P., C. TERTRIN and M. JUSTISZ, Biochim. Biophys. Acta **93**, 40 (1964).

[3] FRITZ, H., K.-H. OPPITZ, M. GEBHARDT, I. OPPITZ and E. WERLE, Z. physiol. Chem. **350**, 91 (1969).

[4] FRITZ, H., E. JAUMANN, R. MEISTER, P. PASQUAY, K. HOCHSTRASSER and E. FINK, this volume, p. 257.

[5] FRITZ, H., M. GEBHARDT, E. FINK, W. SCHRAMM and E. WERLE, Z. physiol. Chem. **350**, 129 (1969); see also ref. [6].

[6] FRITZ, H., B. BREY, M. MÜLLER and M. GEBHARDT, this volume, p. 28.

[7] BAILEY, L., Techniques in Protein Chemistry. Elsevier Publishing Comp., Amsterdam—London—New York 1967, p. 342.

[8] FRITZ, H., E. FINK, M. GEBHARDT, K. HOCHstrasser and E. WERLE, Z. physiol. Chem. **350**, 933 (1969).

[9] LASKOWSKI, Jr., M. and R. W. SEALOCK, in The Enzymes **2** (third edition), in press, paragraph III-C-3.

[10] Cf. ref. [4] and ref. [19—24] given there.

[11] Cf. paragraph "Enzymatic Cleavage" in ref. [6].

[12] Cf. ref. [9], paragraph III-C-5.

[13] MARX, R., unpublished data.

[14] FINK, E., G. KLEIN, F. HAMMER, G. MÜLLER-BARDORFF and H. FRITZ, this volume, p. 225; further references are given there.

[15] ZANEVELD, L. J. D., K. L. POLAKOSKI, R. T. ROBERTSON and W. L. WILLIAMS, this volume, p. 236; further references are given there. Cf. INGRISCH, H., H. HAENDLE and E. WERLE, this volume, p. 244.

[16] FRITZ, H., K.-H. OPPITZ, D. MECKL, B. KEMKES, H. HAENDLE, H. SCHULT and E. WERLE, Z. physiol. Chem. **350**, 1541 (1969).

Inhibitors from *Ascaris Lumbricoides*: Interactions With the Host's Digestive Enzymes

R. J. Peanasky and Ghaleb M. Abu-Erreish

Department of Biochemistry, The University of South Dakota, School of Medicine Vermillion, South Dakota 57069

Ascaris lumbricoides is a round worm parasite that lives in the *intestine* of its host, either man or pig. Although, these worms live and thrive in a most hostile environment — one geared to the breakdown of protein, carbohydrate, fat and nucleic acids their survival is not due to some unique property of their cuticle. If live *Ascaris* are kept at 37° in a salt solution (Baldwin and Moyle 1947), they will survive for days. If one adds a few milligrams of a plant protease like ficin, bromelin or papain, the worm is completely digested in a matter of hours (Robbins 1930, Robbins and Lamson 1934, Berger and Asenjo 1939 and 1940, Walti 1938). In another experiment, if an equal amount of trypsin or chymotrypsin is included the worm continues to survive. Extracts of *Ascaris* indeed inhibit the proteolytic enzymes of the pancreas (Weinland 1903, Mendel and Blood 1910, Collier 1941, Green 1957). To our knowledge however, no one has as yet measured the rate of disappearance of a proteolytic enzyme from such a medium. To measure this parameter, fresh *Ascaris* were collected at the abbatoir and brought to the laboratory in the salt medium. A washed worm (4—5 grams) was placed in 75 m*l* of fresh 37° salt medium to which was added 16 mg of chymotrypsin α. Figure 1 shows that live worms inactivated 10—15% of the chymotrypsin α in 8 hours and by 26 hours in-

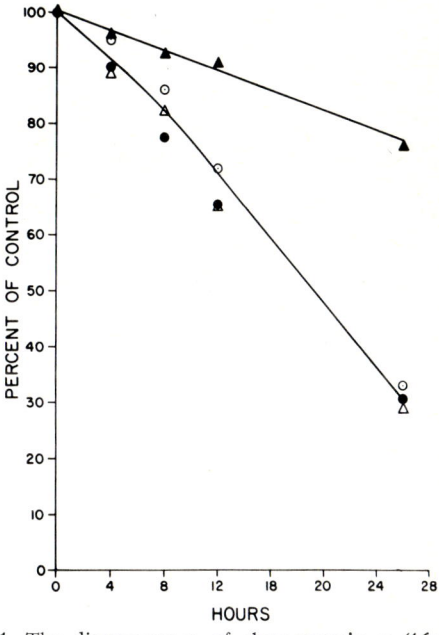

Fig. 1. The disappearance of chymotrypsin α (16 mg) from 75 m*l* of a salt solution (Baldwin and Moyle 1947) containing one *Ascaris* worm. When chymotrypsin α (16 mg) was incubated in this salt solution for 27 hours at 37°C, 14 to 17% of its activity was lost. This reaction mixture was used as 100% control at each time interval. ▲—▲ a worm was killed by lowering and maintaining its temperature at 0°C for 1 hour and then it was placed in the salt solution containing chymotrypsin. ⊙—⊙, △—△, and ●—● live worms incubated in the salt solution at 37°C.

activated 70% or 11 milligrams of chymotrypsin α. A similar experiment with trypsin was performed and the same kind of a curve was obtained when expressed as percent of control. The instability of trypsin (it lost 60% of its activity under these conditions) left the estimate of the amount of trypsin inactivated less certain. RHODES et al. (1963) found chymotrypsin inhibitor in every organ of *Ascaris* that they tested. We did a far more simple-minded experiment. After removal of the urogenital tract and the perienteric fluid the worm was cut into 3 sections (heads, middles, and tails). Table 1 shows that most of the activity for inhibiting chymotrypsin lies in the last 2/3rds of the worm. When these extracts were tested for trypsin inhibition the same distribution was found. However, each of these extracts was found to inhibit pepsin equally well.

These two experiments suggested to us that the inhibition of the host's digestive enzymes is far more complex than the presence of a chemical substance uniquely placed in a strategic position in the body walls of the parasite. In studying the biological function of inhibitors, therefore, one has to consider their location and how this affects their activity. Our studies along these lines are still preliminary. For the remainder of this discussion *soluble* enzymes and inhibitors from *Ascaris in solution* will be considered.

Table 1. Distribution of Chymotrypsin Inhibitor in *Ascaris Lumbricoides*

	Protein mg/ml	μg Inh./ml	specific Activity	Total Units	% of Activity
Heads	7.65	223	29	27,880	20%
Middles	8.50	320	38	52,480	38%
Tails	10.20	459	45	58,300	42%

The urogenital tract of ten inch *Ascaris* (50 in number) was dissected out, the perienteric fluid allowed to escape, and the worm was cut into 3 equal sections (heads, middles, and tails). Each section was extracted into 4 volumes of H_2O and the inhibitor was determined following centrifugation at 105,000 $\times$ g for 1 hour

Chymotrypsin Inhibitors

While RJP was a trainee of Professor M. LASKOWSKI, Sr. the separation of the inhibitory activity in extracts of *Ascaris* body walls towards trypsin and chymotrypsin was achieved by using continuous flow paper curtain electrophoresis and something which was protein and which would inhibit chymotrypsins α and B was crystallized (PEANASKY and LASKOWSKI Sr. 1960). Moving boundary electrophoresis, however, showed that these preparations had more than one component. Chromatography with Sephadex G-25 and G-75, achieved a preparation that had constant specific activity across a peak (PEANASKY and SZUCS 1964) and behaved as a single boundary in sedimentation studies with a coefficient of 1.2 S. Such preparations were devoid of carbohydrate and sialic acid. With a partial specific volume of 0.747 measured by pycnometry, a molecular weight by approach to sedimentation equilibrium of 8,000 $\pm$ 200 was calculated. The specificity of these preparations is shown in Table 2. The reaction with chymotrypsins α and B was complete in 10—12 minutes and on the basis of a 1:1 mixture a K_i for the complex between chymotrypsin α and inhibitor of 6.9 $\times$ 10^{-9} M and for the complex between

Table 2. Specificity of Chymotrypsin Inhibitor

Chymotrypsin α	$K_i = 6.9 \times 10^{-9}$M
Chymotrypsin B	$K_i = 3.2 \times 10^{-8}$M
Carboxypeptidase A	0
Carboxypeptidase B	0
Leucine Amino Peptidase	0
Trypsin	0
Pepsin	0
Ficin	0
Papain	0
Subtilisin	$K_i = 1.2 \times 10^{-6}$M

chymotrypsin B and the inhibitor of 3.2×10^{-8}M was calculated according to GREEN and WORK (1953). Carboxypeptidases A and B, pepsin, leucine amino peptidase and the protease, esterase and chymotryptic activity of trypsin are not inhibited. The plant proteases, ficin and papain, are not inhibited but can digest the inhibitor as evidenced by a slow disappearance of inhibitor activity when measured against chymotrypsin following exposure to these proteases. This effect ought to be expected because whole *Ascaris* are digested by these plant proteases. Subtilisin is inhibited by this material with a K_i of the order of 1.2×10^{-6}M. This is a better inhibitor for subtilisin by some $50 \times$ than the synthetic arsonic acids (GLAZER 1968).

What has continuously irritated us was that every attempt to establish a satisfactory amino acid composition of our material failed. In some earlier work preparations were obtained that were devoid of tyrosine, isoleucine and phenylalanine. Subsequently, other preparations contained detectable amounts (about a half of a residue of isoleucine) and some tyrosine or phenylalanine. We were not prepared to accept the idea that there is more than one chymotrypsin inhibitor even though RHODES et al. (1963) chromatographed a crude extract of *Ascaris* and found more than one chymotrypsin inhibitor fraction and ROLA and PUDLES (1966) reported that *Ascaris* extracted in 10% TCA, precipitated with ammonium sulfate and chromatographed on DEAE and CM-cellulose contained 2 and possibily 3 ill defined chymotrypsin inhibitors.

Therefore, about two years ago we set out to design a preparation of chymotrypsin inhibitor that would eliminate the use of continuous flow paper curtain electrophoresis and hopefully lead to a more satisfactory preparation. An outline of this approach is shown in Figure 2. We found no way to bypass the TCA treatment so it was rigorously controlled. Samples (8 mg/ml) were brought to 7.5% TCA and held there no longer than 20 minutes including time to centrifuge the precipitate. The supernatant was quickly raised to pH 7.5 and precipitated with 0.80 saturation with ammonium sulfate. Following dialysis in 50 mM ammonium acetate buffer at pH 6.0 the sample was applied to a CM cellulose column. Figure 3 shows that better than 70% of the applied activity was recovered from one peak. The material from this peak has relatively constant specific activity. Table 3 shows that the fraction from CM cellulose represents 39% of the total inhibitor present and a 65 fold purification. Attempts at further purification on

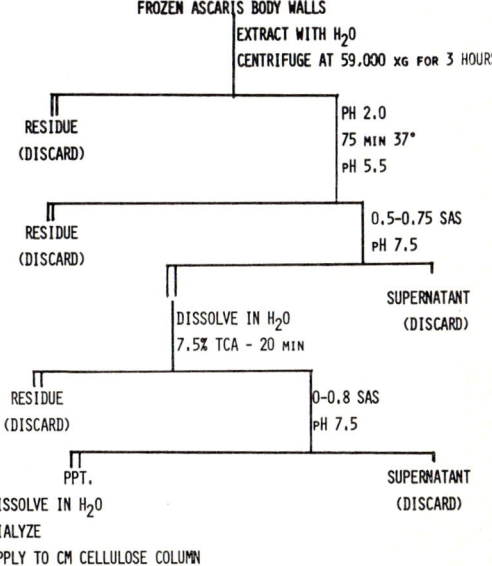

Fig. 2. Fractionation of frozen *Ascaris* body walls for chymotrypsin inhibitor.

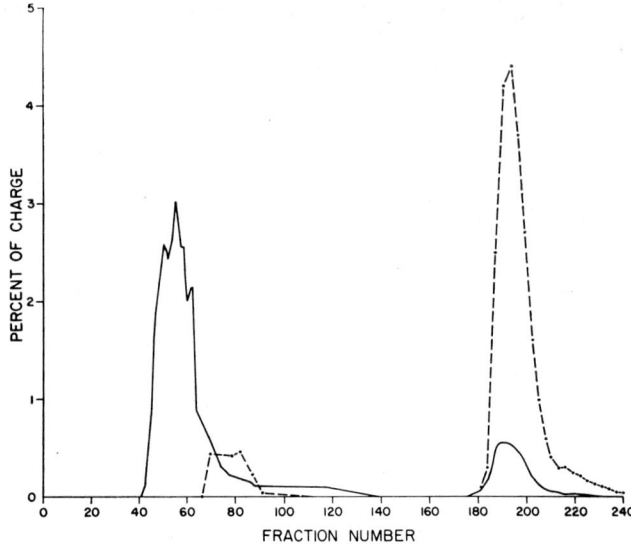

Fig. 3. Chromatography on CM-cellulose of *Ascaris* chymotrypsin inhibitor from the ammonium sulfate precipitate (Fig. 2). Column (4.5 × 46 cm) and sample were equilibrated with 50 mM ammonium acetate pH 6.0. Flow rate 60 ml/hr, each fraction 15 ml. At fraction 140 elution was continued with 200 mM ammonium acetate pH 6.5. Ordinate is percent of the protein —— and of the units of chymotrypsin inhibitor ----- charged onto the column.

Table 3. Recovery of Chymotrypsin Inhibitor from 600 Grams of *Ascaris*

	μg Chymotrypsin Inhibited	Specific Activity: μg Chymotrypsin inhibited per mg of Protein	% Recovery
Crude extract	602,000	40	(100)
After heating	600,000	77	100
0.5—0.8 SAS	534,000	144	89
TCA— 0.8 SAS	331,000	430	55
CM-Cellulose	234,780	2610	39

CM cellulose indicated heterogeneity of the distribution of protein but all tubes had identical specific activity. A sample of this material passed through a 2 meter column of Sephadex G-75 is shown in Figure 4. Such a preparation will behave as a single boundary on sedimentation in an ultracentrifuge. A polyacrylamide gel pattern of this material in 15% cross linked gel at pH 9.0 showed four components, electrophoretically different. Chromatography through a CM cellulose column employing the conditions that were used in the polyacrylamide gel electrophoresis experiment are shown in Figure 5. Four different fractions were obtained, all inhibit chymotrypsin and each has the same specific activity. Polyacrylamide gel electrophoresis of each peak is shown in Figure 6 and it appears that each is different.

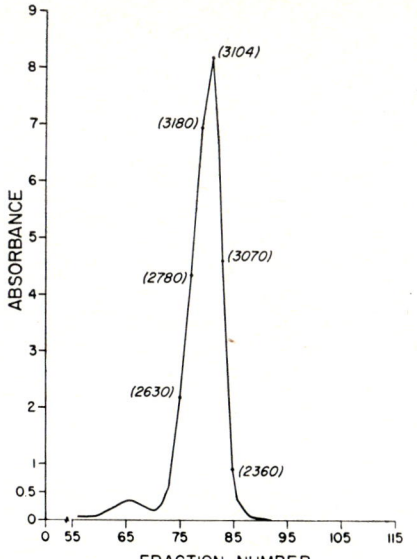

Fig. 4. Chromatography on Sephadex G-75 of *Ascaris* chymotrypsin inhibitor following CM-cellulose. Column (2.4 × 200 cm) was equilibrated with 50 mM acetate pH 6.0. Each fraction 9 ml. Ordinate is the absorbance of each fraction at 280 nm. Numbers in parentheses represent specific activity.

Table 4. Minimum Amino Acid Composition of Each Chymotrypsin Inhibiting Peak in Figure 5

Peak	I	II	III	IV
Lys	5.17	3.96	7.23	5.09
His	1.00	0.94	0.91	0.91
Arg	5.78	4.04	6.09	5.00
Asx	3.91	2.98	3.95	2.77
Thr	3.49	2.69	4.86	4.09
Ser	2.69	1.90	2.00	1.89
Glx	7.23	5.29	8.73	6.28
Pro	8.09	5.67	9.05	6.81
Gly	5.75	4.17	7.14	5.18
Ala	1.09	1.00	2.27	1.00
1/2 Cys	8.11	5.67	10.82	7.71
Val	2.86	2.17	3.59	2.60
Met	2.54	1.77	2.41	2.33
Ileu	—	0.12	0.45	—
Leu	1.86	1.40	1.82	1.76
Tyr	—	—	—	—
Phe	—	0.17	0.86	—
Try	(1.0)	(1.0)	(1.0)	(1.0)
X	0.46	—	0.36	—
Total Residues	62	46	73	55

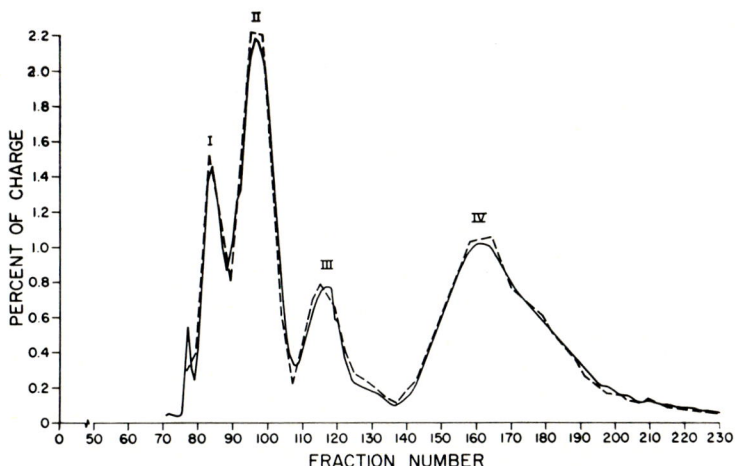

Fig. 5. Second chromatography on CM-cellulose of *Ascaris* chymotrypsin inhibitor after Sephadex G-75. Column (2 × 115 cm) was equilibrated with 50 mM ammonium carbonate pH 8.5. Flow rate was 40 ml/hr., each fraction 5 ml. Ordinate is percent of the protein —— or of the units of chymotrypsin inhibitor ---- charged onto the column. I to IV are chymotrypsin inhibiting peaks.

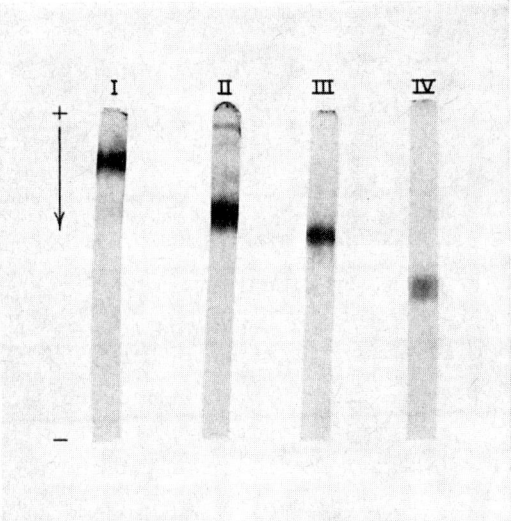

Fig. 6. Polyacrylamide gel patterns of each of the chymotrypsin inhibitor peaks obtained from CM-cellulose chromatography in Fig. 5. A 15% cross-linked gel was used at pH 9.0.

The amino acid composition of each of these 4 peaks are shown in Table 4. Peak IV appears to be the closest to a single component. Its amino acid composition is based on a single residue of alanine. This fraction is completely devoid of phenylalanine, tyrosine and isoleucine and appears to have 55 amino acid residues. Peak I is also devoid of isoleucine, tyrosine and phenylalanine. An unknown peak is obtained, which appears right on the buffer change in the long column as employed in the Beckman 2-column system. We do not know the identity of this peak. This fraction appears to contain 62 amino acids and differs from Peak IV by one residue each of arginine, aspartic acid, serine, glutamic acid, proline, glycine, and methionine. Peaks II and III contain some isoleucine and phenylalanine but probably less than a residue, at least of isoleucine. Fraction III, but not fraction II contains some of the unknown compound at the buffer jump. On the basis of a single alanine, Peak II appears to have 46 amino acids. An estimate is that Peak III contains 73 residues.

Not only do each of the peaks have the same specific activity against chymotrypsin α but upon titrating chymotrypsin α with each of the peaks the same complex is formed so that the peaks cannot be distinguished on this basis as the trypsin inhibitors can.

Some years ago an experiment was performed on chymotrypsin inhibitor which we hoped would throw some light on its mechanism of action. Since tryptophan was recognized as the only aromatic residue which the chymotrypsin inhibitor possessed we wondered if it was the active site of the inhibitor of chymotrypsin. To test this, the chymotrypsin inhibitor was titrated with N-bromosuccinimide. The ability of the tryptophan oxidized inhibitor to react with chymotrypsin α was determined after each increment of NBS and the results are shown in Figure 7. At the point where all of the tryptophan has been destroyed (on the basis of the factor 1.31 × disappearance of spectrum at 278 nm [Patchornik et al. 1958]), the inhibitor reacts with chymotrypsin α and is 75% as good an inhibitor. When the tryptophan oxidized inhibitor was tested against chymotrypsin B a similar curve was obtained. Whether this means that tryptophan is not involved in the inhibitor chy-

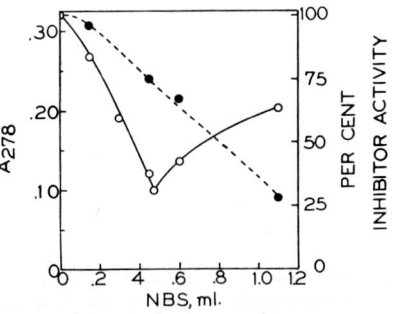

Fig. 7. Titration of *Ascaris* chymotrypsin inhibitor with N-bromosuccinimide (NBS) at pH 4.0 in 0.1M acetate buffer and determination of the chymotrypsin α inhibiting activity remaining. NBS solution was 2×10^{-3}M. After each increment of NBS was added an aliquot of the solution was removed, and unreacted NBS was destroyed by diluting it in formate buffer at pH 4.0 o—o absorbance at 278 nm. ●---● percent of inhibitor activity remaining.

motrypsin combination or whether the peptide bond involving oxidized tryptophan is still a site of chymotryptic activity cannot be concluded from these experiments.

Trypsin Inhibitors

Although the initial work on *Ascaris* was directed at the chymotrypsin inhibiting fractions, the method used to obtain the separation of trypsin and chymotrypsin inhibiting fractions, continuous flow paper curtain electrophoresis, made available some fractions that inhibited trypsin only (PEANASKY and LASKOWSKI, Sr. 1960). Umberto KUCICH, in our laboratory, used this as a starting material in the isolation of trypsin inhibitors. Direct chromatography on DEAE cellulose, CM cellulose or Sephadex yielded little purification unless these fractions were first treated with Dowex-1 at pH 10. When Dowex-1 treated material was equilibrated and chromatographed on CM cellulose beginning at pH 4.7 in 50 mM buffer, two trypsin inhibiting fractions referred to as Peak I and Peak II, were obtained (Fig. 8). Chromatography of Peak I on Amberlite IRC-50 at pH 6.0 showed that it approached constant specific activity under these conditions. Ultracentrifugation in a synthetic boundary cell at pH 4.4 showed a monodisperse preparation with a sedimination coefficient of 1.0 S (KUCICH and PEANASKY 1970). Figure 9 A shows that Peak I forms a complex with pork trypsin for which a K_i of 90 nM can be calculated from the mass action equation and the assumption of a 1:1 complex (GREEN and WORK 1953). What is more important is that even $10 \times$ the equivalent amount of inhibitor will not fully inactivate 257 picomoles of porcine trypsin. Peak II (Fig. 9 B) inactivates pork trypsin (257 picomoles) when $1.5 \times$ the stoichiometric amount of inhibitor is added. Under the above conditions and with the same restrictions as for Peak I, the K_i was calculated to be 13 nM.

The possibility that Peak I engaged pork trypsin as a temporary inhibitor was suggested because

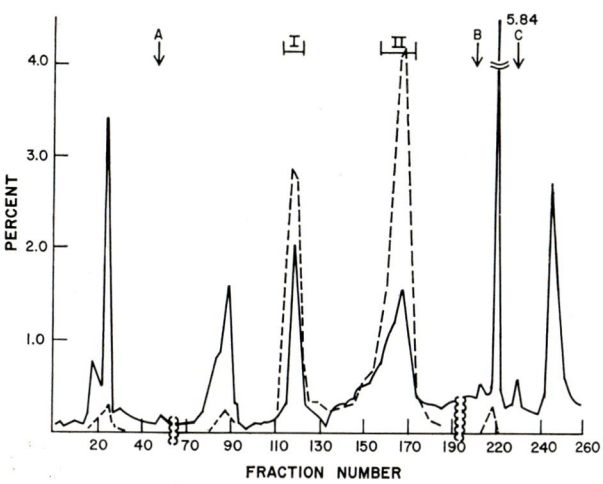

Fig. 8. Chromatography of *Ascaris* trypsin inhibitor on CM-cellulose after treatment with Dowex 1. Column (0.9 cm × 35 cm) was equilibrated with 50 mM ammonium acetate pH 4.7. Flow rate 6—8 ml/hr; each fraction 3 ml. At point A, a linear gradient 50 mM ammonium acetate pH 4.7—100 mM ammonium acetate pH 5.5 was applied. Ordinate is percent of the protein —— and of the units of trypsin inhibitor ---- charged onto the column. I and II are pooled fractions of trypsin inhibitor (KUCICH and PEANASKY 1970).

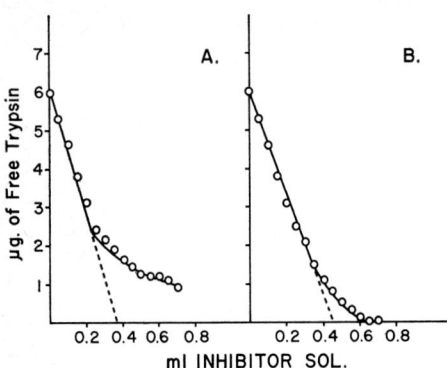

Fig. 9. Reaction of pork trypsin (257 picomoles) with Peak I (A) and Peak II (B) *Ascaris* trypsin inhibitors. Stock trypsin inhibitor solutions had an absorbance of 5×10^{-3} at 280 nm. Reaction mixtures (1.0 ml) made up of pork trypsin and the amounts of inhibitor indicated on the abscissa in 40 mM Tris buffer-10 mM $CaCl_2$ (pH 7.5) were allowed to react at 37°C for 10 minutes. Free trypsin was determined by the rate of hydrolysis of hemoglobin substrate (KUCICH and PEANASKY 1970).

increase in ninhydrin positive material. Since the final equilibrium is achieved at 65% of complex and 35% of free trypsin and inhibitor it cannot be a case of digestion of the inhibitor but it could be a modification of the inhibitor or a modification of trypsin so that it reacts differently with the inhibitor. These crude measurements do not permit speculation beyond this point.

When a synthetic substrate like TAME is used to determine the free trypsin the two inhibitor fractions can also be distinguished (Fig. 10). Peak II — pork trypsin complex dissociated to the extent of 19—22%. Peak I — pork trypsin complex dissociated to the extent of 65%. When the amount of trypsin and inhibitor was halved and the TAME concentration maintained — the extent of displacement reached 76% for Peak I as would be expected from the mass action equation. PUDLES, ROLA and MATIDA (1967) reported the presence of only one trypsin inhibitor and we have no way of comparing our preparations with theirs. PORTMAN and FRAEFEL

it cannot fully inactivate trypsin even at high inhibitor concentrations. It should be emphasized that in these studies the inhibitor from *Ascaris* is being reacted with pork trypsin, its "natural" enemy. When mixtures of trypsin (257 picomoles) and the amount of inhibitor necessary to inactivate up to 65% of the trypsin present are incubated over a period of up to 10 days the amount of free trypsin is unchanged when it is corrected for the slow loss of trypsin. However when mixtures of trypsin and an amount of inhibitor that inactivates between 65% and 95% of 257 picomoles of trypsin are incubated the free trypsin measured is a minimum initially and then there is a slow increase of free trypsin, an amount that eventually becomes 35% of free trypsin 65% of complex.

If the 65% of trypsin inhibited is just exceeded then the reversal to this point may only be a matter of 30 minutes, but if a ratio of inhibitor of trypsin to trypsin of 2 is selected it may take as long as 7 days to achieve this degree of regeneration. Various of these mixtures have been tested with ninhydrin but there is no detectable

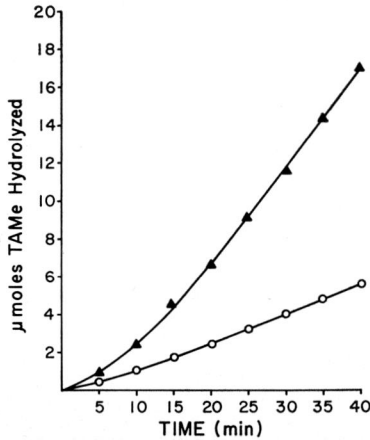

Fig. 10. Displacement of Peak I and Peak II *Ascaris* trypsin inhibitors from pork trypsin by TAME. Pork trypsin (257 picomoles) was reacted with 0.9 ml of the same inhibitor solution as used in Fig. 9 and under the same conditions as employed in that experiment. This is 2—3 times the amount of trypsin inhibitor necessary to stoichiometrically inhibit the pork trypsin. After 10 minutes, 4.0 ml of 10 mM TAME solution was added. ▲—▲, pork trypsin — Peak I; o—o, pork trypsin — Peak II. (KUCICH and PEANASKY 1970).

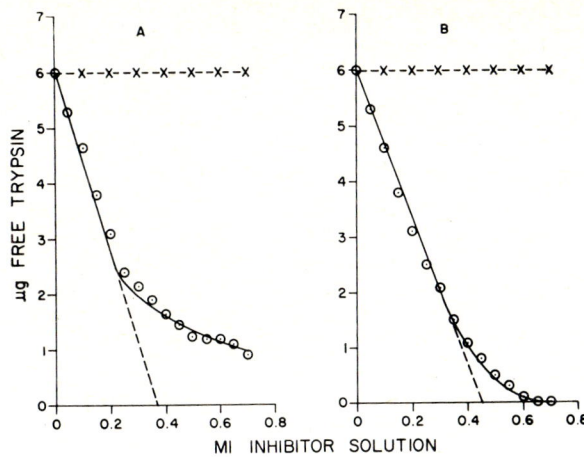

Fig. 11. Reaction of pork trypsin and human trypsin with Peak I (A) and Peak II (B) *Ascaris* trypsin inhibitors. Reaction conditions are the same as in Fig. 9. o—o, pork trypsin — Peak I (A) or Peak II (B) *Ascaris* trypsin inhibitor. x— —x, human trypsin — Peak I (A) or Peak II (B) *Ascaris* trypsin inhibitor.

(1967) found at least two trypsin inhibiting fractions following chromatography on CM cellulose. The amino acid composition and the sequence of one of these fractions has been reported (Fraefel and Acher 1968). From the amino acid composition, our Peak I does not seem to be identical with the sequenced inhibitor. We do not know about Peak II as yet. A very interesting problem relating to the species specificity of *Ascaris* suggested itself to us. Embryonated eggs from *Ascaris* raised in a hog, can be hatched in a human who might injest them; they migrate into the blood stream, undergo several molts over an 8 day period and finally are found in a larval form in the lungs. In the appropriate host the larval form is coughed up, swallowed and soon grows to adulthood. In the wrong host development stops at the lung stage (Faust and Russell 1964). Since we believe that the trypsin and chymotrypsin inhibitors are key substances in the survival of the worm in the intestines, we wondered if the host specificity of the worms could not be based upon the specificity of the inhibitor. If the trypsin inhibitor from pork *Ascaris* could not inhibit human trypsin, this could explain why the development of eggs from the pork *Ascaris* never resulted in adult worms in humans. Dr. Travis kindly supplied us with a sample of human trypsin (Travis and Roberts 1969). Human trypsin equal to 257 picomoles of pork trypsin was not inhibited (Fig. 11). When the amount of *Ascaris* trypsin inhibitor was increased by a factor of 20 (the amount of human trypsin remained the same as above) little evidence of inhibition by Peak I and a barely detectable amount of inhibition by Peak II was observed (Fig. 12). When the reaction of these inhibitors (20× the stoichiometric amount) with human trypsin was allowed to proceed for 10 minutes and the free trypsin determined in a pH stat with TAME as substrate only Peak II showed evidence of a loose complex which was completely displaced by zero order concentration of TAME in 3 minutes (Fig. 13). Allowing the complex 2 hours to form did not change the results.

We haven't been able to obtain human *Ascaris* as yet and try the reverse experiment but the evidence we have strongly suggests that in *Ascaris* host specificity of the intestinal parasite might be controlled by host specificity of proteolytic inhibitors.

Pepsin Inhibitor

Although adult *Ascaris* live in the intestine of their host, an environment in which pepsin is not stable, extracts of adult *Ascaris* still do inhibit pepsin. This activity is probably not essential to the adult *Ascaris* but its presence at the appropriate molting stage might play an important role in the life cycle of the *Ascaris*. The

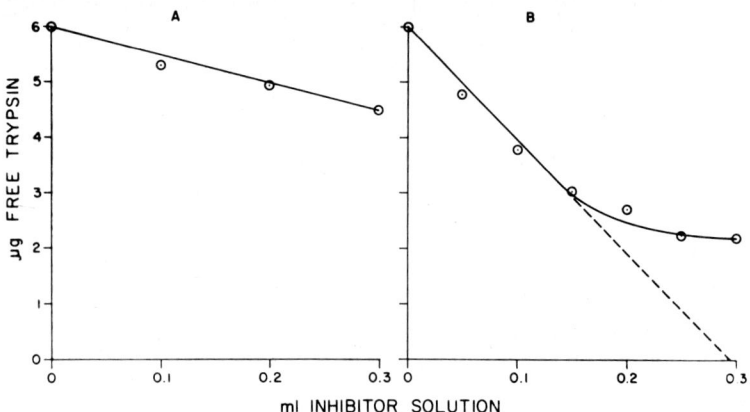

Fig. 12. Reaction of human trypsin with Peak I (A) and Peak II (B) *Ascaris* trypsin inhibitors. Stock trypsin inhibitor solutions had an absorbance of 102×10^{-3} at 280 nm. This is a $20 \times$ greater concentration of inhibitor solution than used in Fig. 9 and 11; all other conditions are the same.

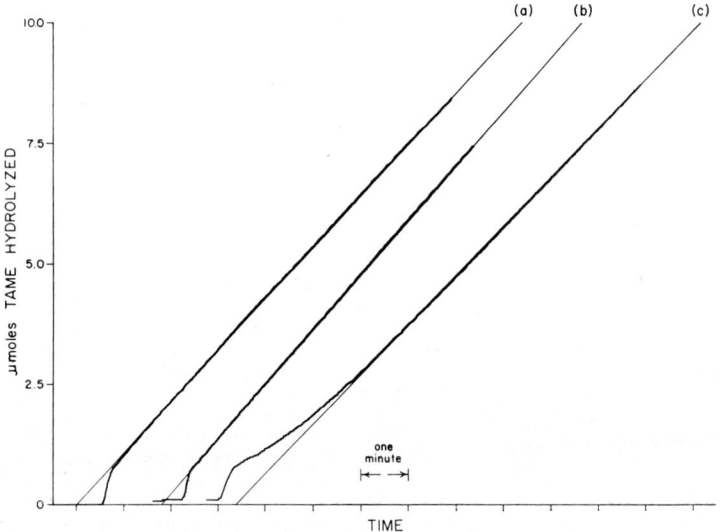

Fig. 13. Rate of reaction of human trypsin (a), of human trypsin + Peak I *Ascaris* trypsin inhibitor (b), of human trypsin + Peak II *Ascaris* trypsin inhibitor (c) on TAME (pH-stat tracing). Human trypsin (257 picomoles) was incubated alone or with an amount of Peak I (a) or Peak II (b) inhibitor (absorbance of each inhibitor was 34×10^{-3} at 280 nm) at 27°C in 100 mM KCl-20 mM $CaCl_2$-5 mM Tris buffer (pH 7.5) for 10 minutes in a volume of 0.5 ml. This is $20 \times$ the amount of either inhibitor necessary to inhibit stoichiometrically an equivalent amount of pork trypsin. 4.0 ml of 10 mM TAME was added to determine free trypsin.

pepsin inhibitor is present in adult *Ascaris* at concentration of only 0.5—1% of the concentration of trypsin and chymotrypsin inhibitors in *Ascaris*. The initial steps in it's purification are shown in Figure 14.

Chromatography of the ammonium sulfate fraction on Biogel P-30 completely removes carbohydrate from pepsin inhibiting fractions (Fig. 15). Such samples, concentrated with an Amicon

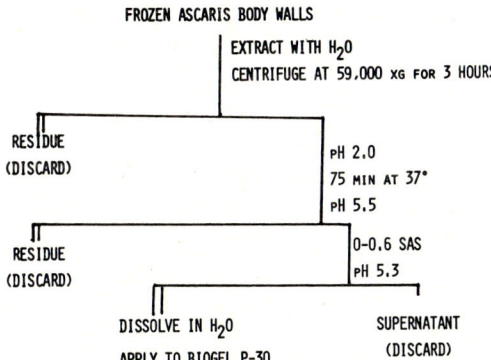

Fig. 14. Initial steps in the purification of pepsin inhibitor from frozen *Ascaris* body walls.

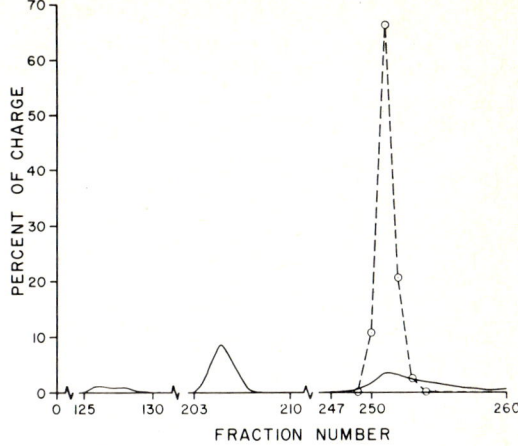

Fig. 16. Chromatography on Cellex SE of *Ascaris* pepsin inhibitor after chromatography on Biogel P-30. Column (2.4 × 45 cm) was equilibrated and eluted with 20 mM TRIS-HCl adjusted to pH 3.65. At tube 30 elution was continued with 20 mM TRIS-HCl pH 4.6. At tube 125 a 5 chamber gradient in which the first 3 chambers contained 20 mM TRIS pH 4.6 and the last 2 chambers contained 20 mM acetate pH 5.0 was used. Each fraction 12 ml. Ordinate is percent of charge of protein ——, or percent of charge of pepsin inhibitor ---.

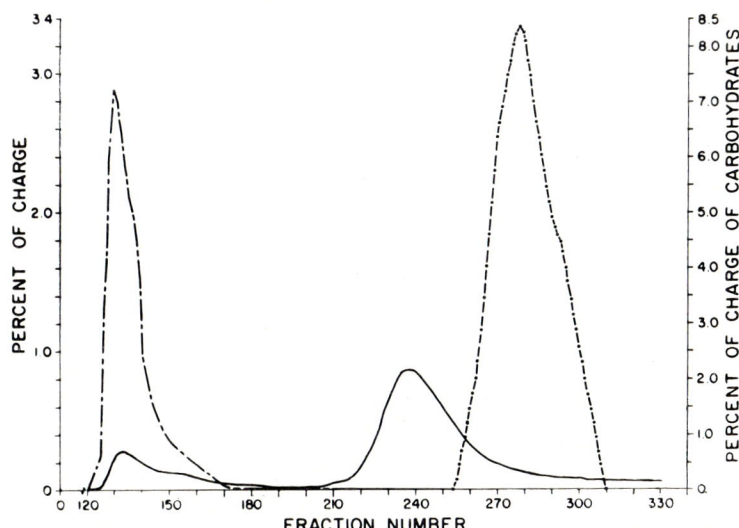

Fig. 15. Chromatography on Biogel P-30 of *Ascaris* pepsin inhibitor precipitated by ammonium sulfate (Fig. 14). Column (10 × 95 cm) was equilibrated and eluted with 0.1M TRIS-HCl adjusted to pH 2.1. Each fraction 17 ml. Left ordinate is percent of charge of protein —— and percent of charge of units of pepsin inhibitor ------. Right ordinate is percent of charge of carbohydrate —·—.

cell were next chromatographed on Cellex SE (Fig. 16). Although recovery of protein at pH 4.7 is poor, a 70% recovery of activity is achieved (the rest of the protein can be removed from Cellex SE at higher pH). This fraction was then chromatographed on Sephadex G-25 (Fig. 17). Table 5 shows that the recovery of activity is about 50% overall — the material

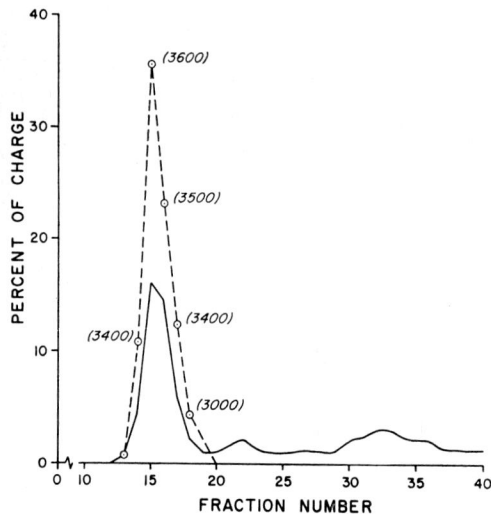

Fig. 17. Chromatography on Sephadex G-25 of *Ascaris* pepsin inhibitor after Cellex SE. Column (0.9 × 45 cm) was equilibrated and eluted with 100 mM Tris buffer pH 8.6. Ordinate is percent of charge of protein ——— and of units of pepsin inhibitor ----. Numbers in parenthesis represent specific activity.

has been purified about 7,000 fold. To our knowledge this is the first purification of a pepsin inhibitor. Reports (e. g. ROLA and PUDLES 1966) of the presence of pepsin inhibitor are in error for this reason: Pepsin inhibitor, although stable to pH 1.0 is not stable in the presence of trichloroacetic acid. If *Ascaris* or a sample of *Ascaris* extract is treated with TCA and then assayed it will inhibit pepsin but the extent of inhibition will be exactly that found after treating pepsin with an equivalent amount of a TCA-H_2O solution. We believe that some reports of the presence of pepsin inhibitor and then that it could not be precipitated or was lost on dialysis reflects the fact that it was the TCA which was mistakenly identified as pepsin inhibitor.

Polyacrylamide gel electrophoresis performed in 7.5% cross linked gel at pH 8.6 indicated that the inhibitor was about 80% pure. This inhibitor reacts with pepsin at pH 2.0 (Fig. 18) and on the assumption of a 1 : 1 complex a K_i of 5.4×10^{-10} M can be calculated (GREEN and WORK 1953). These lines of evidence suggest that the inhibitor has a rather low molecular weight. It passes through a UM-05 membrane unless the pressure is kept very low, it is retarded on a Sephadex G-25 column and its specific activity predicts a molecular weight of 8,000 or less. A preliminary amino acid composition of the preparation from Sephadex indicates that in a molecule of this size the amount of half cystine is not enough to form a single disulfide bridge.

Final Observation

In spite of the great similarity between the catalytic sites of trypsin and chymotrypsin as has been so elegantly reviewed at this symposium,

Table 5. Recovery of Pepsin Inhibitor from 600 Grams *Ascaris*

Step of Purification	µg Pepsin Inhibited	Specific Activity (µg Pepsin Inhibited) per mg of Protein	Recovery (%)
Crude extract	12,600	0,5	(100)
Ammonium sulfate	10,080	6	80
Bio-GEL P-30	8,100	120	65
Cellex SE	6,400	1,600	50
Sephadex G-25	6,000	3,500	50

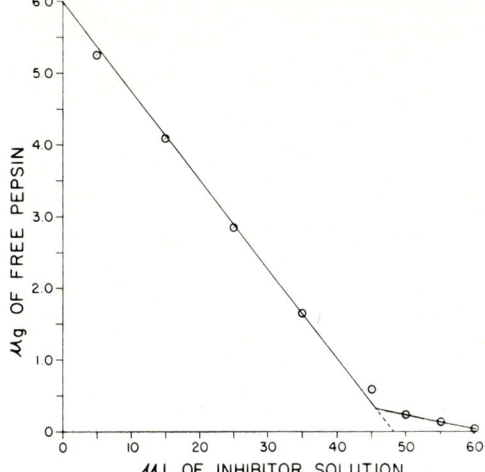

Fig. 18. Reaction of pork pepsin (174 picomoles) with *Ascaris* pepsin inhibitor. Stock pepsin inhibitor solution had an absorbance of 58×10^{-3} at 280 nm. Reaction mixtures (1.0 ml) made up of pork pepsin and the amounts of inhibitor indicated on the abscissa in 10 mM HCl (pH 2.0) were allowed to react at 37°C for 5 minutes. Free pepsin was determined by the rate of hydrolysis of hemoglobin substrate.

Ascaris lumbricoides whose survival depends upon its ability to inhibit both trypsin and chymotrypsin has designed *different* molecules to inhibit trypsin and chymotrypsin. The specificity of these inhibitor molecules is absolute.

Acknowledgements. I am deeply grateful to Dr.-to-be U. Kucich, and to Drs. M. Szucs, L. McEvoy, R. Komorowski, W. Olmsted, C. Moyer, R. Agostinelli who at the time were medical students and labored hard and diligently during one or more of their summer vacations. Mrs. M. Wales and Mr. R. Tiahrt supplied expert technical assistance. Mrs. M. Wales and (G. M. A.) fractionated such large quantities of *Ascaris* extract that they were overcome by an allergic response to *Ascaris*. In spite of this they doggedly continued on the project. Finally, I acknowledge grant GM 16203 from the National Institute of General Medical Sciences and the U. S. Department of Agriculture for financial assistance. Some of the experiments were performed while (R. J. P.) was supported by a U. S. Public Health Service Research Career Program Award K-3-GM-4282.

References

Baldwin, E. and V. Moyle, J. Exptl. Biol. **32**, 277 (1947).
Berger, J. and C. F. Asenjo, Science **90**, 299 (1939).
Berger, J. and C. F. Asenjo, Science **91**, 387 (1940).
Collier, H. B., Can. J. Res. **19 B**, 90 (1941).
Faust, E. C. and P. F. Russell, Clinical Parasitology, Philadelphia, Lea and Febiger pp. 419—429 (1964).
Fraefel, W. and R. Acher, Biochim. Biophys. Acta **154**, 615 (1968).
Glazer, A. N., Proc. Nat. Acad. Sci. USA **59**, 996 (1968).
Green, N. M. and E. Work, Biochem. J. **54**, 257 (1953).
Green, N. M., Biochem. J. **66**, 416 (1957).
Kucich, U. and R. J. Peanasky, Biochim. Biophys. Acta **200**, 47 (1970).
Mendel, L. B. and A. F. Blood, J. Biol. Chem. **8**, 177 (1910).
Patchornik, A., W. B. Lawson and B. Witkop, J. Am. Chem. Soc. **80**, 4747 (1958).
Peanasky, R. J. and M. Laskowski, Sr., Biochim. Biophys. Acta **37**, 167 (1960).
Peanasky, R. J. and M. M. Szucs, J. Biol. Chem. **239**, 2525 (1964).
Portman, P. and W. Fraefel, Helv. Chim. Acta **50**, 2078 (1967).
Pudles, J., F. H. Rola and A. K. Matida, Arch. Biochim. Biophys. **120**, 594 (1967).
Rhodes, M. B., C. L. Marsh and G. W. Kelley, Jr., Exptl. Parasitol. **13**, 266 (1963).
Robbins, B. H., J. Biol. Chem. **87**, 251 (1930).
Robbins, B. H. and P. D. Lamson, J. Biol. Chem. **106**, 725 (1934).
Rola, F. H. and J. Pudles, Arch. Biochem. Biophys. **113**, 134 (1966).
Travis, J. and R. C. Roberts, Biochemistry **8**, 2884 (1969).
Walti, A., J. Am. Chem. Soc. **60**, 493 (1938).
Weinland, E., Z. Biol. **44**, 1 (1903).

Interaction of Human Pancreatic Proteinases With Naturally Occuring Proteinase Inhibitors*

M. H. COAN and J. TRAVIS

Department of Biochemistry, University of Georgia, Athens, Georgia 30601

Naturally occurring proteinase inhibitors have been assumed to play an important, though as yet unknown, physiological role because nearly all specifically inhibit bovine trypsin and, to a lesser extent, bovine α-chymotrypsin. Although a particular inhibitor may play an important role in regulating the proteolytic activity in the organism which synthesizes it, there is no reason to assume that the inhibitor will show similar activity toward proteases with the same specificity, which are isolated from other tissues. Thus, for example, chicken ovomucoid completely inhibits bovine trypsin at an inhibitor to enzyme molar ratio of 1:1, whereas human trypsin is essentially unaffected, even in the presence of a large excess of inhibitor (BUCK et al., 1962).

The results of a recent investigation on the effect of a variety of inhibitors of bovine trypsin on the activity of human trypsin, plasmin, and thrombin showed that there were substantial differences in the extent of inhibition of human trypsin (FEENEY et al., 1969). Quantitative results were not given, however, presumably because of the difficulty in activation of human trypsinogen which was used as a source of enzyme.

In our laboratory we have been investigating the properties of human pancreatic proteases and have succeeded in purifying both human trypsin and the major form of human chymotrypsin present in activated extracts of human pancreas. We wish to report, at this time, the effect of several of the naturally occurring trypsin inhibitors on the activity of each of these enzymes.

Materials and Methods

Human trypsin and human chymotrypsin were prepared by ion-exchange chromatography of activated extracts of human pancreas (COAN et al., 1971). Bovine trypsin, bovine chymotrypsin, lima bean trypsin inhibitor, chicken ovomucoid, and soybean trypsin inhibitor were obtained from the Worthington Biochemical Co. KUNITZ pancreatic trypsin inhibitor was a gift of Dr. L. J. GREENE.

Inhibition experiments were carried out by mixing given quantities of enzyme and inhibitor in 0.05M Tris-HCl buffer, pH 8.0, containing 0.05M $CaCl_2$. After fifteen minutes incubation at room temperature, the samples were assayed for both esterolytic activity, using BAEE or ATEE (SCHWERT and TAKENAKA, 1955), and for caseinolytic activity using the KUNITZ procedure (LASKOWSKI, 1955). Experiments performed using the Kunitz pancreatic trypsin inhibitor were allowed to incubate for sixty mi-

---

* This research was supported in part by Grant No. AM-13169 from the U. S. Public Health Service.

Table 1. Comparative Properties of Human and Bovine Pancreatic Proteases

Property	Human Trypsin	Bovine Trypsin	Human Chymotrypsin	Bovine α-chymotrypsin
Molecular weight	22,900	23,000	27,000	23,000
Disulfide bonds	5	6	4	5
Stability				
a) alkaline pH (8.1, no Ca^{++})	unstable	unstable	unstable	stable
b) alkaline pH (8.1, 0.5M Ca^{++})	increased stability	increased stability	unstable	stable
c) acid pH (3.0)	slowly inactivated	stable	stable	stable
N-Terminus	Ileu	Ileu	1/2 Cys, Ileu	1/2 Cys, Ileu, Ala
Antigenicity				
a) Against anti-human trypsin	+	−	−	−
b) Against anti-human chymotrypsin	+	−	+	−
c) Against anti-bovine trypsin	−	+	−	+
d) Against anti-bovine chymotrypsin	−	+	−	+

nutes prior to assay because of the slow response of the inhibitor to the enzymes tested.

Results

The properties of the human pancreatic proteases differ in many respects from those of their bovine counterparts (Tab. 1), even though bond cleavage specificities of the two pairs of enzymes appear to be identical. The characteristics listed also indicate that the tertiary structure of the human enzymes, although similar to each other, are certainly not similar to those of the bovine species. Such differences could be reflected in changes in both the type and degree of inhibition obtained with naturally occurring trypsin inhibitors, especially if steric factors play an important role in the interaction of large molecules such as the formation of a trypsin-trypsin inhibitor complex.

Effect of Soybean Trypsin Inhibitor

In our early experiments with human trypsin, a lack of interaction between soybean trypsin inhibitor and this enzyme was detected which was based on changes in esterolytic activity (TRAVIS and ROBERTS, 1969). From the results shown in Figure 1, it can be seen that our original conclusions were essentially correct. However,

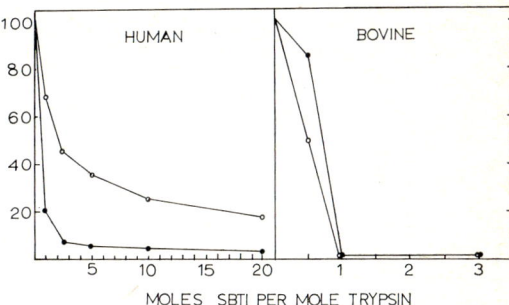

Fig. 1. Effect of soybean trypsin inhibitor on the esterolytic and proteolytic activities of human and bovine trypsins. Substrates used were BAEE and casein. The amount of enzyme preparation used was 0.01 mg in a final volume of 3 ml. o-o, esterolytic activity; ●-●, proteolytic activity. Time of incubation prior to assay, 15 minutes. Ordinate: Remaining activity in %.

there is significant inhibition of proteolytic activity, although soybean trypsin inhibitor is not as effective as with the bovine enzyme.

The protective effect afforded bovine trypsin after complex formation with soybean trypsin inhibitor is well known (GREEN, 1953), and no incorporation of DFP into bovine trypsin in

Table 2. Inhibition of Esterolytic Activity of Human and Bovine Trypsins by SBTI

Molar Ratio SBTI : Trypsin	% Activity Remaining		% C^{14} DFP Incorporated	
	Bovine	Human	Bovine	Human
0:1	100	100	100	100
0.5:1	50	74	51	80
1:1	13	68	15	71
2:1	8	50	5	48
4:1	2	35	5	30
8:1	—	26	—	23

such a complex can be detected. The difference in reactivity of soybean trypsin inhibitor towards human trypsin suggested the possibility of a loose interaction and for this reason attempts were made to incorporate C^{14}-DFP into a human trypsin-soybean trypsin inhibitor complex. As can be seen in Table 2, a strong correlation was obtained between free enzyme present after addition of inhibitor and the amount of C^{14}-DFP which could be incorporated. From these results it can be concluded that a loose binding does not occur and that the differences in reactivity shown in Figure 1 are probably attributable to equilibrium effects, although non-specific binding of soybean trypsin inhibitor to the human enzyme cannot be ruled out.

In Figure 2, the inactivation of bovine and human chymotrypsins by soybean trypsin inhibitor is depicted. There are no major differences in the reactivity of either towards the inhibitor and this is in agreement with the fact that this inhibitor is much more reactive towards bovine trypsin and reacts nonstoichiometrically with bovine chymotrypsin.

Effect of Kunitz Pancreatic Trypsin Inhibitor

KUNITZ pancreatic trypsin inhibitor is a low molecular weight protein isolated from bovine pancreas. This inhibitor causes the total loss in enzymatic activity of both bovine trypsin and chymotrypsin but only after prolonged incubation. When tested against human trypsin this inhibitor showed essentially no difference in either the rate or stoichiometry of inhibition (Fig. 3). However, the effect on the enzymatic

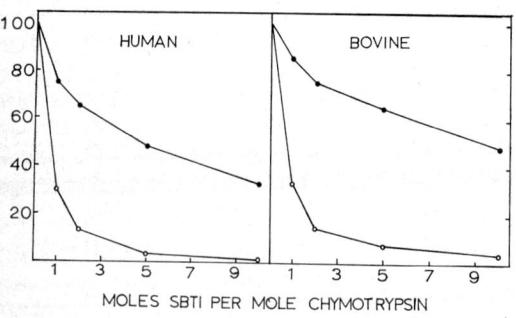

Fig. 2. Effect of soybean trypsin inhibitor on the esterolytic and proteolytic activities of human and bovine chymotrypsins. The substrate used for measuring esterolytic activity was ATEE. All other conditions were the same as described in Fig. 1.

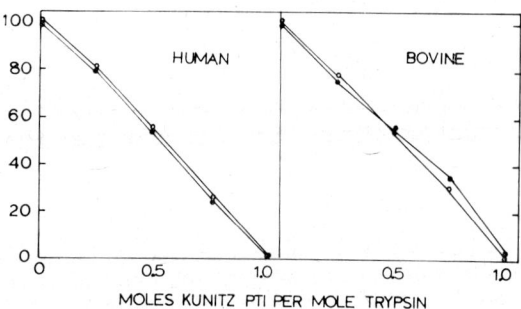

Fig. 3. Effect of KUNITZ pancreatic trypsin inhibitor on the esterolytic and proteolytic activities of human and bovine trypsins. Time of incubation, 60 minutes. All other conditions were the same as described in Fig. 1.

activity of human chymotrypsin was much more significant (Fig. 4), in that the esterolytic activity remained virtually unchanged and the caseinolytic activity was only very slightly altered. Whether or not this lack of inhibition is species specific cannot be established until a

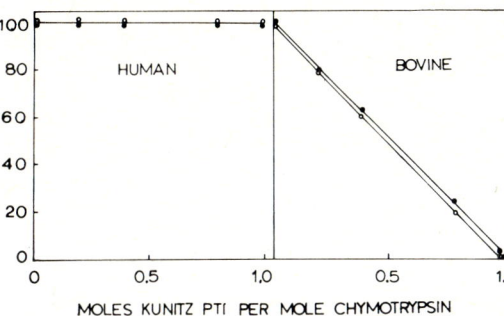

Fig. 4. Effect of KUNITZ pancreatic trypsin inhibitor on the esterolytic and proteolytic activities of human and bovine chymotrypsins. Time of incubation, 60 minutes. The substrate used for measuring esterolytic activity was ATEE. All other conditions were the same as described in Fig. 1.

human KUNITZ pancreatic trypsin inhibitor is isolated. Such a protein has been looked for in this species but has not yet been detected.

Effect of Lima Bean Trypsin Inhibitor

Of all of the inhibitors examined in the current study, lima bean trypsin inhibitor was, by far, the most efficient in inactivating both human trypsin and human chymotrypsin (Fig. 5 and 6). Although neither human chymotrypsin nor bovine chymotrypsin was inhibited stoichio-metrically it should be noted that this was the only example detected where inhibition of a human enzyme was stronger than that from the bovine species.

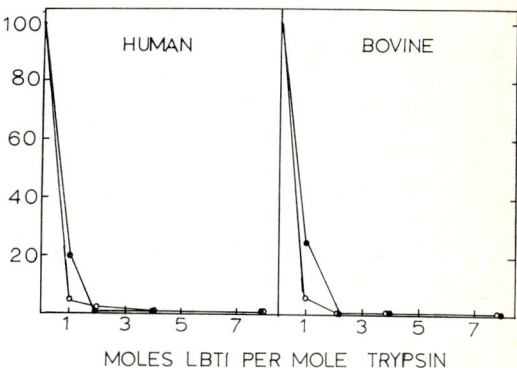

Fig. 5. Effect of lima bean trypsin inhibitor on the esterolytic and proteolytic activities of human and bovine trypsins. All conditions were the same as described in Fig. 1.

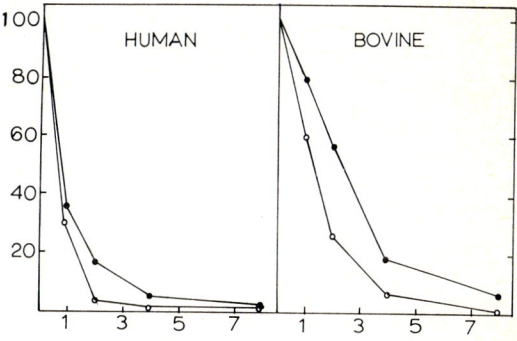

Fig. 6. Effect of lima bean trypsin inhibitor on the esterolytic and proteolytic activities of human and bovine chymotrypsins. The substrate used for measuring esterolytic activity was ATEE. All other conditions were the same as described in Fig. 1.

Table 3. Effect of Ovomucoid on Human and Bovine Trypsin and Chymotrypsin

Molar Ratio Ovomucoid: Enzyme	% Activity Remaining			
	Human		Bovine	
	Trypsin	Chymotrypsin	Trypsin	Chymotrypsin
1:1	100	100	51 (71)	100
2:1	100	100	26 (45)	100
4:1	100	100	11 (25)	100

Effect of Ovomucoid

The complete retention of the proteolytic activity of human trypsin, human chymotrypsin, and bovine chymotrypsin, even in the presence of excess chicken ovomucoid, is demonstrated in Table 3. The lack of inhibition is striking, especially when a comparison is made with bovine trypsin. No reason can, as yet, be given for the stoichiometry which has been observed between bovine trypsin and chicken ovomucoid but which could not be obtained here. Because of the absence of inhibitory activity towards human enzymes, it would seem obvious that there can be no nutritional importance ascribed to chicken ovomucoid, especially with regard to human growth patterns.

Discussion

Of the four inhibitors examined in this study, only one, lima bean trypsin inhibitor, was able to completely repress the proteolytic activity of human trypsin and human chymotrypsin. The general lack of similarity in sensitivity to inhibitors denoted here between the bovine and human species must then be ascribed to differences in the tertiary structure of the two sets of enzymes. Any nutritional significance previously attached to the inhibitors must be re-evaluated in terms of the effect observed on the particular organism examined and cannot be generalized. Furthermore, a proper evaluation of inhibitory capacity cannot be based on loss of esterolytic activity but must, instead, rely on data obtained from experiments on the inhibition of proteolytic activity.

Species specificity is extremely important in determining the function of individual inhibitors and cannot be overemphasized. This is further borne out by the results of both Dr. GREENE and Dr. PEANASKY which are described elsewhere in this volume.

References

BUCK, F. F., M. BIER and F. F. NORD, Arch. Biochem. Biophys. **98**, 528 (1962).
COAN, M. H. and J. TRAVIS, Biochemistry, in press (1971).
FEENEY, R. E., G. E. MEANS and J. C. BIGLER, J. Biol. Chem. **244**, 1957 (1969).
GREEN, N. M., J. Biol. Chem. **205**, 535 (1953).
LASKOWSKI, M., in S. P. COLOWICK, N. O. KAPLAN, Methods in Enzymology, II 36 (1955).
SCHWERT, E. W. and Y. TAKENAKA, Biochim. Biophys. Acta **16**, 570 (1955).
TRAVIS, J. and R. C. ROBERTS, Biochemistry **8**, 2854 (1969).

On the Mechanism of Temporary Inhibition

HARALD TSCHESCHE, HELMFRIED KLEIN and GÜNTER REIDEL

Organisch-Chemisches Laboratorium der Technischen Universität München, Lehrstuhl für Organische Chemie und Biochemie, Germany

The phenomenon that tryptic activity reappeared upon prolonged incubation of an enzyme-inhibitor complex was first noted by GORINI and AUDRAIN [1, 2] and later named "temporary inhibition" and described in detail by LASKOWSKI, Sr. and WU [3]. The physiological significance of this phenomenon is obvious, since it enables the pancreatic gland to inhibit untimely activated trypsinogen immediately and prior to autocatalytic activation of the zymogens within the tubular system. After secretion into the duodenal tract, however, the trypsin may be liberated again and used for digestion, whereas the protein-like inhibitor is degraded and incorporated into the protein metabolism.

In 1967 we started investigations concerned with the nature of the multiple active components of porcine pancreatic secretory trypsin inhibitor found in the active fractions isolated from crude tissue extracts by affinity chromatography and in acid or ethanol extracts [4, 5]. It could be demonstrated that the multiple active (and inactive) components are generated by trypsin during the process of inhibition [6]. Degradation during temporary inhibition was made responsible for this result. The modification of the native (virgin) inhibitor occurred with minor and/or catalytic amounts of trypsin as low as 1 mole percent. The rate of modification and the appearance of electrophoretically new components were found to be strongly depending on the pH used during incubation. Using a constant ratio of inhibitor to trypsin maximal

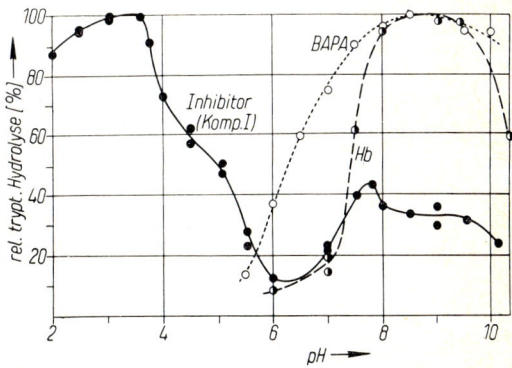

Fig. 1. pH-dependence of tryptic hydrolysis of the substrates: N^α-benzoyl-D, L-arginin p-nitroanilide (BAPA); hemoglobin (Hb). values adopted from NORTHROP [J. gen. Physiol. 5, 263 (1922)]; and PSTI. The values for tryptic hydrolysis of PSTI were taken from Fig. 2 measuring the amount of virgin inhibitor undigested (refers to band I in Fig. 3), (from TSCHESCHE and KLEIN, 1968).

The abbreviations used are: PSTI, pancreatic secretory trypsin inhibitor; BAPNA, N^α-benzoyl-D, L-arginine p-nitroanilide; NPGB, p-nitrophenyl p'-guanidobenzoate; TPCK, p-toluenesulfonamido phenylethyl chloromethyl ketone.

modification occurred in the acid pH range with a strong maximum at pH 3.4 [6]. A second lower optimum was found at pH 7.8 [6] (Fig. 1), i. e. the normal activity range of trypsin. These results were in accordance with the findings of Ozawa and Laskowski, Jr. [7] made on Kunitz soybean trypsin inhibitor for the acid partial proteolysis reaction of the reactive site peptide bond. Since we determined the complete covalent structure of the porcine pancreatic secretory trypsin inhibitor (this volume, page 221), investigations concerning the bonds to be split for inactivation during temporary inhibition were tempting. Now we wish to report on the structure of the first clearly identified inactive degradation product of porcine pancreatic secretory trypsin inhibitor I.

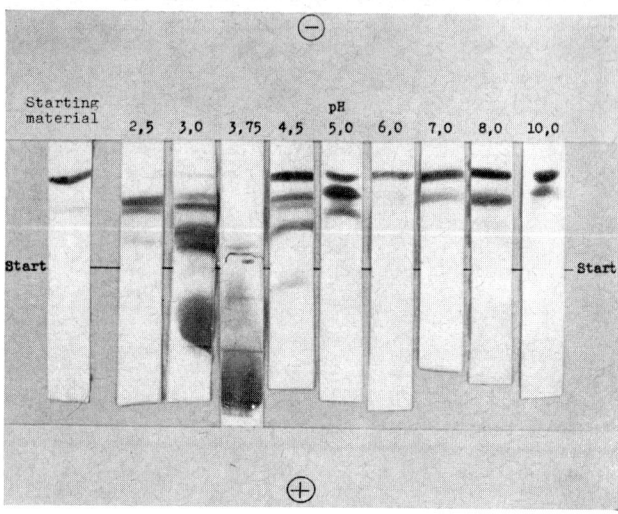

Fig. 2. Electrophoretic band pattern obtained from PSTI modified for 24 hours with 50% molar amount of bovine trypsin at 37°. Conditions: 0.01M veronal-buffer, pH 7.8, 40 V/cm, 110 min on cellulose acetate, stained with Amido Black 10 B (Merck AG, Darmstadt, from Tschesche and Klein, 1968).

Characterization and Separation of Modified Components

In our early experiments we used cellulose acetate electrophoresis [5] for the qualitative separation of virgin inhibitor from modified components. The details of the procedure were given in Tschesche [4] and Tschesche and Klein [6]. The activity of individual components was determined after cutting the folio according to guide strips and elution of the individual bands. Using this procedure we demonstrated that virgin inhibitor is modified consecutively into several distinct electrophoretic components which are still active, but migrate less far towards the cathode (Fig. 2 and 3). One of the components formed could already be isolated and purified from modified inhibitor mixtures [8] and was identified as des-Thr-Ser-Pro-Gln-Arg-inhibitor. This fragment inhibitor is still composed of a single polypeptide chain, exhibits full inhibitory activity, but lacks the amino terminal pentapeptide [8].

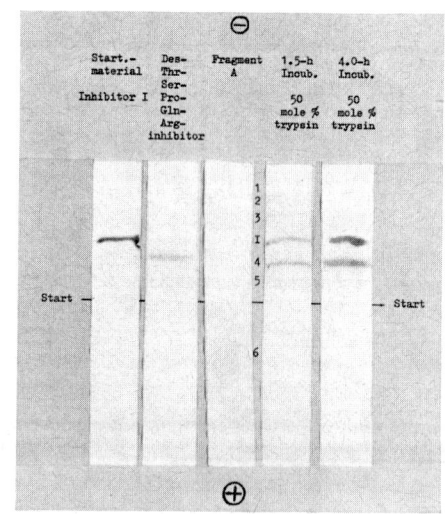

Fig. 3. Electrophoretic band pattern obtained from PSTI, des-Thr-Ser-Pro-Gln-Arg-inhibitor and fragment A, 1.5-hour modified and 4-hour modified PSTI on cellulose acetate folio. Conditions: 0.01M veronal-buffer, pH 7.8, 400 V/cm, 20 min, stained with Amido Black 10 B. For conditions of tryptic modification see text.

Because the positively charged arginine (residue 5) is lost, this component migrates less far towards the cathode as can be seen in the electrophoresis (Fig. 3).

More extensively degraded inhibitor is produced upon incubation at acidic pH, between pH 2.0 and 4.5 (Fig. 2 and 3). This material migrates towards the anode and exhibits no inhibitory activity when tested against bovine trypsin (TPCK-treated trypsin, 2.0 U/mg, Merck AG, Darmstadt) and BAPNA substrate [6] or when tested with NPGB [9]. It fails to give an electrophoretically sharp band (Fig. 2 and 3, band 6) and smears over the corresponding side of the folio towards the anode, indicating the loss of a distinct and ordered protein structure. This behavior and the corresponding explanation were related to the difficulties encountered in staining small amounts of the latter material with Amido-Black 10 B on the folio.

For the separation of modified components on a preparative scale an ion exchange system has been developed which enabled us to isolate distinct components. A column (0.9 × 90 cm) of SE-Sephadex C-25 was equilibrated with 0.05M ammonium acetate buffer pH 4.7 and developed with 45 ml of equilibrating buffer. Then a gradient was employed by mixing 0.05 M ammonium acetate pH 6.1 to 20 ml of equilibrating buffer (Fig. 4).

Identification of Individual Components

For the preparation of modified from virgin inhibitor, inhibitor I (8 mg) was incubated with 50% molar amount of bovine trypsin. The same inhibitor preparation was used as for the determination of the primary structure. The incubation was carried out for different time intervals in (4.8 ml) 0.05M sodium citrate (0.01M in $CaCl_2$) pH 3.4 at 22°. The reaction was stopped by dilution with the same volume of 6% perchloric acid and heating for 1 min to 60°. The solution was neutralized with KOH, centrifuged and lyophilized.

The modified inhibitor mixture was divided and subjected to cellulose acetate electrophoresis, to preparative separation of the individual components on SE-Sephadex and to EDMAN degradation using mass spectral identification via the p-bromophenyl-thiohydantoins [10]. The electrophoretic band patterns of inhibitor I modified for 1.5 and 4 hours at pH 3.4 are shown in Fig. 3. Several bands may be distinguished after

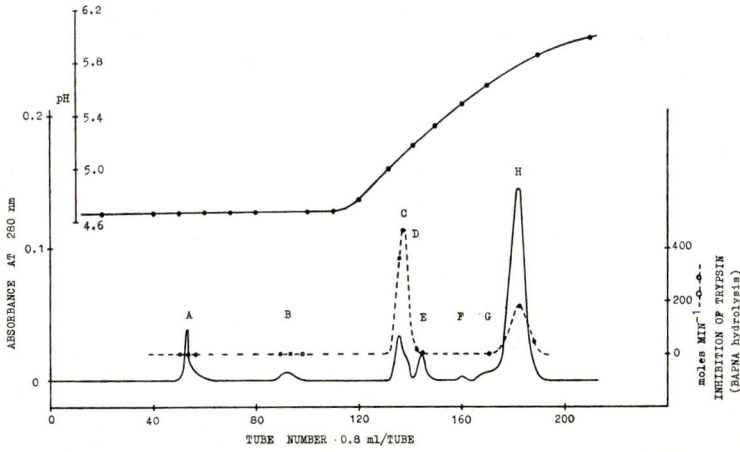

Fig. 4. Elution profile of a 4-hour modified inhibitor mixture obtained by ion exchange chromatography on SE-Sephadex C-25. The column (0.9 × 90 cm) was equilibrated with 0.05M ammonium acetate pH 4.7, developed with 45 ml of equilibrating buffer and then eluted with a gradient obtained by mixing 0.05 M ammonium acetate pH 6.1 to 20 ml of the starting buffer. The flow rate was 4.8 ml/h.

staining with Amido Black 10 B. This finding is in accordance with the results presented by Laskowski, Sr. et al. at this meeting (this volume, page 70). The two predominating bands stained heavily may be correlated to starting material (inhibitor I, i. e. band I, Fig. 3) and to des-Thr-Ser-Pro-Gln-Arg-inhibitor (band 4, Fig. 3) according to their migration distances, to their behavior in ion exchange chromatography on SE-Sephadex C-25 and according to their amino acid compositions and the single peak obtained in gel filtration on Bio Gel P-2 after performic acid oxidation. Besides these two bands several weaker ones (Nos. 1, 2, 3, 5, 6) can be distinguished.

A somewhat different looking picture than that obtained from electrophoresis is presented from the ion exchange chromatography of the 4-hour modified inhibitor mixture (Fig. 4). The same components found in electrophoresis may, however, be easily identified. The starting material (inhibitor I) is found in peak H*, and des-Thr-Ser-Pro-Gln-Arg-inhibitor in peak C. Both components were identified by their amino acid compositions and their elution diagrams obtained from gel filtration after performic acid oxidation. The peaks E, F and G and the more acidic products A and B were devoid of inhibitory activity. All inhibitor fragments were designated according to the peak lettering.

Fragment A was subjected to amino acid analysis and to characterization by electrophoresis. The material migrated towards the anode (Fig. 3, band 6). The amino acid composition is given in the Table and indicates the loss of the amino terminal residues 1 through 5 and of the inner decapeptide, residues 43 through 52. Performic acid oxidation of fragment A and gel filtration of the oxidized peptide mixture on a column of

---

* The specific activity of the recovered material was low in this experiment due to partial disulfide oxidation.

Amino Acid Composition of Porcine PSTI and Modified Components

Amino acid	PSTI		Fragment C		Fragment D		Fragment A	
	16 hrs[a]	Integer	16 hrs[b]	Integer	16 hrs[c]	Integer	16 hrs	Integer
Aspartic acid	3.89	4	4.00	4	3.54	4	4.07	4
Threonine	5.84	6	4.85	5	3.63	4	3.87	4
Serine	5.91	6	4.76	5	5.00	5	5.47	5
Glutamic acid	7.11	7	6.09	6	3.97	4	3.94	4
Proline	4.82	5	3.68	4	3.12	3	3.04	3
Glycine	4.00	4	3.91	4	3.96	4	3.74	4
Alanine	1.00	1	0.93	1	1.68[d]	1	1.32	1
Valine	3.83	4	3.72	4	2.72	3	2.96	3
Half-cystine	5.85	6	5.41	6	2.05	6	4.91	6
Isoleucine	3.00	3	3.00	3	2.12	2	2.12	2
Leucine	1.95	2	1.97	2	1.42	1	1.17	1
Tyrosine	1.97	2	1.70	2	1.23	2	1.47	2
Lysine	3.96	4	3.88	4	3.04	3	2.06	2
Arginine	1.88	2	0.88	1	1.09	1	0.08	0
Total		56		51		43		41

[a] Tschesche, Wachter, Kupfer and Niedermeier, 1969 [11].
[b] Tschesche, Wachter and Kallup, 1969 [8].
[c] Fragment D may be contaminated with fragment C.
[d] Partial oxidation of Cys caused an unknown peak to appear under Ala.

Bio Gel P-2 with subsequent analysis of the isolated peptides revealed that the reactive site peptide bonds Lys[18]-Ile[19] was broken in fragment A.

Fragment D, which was not clearly separated from fragment C (des-Thr-Ser-Pro-Gln-Arg-inhibitor) by its amino acid composition indicated cleavage of the bond Arg[44]-Gln[45] and loss of the corresponding octapeptide, residues 45 through 52.

The same fragments A to H were obtained already after 1.5 hours of tryptic modification (compare Fig. 3), however, in smaller amounts. They were formed as well under physiological conditions of incubation (8 mg inhibitor I, 16 mg TPCK-treated trypsin in 4.8 m/0.05 M triethanolamine-HCl, 0.01M $CaCl_2$, pH 7.0 for 65 hours at 22°). The chromatographic pattern obtained from the SE-Sephadex column under identical conditions of elution was quite similar to the one shown in Figure 4 with only minor changes in the relative peak heights of fragments A, D and G.

Results and Discussion

The results presented here indicate that inactivation of the inhibitor proceeds via formation of several distinct components which are generated by tryptic cleavages of the entire polypeptide chain. One of the early modification reactions, as indicated by electrophoresis experiments, is hydrolysis of the bond Arg[5]-Glu[6], thus generating des-Thr-Ser-Pro-Gln-Arg-inhibitor which still exhibits full inhibitory activity [8]. Experiments are under way in our laboratory to elucidate whether this modification is a perequisite for further degradation and inactivation during temporary inhibition. Cleavage of this amino terminal pentapeptide might be the "switch" for the initial opening of the virgin inhibitor for further tryptic digestion.

During temporary inhibition the same intermediate active and inactive fragments are generated, independent of the pH used during incubation with trypsin. The individual rates of appearance, however, seem to be a function of pH as might be expected and could be demonstrated for the rate of complete inactivation by the experiments performed by LASKOWSKI, Sr. et al. (this volume, page 71).

By five cycles of EDMAN degradation performed on the 4-hour modified inhibitor mixture, cleavage of the peptide bonds Arg[5]-Glu[6], Lys[18]-Ile[19] and Lys[42]-Lys[43] could be deduced from the sequences of the residues identified after each cycle. Hydrolysis of these bonds during inactivation was confirmed by the isolation of fragment A. Fragment A is inactive towards trypsin when tested against BAPNA and NPGB. When compared to the virgin inhibitor (Fig. 5) fragment A has lost the amino terminal penta-

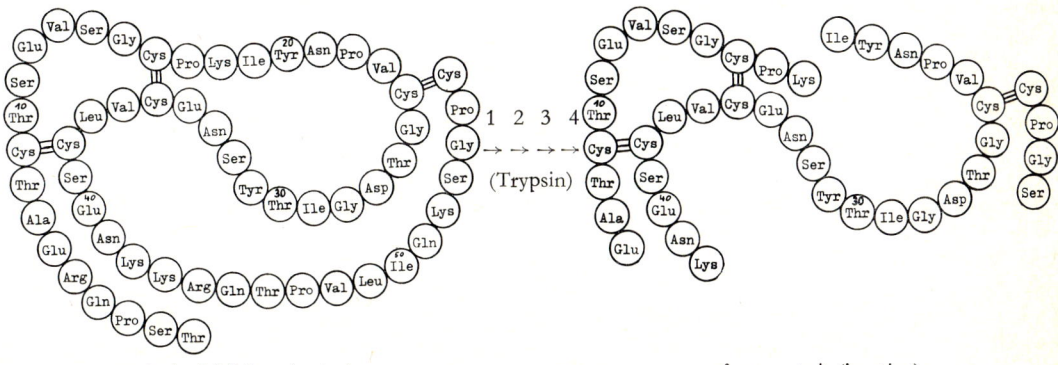

Fig. 5. Schematic diagram of the covalent structure of porcine PSTI and fragment A showing the peptides to be eliminated in the course of temporary inhibition.

peptide, which is insignificant for inhibitory activity, and the internal decapeptide (residues 43 through 52). Therefore, the bond Lys^{51}-Ser^{52} is hydrolyzed as well when the final and inactive degradation product, fragment A, is formed which accumulates during temporary inhibition. No exact information is available at present to the timing of the events, whether the bond Lys^{42}-Lys^{43} or Lys^{51}-Ser^{52} is hydrolyzed first or if no preference exists for either bond. From the acid partial proteolysis reaction, an experiment designed to elucidate the reactive site of the inhibitor (this volume, page 135), it is evident that the cleavage of the reactive site peptide bond Lys^{18}-Ile^{19} precedes the hydrolysis of the two other bonds. The situation is, however, still more complecated by the fact, that fragment D which is a fragment A including the dipeptide Lys^{43}-Arg^{44}, indicates a cleavage of the bond Arg^{44}-Gln^{45} to occur as well.

Therefore, it still remains to be shown, whether one or two further cleavages in addition to the hydrolysis of the reactive site peptide bond are necessary for inactivation. It could, however, be demonstrated that elimination of a decapeptide (residues 43 through 52) far remote from the reactive site sequence destroys the ordered structural features encountered with the inhibitory activity. This is not the case if only the reactive site peptide bond Lys^{18}-Ile^{19} is split.

The positioning of the disulfide bridges in the covalent structure of the inhibitor (this volume, page 221) seems to favour an antiparallel arrangement of the polypeptide chain containing the structural important decapeptide (residues 43—56) and the polypeptide chain including the reactive site disulfide loop (residues 24—35). Another alternative would be parallel arrangement of residues 43—56 to residues 16—24 of the disulfide loop. An important structural feature of this kind would explain why this section of the sequence (residues 44—56) was found to be an area of lower mutability (this volume, page 210). It would be interesting to see whether this expectation could be verified by X-ray crystallography.

Acknowledgements. We wish to thank Mrs. S. KUPFER for her skillful technical assistance and Miss C. FRANK for carrying out the amino acid analyses.

We are indebted to Mr. M. SCHNEIDER for the EDMAN degradations and Mrs. I. MELOCH for the recording of the mass spectra.

We are grateful to the Farbenfabriken Bayer AG, Wuppertal-Elberfeld, Germany, for their assistance and the supply of crude extracts from porcine pancreas.

We express our thanks to the Deutsche Forschungsgemeinschaft, Bad Godesberg, Germany, for their support with grants and materials.

References

[1] GORINI, L. and L. AUDRAIN, Biochim. Biophys. Acta **8**, 702 (1952).
[2] GORINI, L. and L. AUDRAIN, Biochim. Biophys. Acta **10**, 570 (1953).
[3] LASKOWSKI, M., Sr. and F. C. WU, J. Biol.Chem. **204**, 797 (1953).
[4] TSCHESCHE, H., Z. Physiol. Chem. **348**, 1216 (1967).
[5] TSCHESCHE, H., Z. Physiol. Chem. **348**, 1653 (1967).
[6] TSCHESCHE, H. and H. KLEIN, Z. Physiol. Chem. **349**, 1645 (1968).
[7] OZAWA, K. and M. LASKOWSKI, Jr., J. Biol. Chem. **241**, 3955 (1966).
[8] TSCHESCHE, H., E. WACHTER and G. KALLUP, Z. Physiol. Chem. **350**, 1662 (1969).
[9] CHASE, T., Jr. and E. SHAW, Biochem. Biophys. Res. Comm. **29**, 508 (1967).
[10] TSCHESCHE, H. and E. WACHTER, Eur. J. Biochem. **16**, 187 (1970).
[11] TSCHESCHE, H., E. WACHTER, S. KUPFER and K. NIEDERMEIER, Z. Physiol. Chem. **350**, 1247 (1969).